Parasites of North American Freshwater Fishes

Parasites
OF NORTH AMERICAN
FRESHWATER FISHES

GLENN L. HOFFMAN

UNIVERSITY OF CALIFORNIA PRESS
Berkeley, Los Angeles, London

University of California Press
Berkeley and Los Angeles, California
University of California Press, Ltd.
London, England
Copyright © 1967 by The Regents of the University of California

*Standard Book **Number** 520-00565-1*
Library of Congress Catalog Card Number: 67-14063

3 4 5 6 7 8 9 0

Errata and Omissions

P. 4, l. 13: also Israel (Witenberg, G., 1944).

P. 5, l. 5: also Crustacea: *Argulus* as a fortuitous eye parasite (Hargis, 1958).

P. 11, l. 7: add BARBELS—*Henneguya* in catfish.

P. 11, l. 12: add Leeches.

P. 11, l. 31: add NASAL FOSSAE—*Aplodiscus nasalis, Ergasilus megaceros,* larval *Philometra.*

P. 12, no. 14: add URETERS—*Dactylogyrus capoetobramae* (Russia).

P. 15, l. 4: add body cavity.

P. 18, l. 36: also an internal form, *S. invaderis,* was described by Davis and Lazar (1940).

P. 21, col. 2, l. 7: should read Genus *Ambiphrya.* . . . 39.

P. 21, col. 2, l. 8: should read Genus *Apiosoma.* . . . 41.

P. 25, l. 7: add *Gill amoeba*—salmonids and trout (undescribed).

P. 28, l. 14 and P. 29, l. 12: blepharoplast should read kinetoplast.

P. 36: *Balantidium* sp. should read *Nyctotherus* sp. because Fantham and Porter's identification was in error according to R. M. Stabler.

P. 37: *multifilis* should read *multifiliis.*

P. 40, no. 16: should read NYCTOTHERUS; no. 22 should read AMBIPHRYA; no. 23 should read APIOSOMA.

P. 56, bottom: add *M. moxostomatis* Kudo, 1921: in *Moxostoma* sp., N.Y.

P. 57: add *M. serotinum* Kudo, 1943: in gall bladder of frogs.

P. 71: Order ASPIDOCOTYLEA should not be listed under MONOGENEA.

P. 72, l. 3: should read ethyl m-aminobenzoate.

P. 84, no. 67: should read PSEUDOMURRAYTREMA.

P. 90, no. 71a: should read UROCLEIDUS.

P. 102: Order ASPIDOCOTYLEA should not be listed under MONOGENEA.

P. 105: add Genus *Parasitotrema.* . . . 142.

P. 109, no. 11: should read Body spined or not . . . . 12.

P. 110: add Genus CARDICOLA—*C. alseae* Meade and Pratt, 1965: in blood of *Salmo clarki* and *S. gairdneri.*

P. 112, no. 88: should read BUCEPHALOIDES.

P. 119, l. 1 and 15: should read Osborn, 1903.

P. 120, l. 3: should read (Stafford, 1900).

P. 128, par. 2: add Lewis (1935), review and key to species.

P. 142, par. 6: add Genus Parasitotrema—*P. ottawenesis* Miller, 1940: in catfish, Canada.

P. 143, no. 4: should read *Bunodera.*

P. 147, par. 3: should read Faust, 1918.

P. 154, second no. 2: should read Cirrus pouch absent.

P. 159: add HAPLOPORIDAE—*Saccocoelioides sogandaresi* Lumsden, 1963: in *Mollienisia latipinna,* Texas.

P. 171, l. 17: add Becker and Brunson (1966): precocious metas in snail.

P. 173, no. 3: These are not good key characteristics because some authors apparently measured the cavity instead of the entire cotylae. *T. intermedia* is probably misplaced in this key.

P. 177, l. 3: add Sillman (1957): effect on fish.

P. 182, par. 3: occasionally white cysts are found.

P. 186, l. 34: add (Syn. = *Psilostomum*).

P. 187, l. 6: add (Syn. = *R. thomasi*).

P. 198, l. 21: medium should read median.

P. 199, l. 4: *spinetum* should read sp.

P. 199, par. 5: should read Osborn, 1903.

P. 199, par. 7: should read (Stafford, 1900).

P. 201, par. 2: *Hybornchus* should read *Pimephales.*

P. 207: *Isoglaidacris longus* Frederickson and Ulmer, 1965; *I. folius* F and U, 1965; and *I. bulbocirrus* Mackiewicz, 1965 were omitted.

P. 221: *P. filicollis:* add Meggitt, 1914: life cycle and morphology.

P. 226, l. 25: McCloud should read McLeod.

P. 231: *D. salvelini* occurs as an adult in *S. alpinus.*

P. 232, l. 34: add (Syn = *S. dimorphus*).

P. 241, l. 8: should read TRICHURIDEA.

P. 241, l. 47: FILARIIDEA should read PHILOMETRIDEA.

P. 244, no. 11: (fig. 210) should read (fig. 211); (fig. 211) should read (no fig.).

P. 245, no. 1: No intestinal should read Usually no intestinal.

P. 254, l. 4: should read 1917.

P. 255, l. 40: should read 1917.

P. 261, l. 7: add Gustafson (1949), review. Sahay and Prasad (1965), key to spp.

P. 262, l. 10: add Mueller and VanCleave (1932), redescriptions.

P. 267, l. 1: should read PHILOMETRIDEA.

P. 269, l. 20: add Canavan, W. P. N. (1929), Parasitol. 21 (1/2): 63-102.

P. 270: add *Atactorhynchus* . . . 279.

P. 279: add Genus Atactorhynchus—*A. verecundus* Chandler, 1935: in *Cyprinodon variegatus.*

P. 282, no. 235: should read LEPTORHYNCHOIDES.

P. 284, l. 3: insert and intestine.

P. 286: should read *Rhadinorhynchus.*
P. 302, l. 33: add *A. foliaceus* is valid (Stammer, 1959).
P. 305, l. 13: add Neuhaus, E. (1929), Z. Fischerei 27(3): 341-397, life cycle.
P. 306, l. 19: Allison should be Kelley and Allison.
P. 308, 247d: should read copepodid.
P. 316, l. 27: add Clarke and Berg (1959): N.Y.; Coker *et al.* (1921): life cycle and propagation; Howard (1914, 1922): propagation; Ortman (1911): identification; Reuling (1919); Tucker (1928): life cycle.
P. 317, l. 9: add (*Bdukus ichthyius* Holl, 1928).
P. 318: add 7. GORDIACEA—*Chorodes:* In stomach and body cavity of *Salvelinus fontinalis, Salmo gairdneri, Micropterus punctulatus,* and tropicals (Nigrelli, 1943).
P. 318: add 8. CYCLOSTOMATA—lampreys.
P. 319: to salamanders add *Gyrodactylus.* To frogs add **Eustrongylides, Gyrodactylus, Diplostomulum flexicaudum.*
P. 327, l. 16: add *Isoglaridacris bulbocirrus.*
P. 328, l. 23: add *Isoglaridacris bulbocirrus.*
P. 332, col. 2, l. 3: add Cestoda: *Isoglaridacris folius.*
P. 332, col. 2, l. 17: add Cestoda: *Isoglaridacris folius.*
P. 340, col. 2, l. 7: add *Ergasilus arthrosis.*

P. 343, col. 2, l. 27: add Gordiacea: *Chorodes* sp.
P. 349, l. 40, p. 352, l. 5, and p. 355, col. 2. l. 6: add *G. katharineri.*
P. 363, col. 2, l. 42: add Acanthocephala: **Atactorhynchus verecundus.*
P. 371, l. 8: add *E. celestis.*
P. 371, col. 2, l. 42: add *Trypanosoma occidentalis.*
P. 371, col. 2, l. 45: add *Bucephalus polymorphis* and *Bunodera mediovitellata* (USSR).
P. 372, l. 1: add *Diplostomulum gasterostei* (Scotland).
P. 372, l. 9: add *G. alexandria.*
P. 372, col. 2, l. 10: *P. proteus* should read *P. bulbocolli.*
P. 391, l. 29: add *Ceratomyxa shasta.*
P. 393: to *P. williamsoni* add *Henneguya zschokkei, *Bothruocephalus* sp. and *P. laruei.*
P. 395, col. 2, l. 1: add *Pseudochetosoma* (Europe).
P. 397, col. 2, l. 32: *P. proteus* should read *P. bulbocolli.*
P. 398, l. 34: add *Ceratomyxa shasta.*
P. 399, col. 2, l. 8: *proteus* should read *bulbocolli.*
P. 400, l. 37: add *Ceratomyxa shasta.*
P. 401, l. 38: add Gordiacea: *Chorodes.*
P. 404, l. 43: add *Lepocreadium areolatum.*
P. 405, l. 44: add **Posthodiplostomum minimum* (unpub. research).
P. 474: add *Diclybothrium,* 101.
P. 481: add *Parasitotrema,* 142.
P. 484: add *Saccocoelioides,* 159.

Preface

If this manual proves to be of value, much credit belongs to the many people who have helped me, directly or indirectly, during the past twenty-three years.

My major professor, Dr. L. O. Nolf, University of Iowa, encouraged my interest in fish parasites when I was an undergraduate in 1942. He steered me in the proper methods of parasite identification and a method of cataloguing the literature on fish parasites which is scattered in many journals and miscellaneous publications. This book, twenty-three years later, is the first step in an attempt to bring that literature together in one publication.

Mr. E. B. Speaker and the late Mr. Robert Cooper, Iowa State Conservation Commission, assisted me by providing part-time employment during my undergraduate studies and before I entered the Armed Forces during the Second World War. During that time they encouraged my interest in fish parasitology.

The late Dr. Chester Herrick, University of Wisconsin and the Wisconsin Conservation Department, provided working facilities for me at the Woodruff Biological Station, where I was able to do my first fish parasite life-history studies (*Posthodiplostomum minimum*).

Dr. Bangham, Dr. Fischthal, Dr. Hunter, and the late Dr. H. J. Van Cleave have given me much counsel in the study of fish parasites. They also laid the groundwork for fish parasitology in the United States (see Bibliography).

The late Dr. H. S. Davis, U.S. Fish and Wildlife Service, paved the way for government participation in fish parasite research at Leetown.

Dr. George R. LaRue, U.S. Animal Disease and Parasite Research Branch, has given me much kindly counsel on strigeoids of fish during the past fifteen years.

Dr. John Mackiewicz, State University of New York, College of Education, Albany, generously helped with the caryophyllaeid cestode section. Dr. Marvin C. Meyer, University of Maine, gave me much help with the leeches. Dr. Victor Sprague, University of Maryland, Natural Resources Institute, assisted me with the genera *Dermocystidium* and *Ichthyophonus*.

The following have aided me with counsel or specimens which

vii

helped directly or indirectly in the preparation of this book: C. D. Becker, Art Bradford, W. Bullock, R. M. Cable, MaeBelle Chitwood, G. Dubois, J. E. Hall, S. H. Hopkins, E. J. Hugghins, R. R. Kudo, J. Lom, A. McIntosh, G. Malmberg, F. P. Meyer, J. D. Mizelle, J. F. Mueller, P. Osborn, F. Sogandares-Bernal, H. W. Stunkard, L. J. Thomas, J. H. Wales, R. A. Wardle, and S. Yamaguti.

Mr. Paul Thompson, Director, Division of Fishery Research, U.S. Bureau of Sport Fisheries, and Dr. S. F. Snieszko, Director, Eastern Fish Disease Laboratory, have made it possible for me to continue and bring this work to completion. Their understanding of the need for such a book for students and researchers is much appreciated.

My wife, Carolyn, has graciously prepared most of the drawings, most of which were traced from photocopies of the originals.

Mr. Robert Putz, my associate, has very ably carried on our laboratory work during the preparation of the book.

Without the help of the typing and library assistance of Mrs. Juanita Collis, Mrs. Bonnie Knott, Mrs. Millicent Quimby, Mrs. Mary Ann Strider, and Mrs. Florence Wright of this laboratory, the finished copy would not have been possible.

Mr. Ernest Callenbach of the University of California Press, and Genevieve Rogers are to be commended on their gracious and efficient handling of the many editorial details.

G. L. Hoffman

Kearneysville, West Virginia

Contents

Introduction   1
Public Health Aspects of Fish Parasites   4
Methods   6
Some North American Fish Parasites
    Listed by Location in the Fish   11
Brief Descriptions of the Groups
    of Fish Parasites   14
Algae and Fungi   16
Protozoa   21
Monogenetic Trematodes   70
Adult Digenetic Trematodes   105
Metacercarial Trematodes   161
Cestodes   203
Nematodes   241
Acanthocephala   270
Leeches   288
Parasitic Copepods   299
Miscellaneous Parasites   316
Fish Parasites Found on or in Other Animals   319
Predators   320
Fish and Parasite Check-List   321
Bibliography   407
Index   469

"We must possess more knowledge concerning the diseases of the lower animals because it is of prime importance in all conservation programs, in that it may be helpful in preventing great losses of animals which are beneficial to man."—Fasten (1922)

Introduction

The importance of fish parasites is related directly to the importance of the fish that they may affect. As our world becomes more and more crowded with people, all foodstuffs, including fish, become increasingly valuable. As a recreational asset, fish rank at or near the top, both for sport fishing and as one of the attractions of nature. Observing live fish, both in nature and in the display aquarium, is enjoyable to young and old alike.

Chubb (1965), in the introduction to a fish-parasite survey in England, has clearly described the dynamics of fish parasitology:

In natural populations of plants and animals parasites are always present. The parasites are normally in a complex dynamic equilibrium with the free-living communities of plants and animals. Fishes are the apex of the predator-prey pyramid within fresh waters, and therefore tend to be infected by a considerable range of parasites, which may occur in large numbers. This is the normal condition found in any natural environment.

However, if some unusual event occurs in the environment, of natural or human origin, the equilibrium between host and parasite may be disturbed, and an (epizootic) of one or more species of parasites may occur. Regulating mechanisms in the environment soon come into play, and a new equilibrium will be established; but in the intervening period, there may be a serious loss of fishes.

It is thus important from an economic point of view, for fishing as an amenity, or for fish farming, that we have a knowledge of the occurrence of parasites on our freshwater and marine fishes. Once we have a sound background knowledge, it may at least be possible to avoid undesirable human interference in natural waters, and even to control some of the more harmful parasites. Unfortunately, for really effective control the fullest possible details of the biology of the parasites are required. Currently, we have very little information on the biology, or indeed of the occurrence, of the parasites of either the freshwater or marine fishes of the British Isles.

Before treatment or control of fish parasitic diseases can be best achieved, the study of fish parasites should follow a logical pattern:

1. Identifying the parasite.
2. Obtaining a thorough knowledge of its life history, which may be simple or very complicated.

1

3. Learning the ecological requirements of the parasite, such as host specificity, optimum temperature, pH, nutrition, and other metabolic requirements.

4. Mapping the geographic range of the parasite.

5. Determining the effect of immunologic mechanisms of the host on the parasite or vice versa.

6. Studying control and treatment methods.

This manual is intended to be an aid to the identification of the freshwater fish parasites of North America. Very little work has been reported from Alaska and Mexico; so those areas are not well represented. Most of the keys are for generic identification. The known species of each genus are listed along with reference citations to enable the worker to find the literature pertinent to species identification, life cycles, and control. I hope to revise this manual at a later date and include keys to the species.

Most of the identification keys are set up in couplets of contrasting characteristics, but a few of them are set up in triplets and quadruplets to save space. Page numbers are given in the keys only when the reference is not to a nearby page.

In many cases, not enough parasite-host records are available to determine host specificity; insufficient records can be very misleading.

Parasites of brackish-water fishes and anadromous fishes are included because these fishes are sometimes found in fresh water. Euryhaline species are represented by *Ergasilus lizae* (copepod), *Glugea hertwigi* (protozoa), and *Ichthyophonus hoferi* (fungus). The genera *Gyrodactylus* (monogenetic trematode), *Oodinium* (protozoa), *Trichodina* (protozoa), and some of the myxosporideans have representatives in both fresh and marine water, but they are not euryhaline.

Foreign fish parasites are included only when the host exists on both continents or if there is a probability that the parasites will also be present in North America.

Scope.—In our laboratory the work on diseases is divided into virology, bacteriology, parasitology, and histopathology.

Importance of fish parasitology.—No attempt will be made here to place a monetary value on the damage done to fish by parasites. In fish hatcheries this could probably be done with a reasonable amount of accuracy where fish are either killed by parasites or are so adversely affected that they do not grow or reproduce properly. One infection, whirling disease of trouts, caused by *Myxosoma cerebralis,* results in such spectacular deformities in surviving fish that they cannot be marketed or used for stocking purposes. Another disease, caused by *Ichthyophonus hoferi*, brings similar crippling. In nature it has been impossible to determine the damage done by parasites except in serious fish kills such as are caused by *Ichthyophthirius multifilis* and *Lernaea cyprinacea.* Each true fish parasite uses the fish for its home and food, and the total damage is relative to the numbers of parasites present. Some parasites effect much mechanical damage in migrating through tissues and may also cause extensive connective tissue proliferation which impairs growth and reproductive processes. If few parasites are present, the fish is usually not visibly impaired; if large numbers are present, the fish may be killed; and there are various intermediate conditions. The pathology is well covered by Dogiel *et al.* (1958, p. 84).

Prevention, treatment, and eradication.—One of the most important factors leading to the control of fish parasites is the study of the ecology of the parasite. This information usually leads to a weak link where control may easily be effected. Ecology of fish parasites is discussed by Dogiel *et al.* (1958, p. 1) and Bauer *et al.* (1959).

Preventing the introduction of dangerous parasites into hatcheries or lakes and streams is far more effective than the chemical treatment of parasitized fish. In a hatchery or other body of water which does not have serious parasites, any new stock should be carefully examined and rejected or treated if parasitized.

Chemical treatment of externally parasitized fish is usually successful (Hoffman, 1959*b*, 1965; Post, 1965; Snieszko and Hoffman, 1963). Intestinal cestodes can be removed with di-n-butyl tin oxide and kamala (see Hoffman, 1959*b*). Much more research should be done in this field because of the rapidly growing importance of fish culture.

Eradication of external parasites is possible, but is usually only temporary if there are fish somewhere in the water supply to a hatchery. Properly designed spring water supply units or wells will never provide a source of parasites, and if no parasites are brought in with transferred fish, there will be no parasites. Some internal infections, such as whirling disease of trouts, are, however, almost impossible to eradicate because of their resistant spores.

The literature has been surveyed as thoroughly as possible through 1965. The outstanding aids to the bibliographic work were the *Index-Catalogue of Medical and Veterinary Zoology* (U. S. Dept. Agric.), including the latest series, *Trematoda and Trematode Diseases—Parts 1–3, Helminthological Abstracts, Biological Abstracts,* and *Publications on Fish Parasites and Diseases, 330 B.C.—A.D. 1923* (McGregor, 1963). The leading contributor to freshwater fish parasitology of North America, because of his numerous parasite surveys, has been Dr. R. V. Bangham (deceased), formerly of College of Wooster, Wooster, Ohio.

Public Health Aspects
of Fish Parasites

AS PARASITES OF MAN

Some larval parasites of North American fish will infect man. So far there are few records of serious cases in this country, but very serious cases, including fatalities, have been reported from Europe, Asia, and the Philippine Islands. The following parasites have been reported from man in North America or are potentially dangerous because of their lack of host specificity:

TREMATODES: *Apophallus* and other heterophyids* would probably develop in the intestine of man, and their ova would probably be trapped in the villi, work into the circulatory system, and be filtered out in various organs, including vital organs (see Africa *et al.*, 1936). *Clinostomum marginatum*, which usually lives in the mouth of herons, might migrate into the trachea of man; a related species in India and Japan has been so reported (Cameron, 1945). The opisthorchids, *Metorchis* and *Opisthorchis*, exist on this continent and are serious liver parasites of man in Asia. *Plagiorchis* has been reported from man in the Philippines (Africa *et al.*, 1936). *Nanophyetus salmincola* has been reported from man in Russia (Chandler, 1955).

CESTODES: *Diphyllobothrium latum* will develop in man and is present in north central United States and Canada.

NEMATODES: The very serious kidney worm, *Dioctophyma renale*, will develop in man. Larval migrans from *Anisakis*-type larvae causes an acute abdominal syndrome in man.† The latter is usually in marine fish.

All the above are acquired by ingesting raw or undercooked fish. Some are acquired accidentally while tasting fish preparations before

---

* Welberry and Pacetti (1954) reported *Heterophyes heterophyes* eggs from a Florida child. Hutton (1957) noted that identification of the eggs is not possible, and he obtained adult *Phagicola* from animals which were fed the same species of fish from Florida.

† A closely related nematode, *Eustoma*, has been reported as causing acute abdominal syndromes in man in the Netherlands (van Thiel, Kuipers, and Roskam, 1960; Williams, 1965).

cooking, or while cleaning fish. All would be killed by *thorough* cooking, hot smoking, pickling, or freezing. Hutton and Sogandares (1959*b*) found that freezing, hot smoking (4 hrs. at 100 to 125°C), but not brine preparation, killed the metacercariae of *Mesostephanus appendiculatoides*.

AS ESTHETICALLY UNDESIRABLE PARASITES

Most fish parasites would not develop in man even if eaten raw, but some are unsightly. In two instances such diseased fish have been banned from interstate commerce by the U. S. Food and Drug Administration. Many fish are discarded by fishermen because of unsightly parasites, but none are harmful to man if the fish are thoroughly cooked.

Methods

If an examination for all parasites is to be made, it is necessary that the fish be dead no longer than a few minutes, for some of the external protozoa leave the dead fish. Some of the larger parasites can be seen with the naked eye, and almost all other parasites, or their cysts, can be seen with a good 10 to 30× dissection microscope. Some protozoa (blood, intestinal, and some external forms) can be seen only with the compound microscope. One of the most satisfactory routines is the following:

1. Kill the fish by pithing or with anesthetics in the water.
2. If it is small, examine the entire fish submerged in water in a Petri dish with the aid of a 10 and 30× dissection microscope. If the fish is too large, remove the fins and examine them as above.
3. Examine the fins or mucus scrapings from the dorsolateral portion of the fish with 100 and 450× magnification for small external protozoa.
4. Remove the gills, submerge in water, and examine with 10 and 100× dissection microscope.
5. Open the fish and remove a drop or two of blood from the heart, dilute about 50:50 with normal physiological saline, and examine at 100× for motile *Cryptobia* and *Trypanosoma*. Blood smears for staining for blood sporozoa should be made at this time or before the gills are removed. Next remove the viscera. If the fish is small this should be done in saline under the dissection microscope. Examine all organs submerged in saline under the dissection microscope; organs should be teased apart with small forceps. Remove a drop or two from the intestine and examine for protozoa (*Hexamita, Schizamoeba, Eimeria*).
6. Remove the gall and urinary bladders carefully with fine forceps and examine for sporozoa and ciliates. Always squash a small piece of kidney under a cover slip and examine for Myxosporidea and *Cryptobia*.
7. Remove the gastrointestinal tract. If the fish is small it should be opened under the dissection scope. If it is larger, scrape the wall,

dilute the contents with saline, shake, and allow the parasites to settle out in a conical or cylindrical container. After 5 to 10 minutes, aspirate the fluid, leaving the parasites on the bottom; repeat until clean and pour into a small Petri dish for examination under the dissection microscope.

8. Some of the body musculature should be teased apart carefully under observation with a dissection microscope. Larval worms, sometimes encysted, are usually easily seen, but a few are very small and can be seen only with magnification. Many trematode and nematode larvae can be recovered by the digestion technique (see below).

9. Remove and examine the eyes and brain. The head is cut lengthwise to remove the brain, and the inside of the mouth and esophagus should be examined in water under the dissection microscope.

All ectoparasites should be studied alive in chlorine-free tap water, and most internal parasites in physiological saline; 0.8 to 0.9 per cent sodium chloride is adequate for temporary use (rainbow trout blood, e.g., is almost isotonic with mammalian blood; others may be different).

FREEING INTESTINAL NEMATODES FROM MUCUS

Trematodes and cestodes are usually easily freed from mucus, but the longer nematodes often become entangled in it. Shaking them for a few minutes in digest fluid (see below) usually frees them. They must not be left in the solution very long or it will harm them.

DIGESTION TECHNIQUE

When trematode and nematode larvae are not numerous, or if large numbers of them are needed, some species of trematodes and all larval nematodes can be concentrated by removing the fish tissue by digestion. Some species of trematodes and all larval cestodes are rather rapidly destroyed by this method. Dissolve 0.5 per cent of commercial pepsin in water containing 0.5 per cent hydrochloric acid. Cut the fish into very small pieces or grind in a hand-powered kitchen food grinder; individual organs can be digested separately if necessary. Add approximately 1 gram of fish to 20 ml of solution, place in a jar containing a few glass marbles to aid agitation, and digest at 37 to 39°C for one to two hours. Prolonged digestion will destroy some parasites which survive the two hour method. Continuous agitation in a waterbath shaker gives the best results, but it can be done in an incubator or stationary water bath with the aid of a magnetic stirrer, compressed air agitation, or even shaking by hand.

The digested material should be strained through a wire tea strainer to remove bones and undigested particles. Allow the digest to stand for about 15 minutes; the helminths will fall to the bottom and the supernatant can be aspirated. Add saline and repeat until clean. For counting purposes only, water instead of saline may be used. Pour the concentrate into a small Petri dish and examine under the dissection microscope.

FISH PARASITE FIXATION AND PRESERVATION

Many parasites contract badly when placed in cold fixative; therefore the following methods should be used. These methods may not work well on all species and the worker may want to try other procedures.

I.   PROTOZOA
    A. Myxosporidea (often rather large white cysts)
        Cut out the cyst with enough adjacent tissue to prevent rupturing of the cyst. Place in a small vial of 10 per cent formalin and label properly.
    B. Trophozoites of motile forms (external or internal)
        1. Place as many protozoa as possible on a microscope slide in one drop of water, add one drop of PVA-AFA fixative adhesive, mix thoroughly, spread over about half of the slide, and allow to dry for future staining. Wet fixation on a slide yields better results (consult books on techniques).
        2. Place as many protozoa as possible in a small vial of 10 per cent formalin. Protozoa usually shrink greatly.

II.   TREMATODES: Most trematodes contract greatly if killed in in cold fixative; so they should be dropped in a small vial of hot Bouin's fixative. Replace with 70 per cent alcohol next day; hot 10 per cent formalin or AFA may be used. Those that are too thick must be flattened under a cover slip with the fixative drawn under the cover slip with absorbent paper; allow for distortion. For monogenetics, drop infected gills or fins into 10 per cent formalin. See also methods in section on monogenetic trematodes.

III. CESTODES: All cestodes, larval or adult, should be fixed and preserved in 10 per cent formalin. Most cestodes can be fixed in a distended position by dropping them into 85°C water and transferring to formalin (Meyer and Penner, 1962).

IV.   NEMATODES: All nematodes, larval or adult, should be fixed and preserved in hot 70 per cent alcohol. The addition of 2 to 5 per cent glycerine will prevent their air-drying if the alcohol evaporates later.

V.   ACANTHOCEPHALA (thorny-headed worms): All acantho-

cephala should be left in distilled water until hook-bearing proboscis is extruded (about 1 to 4 hours) and then dropped into
hot Bouin's. Transfer to 70 per cent alcohol. Hot formalin is
second choice.

VI.   LEECHES: See section on leeches.

VII. COPEPODS (fish lice): If easily detached (*Argulus*), drop individuals into a small vial of 70 per cent alcohol. If not easily
detached, cut out a small piece of tissue containing the para-
site(s) and drop into a small vial of 70 per cent alcohol, or
carefully remove the tissue from the parasite before preserving.

PERMANENT PREPARATION OF PARASITES

Only a brief outline of methods used will be given here. The techniques can be found in books on general parasitology and histologic
technique.

I.   PROTOZOA: (*a*) Blood flagellates are found more easily alive
in wet mounts, but they and other blood protozoa should be
spread in a thin smear on a microslide and stained with Giemsa's
stain. The stain and wash should be buffered at about pH 7. (*b*)
Trichodinid morphology is best studied with the silver impregnation method (Lom, 1958; Raabe, 1959). (*c*) Nuclear detail
shows up well with iron hematoxylin staining following osmic
acid fixation. (*d*) Myxosporidean spores are best studied fresh,
but can be affixed to slides, fixed while still wet, and stained
with Giemsa's. The vegetative stages are best studied in cross
sections stained with hematoxylin and eosin. If spores are present, Giemsa's is also recommended.

II.   TREMATODES: Several stains yield good results, but worms
vary considerably; so no one stain is recommended. Those most
widely used are (1) Semichon's and borax carmine, counter-
stained with fast green, (2) Harris' or Heidenhain's hematoxylin,
(3) hematin, (4) coelestin blue B, and (5) chlorazol black E.
The latter is selective for certain of the chitinoid hard parts of
Monogenea. Most of the staining involves over-staining and
then destaining under observation until the desired result is at-
tained. Some species stain more readily than others; so it is
difficult to recommend exact staining times. Be certain that specimens stained with hematoxylin are well neutralized before
mounting or they may fade. Also see methods in section on
monogenetic trematodes.

III. CESTODES: Much the same as trematodes.

IV.   NEMATODES: Because of the impervious cuticle, it is very difficult to prepare stained permanent mounts; however, smaller ones can be mounted in glycerine jelly and perhaps other semi-solid material. Most workers make temporary mounts of larger nematodes in lacto-phenol or alcohol-phenol for clearing; the specimens can be returned to alcohol after use.

V.   ACANTHOCEPHALA: Much the same as nematodes.

VI.   LEECHES: Similar to trematodes, but some are too thick for staining (see section on leeches).

VII.   COPEPODS: Store in alcohol and study as temporary mounts (see Harding, 1950). Small *Argulus* could be cleared and mounted permanently.

Some North American Fish Parasites Listed by Location in the Fish

This section is included as an aid to identification because some parasites are found only in very selective organs or tissues. Some of these are probably rare records. For references see specific parasite section.

1. EGGS
   Fungi: *Saprolegnia* and relatives.
   Protozoa: *Carchesium* (walleye and trout eggs; only record). *Epistylis* reported from catfish eggs (Wellborn, 1964).

2. BODY AND FIN SKIN
   Protozoa: All protozoa and fungi of no. 1 above and no. 4 below.
   Trematoda: *Gyrodactylus*, but usually not other monogenetics; metacercariae of many species.
   Copepoda: *Argulus, Lernaea*, occasionally *Salmincola*.

3. CAUDAL FIN: *Philometra carassii* between fin rays of *Carassius auratus*.

4. GILLS
   Fungi: *Dermocystidium* (salmon and trout).
   Protozoa: *Amphileptus, Bodomonas, Chilodonella, Colponema, Costia, Cyclochaeta, Epistylis, Ichthyophthirus*, Microsporidea, Myxosporidea, *Oodinium, Scyphidia, Trichodina*, and *Trichophrya* in gill arches and filaments.
   Trematoda: *Gyrodactylus* and other monogenetics; certain metacercariae.
   Copepoda: *Achtheres, Ergasilus, Lepeoptheirus, Lernaeocera, Salmincola*.

5. BLOOD
   Protozoa: *Trypanosoma* and *Cryptobia* free, *Babesiosoma, Dactylosoma*, and *Haemogregerina* in red cells, rarely *Kudoa* (Myxosporidea).
   Trematoda: *Sanguinicola* in blood vessels, including gill vessels, also some migrating larval forms.
   Nematoda: *Philometra obturans* in gill vessels of pike, Russia.

6. BRANCHIAL CAVITY
   Trematoda: *Syncoelium*.

11

7. MOUTH
Trematoda: Some monogenetics, *Leuceruthrus*.
Copepoda: *Salmincola*, sometimes others.

8. ESOPHAGUS
Trematoda: *Azygia, Cotylaspis, Derogenes, Proterometra*.
Arthropoda: Larval mites.

9. STOMACH
Protozoa: *Schizamoeba* in fingerling trout.
Trematoda: *Allocreadium, Aponurus, Azygia, Caecincola, Centrovarium, Derogenes, Genolinea, Hemiurus, Leuceruthrus*.
Nematoda: *Haplonema*.

10. INTESTINE AND PYLORIC CECA
Protozoa: *Hexamita* trophozoites and cysts in lumen, *Schizamoeba* cysts in fingerling trout, *Eimeria* in goldfish and trout.
Trematoda: Adults of many species.
Cestoda: Adults of many species.
Nematoda: Adults of many species.
Acanthocephala: Adults of many species.

11. BODY CAVITY: MESENTERIES, LIVER, SPLEEN
Protozoa: Many Myxosporidea species; rarely Microsporidea.
Trematoda: Many metacercarial species, including white grub (*Posthodiplostomum*); adult *Paurorhynchus* (bucephalid), adult *Acetodextra* in catfish.
Cestoda: Larval *Diphyllobothrium, Haplobothrium, Ligula, Proteocephalus, Schistecephalus, Triaenophorus*.
Acanthocephala: Larvae of *Echinorhynchus salmonis, Leptorhynchoides thecatum, Pomphorhynchus bulbocolli*.
Nematoda: Adult *Philonema*, many larval species.

12. GALL BLADDER
Protozoa: Myxosporidea, (*Hexamita*? Europe).
Trematoda: *Crepidostomum cooperi, C. farionis, Derogenes, Plagioporus sinitsini*.
Cestoda: *Eubothrium salvelini*; Nematoda: *Rhabdochona*.

13. HEPATIC BILE DUCTS
Trematoda: *Phyllodistomum* sp.

14. KIDNEY
Fungi: *Ichthyophonus hoferi*.
Protozoa: Myxosporidea in renal tubules and also cysts.
Trematoda: Metacercaria of *Posthodiplostomum minimum centrarchi* and *Nanophyetus salmincola*. Adult *Phyllodistomum* in renal tubules and ureters.

15. URINARY BLADDER
Protozoa: Myxosporidea, *Vauchomia* (trichodinid).
Trematoda: *Acolopenteron* (monogenetic), *Phyllodistomum*.

16. OVARY
Protozoa: *Wardia* (Myxosporidea).
Trematoda: Adult *Acetodextra* in catfish, adult *Nematobothrium* in *Ictiobus*.
Nematoda: *Philonema*.

17. TESTES (cf. Moore, E., 1925, p. 93, *Hexamita*).

18. SWIM BLADDER
Trematoda: *Acetodextra*.
Nematoda: *Cystidicola*.

19. HEAD SINUSES
Nematoda: *Philometra*.

20. CARTILAGE
Protozoa: *Myxosoma cerebralis* (Myxosporidea) in trout, *M. cartilaginis* in bluegills, *M. scleropercae* in cartilaginous sclera of eye of perch, *M. hoffmani* in sclera of eye of *Pimephales promelas*, *Henneguya brachyura* from fin ray of *Notropis*, *H. schizura* from sclera of *Esox lucius*, *Henneguya* sp. from branchial arch of *Pomoxis*.

21. EYES
Protozoa: rarely Myxosporidea in eye capsule.
Trematoda: *Diplostomulum* in lens and vitreous humor.

22. NERVOUS SYSTEM
Protozoa: *Myxosoma cerebralis* (Myxosporidea), effect on CNS, but parasites in cartilage, *Myxobolus neurobius* in spinal and peripheral nerves of salmonids in Europe, *Myxobolus encephalica* in brain of carp in Europe.
Trematoda: *Diplostomulum*, *Ornithodiplostomum*, *Parastictodora*, and *Euhaplorchis* metacercariae on brain; *Psilostomum* metacercariae in lateral line canal.

23. MUSCLE AND BODY CONNECTIVE TISSUE
Protozoa: Many Myxosporidea species, *Sarcocystis*.
Trematoda: Many metacercarial species, including yellow grub (*Clinostomum*) and black spot (*Neascus*).
Cestoda: Larval *Triaenophorus*, *Diphyllobothrium*.
Acanthocephala: Larval forms.

Brief Descriptions of the Groups of Fish Parasites

The following list is intended as an aid to identification. It may be difficult to place a parasite in its proper phylum, for some larval forms and at least two adults do not have some of the usual typical morphological characteristics.

1. ALGAE: Unicellular, 10 to 20µ in diameter; green pigment. Usually in clumps as small cysts in gills, eye orbit, occasionally in viscera of small tropical fish.

2. FUNGI: Usually filamentous, about 20µ in diameter, nonseptate. *Ichthyophonus* often occurs as spheres up to 100µ in diameter, but some show hypha-like outgrowths. Filamentous bacteria (Myxobacteria) and actinomycetes only 1 to 3µ in diameter.

3. PROTOZOA: Commonly called single-celled animals. Amoebas, ciliates, and flagellates easily recognized, but not vegetative stages of some sporozoa. Developing Myxosporidea syncytial rather than truly single-celled, but with quite distinctive spores.

4. TREMATODES, monogenetic: Body flattened dorsoventrally; no true suckers; mouth usually opens into muscular pharynx; posterior organ of attachment (haptor) bears chitinoid hooks or clamps; external parasites except some in mouth and one in urinary bladder.

5. TREMATODES, digenetic: Body flattened dorsoventrally; oral and ventral suckers except in gasterostomes and *Sanguinicola*; only a weak oral sucker in *Nematobothrium*. Young metacercariae may resemble adults or cercariae; some developing metacercariae go through a "reorganization" stage which appears inflated and may not resemble typical trematode.

6. CESTODES: Body flattened dorsoventrally; some adults segmented; head (scolex) usually bears suckers, hooks, or suctorial grooves, occasionally no organs of attachment. Larvae contain microscopically conspicuous calcareous concretions.

7. NEMATODES: Body cylindrical with rigid cuticle, one or both ends attenuated; no organs of attachment.

14

8. ACANTHOCEPHALA: Body cylindrical, sometimes slightly flattened; spectacular hook-bearing eversible proboscis present.

9. GORDIACEA: Not true fish parasites; may be recovered from stomach of fish; parasites of insects; resemble nematodes except for extreme length, both ends somewhat blunt rather than attenuated, and body surface ornamented.

10. LEECHES: Some flattened dorsoventrally, some only slightly; body segmented; anterior and posterior suckers may be larger or smaller than body diameter; external parasites.

11. COPEPODS, parasitic: May be louselike (*Argulus*), wormlike (*Lernaea*), or grublike (*Salmincola, Ergasilus*); external parasites.

12. GLOCHIDIA: Larval clams encysted in fins and gills; bivalve shell, usually armed with hooks.

13. INSECT PREDATORS: Not true parasites; some larval and adult Hemiptera and Coleoptera extremely predacious.

Algae and Fungi

ALGAE

Algal parasites of North American freshwater fish are probably rare; there are very few published records of them.

A dinoflagellate, *Oodinium limneticum* Jacobs, was described from fish in Minnesota. This is discussed under protozoa.

A filamentous green alga, *Cladophora* sp., has been described from the opercula of black bass (Vinyard, 1955). The alga had probably started growing on bone that was exposed after spawning activity. I have seen similar growth on the opercula of rainbow trout which had been kept in concrete ponds for two or three years.

Stigeoclonium (filamentous) and *Chlorococcales* (unicellular) were found growing subepidermally within the nasal capsule and below the frontal bones of a kissing gourami (*Helostoma temmincki*) (Nigrelli, McLaughlin, and Jakowska, 1958). The latter alga grew visibly in three months after being injected subdermally into a young gourami.

A unicellular alga was reported, but not named, from the gills and viscera of swordtail (*Xiphophorus helleri*) and kissing gourami (*Helostoma temmineki*) from a tropical fish farm in Florida (Hoffman, Bishop, and Dunbar, 1960). The fish farm was losing many fish, and it is probable that extensive invasion by the alga was a major cause of mortality.

Nine cysts, 0.29 to 1.16 mm in diameter, of *Chlorella* sp. (*Chlorophyta*) were found in the eye orbit of a bluegill (Hoffman, Prescott,

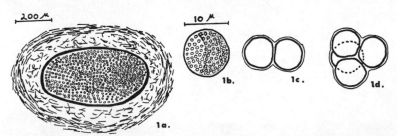

Fig. 1. *Chlorella* sp. from eye orbit of bluegill (from Hoffman, Prescott, and Thompson, 1965). *a*, cyst containing many algal cells; *b*, single cell showing cup-shaped chloroplast and starch bodies; *c* and *d*, dividing stages showing mother cell wall.

and Thompson, 1965). The interior of the cysts was filled predominantly with *Chlorella* (figs. 1a–1d), a unicellular, round, smooth alga, further characterized by having a single cup-shaped green chloroplast, and usually lacking pyrenoids. Multiplication is by autospores, 2, 4, 8, or 16 of which are formed within a cell and liberated by the rupture of the mother cell wall (figs. 1c, 1d). The individual cells were 7 to 10μ in diameter. Within the cysts was a lesser amount of blue-green algal growth, apparently *Phormidium mucicola*. Similar cysts were found in the gills of a bluegill from the same pond two years later. These were successfully transferred to other fish by injection and will be reported later.

Myxonema tenue, a filamentous green alga, was found growing on goldfish in England (Hardy, 1907) and on *Haplochilus latipes* in Japan (Minakata, 1908). Other European algae of fish are discussed in Schäperclaus (1954).

There is no known treatment of algal diseases of fish.

FUNGI

Fungi are plantlike structures lacking chlorophyl; the assimilative phase consists of a true plasmodium or a mycelium, or rarely of separate uninuclear independent cells not amoeboid and at no time uniting as a plasmodium-like structure. Only those concerned with North American freshwater fish will be considered.

Class PHYCOMYCETES

Mycelium, if present, usually continuous throughout in active assimilative phase (nonseptate); if lacking, reproduction not by budding; perfect (sexual) stage usually represented by oospores or zygospores; imperfect (asexual) stage by sporangiospores or modified sporangia, less commonly by true conidia.

Genus **Saprolegnia** and Related Genera
(Fig. 2)

Species of the genus *Saprolegnia* are usually implicated in fungus diseases of fish and fish eggs, although *Achlya, Aphanomyces, Leptomitus*, and *Pythium* have also been reported (Scott and O'Bier, 1962; Scott and Warren, 1964). These fungus diseases of fish are often considered primary or secondary invaders following injury, but once they start growing on a fish the lesions usually continue to enlarge and may cause death unless medication is provided. Fungi often attack dead fish eggs and soon encompass adjacent live eggs, which are attacked and killed; thus they constitute one of the most important egg diseases. These fungi grow on many types of decaying organic matter and are widespread in nature.

Ref. Scott (1964); Scott and O'Bier (1962); Scott and Warren (1964).

MORPHOLOGY, IDENTIFICATION, AND LIFE CYCLE

The presence of fungus on fish or fish eggs is indicated by a white cottony growth which consists of a mass (mycelium) of nonseptate filaments (hyphae) each of which is about 20μ in diameter. Under low magnification the filaments of older infections may be seen to terminate in clublike enlargements which contain the flagellated zoospores (fig. 2). These eventually escape and cause infection of other fish or eggs.

Fish eggs.—During incubation, some eggs die and may soon be invaded by fungus. In time, surrounding eggs are covered by the mycelium, and death results. Unless control measures are exercised, the ever-expanding growth will claim virtually every egg. Kanouse (1932) found circumstantial evidence that under some conditions (probably crowding) fungus filaments could penetrate living eggs. Under experimental conditions, living eggs which were in no way crowded were not invaded. H. S. Davis (1953) states that there is no evidence that *Saprolegnia* can develop on normal eggs unless foreign organic matter is present.

Fish.—Injuries produced by spawning activity and other trauma, or lesions caused by other infections, are often attacked (Hoffman, 1949; Vishniac and Nigrelli, 1957; Scott and O'Bier, 1962; Scott, 1964; Scott and Warren, 1964). Holding warm-water fish in cold water during summer and debilitation caused by other factors probably render fish susceptible to fungus attack.

These fungi belong to the class Phycomycetes and are typified by the nonseptate mycelium. Phycomycetes belonging to the *Saprolegnia* family (Saprolegniaceae) are characterized by having club-shaped sporangia containing zoospores (fig. 2) and round oogonia (sexual reproductive structures). The latter can be obtained only by culturing on special material (Hoffman, 1949; Scott and O'Bier, 1962; Scott, Powell, and Seymour, 1963). Species identification can be made only after studying these structures. The other genera mentioned above can be likewise identified, but Scott (1964) and Scott and O'Bier (1962) should be consulted.

Depending upon the temperature, 24 to 48 hours are required to complete a minimum life cycle of reproductive spore to mycelium to reproductive spore.

TRANSMISSION AND PATHOGENICITY

These fungi grow on many types of decomposing organic matter, and the resulting asexual reproductive spores are present almost everywhere in natural waters.

Dead eggs and fish are generally susceptible. Under favorable conditions healthy eggs resist fungus invasion, and at times dead eggs do not succumb to penetration for many days after turning white. Infertile eggs at times can be incubated a month or longer in the presence of *Saprolegnia* and relatives without being invaded.

Treatment is discussed by Hoffman (1963) and Scott and Warren (1964).

GEOGRAPHIC RANGE

Saprolegnia and relatives are present in all parts of North America.

There are no known seasonal restrictions to infestation of fish eggs by Saprolegniaceae, but fish are more likely to be affected in early spring (temperate zone) and after spawning activity.

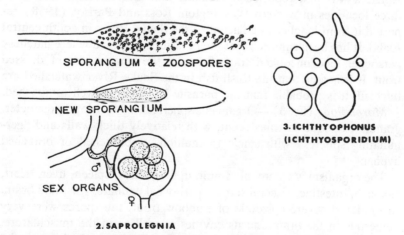

SPORANGIUM & ZOOSPORES

NEW SPORANGIUM

3. ICHTHYOPHONUS
(ICHTHYOSPORIDIUM)

SEX ORGANS

2. SAPROLEGNIA

Fig. 2. *Saprolegnia* sp. TOP: Sporangium with many flagellated zoospores. MIDDLE: An old sporangium with a new one proliferating into it, a characteristic of the genus. BOTTOM: Sex organs, antheridium on left, oogonium on right; usually found only on special media.

Fig. 3. Spores of *Ichthyophonus hoferi* (*Ichthyosporidium h.*) showing variation in size and "hyphal proliferation."

Form Class FUNGI IMPERFECTI
Genus Ichthyophonus
Ichthyophonus hoferi Plehn and Mulsow
(Syn.: *Ichthyosporidium hoferi*)
(Fig. 3)

This fish fungus has been considered to belong to the Phycomycetes by Sproston (1944). Others have avoided trying to place this parasite because it is not a typical member of any class of fungi. Because of this I recommend placing it in the form class Fungi Imperfecti (= neither asci nor basidia known, and relationship not otherwise inferable with reasonable assurance). Sprague (1965, pers. comm.) states that the generic name *Ichthyophonus* has priority over *Ichthyosporidium*, which is a protozoan.

Ichthyophonus hoferi and *Ichthyosporidium* sp. have been reported from Atlantic Ocean fish (cf. Sindermann and Scattergood, 1954), European freshwater fish (cf. Schäperclaus, 1954), aquarium fish (Reichenbach-Klinke, 1957), and North American rainbow trout (Erickson, 1965; Gustafson and Rucker, 1956; Ross and Parisot, 1958). It is not known whether these records concern one species or several.

Ichthyophonus (*Ichthyosporidium*) in rainbow trout in North America was first reported by Rucker and Gustafson (1953) from three localities in western Washington. Ross and Parisot (1958) reported it from hatcheries adjacent to the Snake River in south central Idaho. It has been rumored that the parasite has spread where hatchery personnel feed uncooked trash fish and the raw viscera of dressed trout. If it is true that the trash fish in the Snake River watershed are infected, it is probable that the parasite has become well established.

Morphology (fig. 3)—Organisms spherical, 10 to 145µ in diameter, granular, opaque to translucent, with relatively thick walls and "germinating spheres" with single to multiple, unbranched or branched hyphae.

The organism is commonly found in the kidney, spleen, liver, heart, stomach, intestine, visceral serosa, peritoneal exudate, gills, and brain. In the latest severe epizootic of rainbow trout, the spores were very numerous in the brain and its cavities as well as in the musculature. Central nervous system involvement apparently resulted in partial denervation of the skeletal musculature which caused spinal curvature.

Most of the older spores are encapsulated in small host cysts or granulomas.

Control.—No chemotherapy is known. Prophylaxis consists of sanitation and not feeding raw fish products.

Protozoa

Contents

Key to the Groups of Protozoa   24
Superclass SARCODINA . .   24
  Genus *Schizamoeba* . . .   24
Superclass MASTIGOPHORA
(FLAGELLATA)   25
Key to the Genera of
Mastigophora   25
  Order DINOFLAGEL-
LIDA   27
    Genus *Oodinium*   27
  Order EUGLENIDA   28
    Genus *Euglenosoma* . .   28
  Order "PROTOMONA-
DINA"   28
    Genus *Trypanosoma* . .   28
    Genus *Lamellasoma* . . .   29
    Genus *Cryptobia*   29
    Genus *Colponema*   30
    Genus *Bodomonas*   30
  Order "POLYMASTI-
GINA"   30
    Genus *Costia*   30
    Genus *Urophagus*   31
    Genus *Hexamita*   31
Subphylum CILIOPHORA
(INFUSORIA)   31
Key to the Genera of Cili-
ophora   31
Subclass SUCTORIA   35
  Genus *Trichophrya* . . .   35
Nonsuctorian CILIATES . .   36
Subclass SPIROTRICHIA .   36
  Genus *Balantidium* . . .   36
  Genus *Nyctotherus* . . .   36
Subclass HOLOTRICHIA .   37
  Order HYMENOSTO-
MATIDA   37
    Genus *Ichthyophthirius*   37

    Genus *Glaucoma*   38
  Order GYMNOSTOMA-
TIDA   38
    Genus *Chilodonella* . . .   39
    Genus *Amphileptus* . . .   39
  Subclass PERITRICHIA . .   39
    Genus *Scyphidia*   39
    Genus *Glossatella*   41
    Genus *Epistylis*   41
    Genus *Zoothamnium* . .   41
    Genus *Carchesium*   43
    Family Urceolariidae . . .   43
    Genus *Tripartiella*   43
    Genus *Trichodinella* . .   44
    Genus *Vauchomia*   44
    Genus *Trichodina*   44
SPOROZOA   46
Key to the groups of SPORO-
ZOA   46
  Group ACNIDOSPORIDIA   47
    Genus *Sarcocystis*   47
  Class TELOSPOREA   47
  Subclass COCCIDIA   47
    Genus *Eimeria*   47
    Genus *Haemogregarina*   48
  Order HAEMOSPORI-
DIA   48
    Genus *Dactylosoma* . .   48
    Genus *Babesiosoma* . . .   48
    Genus *Leucocytozoon* .   48
Subphylum CNIDOSPORA . .   49
Class MYXOSPORIDEA . .   49
Key to the Genera of
Myxosporidea   51
  Suborder EURYSPOREA   53
    Genus *Ceratomyxa* . . .   53
    Genus *Myxoproteus* . .   54
    Genus *Leptotheca*   54

21

Genus *Trilospora* 54
Genus *Wardia* 54
Suborder SPHAERO-
SPOREA 54
Genus *Chloromyxum* .. 54
Genus *Kudoa* 55
Genus *Sinuolinea* 55
Genus *Sphaerospora* .. 55
Genus *Unicapsula* 56
Suborder PLATY-
SPOREA 56
Family MYXIDIIDAE . 56
Genus *Myxidium* 56
Genus *Sphaeromyxa* .. 57
Genus *Zschokkella* ... 57
Family COCCO-
MYXIDAE 57
Genus *Coccomyxa* 57
Family MYXOSO-
MATIDAE 57
Genus *Myxosoma* 57

Genus *Agarella* 60
Family MYXO-
BOLIDAE 60
Genus *Myxobolus* 60
Genus *Unicauda* 63
Genus *Henneguya* 63
Genus *Thelohanellus* .. 64
Genus *Hoferellus* 64
Family ELONGIDAE . 64
Genus *Mitraspora* 64
Genus *Myxobilatus* ... 65
Class MICROSPORIDEA .. 65
Key to the genera 65
Genus *Nosema* 67
Genus *Glugea* 67
Genus *Plistophora* 67
Genus *Thelohania* 68
Sporozoa of uncertain classi-
fication 68
Genus *Dermocystidium* 68

At the present time protozoa probably cause more disease in fish culture than any other type of animal parasite. In some instances, perhaps under environmental circumstances unfavorable to the fish, large populations of external protozoa appear on the body and gills of the fish. When present in small numbers, they usually produce no obvious damage; but in large numbers they greatly impair the epithelium, particularly of the gills. Some of the protozoa actually feed on the cells and mucus of the fish; others, which attach to the fish but do not feed on it, may injure it by blockage, particularly to the gills. One external protozoan, *Ichthyophthirius multifilis*, produces severe devastation by burrowing under the epithelium. *Costia* and some species of *Trichodina* cause "blue slime" of the body.

The possible damage done by internal protozoan parasites of fish has not been adequately studied. Myxosporidea have been found in most organs of fish and in some instances are very harmful. Blood flagellates (*Cryptobia*) and sporozoa (*Haemogregarina*) occur in the blood and are transmitted by leeches. Only one flagellate, *Hexamita* (*Octomitus*), one amoeba, *Schizamoeba*, and one sporozoan, *Eimeria*, have been found in the intestine.

Fish in nature are infected with a great variety of protozoa. Disease resulting from these infections has not been reported very often, for several possible reasons: (1) certain stages of the parasites may be dispersed in a large volume of water and therefore fish are not heavily parasitized; (2) the parasites, even if present in large numbers, may do little harm; or (3) the most severely affected fish die, or may be weakened and utilized by natural predators and therefore are not seen.

The most comprehensive works which include protozoa in fish culture are Bychovskaya-Pavlovskaya *et al.* (1962), Davis (1953), Dogiel *et al.* (1958), and Schäperclaus (1954). For general description of protozoa, as well as of specific protozoa, the reader is referred to Kudo (1954).

For technique in studying protozoa refer to the section on methods.

The classification of Honigberg *et al.* (1964) has been followed wherever possible.

Certain protozoa from other parts of the world are included for reference in the event that related forms are discovered in North America.

KEY

TO THE GROUPS OF PROTOZOA
(Also see Wellborn and Rogers, 1966)

1. Trophozoite with pseudopodia (cysts in intestine) . . . .
. . . . .Superclass SARCODINA
1. Trophozoite with flagellum or flagella . . . .
. . . . Superclass MASTIGOPHORA
1. Trophozoite with cilia . . . . .Subphylum CILIOPHORA
(p. 31)
1. Trophozoite with tentacles . . . . .See no. 1 of CILIOPHORA
(p. 35)
1. Usually without cell organs of locomotion, but trophozoites
of some myxosporidea have very sluggish pseudopodia; producing spores . . . . .Subphylum SPOROZOA
(p. 46)

Superclass SARCODINA Hertwig and Lesser, 1874

The members of this group do not possess a thick pellicle and are capable of forming pseudopodia. This is one of the most useful criteria for their identification. The trophozoites of some Myxosporidea have sluggish pseudopodia.

The food of Sarcodina which are parasitic in the intestinal tract is usually bacteria and other organic matter, but at least one amoeba of mammals feeds on blood cells as well.

The life cycle usually consists of a trophozoite in the intestine which encysts and passes out with the feces. The one which is parasitic in fish, however, develops in the stomach. The host is infected when the cysts are accidentally ingested. An undescribed species of amoeba has been reported from the gills of salmonids in Washington.

For a complete description of the Sarcodina see Kudo (1954, p. 417).

Genus **Schizamoeba** Davis, 1926
(Fig. 4)

Trophozoite nucleus vesicular, without endosome, but with large discoid granules along nuclear membrane; one to many nuclei; cyst muclei with large endosome. One species.

Schizamoeba salmonis Davis, 1926. Trophozoite in mucus of stomach of trout; cysts from intestine seen more often. Very common in young hatchery trout but apparently not pathogenic (Davis, 1926; 1946, p. 53). Plehn (1924, p. 424) notes an amoeba in the kidney of brook and rainbow trout in Europe; pathogenicity uncertain (there is a possibility that this was the trophozoite of a myxosporidean).

Superclass MASTIGOPHORA Diesing, 1866
(FLAGELLATA)

The parasitic protozoa of this group possess one to several flagella which are usually visible except on the attached forms of *Costia*, *Oodinium*, and *Euglenosoma*.

Some forms ingest food through a mouthlike opening, the cytostome, and some absorb food in solution through their pellicles.

Most of them have one nucleus, but *Hexamita* has two. The nucleus is vesicular and contains an endosome. Contractile vacuoles are present in some.

Asexual reproduction is by longitudinal fission, and trypanosomes in culture divide in such a fashion as to produce "rosettes" of many individuals which eventually separate.

Ref. Kudo (1954, p. 254).

KEY

TO THE GENERA OF MASTIGOPHORA

1. With chromatophores   2
1. Without chromatophores   3
2. With 2 flagella in free-swimming stage, 1 transverse; parasitic stage without flagella but basal filipodia about 50μ long
   . . . . (fig. 5) *Oodinium*
2. With 2 flagella in free-swimming stage, none transverse; globular on gills of crappies; one species . . . . . (fig. 6) *Euglenosoma*
3. With 1 or 2 flagella . . . . Order "PROTOMONADINA"   4
3. With 3 to 8 flagella which may not show on some attached forms . . . . Order "POLYMASTIGINA"   8
4. With 1 flagellum . . . . Family Trypanosomatidae   5
4. With 2 flagella   6
5. In blood . . . . (fig. 7) *Trypanosoma*
5. On gills; several rows of small refringent rods running lengthwise of body . . . . (fig. 8) *Lamellasoma*
6. Trypanosome-like but with 2 flagella, 1 anterior, 1 posterior; in blood . . . . (fig. 9) *Cryptobia*
6. Not trypanosome-like but body more rounded; 1 anterior flagellum, 1 trailing flagellum; on gills   7

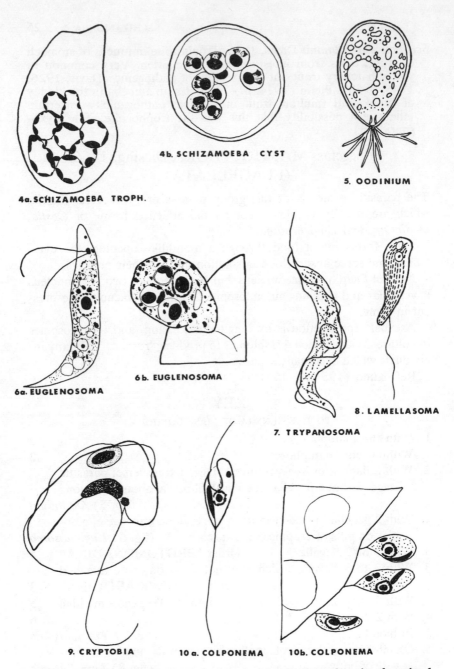

4a. SCHIZAMOEBA TROPH.

4b. SCHIZAMOEBA CYST

5. OODINIUM

6a. EUGLENOSOMA

6b. EUGLENOSOMA

7. TRYPANOSOMA

8. LAMELLASOMA

9. CRYPTOBIA

10a. COLPONEMA

10b. COLPONEMA

Figs. 4–10. Fig. 4. *Schizamoeba salmonis. a*, stained trophozoite; *b*, stained cyst (from Davis, 1926). Fig. 5. *Oodinium limneticum*, parasitic form (from Jacobs, 1946). Fig. 6. *Euglenosoma branchialis. a*, free form; *b*, attached form (from Davis, 1947). Fig. 7. *Trypanosoma percae canadense* (from Fantham, Porter, and Richardson, 1942). Fig. 8. *Lamellasoma*

7. With rod-shaped blepharoplast; on gills of crappie . . . .
. . . . (fig. 10) *Colponema agitans*
7. With round or oval blepharoplast; on gills of crappie and bluegills (may be related to European *Cryptobia branchialis*)
. . . . (fig. 11) *Bodomonas concava*
8. With 1 nucleus and 4 posterior flagella; latter may not show
on attached forms . . . . (fig. 12) *Costia*
8. With 2 nuclei and 8 flagella; in intestine of fish   9
9. With 2 movable posterior appendages . . . . . (fig. 14) *Urophagus*
9. Without movable posterior appendages; from salmonids, perhaps others . . . . (fig. 13) (*Octomitus*) *Hexamita*

Order DINOFLAGELLIDA Bütschli, 1885

With the characteristics of the order (see Kudo, 1954, p. 310). The
most striking are the presence of chromatophores, yellow, brown,
green; a posterior flagellum, and a "girdle" with a transverse flagellum.
Flagella do not show on forms attached to fish. There are many free-
living species, mostly marine, and at least 55 species are parasitic in
or on invertebrates. Only one genus, *Oodinium*, has been found on fish.

Genus **Oodinium** Chatton
(Fig. 5)

Parasitic stage spherical or piriform; 12 to 96µ in greatest diameter
with short stalk; nucleus large; often with yellowish pigment; three
species described from skin of fish, mostly marine (Kudo, 1954, p.
321). One European species, *O. pillularis*, invades skin (Reichenbach-
Klinke, 1956). Life cycle: piriform parasitic stage grows to subsphere
of about 96µ in diameter in about a week, drops off the fish, becomes
spherical cyst which divides 8 times in 3 to 4 days to produce about
256 gymnodinian swarmers capable of attaching to fish.

Ref. Brown (1934); Brown and Hovasse (1946); Jacobs (1946);
Nigrelli (1936*a*).

Oodinium limneticum Jacobs, 1946: skin of freshwater aquarium
fish, Minn.; Hoffman (1959*c*): causes disease at aquarium fish
hatcheries, Fla.

---

bacillaria (from Davis, 1947). Fig. 9. *Cryptobia lynchi*, from stained
blood smear (from Katz, 1951). Fig. 10. *Colponema agitans. a*, free parasite; *b*, attached form (from Davis, 1947).

Order EUGLENIDA Bütschli, 1884

With the characteristics of the order (see Kudo, 1954, p. 293). The main characteristics are the presence of chromatophores, paramylon body, and anterior stigma. The status is not clear in the one species described from fish.

Genus **Euglenosoma** Davis, 1947
(Fig. 6)

Body euglenoid type, long, spindle-shaped, tapering to a sharp point posteriorly; body plastic; anterior end spatulate and flexible; ventral groove and cytostome present; 2 equal flagella arise from cytostome; body twisted spirally.

Euglenosoma branchialis Davis, 1947: reported once; on gills, *Pomoxis annularis*, *P. sparoides*, W. Va.; 20 to 30µ long by 4 to 5µ wide.

Order "PROTOMONADINA"

With the characteristics of the order (see Kudo, 1954, p. 353). Plastic forms with 1 or 2 flagella.

Genus **Trypanosoma** Gruby
(Fig. 7)

Body leaflike; single nucleus; blepharoplast from which single flagellum arises; basal portion of flagellum forms undulating membrane extending along one side of body; parasites in blood of mammals, birds, reptiles, amphibia, and fish. Vectors for fish and amphibian forms are leeches. Pathogenicity in fish not well known. Life cycle: trypanosomes, picked up by leech with blood meal, appear in crop on second day, multiply rapidly, and become crithidia in 4 to 6 days. At 7 days they become slender metacyclic trypanosomes and migrate back to proboscis where they are expelled during next blood meal (Qadri, 1962).

Ref. Kudo (1954, p. 344).

Trypanosoma myoxocephala Fantham *et al.*, 1942: in blood, *Myoxocephalus octodecimspinosus*, Que.
T. percae canadense Fantham *et al.*, 1942: in blood, *Perca flavescens*, Que.
T. remaki Laveran et Mesnil. Kudo (1921): in *Esox reticulatus*, N. Y.
Trypanosoma sp., Fantham and Porter (1947): in *Salvelinus fontinalis*, Que.; Strout (1961): in *Esox niger*, N. H.

Genus **Lamellasoma** Davis, 1947
(Fig. 8)

Resembles trypanosomes; body flattened, with bluntly rounded anterior end, but wider near middle. Long flagellum thicker at base, tapering gradually toward tip. Several rows of small refringent rods run lengthwise of body. A small contractile vacuole in anterior end. Body 15 to 20μ long by 3 to 4μ wide; flagellum up to 30μ. Nucleus about one-third distance from anterior end, rounded or oval with several large peripheral masses of chromatin; a rounded karyosome may be present. Blepharoplast anterior to nucleus, and stains intensely.

Lamellasoma bacillaria Davis, 1947: on gills, *Pomoxis annularis,* *P. sparoides,* W. Va.

Genus **Cryptobia** Leidy
(Syn.: *Trypanoplasma* Laveran and Mesnil)
(Fig. 9)

Biflagellate, trypanosome-like; one flagellum free, other marks outer margin of undulating membrane; blepharoplast (parabasal body) elongated; parasitic in invertebrates and blood of fish; vector and intermediate host is fish leech. Pathogenicity not well understood but considered serious in some western trout hatcheries.
  Ref. Kudo (1954, p. 357).

Cryptobia borreli (Laveran et Mesnil): several European records; Mavor (1915*a*): moribund sucker; Wales and Wolf (1955): in *Oncorhynchus tshawytscha, O. kisutch, Salmo trutta, Catostomus snyderi, Cottus* sp., flagellated parasites in gut of *Piscicola salmositica* (leech), Calif. The latter is considered to be *C. salmositica* by Becker and Katz (1965*a*).
C. carassii (Syn. *Trypanoplasma c.*). Swezy (1919): an ectoparasite.
C. gurneyorum (Minchin): Europe; Laird (1961): in *Esox, Coregonus, Salvelinus,* N. Can.
C. lynchi Katz, 1951 (see *C. salmositica*).
C. salmositica Katz, 1951 (Syn. *C. borreli* Laveran *et* Mesnil, 1901, in part; *C. lynchi*): in blood, *Oncorhynchus kisutch,* Wash.; Davison, Breese, and Katz (1954): in salmon, November–February only, Ore.; Becker (1964): in salmon, *S. gairdneri, Cottus* spp., *Gasterosteus aculeatus, Prosopium williamsoni, Rhinichthys cataractae,* Calif., Ore., Wash.; Becker and Katz (1965*a*, 1966): *O. kisutch, Cottus rhotheus,* vector *Piscicola salmositica,* Wash. *O. gorbuscha, C. aleuticus, R. cataractae.*
Cryptobia sp. Wolf, H., 1960: in blood of *Dorosoma petenense,* Calif.
Cryptobia sp. Wolf, H., 1960: in blood of salmonids, Calif.
Cryptobia sp. Putz and Hoffman, 1965 (unpubl.): in *Exoglossum maxilingua, Rhinichthys* spp., *Semotilus corporalis,* W. Va.; Wenrich (1931): on gills of carp, Pa.

Genus **Colponema** Stein
(Fig. 10)

Body small; rigid; ventral furrow conspicuous, wide at anterior end; one flagellum arises from anterior end, other from middle; 6 to 8µ long; free-living species mostly, two parasitic species.

Colponema agitans Davis, 1947: on gills, *Pomoxis annularis, P. sparoides, Lepomis macrochirus,* W. Va.
Colponema sp. Clemmens and Sneed (1958): on gills, *Ictalurus punctatus,* Okla.

Genus **Bodomonas** Davis, 1947
(Fig. 11)

Body elongate, tapering to point posteriorly, anterior end broad, flattened; one side of body at anterior much thinner, forming shallow groove which gives it a spoon-shape when on gill; 2 flagella arise from anterior end; nucleus central, containing peripheral granules of chromatin; round-oval blepharoplast anterior to nucleus; about 12µ long.

Bodomonas concava Davis, 1947: on gills, *Pomoxis annularis, P. sparoides, Lepomis macrochirus.* Lom (1964) suggests that this may be *Cryptobia branchialis* Nie.

Order "POLYMASTIGINA"

The protozoa of this group possess 3 to 8 flagella (see Kudo, 1954, p. 369). Some have a cytostome (mouth). One to many nuclei are present.

Genus **Costia** Leclerque
(Fig. 12)

Ovoid in front view, piriform in profile; shallow depression tapers into a funnel (?) from which extend 2 long and 2 short flagella (long ones easily seen); cysts reported; all species ectoparasitic on freshwater fish.
   Ref. Davis (1953, p. 204); Kudo (1954, p. 270).

Costia necatrix (Henneguy): on epidermis of many fish; size 5 to 18 by 2.5 to 7µ; uninucleate cyst, spherical, 7 to 10µ in diameter.
C. pyriformis Davis, 1943: on skin and gills, *Salmo gairdneri, Salvelinus fontinalis;* size 9 to 14 by 5 to 8µ.

Genus **Urophagus** Klebs
(Fig. 14)

Similar to *Hexamita*, but with 1 cytostome and 2 movable posterior processes.

Urophagus sp. I have observed an intestinal flagellate in *Carassius auratus* which is tentatively identified as *Urophagus* sp.

Genus **Hexamita** Dujardin
(Fig. 13)

(Syn.: *Octomitus* Prowazek)

Piriform; 2 nuclei near anterior end; 6 anterior and 2 posterior flagella; 2 axostyles (rigid rods through body); cysts formed; species free-living or parasitic in intestine of arthropods and vertebrates.

Ref. Davis (1953, p. 240); Kudo (1954, p. 392); Kulda and Lom (1964).

Hexamita intestinalis Dujardin. In intestine, *Salmo trutta*, in frogs and in rectum of *Motella* in Europe.
H. salmonis (Moore, 1923). In intestine of young trout and salmon, possibly causes enteritis; Uzmann and Jesse (1963) and Uzmann, Paulik, and Hayduk (1965): pathogenicity; Yasutake *et al.* (1961): chemotherapy; Uzmann and Hayduk (1963b): culture method.

Subphylum CILIOPHORA Doflein, 1901
(INFUSORIA)

The Ciliophora possess cilia and, unlike other protozoa, have two kinds of nuclei—macronucleus and micronucleus. The predominant stages of Suctoria possess tentacles and no cilia, but their developmental stages possess cilia. Sexual reproduction of Ciliophora is mainly by conjugation; asexual reproduction is by binary fission or budding.

KEY

TO THE GENERA OF CILIOPHORA

1   Cilia present throughout trophic life; no tentacles
.....................nonsuctorian CILIATES   2
1   Cilia only when young; adult with tentacles........
........Subclass I SUCTORIA (fig. 15) *Trichophrya*
2   (1). Without adoral (posterior) zone of membranellae
................Subclass III HOLOTRICHIA   6

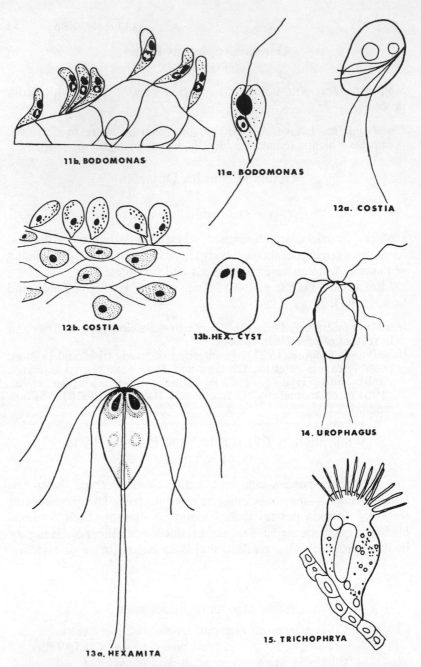

11b. BODOMONAS

11a. BODOMONAS

12a. COSTIA

12b. COSTIA

13b. HEX. CYST

14. UROPHAGUS

13a. HEXAMITA

15. TRICHOPHRYA

Figs. 11–15. Fig. 11. *Bodomonas concava. a*, free form; *b*, attached form (Davis, 1947). Fig. 12. *Costia necatrix. a*, unstained, ventral view showing ventral groove; *b*, attached to epidermis, sectioned and stained (from

2 (1). With adoral zone of membranellae   3
3 (2). Adoral zone of membranellae winds counterclock-
    wise; external parasites. . . . .
    . . . . Subclass IV PERITRICHIA   9
3 (2). Adoral zone of membranellae winds clockwise to
    cytostome; large, oval, intestinal parasites; usually
    active, cilia obvious; rare in fish. . . . .
    . . . . Subclass II SPIROTRICHIA   4
4 (3). Peristome (mouth funnel) sunk in funnel-like hollow
    at anterior end, sometimes difficult to see; 70 to 80μ
    long . . . . (fig. 16) *Balantidium**
4 (3). Peristome not sunk in funnel-like hollow at anterior
    end   5
5 (4). Peristome starts at anterior end, parallels body axis,
    then turns toward right in front of cytostome
    (mouth); up to 150μ long . . . . . (fig. 17) *Nyctotherus*
5 (4). Peristome at anterior end, but is large peristomal field
    with half or completely spiral adoral zone; found in
    intestine of *Barbus* in Europe (cf. Bychovskaya-
    Pavlovskaya *et al.*, 1962) . . . . *Stentoropsis*
6 (2). Cytostome in peristome with inconspicuous mem-
    brane. . . . . Order HYMENOSTOMATIDA   7
6 (2). Cytostome on body surface or in peristome without
    membrane or strong cilia. . . . .
    . . . . Order GYMNOSTOMATIDA   8
7 (6). Cytostome near anterior end, but difficult to determine
    in large specimens; body oval to round; ciliation
    uniform; pellicle longitudinally striated; macronu-
    cleus horseshoe-shaped; multiplication within cyst;
    subepithelial parasite; 50μ to 1 mm in diameter. .
    . . . . (fig. 18) *Ichthyophthirius multifilis*
7 (6). Cytostome at about one-fourth body length near an-
    terior end; body ovoid; in gall bladder and blood;
    about 35μ long . . . . (fig. 19) *Glaucoma*

---

* According to Lom (1964) *Balantidium* is now placed in the order Tricho-
stomatida Bütschli, 1889.

Davis, 1947). Fig. 13. *Hexamita salmonis* (*Octomitus s.*). *a*, free form
(from Davis, 1946); *b*, cyst (orig.). Fig. 14. *Urophagus* sp. from goldfish
(orig.). Fig. 15. *Trichophrya micropteri*, attached to gill (from Davis,
1942).

8   (6). Cytostome ventral, in anterior half; ovoid dorsal sur-
        face without cilia; dorsoventrally flattened; dorsal
        surface convex, ventral surface flat; oral basket con-
        spicuous; macronucleus rounded; 50 to 70μ long
        ....(fig. 20) *Chilodonella* (*Chilodon*)
8   (6). Cytostome a long slit; flask-shaped; somewhat com-
        pressed; ciliation uniform, complete; one species..
        ....(fig. 21) *Amphileptus*
9   (3). Attached; usually no body cilia, though telotroch
        (larva) possesses a posterior ring of cilia........
        ....Suborder SESSILINA   10
9   (3). Free-swimming; low barrel- to saucer-shaped with
        highly developed attaching organellae on posterior
        end consisting of circle of posterior cilia and chiti-
        noid ring attachment disc with radially arranged
        teeth (denticular ring) .... Sub-
        order MOBILINA, Family URCEOLARIIDAE   14
10   (9). Without stalk .... Family SCYPHIDIIDAE   11
10   (9). With stalk (resembles *Vorticella*)   12
11 (10). Cylindrical; posterior end attached, usually by
        widened scopula; body usually cross-striated; nu-
        cleus ribbon-shaped; adoral membrane small....
        .... (fig. 22) *Ambiphrya* (*Scyphidia*)
11 (10). Elongate pear-shaped; macronucleus a solid inverted
        cone; scopula small..........................
        .... (fig. 23) *Apiosoma* (*Glossatella*)
12 (10). Stalk noncontractile .... (fig. 24) *Epistylis*
12 (10). Stalk contractile.............................   13
13 (12). Stalk contractile; entire colony contracts or expands
        simultaneously .... *Zoothamnium*
13 (12). Stalk contractile; colony does not contract simultane-
        ously; recorded from pikeperch egg disease, also
        found on trout eggs .... *Carchesium*
14   (9). Incomplete turn of adoral ciliary spiral; anterior pro-
        jection of tooth present; gill parasites, usually
        "cupped" over gill lamellae, and die soon after re-
        moval from fish.............................   15
14   (9). One or more complete turns of adoral ciliary spiral;
        no anterior projection of tooth present; parasitic on
        body, on gills, or in urinary bladder...........   16
15 (14). Central ray (centripetal thorn) of denticle slender, but
        well developed; blade curved or rod-like........
        .... (fig. 25) *Tripartiella*

15 (14). Central ray of denticle an insignificant little crook. . . .
.... (fig. 26) *Trichodinella*
16 (14). Adoral ciliary spiral less than 2 turns, but at least 1
complete spiral.......................... 17
16 (14). Adoral ciliary spiral of 2 to 3 complete turns; urinary
bladder parasites .... (fig. 27) *Vauchomia**
17 (16). Ciliary girdle with supplementary circlet of cirri. . . .
.... (fig. 28) (*Cyclochaeta*) *Trichodina*
(*Cyclochaeta domerguei* as redescribed by Mac-
Lennan, 1939, is *Trichodina domerquei nomen
nudum*; *cf.* Lom and Hoffman, 1964)
17 (16). Ciliary girdle without supplementary circlet of cirri
.... (fig. 28) *Trichodina**

Subclass I SUCTORIA Haeckel, 1866

Mature stages are without cilia, but have suctorial tentacles.

Genus **Trichophrya** Claparede and Lachmann
(Fig. 15)

Body small; rounded-elongate; without stalk; suctorial tentacles in fas-
cicles, not branching; simple or multiple endogenous budding; attached on invertebrates and vertebrates.

Ref. Culbertson and Hull (1962); Kudo (1954, p. 865).

Trichophrya ictaluri Davis, 1942: on gills, *Ictalurus punctatus*, Iowa.
T. micropteri Davis, 1942: on gills, *Micropterus*, W. Va.
T. piscium (Bütschli, 1889). Culbertson and Hull, 1962 (Synonyms
according to Culbertson and Hull, 1962: *T. ictaluri* Davis, 1947;
T. intermedia Prost, 1952; *T. micropteri* Davis, 1942; *Trichophrya*
sp. from *Salvelinus fontinalis* (Davis, 1942); and probably *T. sinen-
sis* Chen, 1955).
Reported on gills of *Ictalurus punctatus, Micropterus dolomieui,
Oncorhynchus nerka, Salmo gairdneri, Salvelinus fontinalis, S. namay-
cush,* from Iowa, Can. (Manitoba), Md., N. Y., Tex., Vt., W. Va.

This species has been reviewed by Culbertson and Hull (1962),
who placed the above-listed four species in synonomy with it. They
misquote two species: *T. ictalurus* Davis, 1942, is *T. ictaluri* Davis,
1947, and *T. salvelinus* Davis, 1942, is *T.* sp. Davis, 1942. They com-
ment that *T. piscium* is a symbiont, probably feeding on fish ectoparasites; this needs further investigation because the food of *T. piscium*

* The genera *Trichodina* and *Vauchomia* are retained for simplification. Uzmann and Stickney (1954) place them in subgenera of *Trichodina*.

has not been demonstrated and it might be fish mucus instead of ecto-parasites. We have seen *Trichophrya* on fish gills which were not heavily parasitized by other ectoparasites; so it must have been feeding on fish mucus or other fish material. Davis (1942) found *T. micropteri* on *Micropterus dolmieui* only at this station. It was not present on the closely related *M. salmoides* or any of the other fish at the hatchery. F. Meyer (1964) has found a *Trichophrya* sp. on *Lepomis humilis* only, although it was in a pond containing catfish, goldfish, carp, golden shiners, buffalo, largemouth bass, and green sunfish. Apparently there are distinct physiological species or strains of *Trichophrya* in spite of the morphological similarity. Both Davis (1942) and F. Meyer (1964) are of the opinion that the parasite is injurious to the fish.

Trichophrya sp. Davis, 1942: on gills, *Salvelinus fontinalis*.

Nonsuctorian CILIATES

All possess cilia or cirri in the trophozoite stage. Parasitic ciliates of fish vary in size from 20µ to 1 mm. A cytostome is present in all species parasitic in or on fish. Cysts are produced by some species.
    Ref. Corliss (1961).

Subclass II SPIROTRICHIA Bütschli, 1899
Genus **Balantidium*** Claparede and Lachmann
(Fig. 16)

Oval, ellipsoid to subcylindrical; peristome at or near anterior end; cytopharynx not well developed; ciliation uniform; macronucleus elon-gated; micronucleus present; cytopyge (anus) and contractile vacuole terminal; in gut of invertebrates and vertebrates. Many species.
    Ref. Kudo (1954, p. 798).

Balantidium sp. Fantham and Porter, 1947: intestine of brook trout and perch, Can., Europe.

Genus **Nyctotherus** Leidy
(Fig. 17)

Oval; compressed; peristome begins at anterior end, turns to right and ends in cytostome (mouth) midway; cytopharynx runs dorsally and posteriorly, a long tube with undulating membrane; massive macro-nucleus anteriorly; micronucleus present; cytopyge and contractile

---

* According to Lom (1964) *Balantidium* is now placed in the order Tricho-stomatida Bütschli, 1889.

vacuole terminal; in colon of invertebrates, amphibia, and one species in fish.

Nyctotherus pangasia Tripathi, 1954: intestine of freshwater fish, India.

Subclass III HOLOTRICHIA Stein, 1859

These possess uniform ciliation over the entire body. A cytostome is usually present; buccal ciliature, if present, is generally inconspicuous. Asexual reproduction is by transverse fission; sexual reproduction, by conjugation. Some form cysts.

Order HYMENOSTOMATIDA Delage and Herouard, 1896

Buccal cavity ventrally located, contains ciliary apparatus considered fundamentally as tetrahymenal in nature (undulating membrane on right, tripartite, adoral zone on left).

Genus **Ichthyophthirius** Fouquet, 1876
(Fig. 18)

Body oval to round but very plastic; 50μ to 1 mm in diameter; ciliation uniform; pellicle longitudinally striated; vestibulum of cytostome 8 to 20μ in diameter with rotatory beating cilia; near middle of youngest (30 to 50μ) stages, but unequal growth of ciliate produces gradual shift forward until it is nearly anterior in larger (350 to 800μ) individuals; cytostome very difficult to see in larger specimens; cytostome cavity only a little longer than diameter of vestibulum, lined with cilia. Near bottom of cytostome about 5 membranelles can be seen in silver-impregnated preparations. Horseshoe-shaped macronucleus can be seen unstained, micronucleus adhering to it; no division while in skin of fish but multiplication within cyst which is formed after dropping off of fish—up to 1,000 small (30 to 45μ), oval, ciliated tomites with anterior knob (perforatorium) may be produced overnight from each cyst; complete life cycle takes 4 to 40 days depending on temperature; optimum temperature 75 to 80°F. One species.

Ref. Bauer *et al.* (1959); Davis (1953, p. 209); Haas (1933); Kudo (1954, p. 709); MacLennan (1935*a*, 1935*b*, 1943); F. Meyer (1966); Mugard (1949).

Ichthyophthirius multifilis F. (syn. *Enchelys parasitica?*)*, a very se-

---

* *Enchelys parasitica* Dorier (1926) is probably a juvenile stage of *I. multifilis* according to Schäperclaus (1954).

rious pathogen, has been found beneath the epithelium of many species of fish, and it is believed that all species of freshwater fish, including anadromous fish and *Petromyzon* while in freshwater, are vulnerable. Mr. Bob Putz of this laboratory has noted that newly hatched bluegill fry, however, are refractory. Reichenbach-Klinke (1954) noted that "Ich" was present beneath the mucosa of the fore-intestine of small aquarium fish, and I have seen it in the visceral cavity of small stickle-backs. It does not infect other aquatic animals. "Ich" was first described from France (1876), but its origin is not known. It has been reported from all Europe, North America, China, India, Indonesia, and Aus-tralia. It is probable that it is now present in every country which has engaged in foreign fish exchange, particularly goldfish and other aquar-ium fish. Van Duijn (1956) states that it does not exist in nature in England. In Russia it is not found north of 60 to 65° northern latitude, which corresponds in latitude to Anchorage, Alaska.

Death of the fish usually occurs during the growth period of the para-site when much damage is produced by the large active trophozoite. In severe infections the epithelium sloughs off and the fish succumb quickly. The pathogenicity and biology are well covered by Bauer *et al.* (1959) and Dogiel *et al.* (1958).

Previously infected fish retain partial immunity for at least several months (Bauer *et al.*, 1959; Parker, Jack, unpubl.; Putz and Hoffman, unpubl.).

Many methods of treatment have been recommended. Usually only the free-swimming parasites are killed, but Avdos'ev (1962) reported that the trophozoites taken from the skin of carp treated with malachite green did not produce infective tomites. The most reliable methods of treatment are: formaldehyde 1:5000 for one hour at 0 to 18°C, but 1:6000 at warmer temperatures used daily; formaldehyde 15 to 25 ppm daily in aquaria and ponds; malachite green 2 ppm for half an hour daily, or 0.1 ppm in ponds and aquaria; methylene blue 3 ppm in ponds and aquaria. If the water is dirty, higher concentrations of the chemical may be necessary.

Genus **Glaucoma** Ehrenberg
(Fig. 19)

Reported from gall bladder and blood of bream in Europe.
    Ref. Kudo (1954, p. 761); Layman (1949, p. 240).

Order GYMNOSTOMATIDA Bütschli, 1889

No oral ciliature. Cytostome, containing trichites, opens directly to the outside. Ciliature generally simple.

Genus **Chilodonella** Strand
(Syn.: *Chilodon* Ehrenberg)
(Fig. 20)

Ovoid; dorsoventrally flattened; ventral surface with ciliary rows; anteriorly flattened dorsal surface with a cross-row of bristles; cytostome round; oral basket conspicuous, protrusible; macronucleus rounded; on fish and amphipods. Many species.

Ref. Davis (1953, p. 206); Kudo (1954, p. 731); Bauer (1959).

Chilodonella cyprini (Moroff): on gills and skin of trout and many others; 50 to 70μ long; optimum temp. 40 to 50°F.
C. dentatus Fantham and Porter, 1947: on gills, *Micropterus dolomieui*, Que.

Genus **Amphileptus** Ehrenberg
(Fig. 21)

Flask-shaped; slightly compressed; ciliation uniform; slitlike cytostome not reaching middle of body; 2 or more macronuclei.

Amphileptus voracus Davis, 1947: on gills of warm-water fish; considered beneficial, W. Va.

Subclass IV PERITRICHIA Calkins, 1933

These possess an enlarged bell-like anterior end or disclike posterior end surrounded by cilia. Body ciliation is lacking or a single circle on some. Asexual reproduction is by binary fission; the resulting free-swimming young forms of attached species are known as telotrochs. No cysts are produced.

Genus **Ambiphrya** Raabe, 1952
(Fig. 22)
(Syn.: *Scyphidia* Dujardin, in part)

Cylindrical; posterior end (scopula) attached; body usually cross-striated. Scopula broad, flattened; circle of body cilia always present. Macronucleus usually ribbon-shaped.

Ref. Davis (1953, p. 212); Kahl (1935); Kudo (1954, p. 852); Raabe (1952).

Ambiphrya ameiuri (Thompson, Kirkegard, and Jahn, 1946): macronucleus without end branches; on gills of *Ictalurus melas*, Iowa. *Scyphidia macropodia* Davis, 1947, is probably a synonym which

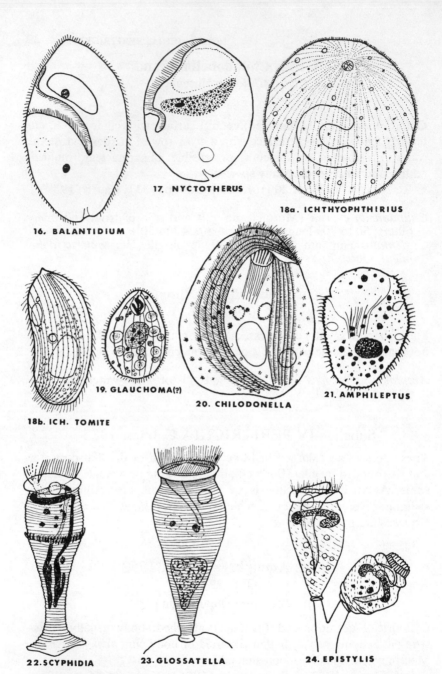

16. BALANTIDIUM

17. NYCTOTHERUS

18a. ICHTHYOPHTHIRIUS

18b. ICH. TOMITE

19. GLAUCHOMA(?)

20. CHILODONELLA

21. AMPHILEPTUS

22. SCYPHIDIA

23. GLOSSATELLA

24. EPISTYLIS

Figs. 16–24. Fig. 16. *Balantidium entozoon* (from Fantham and Porter, 1947). Fig. 17. *Nyctotherus pangasi* (from Tripathi, 1954). Fig. 18. *Ichthyophthirius multifilis. a*, trophozoite (from Bykhovskaya-Pavlovskaya, 1962); *b*, tomite (from MacLennan, 1935). Fig. 19. *Glaucoma* (?)

was reported from *I. nebulosus*, *I. punctatus*, and accidentally on *Lepomis macrochirus*.

A. micropteri Surber, 1940 (see *Apiosoma m.*).
A. tholiformis Surber, 1942: macronucleus with end branches; on gills, *Micropterus dolomieui*, *M. salmoides*, W. Va.

Genus **Apiosoma** Blanchard, 1883
(Fig. 23)
(Syn.: *Glossatella* Bütschli, 1889)

Somewhat similar to *Ambiphrya*, but elongate pear-shaped; macronucleus a compact inverted cone lying in posterior of body; scopula (holdfast) usually small; circle of body cilia present.

Ref. Kahl (1935); Kudo (1954, p. 852); Lom (1966); Schäperclaus (1954, p. 214).

Apiosoma micropteri (Surber, 1940), Syn. *Scyphidia m.*, Surber, 1940): on gills, *Micropterus dolomieui*, *M. Salmoides*, W. Va.
Apiosoma spp. Lom (1966): several from Europe.

Genus **Epistylis** Ehrenberg
(Fig. 24)

Inverted bell form; usually on branched noncontractile stalk, forming colonies. Many species.

Ref. Davis (1953, p. 214); Fischthal (1949b); Kahl (1935); Kudo (1954, p. 853); Lom and Vavra (1961).

Epistylis sp. Fischthal (1949b): on trout and darters, Wis.; Hoffman (1959): on brook trout, N. H., N. C.; Leitritz (1960): on rainbow trout, Calif.

Genus **Zoothamnium** Bory

Similar to *Vorticella*, but colonial; all stalks connected and entire colony contracts at same time. Many species.

Ref. Kudo (1954, p. 857).

Zoothamnium sp. Khajuria and Pillay (1950): from fish, India.

Tetrahymena (from Bychovskaya-Pavlovskaya 1962). Fig. 20. *Chilodonella* sp. (from Davis, 1953). Fig. 21. *Amphileptus voracus* (from Davis, 1947). Fig. 22. *Scyphidia macropodia* (from Surber, 1942). Fig. 23. *Glossatella micropteri* (*Scyphidia m.*) (from Surber, 1940). Fig. 24. *Epistylis niagarae* (from Bishop and Jahn, 1941).

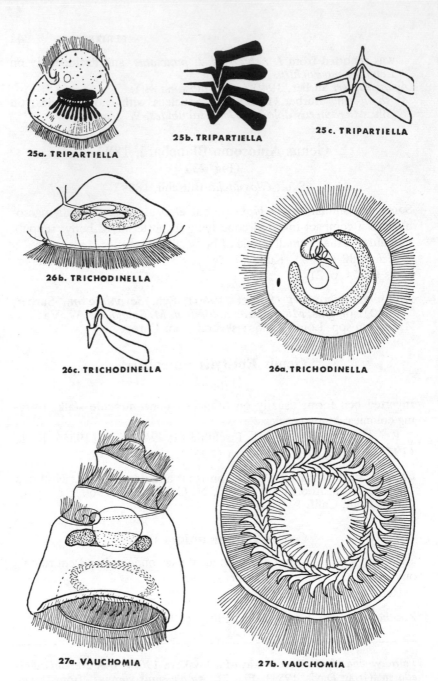

25a. TRIPARTIELLA

25b. TRIPARTIELLA

25c. TRIPARTIELLA

26b. TRICHODINELLA

26c. TRICHODINELLA

26a. TRICHODINELLA

27a. VAUCHOMIA

27b. VAUCHOMIA

Figs. 25–27. Fig. 25. *Tripartiella. a, T. bulbosa* (from Davis, 1947); *b, T. bursiformis*, part of denticulate ring (from Davis, 1947); *c, Tripartiella* denticles (from Lom, 1959). Fig. 26. *Trichodinella epizootica. a*, aboral view (from Lom, 1956); *b*, denticles (from Lom, 1959). Fig. 27. *Vau-*

Genus **Carchesium** Ehrenberg

As *Zoothamnium* but stalks contract independently.
Ref. Davis (1953, p. 214); Kudo (1954, p. 856).

Carchesium sp. On eggs of pikeperch and trout.

Genus Unknown

An unknown stalked peritrich was found on the commercially important mullet, *Mugil cephalus*, in fresh water by Sogandares and Hutton (1958). The visible fungus-like growths were present as large colonies confined to the posterior border or entire exposed portions of the scales. Scale erosion was present at the site of attachment of the colonies.

Family URCEOLARIIDAE Dujardin, 1841

Without stalk; motile. Oral-aboral axis shortened, with prominent basal disc common at aboral pole (posterior) of body. Often found as episymbionts, occasionally causing considerable harm, on a variety of aquatic hosts. The host specificity needs further study. J. Lom (Pers. Comm.) has found a lack of host specificity, but T. Wellborn (Pers. Comm.) has found quite striking host specificity—perhaps both extremes and intermediates are present in the group.
    Ref. Bykhovskaya-Pavlovskaya *et al.* (1964, p. 180); Haider (1964); Lom (1963).

Genus **Tripartiella** Lom, 1959
(Syn.: *Trichodina*, in part)
(Fig. 25)

Buccal apparatus as in *Trichodinella*. Denticles with outer blade, central cone, and well-developed central ray (centripetal thorn). Usually "cupped" over gill lamella and die soon after removal from fish.
    Ref. Lom (1963).

Tripartiella bulbosa (Davis, 1947): on gills, *Semotilus margarita*, W. Va.
T. bursiformis (Davis, 1947), (Syn. *Trichodina b.*): on gills, *Amblo-plites rupestris*, W. Va.
T. symmetricus (Davis, 1947), (Syn. *Trichodina s.*): on gills, *Icta-lurus punctatus*, *Semotilus margarita*, *Rhinichthys atronasus*, Iowa, W. Va.

---

chomia nephritica. a, lateral view; *b*, aboral end (sucking disc), (from Mueller, 1938).

Genus **Trichodinella** Šrámek-Hušek, 1953

(Syn.: *Brachyspira* Raabe, 1950;
Trichodina, in part)
(Fig. 26)

Denticles with outer flat blade and central cone; inner ray (thorn) a small crook. Aboral ciliature with 3 ciliary girdles as in *Trichodina*. Buccal ciliary apparatus of 2 rows, making turn of about 180° on anterior, oral face of ciliate before they plunge into infundibulum. Body usually biscuit-shaped. Usually "cupped" over gill lamella and die soon after removal from fish.
Ref. Lom (1963).

Trichodinella myakkae (Mueller, 1937), (Syn. *Trichodina m.*): on gills, *Micropterus salmoides, Carpiodes carpio, Ictiobus bubalus, Salvelinus fontinalis*, Fla.; Davis (1947): Iowa, N. Y., W. Va.
T. (*Foliella*) *subtilis* Lom, 1959: on crucian carp, Europe; Lom and Hoffman (1964): on *Carassius auratus*, W. Va.

Genus **Vauchomia** Mueller, 1938

(Fig. 27)

As *Trichodina*, except buccal ciliary spiral makes 2 to 3 complete turns; parasites in urinary system.

This genus is retained on the advice of Lom (1964), although Uzmann and Stickney (1954), recommend that it be reduced to a subgenus.

Vauchomia nephritica Mueller, 1938: in urinary bladder, *Esox masquinongy*, N. Y.
V. renicola Mueller, 1932: in urinary tract, *Esox* spp., N. Y.; Bangham (1944) and Fischthal (1947): in *Esox masquinongy*, Wis.; Hunter and Rankin (1939): in *Esox* spp., Conn.; Sindermann 1953): in *E. niger*, Mass.; Fantham and Porter (1947): *Trichodina* sp. in urinary tract, *E. niger*, Que.

Genus **Trichodina** Ehrenberg

(Fig. 28)

Low barrel- to saucer-shaped. Aboral ciliature with 3 ciliary girdles: first made of short, simple cilia, closely associated with border membrane; second a girdle of strong, long locomotor pectinelles not separated by a septum; third with tactile marginal cilia inserted on border of velum. Chitinoid ring of attaching disc with radially arranged hooked

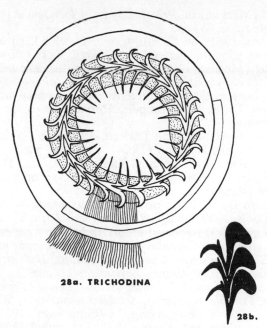

28a. TRICHODINA

28b.

Fig. 28. *Trichodina. a, T. truttae,* aboral disc (from Mueller, 1937); *b, T. fultoni,* denticles (from Davis, 1947).

teeth, each with outer flat blade, central cone, and inner ray (thorn). Buccal ciliary spiral makes more than 1 but less than 2 complete turns. Many species.

Ref. Uzmann and Stickney (1954), latest American review.

Trichodina californica Davis, 1947: on gills, *Oncorhynchus tscha-wytscha,* Calif.

T. discoidea Davis, 1947: on gills, *Ambloplites rupestris, Ictalurus punctatus, Lepomis macrochirus, Pomoxis sparoides,* W. Va., Iowa.

T. domerguei (Wallengren), (Syn. *Cyclochaeta d.*): low saucer-shape (see MacLennan, 1939, for description of *Cyclochaeta domerguei*). Frank (1962): histopathology, *Carassius auratus,* Germany; Lom and Hoffman (1964) consider this to be a *Trichodina* because the marginal cilia (cirri) of many other *Trichodina* spp. are equally long. They consider this species *nomina nuda* until it can be compared with the European *T. domerguei.* Also see Lom and Stein (1966).

T. fultoni Davis, 1947: on gills, *Ambloplites rupestris, Lepomis macrochirus, Micropterus dolomieui, M. salmoides, Salmo gairdneri,* W. Va.; Lom and Hoffman (1964): on *Lepomis cyanellus, M. salmoides, Rhinichthys atratulus,* W. Va.; Davis (1953): also on salamander (*Necturus maculatus*).

T. guberleti (MacLennan, 1939), (Syn. *Cyclochaeta d.*): on *Richard-sonius, Apocape*, Wash.

T. nigra Lom, 1961: on *Cobitis taenia*, Czech.; Lom and Hoffman (1964): on gills, *Huro salmoides, Lepomis macrochirus*, W. Va.

T. pediculus (Mueller, 1786) Ehrb. 1838: on *Esox lucius*, Europe; Mueller (1937): on gills, *Micropterus salmoides*, Fla. Davis (1947): believes that the latter Mueller's material was *T. fultoni*.

T. platyformis Davis, 1947: on gills, *Margariscus margarita, Rhinich-thys atronasus*, W. Va.

T. reticulata Hirschmann and Partsch, 1955: on *Carassius auratus*, Europe; Lom and Hoffman (1964); same, W. Va.

T. truttae Mueller, 1937: on gills, *Salmo clarki*, Ore.; Davis (1947): on *S. clarki*; Bogdanova and Stein (1963): on salmon, Russia (Sakhalin Island).

T. tumefaciens Davis, 1947: on gills, *Cottus bairdi*, W. Va.

T. vallata Davis, 1947: on gills, *Ictalurus punctatus*, Iowa.

Trichodina spp. (?). Reported from *Ambloplites rupestris, Campostoma anomalum, Cyprinus carpio, Etheostoma nigrum, Esox lucius, E. masquinongy, Eucalia inconstans, Hyborhynchus notatus, Hypentelium nigricans, Lepomis gibbosus, L. macrochirus, Micropterus dolomieui, Noturus gyrinus, Perca flavescens, Percina caprodes, Poecilichthys exilis, Salvelinus fontinalis, Semotilus atromaculatus, Stizostedion vitreum*, by Richardson (1938), Bangham (1944), Fantham and Porter (1947), Fischthal (1947a, 1950c), Bangham and Adams (1954), Anthony (1963).

SPOROZOA

These are protozoa which do not move by means of cilia, flagella, or pseudopodia. The trophozoites of some exhibit amoeboid shapes and may move slowly but are not usually seen in motion. All develop spores which are sometimes very resistant to adverse environments. Asexual division is known as schizogony and sexual division as sporogony. Identification is based primarily on spore morphology and size.

KEY

TO THE GROUPS OF SPOROZOA

1. Spore without polar filament 2
1. Spore with polar filament . . . . Subphylum CNIDOSPORA
2. Spore with or without membrane; with many sporozoites; in
    intestinal mucosa and blood cells . . . . . . Class TELOSPOREA
2. Spore with membrane; with one sporozoite . . . .
    . . . . Group ACNIDOSPORIDIA

Group ACNIDOSPORIDIA

This group is not well understood. Simple spores are produced.

Spores banana-shaped, arranged in "tubes" within cyst . . . .
. . . . (fig. 29) *Sarcocystis*
Spores usually oval, with "opercula"; marine . . . .
. . . . Order HAPLOSPORIDIA
Spores small, spherical, with large vacuoles . . . .
. . . . (possibly fungus) *Dermocystidium*

Genus **Sarcocystis** Lankester
(Fig. 29)

Muscle parasites of vertebrates; spores found in opaque white cysts (Meischer's tubes), cylindrical to ovoid. When mature, banana-shaped spores fill cyst; spores contain nucleus and granules. Many species.

Sarcocystis salvelini Fantham and Porter, 1943: in muscle, *Salvelinus fontinalis*, Que.

Class TELOSPOREA Schaudinn, 1900

The spore possesses no polar capsule or filament, but has one to several sporozoites at the end of its development. New hosts are infected when the sporozoites are ingested (in Coccidia) or injected (in blood sporozoa). The sporozoites penetrate cells, and, in Coccidia and Haemosporidia, continue development intracellularly.

Sporozoites enveloped; development in intestinal and blood cells;
zygote not motile . . . . Subclass COCCIDIA
Sporozoites naked; in blood cells; zygote motile . . . .
. . . . Order HAEMOSPORIDIA

Subclass COCCIDIA Leuckart, 1879
Genus **Eimeria**
(Fig. 30)

Parasite develops in epithelial cells of intestine; sporoblast freed in intestine, becomes oocyst containing 4 spores each with 2 sporozoites.

Eimeria aurata Hoffman, 1965: intestine, *Carassius auratus*, Penn.
E. carassii Yakimoff and Gousseff, 1935: intestine, *Carassius carassius*, Russia; very similar to *E. aurata*.
E. nicollei Yakimoff and Gousseff, 1935: intestine, *Carassius carassius*, Russia; very similar to *E. aurata*.
Eimeria sp. Davis, 1946, p. 55: intestine of brook trout; oocyst 10 to 12µ in diameter; Becker (1956, pp. 105, 123) and Schäperclaus (1954, p. 359): in other fish; many species, mostly marine.

Genus **Haemogregarina** Danilewsky
(Fig. 31)

Schizogony (asexual multiplication) in red blood cells of vertebrates; sporogony (sexual development) in gut of leeches. Mature parasite fills red cells and displaces nucleus. Cytoplasm and nucleus stain densely.

Ref. Schäperclaus (1954, p. 361).

Haemogregarina catostomi Becker, 1962: in red cells, *Catostomus macrocheilus*, Wash.
H. irkalukpiki Laird, 1961: in *Salvelinus alpinus*, N. Can.
H. myoxocephali Fantham *et al.*, 1942: in blood cells, *Myoxocephalus octodecimspinosus*, Que.

Order HAEMOSPORIDIA
Genus **Dactylosoma** Labbe
(Fig. 32)

Small parasites which do not fill red cells. Cytoplasm and nucleus stain faintly, no pigment; 4 to 16 nuclei present; gametocytes with karyosomes; 4 to 16 merozoites arranged fanlike. Invertebrate host unknown.

Dactylosoma salvelini Fantham *et al.*, 1942: in red blood cells of brook trout, Que.

Genus **Babesiosoma** Jakowska and Nigrelli, 1956
(No fig.)

Small parasites which do not fill red cells. Cytoplasm and nucleus stain faintly, no pigment; cytoplasm less granular but more vacuolated than *Dactylosoma*; 1 to 4 *Babesia*-like nuclei, no definite karyosome. Reproduces by schizogony, fission, and budding. Merozoites 4 or fewer, arranged in rosette or cross. Invertebrate host unknown.

Babesiosoma tetragonis Becker and Katz, 1965*b*: in *Catostomus* sp., Calif.

Genus **Leucocytozoon** Danilewsky
(Fig. 33)

Large gametocytes with pigment granules in red cells; schizogony in endothelial and visceral cells of vertebrate; in bird forms, life cycle completed in blood-sucking insects; invertebrate host unknown for fish forms.

Leucocytozoon salvelini Fantham *et al.*, 1942: in red blood cells of brook trout, Que.

Subphylum CNIDOSPORA Doflein, 1901

Spores comparatively large (6 to 65μ); polar capsules easily seen
. . . . Class MYXOSPORIDEA
Spores comparatively small (3 to 6μ); no polar capsule, filament usually not easily visible, but has been demonstrated by polar filament extrusion and electron microscopy . . . .
. . . . Class MICROSPORIDEA

Class MYXOSPORIDEA* Bütschli, 1881

Myxosporidea are usually seen when the parasite is in the spore stage, at which time an opaque white cyst containing the spore may often be seen with the naked eye. The histozoic species usually form cysts, sometimes very large, and the identification is customarily based on the spore morphology. Species which develop in the gall bladder, kidney tubules, and urinary bladder do not form such cysts, and the trophozoite is the stage usually detected; identification is based on the spore, however. The resistant proteinaceous spore wall is usually bivalve. Lom and Vavra (1963) have demonstrated mucous envelopes on the spores; species differences were easily recognized. Each spore contains 1 sporoplasm and 1 to 4 polar capsules containing coiled filaments.

The life cycle is direct from fish to fish; this has been experimentally verified for *Myxidium, Chloromyxum,* and *Leptotheca,* but not for other genera (see Kudo, 1930a, p. 313). After ingestion, the sporoplasm leaves the spore, presumably penetrates the intestine, and migrates to the final site, which is often very specific for a given species. The sporoplasm grows into a trophozoite, the nuclei divide, and the structure usually grows to produce finally many spores in a cyst or in a single trophozoite. If the cyst is near the surface, it may rupture and the spores will be freed in the water. If the cysts are internal, the fish must die and disintegrate to free the spores, or perhaps the fish must be eaten by a susceptible fish species; however, recent evidence indicates

---

* Classification of Honigberg *et al.* (1964).

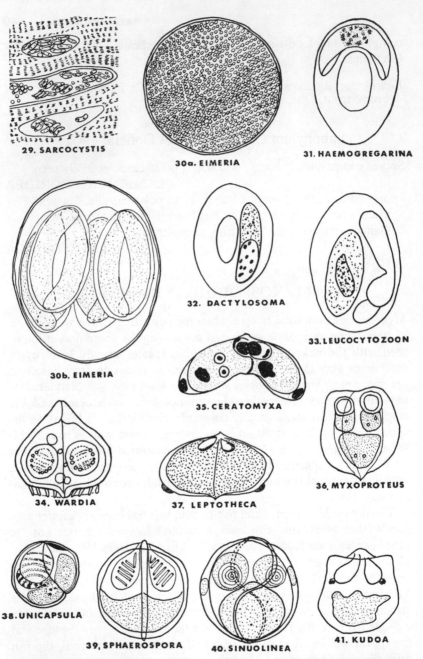

29. SARCOCYSTIS

30a. EIMERIA

31. HAEMOGREGARINA

32. DACTYLOSOMA

33. LEUCOCYTOZOON

30b. EIMERIA

35. CERATOMYXA

34. WARDIA

37. LEPTOTHECA

36. MYXOPROTEUS

38. UNICAPSULA

39. SPHAEROSPORA

40. SINUOLINEA

41. KUDOA

Figs. 29–41. Fig. 29. *Sarcocystis salvelini,* cross section through cysts in muscle (from Fantham and Porter, 1943). Fig. 30. *Eimeria aurati. a,* oocysts; *b,* sporulated oocysts (from Hoffman, 1965). Fig. 31. *Haemogregarina catostomi,* in red blood cell (from Becker, 1962). Fig. 32. *Dac-*

that some spores are voided through the intestine and gills (Uspenskaya, 1957; Hoffman *et al.*, 1965). If the spores are produced in the gall bladder or the urinary bladder, they are voided with excretory products.

Ref. Kudo (1920), synopsis and keys; Kudo (1930*a*), methods. Shulman (1964): evolution and phylogeny.

KEY

TO THE GENERA OF MYXOSPORIDEA

1 Sutural line coincides with or is at acute angle to largest diameter of spore (spore usually flattened slightly); 1, 2, or 4 polar capsules; sporoplasm with or without iodinophilous vacuole* ....
...............Suborder PLATYSPOREA 9

1 Sutural line does not coincide with largest diameter of spore 2

2 (1). Spore not spherical; sutural line at right angles to largest diameter of spore with 1 polar capsule on each side; sporoplasm without iodinophilous vacuole* ..
.................Suborder EURYSPOREA 3

2 (1). Spore spherical or subspherical; 1, 2, or 4 polar capsules; sporoplasm without iodinophilous vacuole* .
................Suborder SPHAEROSPOREA 7

3 (2). Spores laterally expanded; usually coelozoic (gall bladder) in marine fish except one sp. ....
...............Family CERATOMYXIDAE 4

---

* Add an equal amount of iodine solution to wet mount—"glycogen" vacuole stains reddish brown; is less evident in preserved material.

---

tylosoma salvelini, in red blood cell (from Fantham *et al.*, 1942). Fig. 33. *Leucocytozoon salvelini*, in red blood cell; parasite is stippled body at left (from Fantham *et al.*, 1942). Fig. 34. *Wardia ovinocua*, spore (from Kudo, 1920). Fig. 35. *Ceratomyxa shasta*, spore (from Noble, 1950). Fig. 36. *Myxoproteus cordiformis*, spore (from Davis, in Kudo, 1954). Fig. 37. *Leptotheca perlata*, spore (from Balbiani, in Kudo, 1920). Fig. 38. *Unicapsula muscularis*, spore (from Davis, 1924). Fig. 39. *Sphaerospora carassii*, spore (from Kudo, 1920). Fig. 40. *Sinuolinea dimorpha*, spore (from Davis, in Kudo, 1920). Fig. 41. *Kudoa clupeidae*, spore (from Meglitsch, 1947).

Suborder EURYSPOREA

3 (2). Spores less laterally expanded to form isosceles triangle
with 2 convex sides; oval in profile; 2 large polar
capsules; coelozoic in freshwater fish . . . .
. . . . Family WARDIIDAE (fig. 34) *Wardia*

4 (3). Shell valves conical and hollow, attached on bases;
sporoplasms usually not filling intrasporal cavity . .
. . . . (fig. 35) *Ceratomyxa*

4 (3). Shell valves not conical and hollow   5

5 (4). Spores pyramidal, with or without distinct processes
at base of pyramid; in urinary bladder of marine fish
. . . . (fig. 36) *Myxoproteus*

5 (4). Spores not pyramidal   6

6 (5). Shell valves hemispherical; in gall bladder of marine
fish, one amphibian . . . . (fig. 37) *Leptotheca*

6 (5). Spores triangular with concave sides in anterior end
view; profile ellipsoidal; 3 polar capsules and 3 shell
valves; in gall bladder of marine fish . . . . . . *Trilospora*

Suborder SPHAEROSPOREA

7 (2). Spore with 1 polar capsule; between muscle fibers
("wormy" halibut); marine . . . . Family UNICAPSULIDAE (fig. 38) *Unicapsula muscularis*

7 (2). Spore with more than 1 polar capsule   8

8 (7). Spore with 2 polar capsules; in gall and urinary blad-
der, kidney of marine and European fish . . . .
. . . . Family SPHAERO-
SPORIDAE (figs. 39, 40) *Sphaerospora* and *Sinuolinea*

8 (7). Spore with 4 polar capsules; in gall bladder and tissues
of marine and freshwater fish . . . Family CHLOR-
OMYXIDAE (figs. 41, 42) *Kudoa* and *Chloromyxum*

Suborder PLATYOSPOREA

9 (1). 1 polar capsule   10

9 (1). 2 polar capsules, 1 in each end of spore   11

9 (1). 2 polar capsules in one end of spore   12

9 (1). 4 polar capsules in one end of spore; shell prolonged
posteriorly into long processes . . . . . . (fig. 43) *Agarella*

10 (9). Iodinophilous vacuole present; spore ellipsoidal . . . .
. . . . (fig. 44) *Coccomyxa*

10 (9). No iodinophilous vacuole; spore piriform . . . .
. . . . (fig. 45) *Thelohanellus*

11 (9). Spore subspherical . . . . (fig. 46) *Zschokkella*
11 (9). Spore fusiform with pointed or rounded ends; polar
      filament long; histozoic or coelozoic . (fig. 47) *Myxidium*
11 (9). Spore fusiform but ends truncate; polar filament short;
      coelozoic in marine fish . . . . . (fig. 48) *Sphaeromyxa*
12 (9). Iodinophilous vacuole present 13
12 (9). No iodinophilous vacuole 14
13 (12). Spore oval; no posterior processes; histozoic . . . .
      . . . . (fig. 49) *Myxobolus*
13 (12). Body of spore biconvex and compressed parallel to
      sutural plane; tail-like extensions of shell valves
      present; histozoic . . . . (fig. 50) *Henneguya*
13 (12). Spore similar to *Myxobolus*, but single tail-like pro-
      cess not extension of shell valves . . (fig. 51) *Unicauda*
13 (12). Spore pyramidal, with 2 posterior processes from lat-
      eral spaces; histozoic . . . . (fig. 52) *Hoferellus*
14 (12). Sutural line does not bisect polar capsules; spore oval
      to subspherical; histozoic in marine and freshwater
      fish . . . . (fig. 53) *Myxosoma*
14 (12). Sutural line difficult to see, bisects the 2 polar cap-
      sules; shell valves longitudinally striated 15
15 (14). With tail-like processes are extensions of shell valves;
      coelozoic . . . . (fig. 54) *Myxobilatus*
15 (14). With no tail-like extensions; histozoic in kidneys . . .
      . . . . (fig. 55) *Mitraspora*

Suborder EURYSPOREA Kudo, 1920

Largest diameter of spore at right angles to sutural plane; 2 polar capsules, 1 on each side of plane; no iodinophilous vacuole.

Genus **Ceratomyxa** Thélohan
(Fig. 35)

Shell valves conical and hollow, attached on bases; sporoplasm usually
not filling intrasporal cavity; usually in gall bladder.

    Ref. Kudo (1954, p. 649); Noble (1950); Schäperclaus (1954,
p. 348).

Ceratomyxa shasta Noble, 1950: in viscera of fingerling *Salmo gairdneri*, Calif.; Wales and Wolf (1955): heavy mortality of *S. gairdneri*.

Genus **Myxoproteus** Doflein
(Fig. 36)

Spores pyramidal, with or without distinct processes at bases; in urinary bladder of marine fish. Three species.

Ref. Kudo (1954, p. 651).

Genus **Leptotheca** Thélohan
(Fig. 37)

Shell valves hemispherical; in gall bladder of marine fish; one in amphibians. Many species.

Ref. Kudo (1954, p. 651).

Genus **Trilospora** Noble

Spores triangular with concave sides in anterior view; profile ellipsoid; 3 polar capsules, 3 shell valves; in gall bladder of marine fish. One species.

Genus **Wardia** Kudo
(Fig. 34)

Spores isosceles triangle with 2 convex sides; oval in profile; 2 large polar capsules; tissue parasites of freshwater fish. Two species.

Wardia ovinocua Kudo, 1920: in ovary, *Lepomis humilis*, Ill.

Suborder SPHAEROSPOREA Kudo, 1920

Spore spherical; 1, 2, or 4 polar capsules; no iodinophilous vacuole.

Genus **Chloromyxum** Mingazzini
(Fig. 42)

Spore with 4 polar capsules, grouped at anterior end; shell surface often striated or ridged; histozoic or coelozoic in marine and freshwater fish and amphibians. Many species.

Ref. Kudo (1954, p. 654); Schäperclaus (1954, p. 351).

Chloromyxum catostomi Kudo, 1920: in gall bladder, *Catostomus commersoni*.
C. externum Davis, 1947: on gills, *Semotilus margarita*, *Rhinichthys atronasus*, W. Va. This is the only known external myxosporidean.
C. funduli Hahn, 1913: in lesions, *Fundulus*.

C. gibbosum Herrick, 1941: in gall bladder, *Lepomis gibbosus*, Lake Erie; Hoffman (1959): also W. Va.

C. majori Yasutake and Wood, 1957: in glomeruli, *S. gairdneri, Oncorhynchus tshawytscha*, Wash.

C. opladeli Meglitsch, 1942: in gall bladder, *Opladelus olivaris*, Ill.

C. renalis Meglitsch, 1947: in kidney, *Fundulus majalis*, N. C.

C. thompsoni Meglitsch, 1942: in gall bladder, *Ictiobus bubalus*, Ill.

C. trijugum Kudo 1920: in gall bladder, *Pomoxis sparoides, Xenotis megalotis*; Jameson (1931): Calif.; Otto and Jahn (1943): Iowa.

C. truttae Léger, 1906: in gall bladder, *Salmo trutta fario*, France; Davis (1947): *Salvelinus fontinalis*, Vt.

C. wardi Kudo, 1920. Uzmann (1961): gall bladder and intestine of salmonids, Wash.

Chloromyxum sp. Hoffman and Putz (1963): *S. fontinalis*, Pa.

Genus **Kudoa** Meglitsch
(Fig. 41)

Resembles *Chloromyxum*, but spores stellate or quadrate in anterior end view; spore membrane delicate, sutures indistinct; 4 shell valves; histozoic. Several species.

Ref. Kudo (1954, p. 655).

Kudoa sp. Bullock (1958): in blood, *Fundulus heteroclitis, Pseudopleuronectes*, N. H.

Kudoa (?). Rigdon and Hendricks (1955): in *Mugil cephalus*, Tex.

Genus **Sinuolinea** Davis
(Fig. 40)

Spore spherical-subspherical; sutural line sinuous; with or without lateral processes; 2 spherical polar capsules; in urinary bladder; marine fish.

Ref. Kudo (1954, p. 654).

Genus **Sphaerospora** Thélohan
(Fig. 39)

Spore spherical-subspherical; sutural line straight; 2 polar capsules at anterior end; coelozoic or histozoic in marine fish and European freshwater fish.

Ref. Kudo (1954, p. 653).

Sphaerospora carassii Kudo, 1920: in gill filaments, *Carassius carassius*, Japan.

Genus **Unicapsula** Davis
(Fig. 38)

Spore spherical with 1 polar capsule; shell valves asymmetrical; sutural line sinuous; marine; cause of "wormy" halibut. One species.
Ref. Kudo (1954, p. 652).

Unicapsula muscularis Davis, 1924: in muscle fibers of halibut, N. A. (Pacific coast).

Suborder PLATYSPOREA Kudo, 1920

Sutural plane of spore coincides, or is at acute angle, to longest diameter; 1, 2, or 4 polar capsules; with or without iodinophilous vacuole.

Family MYXIDIIDAE Thélohan, 1892

Spore fusiform with pointed or rounded ends; polar filament comparatively long, fine; coelozoic or histozoic in fishes, also in amphibians and reptiles.

Genus **Myxidium** Bütschli
(Fig. 47)

Spores fusiform with pointed or rounded ends; polar capsule in each end; polar filament comparatively long, fine; coelozoic or histozoic in fishes mainly, some in amphibians and reptiles.
Ref. Kudo (1954, p. 655).

Myxidium americanum Kudo, 1920: in kidney tubules, *Trionyx spinifera*, Ill.
M. aplodinoti Kudo, 1934: in gall bladder, *Aplodinotus grunniens*, Ill.
M. bellum Meglitsch, 1937: in gall bladder, *Ictalurus punctatus*, Ill.
M. folium Bond, 1938: in hepatic ducts and gall bladder, *Fundulus heteroclitus*, Md.
M. gasterostei Noble, 1943: in gall bladder, *Gasterosteus aculeatus*, Calif.
M. illinoisense Meglitsch, 1937: in kidney, *Anguilla bostoniensis*, Ill.
M. kudoi Meglitsch, 1937: in gall bladder, *Ictalurus furcatus*, Ill.
M. lieberkuhni Bütschli, 1882: in urinary bladder, *Esox niger, E. lucius*; Noble (1943): in *Lota lota*, Can., Wis., Europe; Guilford (1965): *E. lucius*, L. Mich.
M. macrocapsulare Auerbach, 1910: in gall bladder, *Scardinius*; Otto and Jahn (1943): in *Aplodinotis grunniens*, Iowa.
M. mellum Otto and Jahn, 1943: in gall bladder, *Ictalurus melas, Pomoxis sparoides*, Iowa.
M. minteri Yasutake and Wood, 1957: in renal tubules of chinook and silver salmon, rainbow, steelhead, and brook trout, Wash.

M. myoxocephali Fantham *et al.*, 1940: in gall bladder, *Myoxocephalus octodecemspinosus*, Can.

M. oviforme Parisi, 1912: in gall bladder, *Salmo salar*, Norway; Jameson (1931): in *Salmo gairdneri, Phanerodon furcatus*, Calif.

M. percae Fantham *et al.*, 1939: subdermal, *Perca flavescens*, Can.

M. umbri Guilford, 1965: renal tubules, L. Mich.

Myxidium sp. In gall bladder, kidney tubules, trout (Davis, 1947; Gauthier, 1926; Yasutake and Wood, 1957); in gall bladder, *Micropterus dolomieui* (Fantham and Porter, 1947); in gall bladder of *Umbra limi* (Guilford, 1965).

Genus **Sphaeromyxa** Thélohan
(Fig. 48)

Spore fusiform, but ends usually truncate; polar filament short, thick; trophozoites large, discoid; coelozoic in marine fish.
Ref. Kudo (1954, p. 656).

Genus **Zschokkella** Auerbach
(Fig. 46)

Spore semicircular in front view; fusiform in profile; circular in cross section; ends pointed obliquely; polar capsules large, spherical; sutural line usually in S-form; coelozoic in fish and amphibians.

Zschokkela salvelini Fantham *et al.*, 1939: in kidney capsule, *Salvelinus fontinalis*, Can.

Family COCCOMYXIDAE Léger et Hesse
(Fig. 44)

Spore ellipsoidal; 1 polar capsule at end; circular in cross section.

Coccomyxa morovi Laveran and Hesse, 1910: in gall bladder, *Clupea pilchardus*, marine.

Family MYXOSOMATIDAE Poche

Polar capsules 2 or 4 at anterior end; no iodinophilous vacuole.

Genus **Myxosoma** Thélohan
(Syn.: *Lentospora* Plehn)
(Fig. 53)

Spore ovoidal (front view); usually lenticular in profile; 2 piriform polar capsules at anterior end. Sporoplasm without iodinophilous vacuole; histozoic in freshwater and marine fish. Trophozoites usually

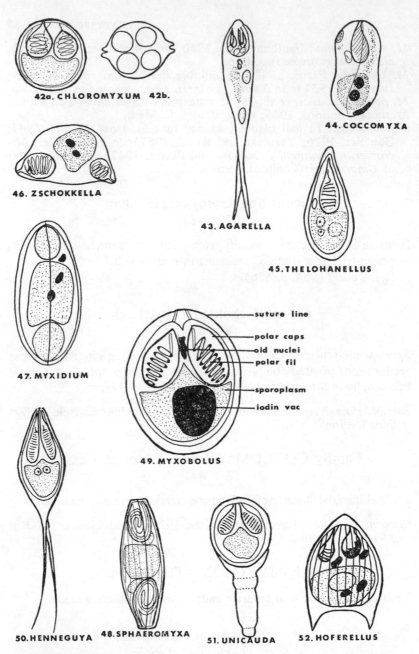

42a. CHLOROMYXUM 42b.

44. COCCOMYXA

46. ZSCHOKKELLA

43. AGARELLA

45. THELOHANELLUS

47. MYXIDIUM

suture line
polar caps
old nuclei
polar fil
sporoplasm
iodin vac

49. MYXOBOLUS

50. HENNEGUYA 48. SPHAEROMYXA 51. UNICAUDA 52. HOFERELLUS

Figs. 42–52. Fig. 42. *Chloromyxum gibbosum*, spore. *a*, front view; *b*, side view (from Herrick, 1941). Fig. 43. *Agarella gracilis*, spore (from Dunkerly, in Kudo, 1954). Fig. 44. *Coccomyxa morovi*, spore (from Léger et Hesse, in Kudo, 1954). Fig. 45. *Thelohanellus notatus*, spore (from Kudo, 1934). Fig. 46. *Zschokkella salvelini*, spore (from Fantham

produce numerous disporoblastic pansporoblasts, rarely monosporo-
blastic or quadrasporoblastic.

Ref. Hoffman, Putz, and Dunbar (1965), synopsis; Davis (1923),
development.

Myxosoma bibullatum Kudo, 1934: in skin, *Catostomus commersoni,*
Ill.; Nigrelli (1943): in gills, *C. commersoni,* N. Y.

M. *cartilaginis* Hoffman, Putz and Dunbar, 1965: in cartilage of head
and spines, *Lepomis macrochirus, L. cyanellus, Micropterus salmoides,* W. Va., Md.

M. *catostomi* Kudo, 1923 (cf. Kudo, 1926): in muscle, connective
tissue, *Catostomus commersoni,* Mich.; Fantham *et al.* (1939);
C. commersoni, Que., Can.

M. *cerebralis* (Hofer), (Plehn, 1905) Kudo, 1933 (Syn. *Lentospora
c.*): in cartilage of head and spinal column, causes "whirling disease" in salmonids; Hoffman, Dunbar and Bradford (1962): reported from N. America.

M. *commersonii* Fantham *et al.,* 1939: in skin, *Catostomus commersoni,* Que.

M. *cuneata* Bond, 1939: in gill arch, *Esox masquinongy,* N. Y.

M. *diaphana* Fantham *et al.,* 1940: in *Fundulus diaphanus,* Nova
Scotia.

M. *ellipticoides* Fantham *et al.,* 1939: in *Catostomus commersoni,*
Can.

M. *endovasa,* Davis, 1947: in gills, *Ictiobus bubalus,* Iowa.

M. *eucalii* Guilford, 1965: in skeleton of *Eucalia inconstans,* L. Mich.

M. *funduli* Kudo, 1920: in *Fundulus heteroclitus, F. majalis, F. diaphanus,* Mass., Md.

M. *grandis* Kudo, 1934: in liver, *Ericymba buccata, Notropis hudsonius, Rhinichthys atronasus,* N. Y.

– M. *hoffmani* Meglitsch, 1963: in cartilaginous sclera of eye, *Pimephales promelas,* N. Dak., Hoffman and Putz (1965): additional
data, same collection.

M. *hudsonis* Bond, 1938: in skin, *Fundulus heteroclitus,* N. Y.

M. *media* Fantham *et al.,* 1939: large abdominal cyst, *Notropis Cornutus,* Que.

M. *microthecum* Meglitsch, 1942: in mesenteries, *Minytrema melanops,* Ill.

M. *mulleri* Bond, 1939: in gills, *Esox masquinongy,* N. Y.

---

et al., 1939). Fig. 47. *Myxidium kudoi,* spore (from Meglitsch, 1937).
Fig. 48. *Sphaeromyxa balbianii,* spore (from Kudo, 1954). Fig. 49.
Myxobolus pfeifferi, spore, stained with iodine showing dark iodinophilous
vacuole (from Keysselitz, in Kudo, 1920). Fig. 50. *Henneguya doori,*
spore (from Guilford, 1963). Fig. 51. *Unicauda crassicauda,* spore (from
Kudo, 1934). Fig. 52. *Hoferellus cyprini,* spore (from Doflein, in Kudo,
1920).

M. multiplicatum (Reuss) Rice and Jahn, 1943: in gills, *Ictiobus bubalus*, Iowa.

M. neurophila Guilford, 1963: in brain, *Perca flavescens, Etheostomum nigrum*, L. Mich., Wis.

M. notropis Fantham *et al.*, 1939: large cyst in abdomen, *Notropis cornutus*, Que.

M. okobojiensis Rice and Jahn, 1943: in gills, *Ictiobus bubalus*, Iowa.

M. orbitalis Fantham *et al.*, 1939: in eye orbit, *Notropis cornutus*, Que.

M. ovalis (Davis, 1923) Kudo, 1933 (Syn. *Lentospora o.*): in gills, *Ictiobus bubalus. I. cyprinella*, Iowa; Rice and Jahn (1943): Iowa; Wagh (1961): Ill.

M. parellipticoides Fantham *et al.*, 1939: large abdominal cyst, *Pfrille neogaeus*, Que.

M. pfrille Fantham *et al.*, 1939: large abdominal cyst, *Pfrille neogaeus*, Que.

M. procerum Kudo, 1934: in skin, *Percopsis guttatus*, Ill. Guilford (1965): skin, *P. omiscomaycus*, L. Mich.

M. robustum Kudo, 1934: in integument, *Notropis cornutus*, Ill.

M. rotundum Meglitsch, 1937: in gills, *Carpiodes cyprinus*, Ill.

— *M. scleroperca* Guilford, 1963: in cartilaginous sclera of eye, *Perca flavescens, Percina caprodes*, L. Mich., Wis.; Dechtiar (1965) L. Erie.

M. squamalis Iversen, 1954: in scales, *Salmo gairdneri, Oncorhynchus kisutch, O. keta*, Wash.

M. subtecalis Bond, 1938: in viscera, fat of cranial cavity, and kidney, *Fundulus heteroclitus*.

Myxosoma (sp. unknown) Rigdon and Hendricks (1955): in *Fundulus similus*, Tex.

Genus **Agarella** Dunkerly

Spore elongate oval; 4 polar capsules at anterior end; shell prolonged posteriorly into long processes.

Ref. Kudo (1954, p. 658).

Agarella gracilis Dunkerly, 1915: in testis of South American lung fish.

Family MYXOBOLIDAE Thélohan

Polar capsules 1, 2, or 4 at anterior end; iodinophilous vacuole present.

Genus **Myxobolus** Bütschli
(Fig. 49)

Spores ovoidal or ellipsoidal, flattened; 2 polar capsules at anterior end; sporoplasm with iodinophilous vacuole; sometimes with posterior prolongation of shell; exclusively histozoic in freshwater fish and amphibians.

Ref. Kudo (1934; 1954, p. 658).

Myxobolus angustus Kudo, 1934: in gills, *Cliola vigilax*, Ill.

M. aureatus Ward, 1919: in fins, *Pimephales notatus, Notropis anogenus*, L. Erie.

M. bellus Kudo, 1934: in skin, *Cyprinus carpio*, Ill.

M. bilineatum Bond, 1938: in brain and viscera, *Fundulus heteroclitus*, Md.

M. bubalis Otto and Jahn, 1943: in gall bladder, *Ictiobus bubalus*, Iowa.

M. capsulatus Davis, 1917: in visceral connective tissue, *Cyprinodon*, N. C.

M. catostomi Fantham *et al.*, 1939; in mouth subepithelium, muscle, *Catostomus commersoni*, Que.

M. compressus Kudo, 1934: in skin, *Notropis blennius*, Ill.

M. congesticius Kudo, 1934: in fins, *Moxostoma anisurum*, Ill.

M. conspicuus Kudo, 1929: in skin, *Moxostoma breviceps*, Ill.; Nigrelli (1943): in *M. aureolum*; Fantham *et al.*, 1939: *M. aureolum*, Que.

M. couesii Fantham *et al.*, 1939: in eye (posterodorsal corner of anterior chamber, on iris), *Couesius plumbeus*, Que.

M. dentium Fantham *et al.*, 1939: sore mouth, *Esox masquinongy*, Can.

M. discrepans Kudo, 1920: in gills, *Carpiodes difformis*, Ill., Rice and Jahn (1943): on gills, *Pomoxis sparoides*, Iowa.

M. funduli (Hahn) Kudo, 1930: in branchiae and muscle, *Fundulus, Cyprinodon*, Woods Hole, Mass.

M. gibbosus Herrick, 1941: in connective tissue of gill arches, *Lepomis gibbosus*, L. Erie.

M. globosus Gurley, 1894; in gills, *Erimyzon oblongus*, N. D., S. C.

M. grandis Fantham *et al.*, 1939: in body cavity, *Notropis cornutus*, Can.

M. gravidus Kudo, 1934: in *Moxostoma anisurum*, Ill.

M. hyborhynchi Fantham *et al.*, 1939: in bone at end of mandible, *Hyborhynchus notatus*, Can.

M. inornatus Fish, 1939: in flesh, *Micropterus salmoides*.

M. insidiosus Wyatt and Pratt, 1963: in muscle, *Oncorhynchus tshawytscha*, Ore.

M. intestinalis Kudo, 1929: in intestinal wall, *Pomoxis sparoides*, Ill.

M. iowensis Otto and Jahn, 1943: in gills, *Pomoxis sparoides*, Iowa.

M. kisutchi Yasutake and Wood, 1957: in spinal column of coho salmon, Wash.

M. koi Kudo, 1920: in gills, *Cyprinus carpio*, Japan.

M. kostiri Herrick, 1936: subcutaneous, *Micropterus dolomieui*, Lake Erie.

M. lintoni Gurley, 1893: subcutaneous, *Cyprinodon variegatus*, Woods Hole, Mass.

M. mesentericus Kudo, 1920: in viscera, *Lepomis cyanellus*, Ill.

M. moxostomi Nigrelli, 1948: in snout (prickle cell hyperplasia), *Moxostoma aureolum*, N. Y.

M. musculi Hahn, 1913: in skin, *Fundulus.*

M. mutabilis Kudo, 1934: in skin, *Pimephales notatus*, Ill.

M. neurobius Schuberg and Schroder, 1905: in nervous tissue, *Salmo trutta*, Europe; also Reichenbach-Klinke (1954, p. 597).

M. nodosus Kudo, 1934: in skin, *Pimephales notatus*, Ill.

M. notatus, Mavor, 1916: intermuscular in body, *Pimephales notatus*, Can.

M. notemigoni Lewis and Summerfelt, 1964: in skin, *Notemigonus crysoleucas*, Ark.

M. notropis Fantham *et al.*, 1939: in skin, *Notropis cornutus*, N. *heterolepis*, Can.

M. okobojiensis Otto and Jahn, 1943: in intestine, *Pomoxis sparoides*, Ia.

M. oblongus Gurley, 1893: beneath skin, *Erimyzon succetta*, Fox R., Kinston.

M. obliquus Kudo, 1934: in muscle, *Carpiodes velifer*, Ill.

M. orbiculatus Kudo, 1920: in muscle, *Notropis gilberti*, Ill.

M. ovatus Kudo, 1934: in skin, *Ictiobus bubalis*, Ill.

M. osburni Herrick, 1936: in mesenteries and peritoneum, *Micropterus dolomieui*, *Lepomis gibbosus*, L. Erie; Otto and Jahn (1943): in gall bladder, *L. macrochirus*, *Pomoxis sparoides*, Iowa.

M. ovoidalis Fantham, 1930: in skin, *Salvelinus fontinalis*, Can.

M. percae Fantham *et al.*, 1939: at base of pectoral fin, *Perca flavescens*, Can.

M. poecilichthidis Fantham *et al.*, 1939: in intestine, *Etheostoma exile*, Can.

M. rhinichthidis Fantham *et al.*, 1939: in skin, *Rhinichthys atronasus*, Can.

M. sparoidis Otto and Jahn, 1943: in gall bladder and intestine, *Pomoxis sparoides*, Iowa.

M. squamae Keysselitz, 1908. Shaw (1947): in skin of *Oncorhynchus kisutch* and bass, Ore.

M. squamosus Kudo, 1934: subepithelium, *Hybopsis kentuckiensis*, Ill.

M. subcircularis Fantham *et al.*, 1939: in fin muscle, *Catostomus commersoni*, Que.

M. symmetricus Rice and Jahn, 1943: in gills, *Ictiobus bubalis*, Iowa.

M. teres Kudo, 1934: in muscle, *Notropis whipplii*, Ill.

M. transovalis Gurley, 1893: under scales, *Clinostomus funduloides*, Va., Rice and Jahn (1943): in gills, *Ictiobus bubalis*, Iowa.

M. transversalis Fantham *et al.*, 1939: in muscle, *Notropis cornutus*, Canada.

M. vastus Kudo, 1934: in corium, *Moxostoma aureolum*, Ill.

Myxobolus sp. Rigdon and Hendricks, 1955: in *Cyprinodon variegatus*, Tex.

Genus **Unicauda** Davis, 1944
(Fig. 51)

Spore similar to that of *Henneguya*, but single caudal appendage not extension of shell valves.

Ref. Davis (1944); Kudo (1934; 1954, p. 660).

Unicauda brachyura (Syn. *Henneguya b.* Ward, 1919): in fin ray, *Notropis anogenus,* L. Erie.

U. clavicauda (Syn. *Henneguya c.* Kudo, 1934): subdermal, *Notropis blennius,* Ill.

U. crassicauda (Syn. *Henneguya c.* Kudo, 1934): in skin and fins, *Campostoma anomalum,* Ill.

U. fontinalis (Syn. *Henneguya f.* Fantham *et al.,* 1939): in skin, *Salvelinus fontinalis,* Que.

U. plasmodia (Syn. *Henneguya p.* Davis, 1922): in gills, *Ictalurus punctatus.*

Genus **Henneguya** (Thélohan) Davis, 1944
(Fig. 50)

Spores ovoid with 2 polar capsules at anterior end. Posterior end of shell valves prolonged into more or less extended processes. Body of spore biconvex, compressed parallel to sutural plane (sutural plane not bisecting polar capsules). Sporoplasm with iodinophilous vacuole. Mostly histozoic, usually forming cysts. Polysporous.

Ref. Kudo (1920, 1929, 1934; 1954, p. 660).

Henneguya acuta Bond, 1939: in gills, *Esox masquinongy,* N. Y.

H. ameiurensis Nigrelli and Smith, 1940: in barbels, *Ictalurus nebulosus,* N. Y.

H. amiae Fantham *et al.,* 1940: in gills, *Amia calva,* Canada.

H. doori Guilford, 1963: in gill filament, *Perca flavescens,* Wis.

H. esocus Fantham *et al.* 1939: in gills, *Esox niger,* Que.

H. exilis Kudo: in gills, *Ictalurus punctatus, I. melas,* Ill; Rice and Jahn (1943): Iowa.

H. fontinalis var. *notropis* Fantham *et al.,* 1939: in skin, *Notropis cornutus, N. heterolepis,* Que.

H. gurleyi Kudo, 1920: base of spines, *Ictalurus melas,* Iowa.

H. limatula Meglitsch, 1937: in gall bladder, *Ictalurus furcatus,* Ill.; Guilford (1965): *I. melas,* Mich.

H. macrura (Gurley) Thélohan, 1895: connective tissue of head, *Hybognathus nuchalis,* Tex.

H. magna Rice and Jahn, 1943: in gills, *Roccus chrysops,* Iowa.

H. monura (Gurley) Labbe, 1899: in subcutaneous intermuscular tissue, *Aphredoderus sayanus,* N. J.

H. nigris Bond, 1939: in gills, *Esox niger,* Md., *E. masquinongy,* N. Y.

H. percae Fantham and Porter, 1947: in gills, *Perca flavescens,* Que.

H. salminicola Ward, 1919: in tissue, *Oncorhynchus kisutch,* Alaska;

Shaw (1947): in *O. tschawytscha*, Ore.; Fish (1939): *O. kisutch, O. tschawytscha, O. gorbusca*, Alaska.
H. salmonis Fantham *et al.*, 1939: in subcutaneous tissue, *Salmo salar*, Can.
H. schizura (Gurley) Labbe, 1899: in eye muscle, sclera, and between sclera and choroid, *Esox lucius*, Germany, U. S. A.

Genus **Thelohanellus** Kudo
(Fig. 45)

Piriform spores, each with 1 polar capsule; sporoplasm with iodinophilous vacuole; histozoic in freshwater fish.
Ref. Kudo (1934; 1954, p. 660).

Thelohanellus notatus (Mavor) Kudo, 1934: in subdermal cysts, *Pimephales notatus, Cliola vigilax, Notropis cornutus, N. blennius, Leuciscus rutilus*, Ill.; Fantham *et al.* (1939): in *Notropis cornutus, Catostomus commersoni, Hyborhynchus notatus, Pfrille neogaeus*, Que.
T. pyriformis (Thélohan) Kudo, 1933 (Syn. *Myxobolus p.*). Kudo (1918): in spleen, *Perca flavescens*, Mass., probably *T. pyriformis*.

Genus **Hoferellus** Berg, 1898
(Fig. 52)

Spore pyramidal with 2 posterior spinous processes arising from lateral faces; in kidney of freshwater fish.

Hoferellus cyprini Doflein, 1898: in renal tubules, *Cyprinus carpio*, Europe.
Hoferellus sp. Fantham, 1919: S. Africa.

Family ELONGIDAE N.N.

Spore elongate; spherical or subspherical in cross section; sutural line passes between polar capsules; shell valves with or without posterior processes similar to *Henneguya*; 2 polar capsules at anterior end; no iodinophilous vacuole; histozoic and coelozoic in freshwater fish.

Genus **Mitraspora*** Fujita
(Fig. 55)

Spores circular or ovoidal in front view; slightly flattened in profile; 2 polar capsules; shell striated; with or without posterior filaments;

---

* This genus was formerly in the suborder Eurysporea, but the position of the sutural plane fits the suborder Platysporea; a new family, Elongidae, is proposed to contain it and *Myxobilatus*.

sutural line passes between polar capsules; in kidneys of freshwater fish. Three species.

Mitraspora elongata Kudo, 1920: in kidney, *Lepomis cyanellus*; Hoffman (1959): *L. cyanellus, L. macrochirus, Micropterus salmoides,* W. Va.

Genus **Myxobilatus** Davis, 1944
(Fig. 54)

Spores bilaterally symmetrical with 2 polar capsules at anterior end. Body of spore flattened on one side, strongly convex on opposite side. Plane of flattening at right angles to sutural plane. Shell valves prolonged into tail-like processes which are separate throughout entire length. No iodinophilous vacuole (Davis, 1944, states that one is present). Coelozoic. Trophozoites irregular in shape or greatly elongated. Monosporous, disporous, or polysporous.
Ref. Davis (1944).

Myxobilatus asymmetricus Davis, 1944: in urinary bladder, *Stizostedion vitreum,* Iowa.
M. caudalis Davis, 1944: in urinary bladder, *Aplodinotus grunniens,* Iowa.
M. mictosporus (Henneguya m. Kudo, 1920) Davis, 1944: in urinary bladder, *Lepomis* spp., *Micropterus salmoides.*
M. ohioensis (Henneguya o. Herrick, 1941) Davis, 1944: in urinary bladder, *Lepomis gibbosus,* L. Erie.
M. rupestris (Henneguya r. Herrick, 1941) Davis, 1944: in urinary bladder, *Ambloplites rupestris.*
M. wisconsinensis (Henneguya w. Mavor and Strasser, 1916) Davis, 1944: in urinary bladder, *Perca flavescens,* Wis.

Class MICROSPORIDEA Corliss and Levine, 1963

Microsporidea invade and undergo asexual division and sporogony within the host cell. Infected cells often show enormous hypertrophy of cytoplasm and nucleus. The spore is small, 3 to 6µ long, rarely 8µ, and the spore membrane is a single piece unlike the myxosporidea. A single polar filament is present, but is not easily seen. One vacuole is usually present at each end.
Ref. Kudo (1924, 1930b, 1954); Kudo and Daniels (1963); Putz, Hoffman, and Dunbar (1965).

KEY
TO THE GENERA OF MICROSPORIDEA

1. Each sporont develops into 1 spore . . . . (fig. 56) *Nosema*
1. Each sporont develops into 2 or more spores 2

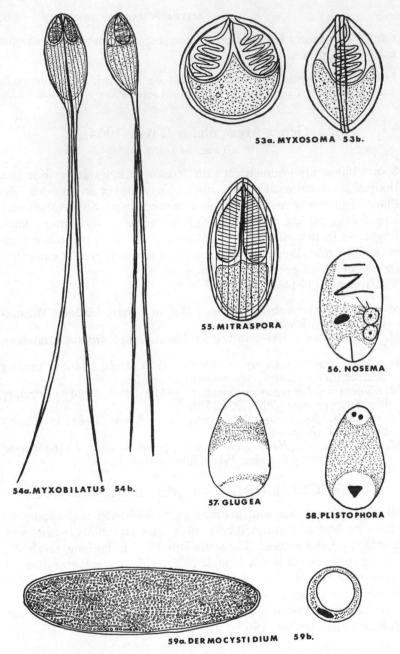

53a. MYXOSOMA 53b.

55. MITRASPORA

56. NOSEMA

54a. MYXOBILATUS 54b.

57. GLUGEA

58. PLISTOPHORA

59a. DERMOCYSTIDIUM 59b.

Figs. 53–59. Fig. 53. *Myxosoma ovalis*, spore. *a*, front view; *b*, side view (from Davis, 1953). Fig. 54. *Myxobilatus caudalis*, spore. *a*, front view; *b*, side view (from Davis, 1944). Fig. 55. *Mitraspora elongata*, spore (orig.). Fig. 56. *Nosema pimephales*, spore (from Fantham *et al.*, 1939).

2. Each sporont develops into 2 spores; infected host cells become extremely hypertrophied, and transform into *Glugea* cysts; polar filament can be extruded; in marine and freshwater fish .... (fig. 57) *Glugea*

2. Each sporont develops into many sporoblasts, each becoming a spore; in marine and freshwater fish .. (fig. 58) *Plistophora*

Genus **Nosema** (Nageli, 1857) Perez, 1905
(Fig. 56)

Differs from *Glugea* and *Plistophora* in that each sporont develops into a single spore.

Ref. Kudo (1954, p. 670); Putz, Hoffman, and Dunbar (1965).

Nosema pimephales Fantham *et al.*, 1941: from abdomen, *Pimephales promelas*, Que.

Genus **Glugea** (Thélohan) Weissenberg, 1913
(Fig. 57)

Each sporont develops into 2 spores; infected cells become extremely hypertrophied, and transform themselves into "*Glugea*" cysts. Usually in marine and brackish fish.

Ref. Kudo (1924, p. 110; 1954, p. 672); Putz, Hoffman, and Dunbar (1965).

Glugea hertwigi Weissenberg, 1911. In North America—Kudo (1924); Fantham *et al.* (1941); Haley (1954a, 1954b): marine and fresh water in many organs of *Osmerus eperlanus* and *O. mordax*; Dechtiar (1965): L. Erie.

Genus **Plistophora** Gurley, 1893
(Fig. 58)

Each sporont develops into more than 16 spores.

Ref. Kudo (1924, p. 167; 1954, p. 676); Putz, Hoffman, and Dunbar (1965).

Plistophora cepedianae Putz, Hoffman, and Dunbar, 1965: large cysts in viscera, *Dorosoma cepedianum*, Ohio. Probably causes mortalities.

Fig. 57. *Glugea hertwigi*, spore (from Schrader, in Kudo, 1954). Fig. 58. *Plistophora cepediana*, spore (from Putz and Hoffman, 1965). Fig. 59. *Dermocystidium* sp. *a*, cyst; *b*, spore (orig.).

P. ovariae Summerfelt, 1964: in ovaries, liver, kidney, *Notemigonus crysoleucas*, Ark., Ill., Ky., Mo.
P. salmonae Putz, Hoffman, and Dunbar, 1965: small cysts in gills, *Salmo gairdneri*, Calif., Wales and Wolf (1955): causes heavy mortalities.
Plistophora sp. Bond, 1937: in muscle, *Fundulus heteroclitus*, Md. This may be a *Nosema* sp.

Genus **Thelohania** Henneguy, 1892
(No fig.)

Each sporont develops into 8 sporoblasts and ultimately into 8 spores. Sporont membrane may degenerate. Spores similar to *Nosema*, *Plistophora*, and *Glugea*. Reported from freshwater fish in Europe.

Ref. Kudo (1924, p. 130; 1954, p. 674); Putz, Hoffman, and Dunbar (1965).

SPOROZOA OF UNCERTAIN CLASSIFICATION

Genus **Dermocystidium** Perez, 1907
(Fig. 59)

This parasite has been considered a fungus by some and a haplosporidean protozoan by others. To my knowledge no one has proved whether *Dermocystidium* of lower vertebrates is protozoa or fungus. The vegetative stage, as shown by Elkan (1962), resembles that of myxosporidean protozoa. The spores, however, have no characteristics of Myxosporidea or Haplosporidea; so further classification remains uncertain. *Dermocystidium marinum* of oysters, which has been studied more extensively (see Macklin, 1961), does possess some fungal characteristics, notably the spore wall which reacts with iodide, and tubular outgrowths of the spore.

Immature stages: cysts with hyaline walls (probably of parasite origin) of varying thickness filled with amorphous protoplasmic mass in which chromatin granules are more or less evenly distributed. Granules coalesce to form nuclei, splitting up protoplasmic mass to form spherical portions 8 to 10μ in diameter which produce spores 3 to 12μ in diameter. *Spore stage*: spores enclosed in elongate oval or round cyst which is very rigid. Hyaline cyst wall probably of parasite origin. Spores round, 3 to 12μ diameter, containing nucleus and large conspicuous eccentric vacuole.

Parasites of fishes and oysters.

Dermocystidium branchialis Léger, 1914: small round cysts in gills, *Salmo trutta*, Europe.

D. gasterostei Elkans, 1962: in skin, *Gasterosteus aculeatus*, England.
D. koi Hoshina and Sahara, 1950: in skin and muscle, *Cyprinus carpio*, Japan.
D. percae Reichenbach-Klinke, 1950: in skin, *Perca fluviatilis*, Europe; Chen (1956): on black Amur fish and sheatfish.
D. salmonis Davis, 1947: small round cysts in gills, *Onchorhynchus tschawytscha*, Calif.
D. vejdovskyi Jirovec, 1939: small round cysts in gills, *Esox lucius*, Europe; Reichenbach-Klinke (1954): same.
Dermocystidium sp. I have observed cysts on fins of *Carassius auratus*, *Lepomis macrochirus*, and *Salvelinus fontinalis*.

Monogenetic Trematodes
(Platyhelminthes: Trematoda: Monogenea)*

Contents

Monogenetic trematodes
(Class Trematodes, Order
Monogenea) 71
Methods 71
Maintaining *Gyrodactylus* in
the laboratory 73
Key to suborders 74
Suborder MONOPISTHO-
COTYLEA 74
Key to the families of
MONOPISTHOCOTYLEA . 74
Family GYRODAC-
TYLIDAE 75
   Genus *Gyrodactylus* 75
Family CALCEOSTO-
MATIDAE 76
   Genus *Acolpenteron* 77
   Genus *Anonchohaptor* ... 77
Family DACTYLO-
GYRIDAE 77
Key to the subfamilies of
DACTYLOGYRIDAE 77
   Subfamily Dactylogyrinae.. 79
   Key to the genera 79
   Genus *Dactylogyrus* 79
   Genus (*Neodactylogyrus*) 82
   Genus *Pellucidhaptor* 83
   Subfamily Diplectaninae .. 83
   Genus *Diplectanum* 83
   Subfamily TETRAON-
CHINAE (ANCYRO-
CEPHALINAE) 85
Key to the genera and synony-
mous genera of TETRAON-
CHINAE 85

Genus *Pseudo-
murraytrema* 87
Genus *Tetraonchus* 87
Genus *Cleidodiscus* 88
Genus *Urocleidoides* 89
Genus *Urocleidus* 91
Genus "*Haplocleidus*" ... 93
Genus "*Pterocleidus*" 95
Genus *Clavunculus* 95
Genus *Anchoradiscus* ... 96
Genus *Actinocleidus* 96
Genus (*Ancyrocephalus*). 97
Genus (*Haplocleidus*) ... 97
Genus (*Leptocleidus*) ... 98
Genus (*Onchocleidus*) ... 98
Genus (*Pterocleidus*) 98
Genus (*Tetracleidus*) 98
Suborder POLYOPISTHO-
COTYLEA 98
Key to the genera of
POLYOPISTHOCOTYLEA 99
   Genus *Microcotyle* 99
   Genus *Lintaxine* 101
   Genus *Diclybothrium* 101
   Genus *Mazocraeoides* ... 101
   Genus *Octomacrum* 102
   Genus *Discocotyle* 102
Order ASPIDOCOTYLEA .. 102
Family ASPIDOGASTERI-
DAE 104
   Genus *Cotylaspis* 104
   Genus *Cotylogaster* 104

* Synonyms are included in this table of contents and are indicated by quotation marks or parentheses.

(Class TREMATODA, Order MONOGENEA)

Monogenea are small (microscopic) to medium-sized (5mm) trematodes which complete the life cycle on one host. Usually the immature worms are morphologically similar to the mature forms. The chief organ of attachment and best identification aid is the haptor, which is posterior (lateral in one freshwater species). Nearly always there are 12 to 16 hooklets around the margin of the haptor; and in most freshwater genera 2 or 4 larger hooks (anchors) are centrally located in the haptor. In the Polyopisthocotylea, however, the larval hooklets disappear and 6 to many cuticular adhesive units (clamps and/or suckers) develop. Anterior adhesive organs (head organs), suckers, or pseudosuckers may be present. The eggs are comparatively few and large, often with polar prolongations. Vitellaria are nearly always minutely follicular, coextensive with intestinal crura.

Some species live on the gills only, some on the body and fins only, and some on both. One genus is found in the urinary system. Most are capable of moving around on the host, but some *Dactylogyrus* spp. apparently remain in one spot on the gills. The food of monogenetic trematodes is host mucus, epithelium, and sometimes blood. The feeding organ (pharynx) of at least one species emits a proteolytic substance which erodes epidermis (Kearn, 1963).

Some monogenetic trematodes are serious pests in fish culture on occasion. Certain species of *Gyrodactylus* may become so numerous on cultured trout, goldfish, and bluegills that the fish suffer great distress and must be treated if they are to survive. Some *Dactylogyrus* species cause great damage to the gill filaments of carp and goldfish in hatcheries. Other Monogenea probably are potential threats to fish culture, but have not been adequately studied. In fish populations that have become crowded in nature, similar hardships have been known to occur.

METHODS

It is preferable that fish be examined immediately after killing, which can be accomplished by severing the spinal cord immediately behind

71

the head with a bone cutter, pithing or narcotizing; cutting with a sharp instrument causes excessive bleeding. Anesthetics, such as Sandoz MS 222 (meta-aminobenzoic acid ethylester) can be used to immobilize the fish that must be kept alive. Preserved fish can be used, but are far less satisfactory than freshly killed ones.

If the fish is small it should be immersed in chlorine-free water and examined at 10× with a dissection microscope; monogenetic trematodes are usually very active, a fact which aids greatly in their detection. The entire body, the fins, the inside of the mouth, and particularly the gills should be observed. After examining the intact fish, the opercula should be carefully cut off, exposing the gills, which can then be excised as needed for thorough examination. *Acolpenteron* spp. should be sought in the urinary system.

If the fish is too large to examine as described above, excised gills and fins, mucus scrapings, and the inside of the mouth should be examined individually.

Flukes may be removed from fish mechanically, but attendant mucus often makes this difficult. *Gyrodactylus* can be removed dead, but fresh enough for making permanent preparations, by immersing the host in formalin (1:4000) for 15 to 45 minutes. The flukes leave the fish and seldom become entangled in mucus. Other flukes may be removed likewise, but do not become free from the gills as readily. Flukes so removed may be studied fresh or fixed for permanent preparations. Monogenetic trematodes can be studied advantageously while fresh, or better yet if alive. Such material can be studied under a cover slip with adequate water to prevent flattening. Later the fluid may be removed sufficiently with pieces of paper toweling or filter paper to flatten the specimen and permit optimum observation of the haptoral armature and copulatory complex.

Freezing the gills for 6 to 24 hours will also free flukes from the gills (Mizelle, 1936). Chloretone (Hargis, 1953) and nembutal (Ikezaki and Hoffman, 1957) have also been used.

Permanent preparations can be made after fixing worms removed by the foregoing methods, providing the worms are fixed very soon after they are immobilized. Freezing sometimes destroys the architecture of the soft parts, but the hard parts, which are usually the most important, will be satisfactory. Mizelle (1936) describes a method for affixing monogenetic trematodes to a slide: "the specimens were transferred to slides coated with Mayer's albumin. They were then placed in a finger bowl on inverted stender dishes. Alcohol was poured in the bottom of the container to a level below the top of the stender dishes, the top of the bowl was covered and the apparatus set in an incubator. It was found that an exposure for one to five minutes at 55°C was sufficient to coagulate the albumin and securely attach the specimens to the slide. After attachment of the specimens to the slides they were fixed in Gilson's fluid and treated with iodine for removal of mercuric salts." Hematoxylin and carmine stains are usually employed for staining monogenetic trematodes.

Unstained semipermanent mounts can be made by mounting in a very small amount of polyvinyl alcohol, Turtox CMC or CMC-S (Turtox Biological Supply House), and sealing with lacquer or permanent mounting medium. The two cover glass method also can be used: place the specimen in a very small amount of water on a 22 mm square cover glass, add a small amount of polyvinyl alcohol, place a smaller cover glass on the specimen, place 4 drops of permanent mounting medium on a slide, invert the cover glass preparation and place it on the mounting medium, which will seal the mount. Glycerine jelly may be used for preserved or fixed specimens (Mizelle and Seamster, 1939). The usefulness of such mounts will depend on how well the anchors are separated in the flattening process.

MAINTAINING GYRODACTYLUS IN THE LABORATORY

Life cycle.—*Gyrodactylus bullatarudis* Turnbull, 1956, of the guppy, completes its life cycle from larva to larva-producing adult in about 60 hours at 25 to 27°C (Turnbull, 1956). It probably takes two weeks for a population of one *Gyrodactylus* to build up to 100.

Maintaining on trout.—Usually small fingerlings (1 to 2 in.) are most heavily infected. As fish become older, the population of *Gyrodactylus* gradually diminishes. Such yearling (9 to 12 in.) usually are still infected, but with a very small number of parasites.

Minimal water supply and crowding is conducive to increasing the *Gyrodactylus* population.

Gyrodactylus can be transferred to uninfected fish by placing them in the same tank with infected fish. They can also be transferred by skin scrapings (Hoffman and Putz, 1964).

Maintaining on bluegills.—*Gyrodactylus macrochiri* (Hoffman and Putz, 1964) can be maintained at 54°F in a manner similar to the above. However, in aquaria at room temperature they disappear.

Maintaining other species of Gyrodactylus.—Goldfish *Gyrodactylus* populations follow the same pattern as above but can be maintained at room temperature. Turnbull (1956) maintained the guppy *Gyrodactylus* at room temperature with no difficulties. Malmberg (1956) was not able to maintain *Gyrodactylus* spp. in his lab in Sweden.

To assure large populations of *Gyrodactylus*, it is probably best to have young or small uninfected fish available, and to infect new lots from time to time, possibly monthly.

The keys herein have been modified from Hargis (1952), Mizelle and Hughes (1938), Sproston (1946), and Yamaguti (1963).

KEY

TO THE SUBORDERS OF MONOGENETIC TREMATODES

Haptor as a single unit; larval haptor retained, more or less unchanged, in adult; anterior end with gland organs; mouth not surrounded by oral sucker, and paired suckers within mouth not present; genitointestinal canal absent; haptor usually bears 1 or 2 pairs of anchors (hooks) and 12 to 16 hooklets . . . . . .
. . . . MONOPISTHOCOTYLEA
Functional haptor (posterior attachment organ) of adult developed anterior to larval haptor in form of lateral rows of suckers or clamps always supported by cuticular sclerites, usually acting as clamps; larval haptor vestigial; anterior end without adhesive glands . . . . POLYOPISTHOCOTYLEA
(p. 98)

Suborder MONOPISTHOCOTYLEA Odhner, 1912

Haptor a single unit derived from larval haptor; armature usually consists of large anchors (hooks) and marginal hooklets. Anterior end frequently with adhesive organs in form of "head organs"; genitointestinal canal absent. Most species highly host-specific.
    Ref. Sproston (1946, p. 192); Yamaguti (1963, p. 7).

KEY

TO THE FAMILIES OF MONOPISTHOCOTYLEA

1. Viviparous; anchor hooks of embryo usually can be seen about
    midway in parent worm . . . . GYRODACTYLIDAE
1. Oviparous   2
2. Cephalic glands not concentrated in head organs; anterior end
    may be expanded into head lappets; usually no anchors,
    but marginal hooklets present; eye spots present or absent
    . . . . CALCEOSTOMATIDAE
2. Cephalic glands concentrated in head organs; 2 or 4 anchors,
    14 marginal hooklets (plus 2 very small ones in some
    genera) on haptor; eye spots present . . DACTYLOGYRIDAE

Family GYRODACTYLIDAE Cobbold, 1864
Subfamily GYRODACTYLINAE Monticelli, 1892

Small elongated Monogenea with anterior end bilobed, each lobe with head organ. Haptor well developed, bearing 1 pair of large hooks (anchors) and 16 marginal hooklets. Intestine bifurcate, the 2 limbs not uniting posteriorly. Eye spots absent. Male copulatory organ (cirrus) armed with row of minute "spines" and usually with triangular cuticular plaque. Ovary V-shaped or lobed, posterior or ventral to testis. Vitellaria absent or united with ovary. Vagina absent. Viviparous. Only one genus, *Gyrodactylus*, represented in North America.

Genus **Gyrodactylus** Nordmann, 1832
(Fig. 60)

Viviparous; larva, including anchors, usually present in parent in-utero. Very host-specific in nature except possibly *G. elegans*. Many species.

Ref. Sproston (1946, p. 195); Malmberg (1956); Putz and Hoffman (1963); Yamaguti (1963, p. 10).

Development is discussed by Hoffman and Putz (1964).

Gyrodactylus aphredoderi Rogers and Wellborn, 1965: on *Aphredoderus sayanus*, Ala.

G. atratuli Putz and Hoffman, 1963: on *Rhinichthys atratulus*, W. Va.

G. bairdi Wood and Mizelle, 1957: on *Cottus b. bairdi*, Ind.

G. bullatarudis Turnbull, 1956: on *Lebistes reticulatus*, Can., original source unknown; Rogers and Wellborn (1965): same, Ala.

G. couesius Wood and Mizelle, 1957: on *Couesius plumbius dissimilis*, Brit. Col.

G. cylindriformis Mueller and Van Cleave, 1932: on *Umbra limi*, N. Y.

G. egregius Wood and Mizelle, 1957: on *Richardsonius egregius*, Nev.

G. elegans Nordmann, 1832: a common species in Europe. Occurrence in North America needs more study. Seamster (1938): on *Ictalurus melas*; Mueller (1936a): var. "A" on *C. carassius*, var. "B" on trout, N. Y.; Haderlie (1953): on *Salmo gairdneri*, Calif.; Curtin (1956): on *C. carpio*, Penn.; Lewis and Lewis (1963): on *Notemigonus crysoleucas*, Ill.; Malmberg (1958) believes that *G. elegans* Nordmann, 1832, from *Abramis brama* is a true species, that *G. elegans* from North American goldfish is probably Wagener's species, and that reported from trout by Mueller (1936) is a different species.

G. elegans muelleri (*G. elegans A*. Mueller, 1936) Yin and Sproston, 1948: on *Carassius auratus*, N. America.

G. elegans salmonis (*G. elegans B*. Mueller, 1936) Yin and Sproston, 1948: on trout, N. America.

G. eucaliae Ikezaki and Hoffman, 1957: on *Eucalia inconstans*, N. Dak.

G. fairporti Van Cleave, 1921: on *Ictalurus melas, Cyprinus carpio,* Iowa.

G. funduli Hargis, 1955: on *Fundulus similis,* Fla.; Hutton (1964): same; Sogandares (pers. com.), same, La.

G. gambusiae Rogers and Wellborn, 1965: on *Gambusia affinis,* Fla.

G. gurleyi Price, 1937: on *Carassius auratus,* Tex.

G. heterodactylus Rogers and Wellborn, 1965: on *Elassoma zonatum,* Ala.

G. limi Wood and Mizelle, 1957: on *Umbra limi,* Ind.

G. macrochiri Hoffman and Putz, 1964 (Syn. *G. elegans* of Hargis, 1953): on *Lepomis macrochirus, L. cyanellus,* and experimentally on *Micropterus salmoides,* W. Va.; Rogers and Wellborn (1965): on *L. macrochirus,* Ala., *L. punctatus,* Fla., *Chaenobryttus gulosua,* Fla.

G. margaritae, Putz and Hoffman, 1963: on *Semotilus margarita,* W. Va.

G. medius Kathariner, 1894: on *Micropterus dolomieui* (questioned by Price, 1937); Meyer (1954): on *Salvelinus fontinalis,* Me.

G. micropogonus Wood and Mizelle, 1957: on *Micropogon undulatus,* Tex.

G. percinae Rogers and Wellborn, 1965: on *Percina nigrofasciata,* Ala.

G. prolongis Hargis, 1955: on *Fundulus grandis,* Fla.; Sogandares (pers. com.): will transfer to *Cyprinodon variegatus* in aquaria, La.

G. protuberus Rogers and Wellborn, 1965: on *Notropis uranoscopus,* Ala.

G. rhinichthius Wood and Mizelle, 1957: on *Rhinichthys osculus,* Nev.

G. richardsonius Wood and Mizelle, 1957: on *Richardsonius egregius,* Nev.

G. spathulatus Mueller, 1936: on *Catostomus commersoni;* Krueger (1954): same, Ohio.

G. stegurus Mueller, 1937: on *Fundulus diaphanus,* N. Y.

G. stephanus Mueller, 1937: on *Fundulus heteroclitus,* Md.; Hargis (1955): on *F. grandis,* Fla.

Gyrodactylus sp. Hargis, 1953: on *Ictalurus nebulosus,* Va.

Gyrodactylus sp. Hargis, 1955: on *Cyprinodon variegatus,* Fla.

Family CALCEOSTOMATIDAE

Cephalic glands not concentrated into head organs, but scattered over a considerable area of anterior of body, which may be expanded into head lappets. Haptor sucker-like, with or without anchors, marginal hooklets usually present. Cirrus simple, cuticularized.

Ref. Sproston (1946, p. 257); Yamaguti (1963, p. 89).

Head lappets absent; no eye spots; intestinal crura united posteriorly .... (fig. 61) *Acolpenteron*

Head lappets present; 4 eye spots present; intestinal crura not united posteriorly.................(fig. 62) *Anonchohaptor*

Genus **Acolpenteron** Fischthal and Allison, 1940
(Fig. 61)

No head lappets, eyes,* or anchors; haptor with 14 marginal hooklets. Body covered with sensory hairs. Testis single, elongated. Ovary elliptical, median. Intestinal crura without diverticula but confluent posteriorly.

Ref. Sproston (1946, p. 259); Yamaguti (1963, p. 90).

Acolpenteron catostomi Fischthal and Allison, 1942: in ureters and urinary bladder. *Hypentelium nigricans, Catostomus commersoni,* Mich.; Fischthal (1950, 1956): *C. commersoni,* N. Y., Wis.

A. ureteroecetes Fischthal and Allison, 1940: in ureters and bladder, *Micropterus dolomieui. M. salmoides,* Mich.; Fischthal (1956): *M. dolomieui,* N. Y.; Du Plessis (1948): *M. salmoides,* S. Africa.

Genus **Anonchohaptor** Mueller, 1938
(Fig. 62)

Head lappets prominent. Haptor disclike, no anchors but 14 marginal hooklets (2 are central); 4 eye spots. Intestine without diverticula, but crura united posterior to testis. Testis small. Ovary looping around right intestinal limb.

Ref. Sproston (1946, p. 259); Yamaguti (1963, p. 91).

Anonchohaptor anomalum Mueller, 1938: on gills, *Catostomus commersoni, Hypentelium nigricans, Moxostoma duquesni,* N. Y.; Fischthal (1947): *C. commersoni,* Wis.

Family DACTYLOGYRIDAE Bychowsky, 1933

Anterior end with head organs. Cephalic glands lateral or throughout pre-oral regions. Haptor with or without accessory structures, but with 1 or 2 pairs of anchors and usually 14 marginal hooklets; *Dactylogyrus* with 14 plus one very small pair (Mizelle and Price, 1963). *Pellucidhaptor* with 16 hooks of approximately same size. Eye spots present. Ovary globular, pre-testicular.

Ref. Sproston (1946, p. 201); Mizelle *et al.* (1956).

KEY
TO THE SUBFAMILIES OF DACTYLOGYRIDAE

1. Haptor with 1 pair of anchors . . Subfamily DACTYLOGYRINAE
1. Haptor with 2 pairs of anchors   2

---

* The original generic description states "no eyes," but the species description of *A. ureteroecetes* states "four eye spots present." I have seen the four eye spots in *A. ureteroecetes* from Shenandoah R. bass.

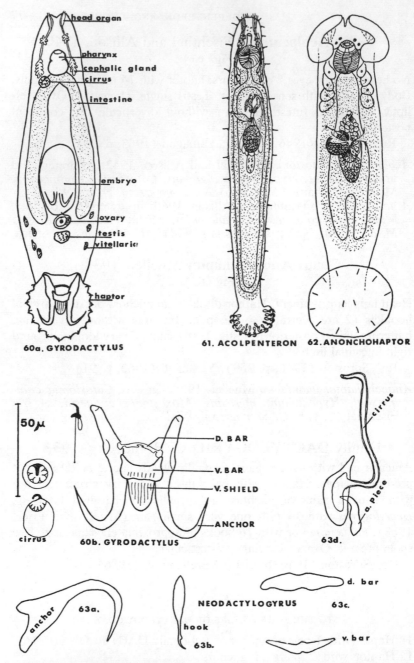

Figs. 60–63. Fig. 60. *Gyrodactylus. a, G. eucaliae* (from Ikezaki and Hoffman, 1957); *b, G. macrochiri*, haptor (from Hoffman and Putz, 1964). Fig. 61. *Acolpenteron ureteroecetes* (from Fischthal and Allison, 1941).

2. Haptor with pair of squamodiscs.......................
   (accessory plaques)......... Subfamily DIPLECTANINAE
   One genus, on brackish-water fish...... (fig. 66) *Diplectanum*
2. Haptor without squamodiscs; 12, 14, or 16 marginal hooklets...................... Subfamily TETRAONCHINAE

Subfamily DACTYLOGYRINAE Bychowsky, 1933

Dactylogyridae: Haptor without accessory sclerotized structures and with one pair of anchors supported by a cuticular bar, vestigial ventral bar may be present; 16 pairs of marginal hooklets, one pair usually very small. Intestine bifurcate, crura confluent posteriorly. Eyes present. Testes and ovary rounded; ovary pre-testicular. Vagina present, with or without supporting structures.
   Ref. Bychowsky (1957), and Putz and Hoffman (1964), life cycles.

KEY

TO THE GENERA OF DACTYLOGYRINAE

1. Anchors supported by 1 cuticular bar.................   2
1. Anchors supported by 1 bar, accessory ventral bar present but
   may not be visible in preserved specimens. Mizelle (1955)
   considers this a synonym of *Dactylogyrus*..............
   .... (fig. 63) *Neodactylogyrus*
2. Haptor with 16 hooklets, one pair much smaller than rest; ceca
   confluent posteriorly .... (fig. 64) *Dactylogyrus*
2. Haptor with 16 hooklets of same size; ceca not confluent posteriorly........................ (fig. 65) *Pellucidhaptor*
                                                   (p. 83)

Genus **Dactylogyrus** Diesing, 1850
(Fig. 64)

Haptor without accessory cuticular structures and with one pair of anchors supported by one cuticular bar; 16 marginal hooklets, one pair of which is very small and can be seen with phase microscopy (Mizelle and Price, 1963). Intestine bifurcate, crura confluent pos-

---

Fig. 62. *Anonchohaptor anomalum* (from Mueller, 1938). Fig. 63. *Neodactylogyrus cornutus* (from Mueller, 1938). *a*, anchor; *b*, hook; *c*, haptoral bars; *d*, cirrus and accessory piece.

teriorly. Eyes present. Testes and ovary rounded; ovary pre-testicular. Vagina present, with or without supporting structures.

Mizelle (1955) considers *Neodactylogyrus* a synonym because the smaller haptoral bar is not always visible, but we have retained the genus tentatively to aid in identification.

Ref. Sproston (1946, p. 202); Mizelle (1938, 1955); Mizelle and Donahue (1944).

Dactylogyrus acus (see *Neodactylogyrus acus*).

D. amblops (see *Neodactylogyrus amblops*).

D. anchoratus (Dujardin, 1845, Wagener, 1857): Europe; Mueller (1936): *Carassius auratus*, N. Y.; Price and Mizelle (1964): same, Calif.

D. apos (see *Neodactylogyrus apos*).

D. atromaculatus (Mizelle, 1938): gills, *Semotilus atromaculatus*, Ill.

D. attenuatus (see *Neodactylogyrus attenuatus*).

D. aureus Seamster, 1948: gills, *Notemigonus crysoleucas*, Okla.; Hargis (1953): on *N. crysoleucas, Chaenobryttus coronarius, Lepomis macrochirus*, Fla.

D. banghami Mizelle and Donahue, 1944: gills, *Notropis cornutus*, Ont.; Monaco and Mizelle (1955): on *Richardsonius balteatus, Couesius plumbeus, Rhinichthys cataractae, Hybognathus placita, Notropis cornutus, N. volucellus, N. lutrensis, N. deliciosus, N. girardi, N. percobromus*, Brit. Col., Ont., Okla.

D. bifurcatus Mizelle, 1937: gills, *Hyborhynchus notatus*, Ill.

D. bulbosus Mizelle and Donahue, 1944: gills, *Notropis cornutus*, Ont.

D. bulbus Mueller, 1938: gills, *Notropis cornutus*, Ont.; Mizelle and Donahue (1944): same, Wis.; Mizelle and Klucka (1953): Wis.

D. bychowskyi Mizelle, 1937: gills, *Hyborhynchus notatus*, Ill.

D. californiensis Mizelle, 1962: gills, *Ptychocheilus grandis*, Calif.

D. claviformis Mizelle and Klucka, 1953: gills, *Semotilus atromaculatus*, Wis.

D. columbiensis Monaco and Mizelle, 1955: gills, *Ptychocheilus oregonensis*, Brit. Col.

D. confusus (see *Neodactylogyrus confusus*).

D. cornutus (see *Neodactylogyrus cornutus*).

D. corporalis Putz and Hoffman, 1964: on *Semotilus corporalis*, W. Va.

D. distinctus (see *Neodactylogyrus distinctus*).

D. dubius Mizelle and Klucka, 1953: gills, *Notropis cornutus frontalis*, Wis.

D. duquesni (see *Neodactylogyrus duquesni*).

D. egregius Price and Mizelle, 1964: gills, *Richardsonius egregius*, Calif.

D. eucalius Mizelle and Regensberger, 1945: gills, *Eucalia inconstans*, Wis.

D. extensus Mueller and Van Cleave, 1932: probably synonym of *D. solidus* of Europe (see Bauer, 1956); gills, *Cyprinus carpio*, N. Y.; Mizelle and Klucka (1953): Wis.; Mizelle and Webb (1953): Wis.; Bangham (1955): Mich.; Haderlie (1953): Calif.; Krueger

(1954): Ohio; Roberts (1957): Okla.; Fantham and Porter (1947): *C. carpio, Micropterus dolomieui,* Que.

D. fulcrum (see *Neodactylogyrus fulcrum*).

D. hybognathus Monaco and Mizelle, 1955: gills, *Hybognathus placita,* Okla.

D. leptobarbus Mizelle and Price, 1964: on *Leptobarbus hoevenii,* Calif. (orig. source Thailand).

D. lineatus Mizelle and Klucka, 1953: gills, *Semotilus atromaculatus,* Wis.

D. maculatus Mizelle and Price, 1964: gills, *Rhinichthys osculus,* Calif.

D. microlepidotus Price and Mizelle, 1964: on *Orthodon microlepidotus, Lavinia exilicauda,* Calif.

D. microphallus Mueller, 1938: gills, *Semotilus atromaculatus,* N. Y.; Mizelle and Klucka (1953): Wis.

D. moorei Monaco and Mizelle, 1955: gills, *Notropis deliciosus,* Okla.

D. mylocheilus Monaco and Mizelle, 1955: gills, *Mylocheilus caurinus, Couesius plumbeus,* Can.

D. nuchalis Wood and Mizelle, 1957: gills, *Hybognathus nuchalis,* Tex.

D. occidentalis Mizelle and Price, 1964: on *Hesperoleucas symmetricus,* Calif.

D. orchis (see *Neodactylogyrus orchis*).

D. orthodon Price and Mizelle, 1964: on *Orthodon microlepidotus,* Calif.

D. osculus Wood and Mizelle, 1957: gills, *Rhinichthys osculus,* Nev.

D. parvicirrus Seamster, 1948: gills, *Notemigonus crysoleucus,* Okla.; Hargis (1953): same, Fla.

D. percobromus Monaco and Mizelle, 1955: gills, *Notropis percobromus,* Okla.

D. perlus (see *Neodactylogyrus perlus*).

D. photogenis (see *Neodactylogyrus photogenis*).

D. pollex, Mizelle and Donahue 1944: gills, *Notropis cornutus,* Ont.

D. ptychocheilus Monaco and Mizelle, 1955: on *Ptychocheilus oregonensis,* Can.

D. pyriformis (see *Neodactylogyrus pyriformis*).

D. rhinichthius Wood and Mizelle, 1957: on *Rhinichthys atratulus,* Ind.

D. richardsonius Monaco and Mizelle, 1955: on *Richardsonius balteatus,* Can.; Price and Mizelle (1964): same, Calif.

D. rubellus (see *Neodactylogyrus rubellus*).

D. scutatus (see *Neodactylogyrus scutatus*).

D. semotilus Wood and Mizelle, 1957: on *Semotilus atromaculatus,* Mich.

D. simplex (see *Neodactylogyrus simplex*).

D. tenax Mueller, 1938: on *Semotilus atromaculatus,* N. Y.

D. texomonensis Seamster, 1960: on *Hybopsis storeriana,* Okla.

D. tridactylus Monaco and Mizelle, 1955: on *Ptychocheilus oregonensis,* Can.

D. ursus (see *Neodactylogyrus ursus*).

D. vancleavei Monaco and Mizelle, 1955: on *Ptychocheilus oregonensis, Acrocheilus alutaceus,* Can.

D. vannus (see *Neodactylogyrus vannus*).
D. vastator Nybelin, 1924: Europe; Price and Mizelle (1964): on *Carassius auratus*, Calif.
D. wegeneri Kulwiec, 1927: Europe; Price and Mizelle (1964): on *C. auratus*, Calif.

Genus (**Neodactylogyrus**) Price, 1938
(Fig. 63)

As *Dactylogyrus* but with ventral cuticular bar present. Considered synonymous with *Dactylogyrus* by Mizelle and Donahue (1944) because the ventral bar is not always visible. It is included as a genus here to facilitate identifying species.

Ref. Sproston (1946, p. 208); Yamaguti (1963, p. 34).

Neodactylogyrus acus (Syn. *Dactylogyrus a.*), (Mueller, 1938) Price, 1938: on gills, *Notropis cornutus, Campostoma anomalum*, N. Y.
N. amblops (Syn. *Dactylogyrus a.*), (Mueller, 1938) Price, 1938: on gills, *Hybopsis amblops*, N. Y.
N. apos (Syn. *Dactylogyrus a.*), (Mueller, 1938) Price, 1938: on gills, *Hypentelium nigricans*, N. Y.
N. attenuatus (Syn. *Dactylogyrus a.*), (Mizelle and Klucka, 1953) Yamaguti, 1963: on gills, *Semotilus atromaculatus*, Wis.
N. bifurcatus (Syn. *Dactylogyrus b.*), (Mizelle, 1937) Price, 1938: on gills, *Hyborhynchus notatus*, Ill.
N. bulbus (Syn. *Dactylogyrus b.*), (Mueller, 1938) Price, 1938: on gills, *Notropis cornutus*, N. Y.; Mizelle and Donahue (1944): *N. cornutus*, Ont.; Mizelle and Klucka (1953) and Mizelle and Webb (1963): *N. cornutus*, Wis.
N. confusus (Syn. *Dactylogyrus c.*), (Mueller, 1938) Price, 1938: on gills, *Clinostomus elongatus*, N. Y.
N. cornutus (Syn. *Dactylogyrus c.*), (Mueller, 1938) Price, 1938: on gills, *Notropis cornutus*, N. Y., Ont.; Mizelle and Klucka (1953): same, Wis.; Mizelle and Webb (1953): same, Wis.
N. distinctus (Syn. *Dactylogyrus d.*), (Mizelle and Klucka, 1953) Yamaguti, 1963: on gills, *Notropis volucellus*, Wis.
N. duquesni (Syn. *Dactylogyrus d.*), (Mueller, 1938) Price, 1938: on gills, *Moxostoma duquesni*, N. Y.
N. fulcrum (Syn. *Dactylogyrus f.*), (Mueller, 1938): on gills, *Notropis cornutus*, N. Y.
N. orchis (Syn. *Dactylogyrus o.*), (Mueller, 1938), Price, 1938: on gills, *Notropis rubellus*, N. Y.
N. perlus (Syn. *Dactylogyrus p.*), (Mueller, 1938) Price, 1938: on gills, *Notropis cornutus*, N. Y., Ont.; Mizelle and Donahue (1944): vestigial ventral bar not observed (= *Dactylogyrus p.* ?).
N. photogenis (Syn. *Dactylogyrus p.*), (Mueller, 1938) Price, 1938: on gills, *Notropis photogenis*, N. Y.
N. pyriformis (Mizelle and Klucka, 1953) Yamaguti, 1963: on gills, *Notropis cornutus*, Wis.

N. rubellus (Syn. *Dactylogyrus r.*), (Mueller, 1938) Price, 1938: on gills, *Notropis rubellus,* N. Y.

N. scutatus (Syn. *Dactylogyrus s.*), (Mueller, 1938) Price, 1938: on gills, *Exoglossum maxilingua, perexoglossum laurae,* N. Y.

N. simplex (Syn. *Dactylogyrus s.*), (Mizelle, 1937) Price, 1938: on gills, *Hyborhynchus notatus,* Ill.

N. ursus (Syn. *Dactylogyrus u.*), (Mueller, 1938) Price, 1938: on gills, *Moxostoma anisurum,* N. Y.

N. vannus (Syn. *Neodactylogyrus v.*), (Mizelle and Klucka, 1953) Yamaguti, 1963: on gills, *Notropis cornutus,* Wis.

Genus **Pellucidhaptor** Price and Mizelle, 1964
(Fig. 65)

Dactylogyrinae: Smooth cuticle; 2 pairs of eye spots, posterior pair larger. Haptor set off from body by distinct peduncle, having one pair of dorsal anchors with doubly recurved points projecting from surface of haptor. Anchor wings (filaments) arise on anchor bases. Hooklets 16, of similar shape and size. Ovary pre-testicular, near midlength of trunk; vagina present or absent. Vitellaria extend from pharynx to or into haptor; often distributed as two lateral bands, confluent anteriorly and posteriorly. Gut bifurcate; crura nonconfluent, without lateral ceca. Parasites of freshwater fishes.

Pellucidhaptor pelliucidhaptor Price and Mizelle, 1964: on gills, *Richardsonius egregius,* Calif.

Subfamily DIPLECTANINAE
Genus **Diplectanum** Diesing, 1858
(Fig. 66)
(Syn.: *Lepidotes*)

Dactylogyridae: Body covered with anteriorly directed scalelike spines. Haptor bearing paired squamodiscs, dorsal and ventral; subsessile or sessile discs covered with concentric rows of scalelike spines or lamellae, without accessory hooks. Two pairs of anchors with 3 supporting bars; 14 marginal hooklets probably always present. Intestine saclike, or crura without diverticula. Two pairs of eyes present. Cirrus with ejaculatory bulb. Testis and ovary without lobes. Vagina present. Ref. Sproston (1946, p. 250); Yamaguti (1963, p. 95).

Diplectanum collinsi (Mueller, 1936) Price, 1937 (Syn. *Lepidotes c.*): on *Roccus lineatus,* Fla.

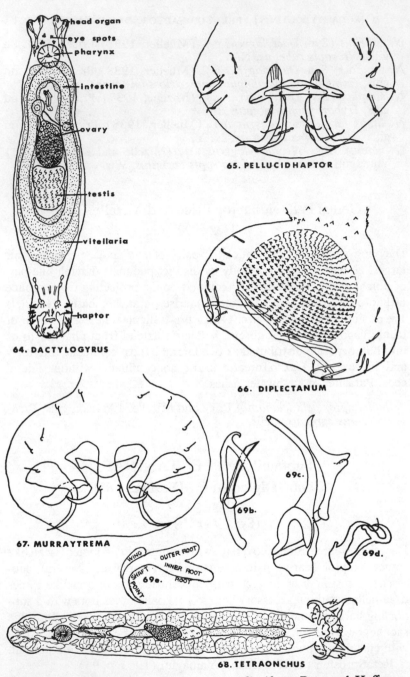

Figs. 64–69. Fig. 64. *Dactylogyrus corporalis* (from Putz and Hoffman, 1964). Fig. 65. *Pellucidhaptor pellucidhaptor*, haptor (from Mizelle and Price, 1964). Fig. 66. *Diplectanum collinsi*, haptor (from Mueller, 1936). Fig. 67. *Murraytrema copulata*, haptor (from Mueller, 1938). Fig. 68.

Subfamily TETRAONCHINAE* (Monticelli, 1903)
Price, 1937

Dactylogyridae: 2 pairs of anchors, 14 marginal hooklets. Intestine single or double. Eyes present or absent. Testes and ovary unlobed. Vagina present or absent.

There are 8 genera in this subfamily; however, the genera *Cleidodiscus* and *Urocleidus* contain species which were originally described in different genera: *Haplocleidus, Pterocleidus, Ancyrocephalus, Leptocleidus, Onchocleidus, Actinocleidus,* and *Tetracleidus.* There is some controversy over the synonymy involved; I am not qualified to attempt to resolve it here, and so have tried to include all generic names. Where a genus is considered a synonym by some or all workers, I have so indicated. The key to the genera includes all such genera as an aid to identification.

Ref. Hargis (1952); Mizelle (1938, 1955); Mizelle and Hughes (1938); Mizelle *et al.* (1956); Sproston (1946); Yamaguti (1963).

KEY

TO THE GENERA AND SYNONYMOUS GENERA† OF TETRAONCHINAE‡

(Modified from Sproston, 1946; Mizelle, 1938; Mizelle and Hughes, 1938; Hargis, 1952)

1. Haptor with 1 transverse bar; gut rhabdocoelate; 16 marginal hooklets . . . . (fig. 68) *Tetraonchus*
1. Haptor with 2 separate non-articulated transverse bars; gut bifurcate   2

---

* This group was placed in the Ancyrocephalinae by Bychowsky (1957). I have retained Price's system for simplification and do not dispute Bychowsky's revision.

† These synonyms have been included to help with the identification of species.

‡ See the key to the genera of Ancyrocephalinae of Mizelle and Price (1964a), which includes all genera of the world. The genus *Trianchoratus* from the kissing gourami (*Helostoma rudolphi*), possessing only 3 anchors, was described after this manuscript was prepared (Price & Berry, 1966).

Tetraonchus monenteron (from Van Cleave and Mueller, 1934). Fig. 69. *a, Cleidodiscus aculeatus* (from Van Cleave and Mueller, 1932—(find on next plate); *b, C. stentor,* cirrus and accessory piece (from Mueller, 1937); *c, C. robustus,* cirrus and accessory piece (from Mizelle and Regensberger, 1945); *d, C. chelatus,* cirrus and accessory piece (from Mizelle and Jaskoski, 1942); *e, Anchor nomenclature* (from Mueller, 1936).

1. Haptor with 2 articulated bars; gut bifurcate   4
1. Haptor with 3 separate transverse bars; gut bifurcate . . . . . .
. . . . (fig. 67) *Pseudomurraytrema*
2. Cirrus and accessory piece basally articulate (possibly not in
    C. alatus); vagina, if present, on left margin [Syn. in part
    according to Mizelle *et al.* (1956): *Leptocleidus, Tetraclei-
    dus, Ancyrocephalus*] . . . . (fig. 69) *Cleidodiscus*
    a. Cirrus relatively simple. . . . .   b
    a. Cirrus in a large coil . . . . *Cleidodiscus megalonchus*
    b. Vagina on right margin of body, old genus in part . . . .
        . . . . (*Tetracleidus*)*
                                                              (p. 98)
    b. Vagina on left margin of body . . . . *Cleidodiscus*
2. Accessory piece never basally articulated with cirrus   3
3. Vagina on left . . . . (fig. 70) *Urocleidoides*
3. Vagina, if present, on right margin [Syn. in part according to
    Sproston (1946, p. 242), and Mizelle *et al.* (1956): *Aristo-
    cleidus, Haplocleidus, Pterocleidus, Tetracleidus, Oncho-
    cleidus*]. . . . . (fig. 71) *Urocleidus*
    a. Anchors dissimilar in shape . . . . . in part (*Aristocleidus*)*
                                                              (p. 91)
    a. Anchors similar in shape . . . .   b
    b. Anchors markedly dissimilar in size . . . .
        . . . . (fig. 72) in part (*Haplocleidus*)*
                                                              (p. 93)
    b. Anchors similar in size   c
    c. Vagina wanting . . . . *Urocleidus*
    c. Vagina present . . . .   d
    d. Anchor shafts with spurs . . (fig. 73) in part (*Pterocleidus*)*
                                                              (p. 95)
    d. Anchor shafts without spurs; cirrus accessory piece
        usually absent or immovable . . (fig. 74) (*Onchocleidus*)*
                                                              (p. 98)
4. Haptor distinct, umbrella-like with scalloped margin, and with
    2 pairs of anchors on small protuberance in center of ventral
    surface; otherwise very similar to *Actinocleidus* [Syn. in
    part according to Mizelle *et al.* (1956): *Actinocleidus,
    Ancyrocephalus*] . . . . (fig. 75) *Clavunculus*
                                                              (p. 95)
4. Haptor distinct, not on central protuberance   5

* These synonyms have been included to help with the identification of species.

5. Haptoral bars similar; anchor shafts vestigial; anchor bases
   occupy most of frontal plane through haptor . . . .
   . . . . (fig. 76) *Anchoradiscus*
   (p. 96)
5. Haptoral bars dissimilar; anchor shafts not vestigial; anchor
   bases of moderate size . . . . (fig. 77) *Actinocleidus*
   (p. 96)

Genus **Pseudomurraytrema** (Price, 1937) Bychowsky, 1957
(Fig. 67)

Cephalic glands open to exterior through 4 pairs of head organs.
Haptor contains 2 pairs of large anchors (hooks) separated by 3
transversely placed nonarticulate bars; ventral pair of anchors connects with ventral bar; each dorsal anchor connects with correspond-
ing lateral, dorsal bar; 14 marginal hooklets. Intestinal branches not
uniting posteriorly. Eyes present. Testis and ovary in equatorial zone.
Cirrus with accessory piece. Vagina present, opening ventrally and
medially.

Ref. Sproston (1946, p. 239); Mizelle *et al.* (1956, p. 169); Bychowsky (1957, p. 287).

Pseudomurraytrema copulata (Mueller, 1938): on *Catostomus commersoni, Hypentelium nigricans, Moxostoma anisurum, M. erythrurum*, N. Y.; Mizelle and Klucka (1953): on *C. commersoni, M.
   anisurum*, Wis.; Mizelle and Webb (1953): on *C. fecundus*, Wyo.

Genus **Tetraonchus** Diesing, 1858
(Fig. 68)

Tetraonchinae: Cephalic glands open in 3 or more pairs of head organs.
Haptor more or less distinctly set off from body, bearing a butterfly-
shaped transverse bar articulating directly with both pairs of anchors,
and 16 marginal hooklets. Intestine unbranched, saclike, without diverticula. Eyes present. Testes and ovary in midbody region.

Ref. Sproston (1946, p. 216); Mizelle (1938, p. 344), key to spp.;
Mizelle *et al.* (1956).

Tetraonchus alaskensis Price, 1937: on *Salmo mykiss, Onchorhynchus
   kisutch, Salvelinus malma*, Alaska.
T. momenteron (Wagener, 1857) Diesing, 1858: Europe; Van Cleave
   and Mueller (1934): on *Esox lucius*, N. Y.; Mizelle and Regensberger (1945): same, Wis.
T. rauschi Mizelle and Webb, 1953: on *Thymallus signifer*, Alaska.
T. variabilis Mizelle and Webb, 1953: on *Prosopium williamsi*, Wyo.,
   and *P. cylindraceum*, Alaska. This species was placed in *Salmonchus*
   by Spassky and Roytman (1958).

Genus **Cleidodiscus** Mueller, 1934
(Fig. 69)

(Syn.: *Leptocleidus* Mueller, 1936, in part; *Tetracleidus* Mueller, 1936, in part; Mizelle (1938): redescribed.)

Tetraonchinae: Somewhat flattened dorsoventrally, trunk narrowly elliptical in outline. Eyes 4, posterior pair larger. Gut bifurcate, without diverticula; rami confluent posteriorly. Gonads near middle of body. Cirrus usually a simple cuticularized tube. Accessory piece always present, generally articulated basally with cirrus. Vagina usually present, opening on left margin near middle of trunk. Vitellaria of numerous small, discrete follicles arranged in pair of lateral bands extending from pharyngeal region to, or into, peduncle; bands always confluent posteriorly and sometimes anteriorly. Haptor generally distinct, discoidal or subhexagonal; armed with 2 pairs of bars and 7 pairs of hooklets. Anchors with superficial roots of each pair connected by transverse bar; bars nonarticulate with each other.

Ref. Mizelle *et al.* (1961), life cycle.

Cleidodiscus aculeatus (Van Cleave and Mueller, 1932) Mizelle and Regensberger, 1945 (Syn. *Ancyrocephalus a.* and *Urocleidus a.*): on *Stizostedion vitreum*, N. Y., Wis.

C. alatus Mueller, 1938: on *Ambloplites rupestris*, N. Y.; Mizelle and Regensberger (1945): same, Wis.

C. articularis (Syn. for *Actinocleidus a.*).

C. banghami (Mueller, 1936) Mizelle, 1940 (Syn. *Tetracleidus b.*, *Urocleidus b.*): on *Micropterus dolemieui*, *M. punctulatus*, Tenn., Ont.

C. bedardi Mizelle, 1936: on *Lepomis megalotis*, Ill.; Mizelle (1940): same, Tenn.

C. brachus Mueller, 1938: on *Semotilus atromaculatus*, *S. margarita*, N. Y.; Mizelle and Klucka (1953): same, Wis.

C. capax Mizelle, 1936: on *Pomoxis annularis*, *P. nigromaculatus*, Ill.; Mueller (1937): *P. nigromaculatus*, N. Y.; Mizelle, La Grave, and O'Shaughnessy (1943): *P. nigromaculatus*, Tenn., Ill.; Mizelle and Regensberger (1945): same, Wis.; Seamster (1948): same, La.; Hargis (1952): same, Va.

C. chelatus Mizelle and Jaskoski, 1942: on *Lepomis miniatus*, Tenn.

C. diversus Mizelle, 1938: on *Lepomis cyanellus*, Ill., Okla.; Seamster (1938): Okla.; Krueger (1954): Ohio R.; McDaniel (1963): *L. cyanellus*, *L. macrochirus*, *L. megalotis*, Okla.

C. floridanus Mueller, 1936 (Syn. *C. mirabilis* Mueller, 1937, in part): on *Ictalurus punctatus*, Fla.; Mueller (1937): *Pylodictus olivaris*, Miss. R.; Mizelle and Cronin (1943): *I. furcatus*, *I. melas*, Tenn.; Mizelle and Klucka (1953): *I. punctatus*, Wis.; Mizelle and Webb (1953): *I. punctatus*, Wis.; Harms (1959): *I. punctatus*, *I. melas*, *I. natalis*, Kans.

C. fusiformis (Syn. for *Actinocleidus f.*).

C. incisor (Syn. for *C. robustus*).

C. longus Mizelle, 1936: on *Pomoxis annularis,* Ill.; Krueger (1954): *Ictalurus melas, I. nebulosus,* Ohio; Mizelle (1938): Okla.; Mizelle, LaGrave, and O'Shaughnessy (1943): Tenn., Ill.

C. malleus (Syn. for *Urocleidus m.*).

C. megalonchus (Mueller, 1936) Mizelle and Hughes, 1938 (Syn. *Tetraonchus unguiculatus, Ancyrocephalus paradoxus, Leptocleidus megalonchus*): on *Micropterus dolomieui,* Ohio.; Cooper (1915): same; Stafford (1905): on *Ambloplites rupestris, L. gibbosus,* Can.

C. mirabilis (Syn. of *C. floridanus*).

C. nematocirrus Mueller, 1937: on *Lepomis gibbosus,* Fla.; Summers and Bennett (1938): on *L. macrochirus,* La.; Mizelle (1941a): on *L. microlophus,* Fla.; Seamster (1948): on *L. megalotis,* La.

C. oculatus (Syn of *Actinocleidis o.*).

C. pricei Mueller, 1936: on *Ictalurus punctatus, I. natalis, I. nebulosus,* Fla.; Seamster (1938): *I. punctatus, I. melas,* Okla., La.; Mizelle and Cronin (1943): *I. punctatus, I. melas, I. natalis, I. furcatus,* Tenn.; Mizelle and Donahue (1944): *I. melas, I. nebulosus,* Ont.; Seamster (1948): *I. melas,* La.; Mueller (1937): *I. nebulosus,* Fla.; Mizelle and Regensberger (1945): *I. nebulosus,* Wis.; Hargis (1952b): *I. nebulosus,* Va.; Mizelle *et al.* (1961): life cycle, *I. catus,* Calif.

C. rarus Mizelle, 1940: on *Micropterus punctalatus,* Tenn.

C. robustus Mueller, 1934 (Syn. *Cleidodiscus incisor, Actinocleidus i.*): on *Lepomis gibbosus, L. cyanellus, L. macrochirus*; Krueger (1954): *L. cyanellus,* Ohio.

C. stentor Mueller, 1937: on *Ambloplites rupestris,* N. Y.; Mizelle and Regensberger (1945): *Pomoxis nigro maculatus,* Wis.; Hargis (1952): same, Va.

C. uniformis Mizelle, 1936: on *Pomoxis annularis*; Ill.; Krueger (1954): same; Ohio; Mizelle (1938): same, Ill., Okla.; Mizelle *et al.* (1943): same, Tenn., Ill.

C. vancleavei Mizelle, 1936 (Syn. *Onchocleidus formosus, Cleidodiscus f.*): on *Pomoxis annularis,* Ill.; Mueller (1936): *P. nigromaculatus,* Fla.; Mizelle (1938): *P. annularis, P. nigromaculatus,* Ill., Okla.; Seamster (1938): same, Okla.; Summers and Bennett (1938): *P. nigromaculatus,* La.; Mizelle *et al.* (1943): *P. annularis, P. nigromaculatus,* Tenn., La.; Hargis (1952b): *P. nigromaculatus,* Va.; Mizelle and Webb (1953): *P. annularis,* Wis.; Krueger (1954): *P. annularis, P. nigromaculatus,* Ohio.

C. venardi Mizelle and Jaskoski, 1942: on *Lepomis miniatus,* Tenn.

Genus **Urocleidoides** Mizelle and Price, 1964
(Fig. 70)

Tetraonchinae: Cuticle smooth; 2 pairs of eye spots, posterior pair larger. Intestinal crura confluent. Haptor usually poorly set off from trunk, with 2 pairs of similar anchors, each pair connected by a hap-

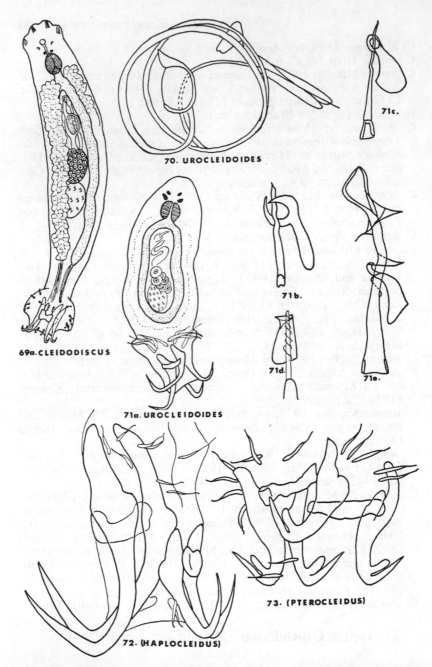

70. UROCLEIDOIDES

71c.

69a. CLEIDODISCUS

71b.

71d.

71e.

71a. UROCLEIDOIDES

73. (PTEROCLEIDUS)

72. (HAPLOCLEIDUS)

Figs. 70–73. Fig. 70. *Urocleidoides reticulatus*, cirrus (from Mizelle and Price, 1964). Fig. 71. *a, Urocleidus ferox* (from Mueller, 1934); *b, U. parvicirrus*, cirrus and accessory piece (from Mizelle and Jaskoski, 1942); *c, U. mucronatus*, cirrus and accessory piece (from Mizelle and Brennan,

toral bar; bars nonarticulated. Hooklets 14, similar in shape and size. Testes ovate, post-ovarian. Accessory piece basally nonarticulate with cirrus. Vagina on left, near midlength of trunk. Vitellaria distributed from level of eye spots or pharynx to or into haptor, infrequently forming lateral bands confluent anteriorly and posteriorly.

Urocleidoides reticulatus Mizelle and Price, 1964: on gills, *Lebistes reticulatus*, Calif.; original source probably Trinidad.

Genus **Urocleidus** (Mueller, 1934) Mizelle and Hughes, 1938
(Fig. 71)
(Syn.: *Aristocleidus, Haplocleidus, Onchocleidus, Pterocleidus, Tetracleidus* in part)

Tetraonchinae: Trunk, eyes, gut, gonads, vitellaria, and haptoral bars as described for *Cleidodiscus*. Cirrus a cuticularized tube, straight or undulate with or without a cirral thread or fin, infrequently corkscrewlike. Accessory piece never basally articulated with cirrus. Vagina, when present, opening on right margin near midlength of trunk. Haptor generally distinct, subhexagonal; armed with 2 pairs of anchors and 7 pairs of hooks.

Mizelle *et al.* (1956) have placed the genera *Pterocleidus* and *Haplocleidus* in the genus *Urocleidus*. I concur, but, since the key characteristics of the old genera (see p. 85) are useful in identifying species, I have here grouped the species in *Urocleidus*, old "*Pterocleidus*," and old "*Haplocleidus*."

Ref. Sproston (1946, p. 242); Mizelle and Hughes (1938), key to spp.; Mizelle *et al.* (1956).

"Urocleidus"

Urocleidus aculeatus (Syn. for *Ancyrocephalus a.*).
U. adspectus Mueller, 1936: on *Perca flavescens*, N. Y.; Mizelle and Donahue (1944):same, Ont.; Mizelle and Regensberger (1945): same, Wis.
U. angularis Mueller, 1934 (Syn. *Ancyrocephalus angularis* Mueller, 1934, 1936): on *Fundulus diaphanus*, N. Y.

1942); *d, U. torquatus,* cirrus and accessory piece (from Mizelle and Cronin, 1943); *e, U. Chaenobryttus,* cirrus and accessory piece (from Mizelle and Seamster, 1939). Fig. 72. (*Haplocleidus dispar*) = *Urocleidus d.,* anchors and hooks (from Mueller, 1936). Fig. 73. (*Pterocleidus acer*) = *Urocleidus a.,* anchors and hooks (from Mueller, 1936).

U. attenuatus Mizelle, 1941: on *Lepomis microlophus*, Fla.; Mizelle and Brennan (1942): *L. microlophus, L. macrochirus*, Fla.; Mizelle and Cronin (1943): *L. microlophus*, Tenn.; Mizelle and Jaskoski (1942): *L. miniatus*, Tenn.

U. banghami (Syn. for *Cleidodiscus b.*).

U. biramosus (Mueller, 1937), (Syn. *Pterocleidus b.*): on *Lepomis macrochirus*, Fla.

U. chaenobryttus Mizelle and Seamster, 1939: on *Chaenobryttus coronarius*, Fla.; Mizelle and Jaskoski (1942): *Lepomis miniatus*, Tenn.; Mizelle and Cronin (1943): *C. coronarius*, Tenn.; Hargis (1952c): *C. coronarius, L. macrochirus*, Va.; McDaniel (1963): *L. cyanellus*, Okla.

U. chautauquensis (Mueller, 1938) Mizelle and Hughes, 1938 (Syn. *Tetracleidus c., Cleidodiscus c.*): on *Ambloplites rupestris*, N. Y.; Mizelle and Regensberger (1945): same, Wis.

U. chrysops Mizelle and Klucka, 1953: on *Roccus chrysops*, Wis.; Mizelle and Webb (1953): same.

U. cyanellus (Mizelle, 1938) Mizelle and Hughes, 1938 (Syn. *Onchocleidus c.*): on *Lepomis cyanellus*, Ill.; Seamster (1948): same, Okla.; Krueger (1954): same, Ohio.

U. distinctus (Mizelle, 1936) Mizelle and Hughes, 1938 (Syn. *Onchocleidus d.*): on *Lepomis megalotis*, Ill.

U. doloresae Hargis, 1952: on *Chaenobryttus coronarius*, Va.

U. ferox Mueller, 1934 (Syn. *Onchocleidus f., O. mucronatus, Urocleidus m.*): on *Lepomis gibbosus, L. macrochirus* (and hybrids), *L. humilis*, N. Y., Ill.; Mizelle (1938): on *L. macrochirus*, Okla., Ill.; Seamster (1938): *L. humilis, L. macrochirus*, Okla.; Summers and Bennett (1938): *L. macrochirus*, La.; Mizelle and Brennan (1942): *L. macrochirus*, Fla.; Mizelle and Donahue (1944): *L. gibbosus*, Ont.; Mizelle and Regensberger (1945): *L. macrochirus*, Wis.; Hargis (1952c): *L. gibbosus, L. macrochirus, Chaenobryttus coronarius*, Va.; McDaniels (1963): *L. macrochirus, L. megalotis*, Okla.

U. hastatus (Mueller, 1936) Mizelle and Hughes, 1938 (Syn. *Aristocleidus h.*): on *Roccus lineatus*, Fla.; Price, 1937: 14, not 12 hooklets.

U. interruptus (Mizelle, 1936) Mizelle and Hughes, 1938 (Syn. *Onchocleidus i.*): on *Roccus interrupta*, Ill.

U. macropterus Harrises, 1962: on *Centrarchus macropterus*, Miss.

U. malleus (Mueller, 1938) Mizelle and Hughes, 1938 (Syn. *Cleidodiscus m.*): on *Percina caprodes, Hadropterus maculatus*, N. Y.

U. mimus (Mueller, 1936) Mizelle and Hughes, 1938 (Syn. *Onchocleidus m.*): on *Esox reticulatus, Roccus chrysops*, Ohio; Mueller (1937): *R. chrysops*, N. Y.; Mizelle and Klucka (1953): *E. reticulatus, R. chrysops*, Wis.; Mizelle and Webb (1953): Wis.

U. moorei Mizelle, 1940: on *Etheostoma flabellaris*, Tenn.

U. mucronatus (Mizelle, 1936) Mizelle and Hughes, 1938 (Syn. *Onchocleidus m.*). Mizelle *et al.* [1956] consider this a synonym of *U. ferox*. Krueger (1954): on *Lepomis cyanellus, L. macrochirus*, Ohio.

U. nigrofasciatus Harrises, 1962: on *Percina nigrofasciata*, Miss.
U. perdix (Mueller, 1937) Mizelle and Hughes, 1938 (Syn. *Oncho-cleidus p.*): on *Lepomis macrochirus*; Krueger (1954): same, Ohio.
U. principalis (Mizelle, 1936) Mizelle and Hughes, 1938 (Syn. *Oncho-cleidus p., O. contortus*): on *Micropterus punctulatus*, Ill.; *Mueller*, (1937): *M. salmoides*, Fla.; Seamster (1938): *M. dolomieui*, Okla.; Summers and Bennet (1938): *M. salmoides*, La.; Mizelle (1938): same, Ill.; Mizelle (1940): *M. dolomieui, M. punctulatus, M. salmoides*, Tenn.; Mizelle and Cronin (1943): *M. salmoides*, Tenn.; Mizelle and Regensberger (1945): same, Wis.; Hargis (1952): same, Va.
U. procax Mizelle and Donahue, 1944: on *Lepomis gibbosus*, Ont.; Hargis (1952): same, Va.
U. seculus Mizelle and Arcadi, 1945: on *Gambusia affinis*, Calif.; Seamster (1948): same, La.
U. similis (Mueller, 1936) Mizelle and Hughes, 1938 (Syn. *Oncho-cleidus s.*): on *Lepomis gibbosus*, N. Y.; Vojtek (1958): same, Czech. (imported from U. S.).
U. spiralis (Mueller, 1937) Mizelle and Hughes, 1938 (Syn. *Oncho-cleidus s.*): on *Lepomis gibbosus*, Fla.
U. torquatus Mizelle and Cronin, 1943: on *Lepomis microlophus*, Tenn.
U. umbraensis Mizelle, 1938: on *Fundulus notatus*, Ill.; Summers and Bennett (1938): *F. dispar*, La.
U. variabilis Mizelle and Cronin, 1943: on *Lepomis microlophus*, Tenn.

"Haplocleidus"
(= *Urocleidus*)
(Fig. 72)

Urocleidus affinis (Mueller, 1937) Mizelle and Hughes, 1938 (Syn. *Haplocleidus a.*): on *Lepomis gibbosus*, Fla.
U. dispar (Mueller, 1936) Mizelle and Hughes, 1938 (Syn. *Oncho-cleidus o., Haplocleidus o.*). Mueller (1936): on *Lepomis gibbosus*, N. Y.; Seamster (1938): *L. humilis, L. macrochirus*, Okla.; Mizelle (1938): *L. macrochirus*, Ill.; Mizelle and Donahue (1944): *L. gibbosus*, Ont.; Mizelle and Cronin (1943): *M. salmoides*, Tenn.; Mizelle and Regensberger (1945): *L. macrochirus*, Wis.; Hargis (1952): *L. gibbosus, L. macrochirus, Chaenobryttus coronarius*, Va.
U. furcatus (Mueller, 1937) Mizelle and Hughes, 1938 (Syn. *Haplo-cleidus f.*): on *Micropterus salmoides*, Fla.; Summers and Bennett (1938): same, La.; Mizelle (1940): *M. salmoides, M. punctulatus*, Tenn.; Mizelle and Regensberger (1945): *M. salmoides*, Wis.; Hargis (1952): same, Va.; Krueger (1954): *Lepomis cyanellus, L. macrochirus*, Ohio.
U. parvicirrus Mizelle and Jaskoski, 1942: on *Lepomis miniatus*, Tenn.; Mizelle and Cronin (1943): *L. microlophus*, Tenn.

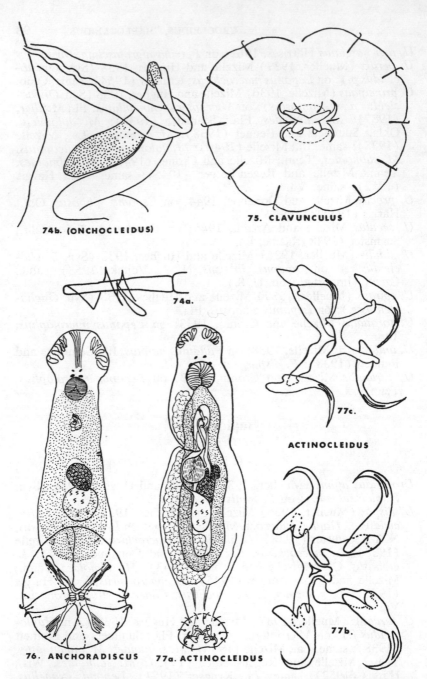

74b. (ONCHOCLEIDUS)

75. CLAVUNCULUS

74a.

77c.

ACTINOCLEIDUS

76. ANCHORADISCUS

77a. ACTINOCLEIDUS

77b.

Figs. 74–77. Fig. 74. *a, (Onchocleidus cyanellus) = Urocleidus c.,* cirrus and accessory piece (from Mizelle, 1938); *b, (O. ferox) = U. ferox,* cirrus and cirrus gland (from Mueller, 1936). Fig. 75. *Clavunculus bursatus,* haptor (from Mueller, 1936). Fig. 76. *Anchoradiscus anchoradiscus* (from

"Pterocleidus"
(= *Urocleidus*)
(Figs. 73, 74)

Urocleidus acer (Mueller, 1936) Mizelle and Hughes, 1938 (Syn. *Onchocleidus a.*, *Pterocleidus a.*): on *Lepomis gibbosus*, N. Y.; Mizelle (1936): *L. gibbosus*, *L. humilis*, Ill.; Mizelle (1938): *L. macrochirus*, Ill.; Seamster (1938): same, Okla.; Mizelle and Brennan (1942): same, Tenn.; Mizelle and Regensberger (1945): same, Wis.; Seamster (1948): same, La.

U. acuminatus (Mizelle, 1936) Mizelle and Hughes, 1938 (Syn. *Onchocleidus a.*, *Pterocleidus a.*): on *Lepomis megalotis*, Ill.; Seamster (1948): same, La.

U. biramosus (Mueller, 1937) Mizelle and Hughes, 1937 (Syn. *Pterocleidus b.*): on *Lepomis macrochirus*, Fla.; Mizelle and Brennan (1942): same, Fla.; Krueger (1954): *Pomoxis nigromaculatus*, Ohio.

U. grandis Mizelle and Seamster, 1939: on *Chaenobryttus coronarius*, Fla.; Mizelle and Cronin (1943): same, Tenn.

U. miniatus Mizelle and Jaskoski, 1942: on *Lepomis miniatus*, Tenn.

U. wadei Seamster, 1948: on *Centrarchus macropterus*, La.

Genus **Clavunculus** Mizelle, Stokely, Jaskoski, Seamster, and Monaco, 1956
(Fig. 75)

Tetraonchinae: Trunk, eyes, gut, gonads, cirrus, accessory piece, vagina (when present), and vitellaria as in *Cleidodiscus*. Haptor distinct, umbrella-like with scalloped margin, and with 2 pairs of anchors on small protuberance in center of ventral surface. Bases of each pair of anchors connected by a bar; bars articulate with each other; 7 pairs of haptoral hooks; each marginal indentation of haptor usually with 1 hook.

Ref. Mizelle *et al.* (1956).

Clavunculus bifurcatus (Mizelle, 1941) Mizelle *et al.*, 1956 (Syn. *Actinocleidus b.*): on *Lepomis microlophus*, Fla.

C. bursatus (Mueller, 1936) Mizelle *et al.*, 1956 (Syn. *Ancyrocephalus b.*, *Actinocleidus b.*): on *Micropterus dolomieui*, *M. salmoides*, *M. punctulatus*, *Lepomis macrochirus*, Ohio, Tenn., Fla.; Meyer (1954): *M. dolomieui*, Me.

C. unguis (Mizelle and Cronin, 1943) Mizelle *et al.*, 1956 (Syn. *Actinocleidus u.*): on *Micropterus salmoides*, Tenn., Wis.; Hargis (1952): same, Va.

Mizelle, 1941). Fig. 77. *Actinocleidus. a, A. fusiformis* (from Van Cleave and Mueller, 1934); *b, A. fusiformis*, anchors and bars (from Mizelle, 1940); *c, A. longus*, anchors and bars (from Mizelle, 1938).

Genus **Anchoradiscus** Mizelle, 1941
(Fig. 76)

Tetraonchinae: Anchor base (roots) parallel to and developed so as to extend through greater part of frontal plane of discoidal haptor. Anchor shafts vestigial or absent; anchor points recurved and directed laterally. Haptoral bars 2, attached to ventral (or lateral) surfaces of anchor bases, and articulating with each other at their midregions. Each of 14 marginal hooklets on haptoral margins and differentiated into base, a solid anchor shaft, a sickle-shaped termination, and an opposable piece. Copulatory complex of cirrus and accessory piece. Vagina on left. Eye spots 4, posterior pair larger.

Ref. Sproston (1946, p. 223); Mizelle *et al.* (1956).

Anchoradiscus anchoradiscus Mizelle, 1941: on gills, *Lepomis macrochirus, L. microlophus,* Fla.
A. triangularis (Summers, 1937) Mizelle, 1941 (Syn. *Actinocleidus t.*): on *Lepomis symmetricus,* La.

Genus **Actinocleidus** Mueller, 1937
(Fig. 77)

Tetraonchinae: Trunk, eyes, gut, gonads, cirrus, accessory piece, vagina (when present), and vitellaria as in *Cleidodiscus.* Haptor distinct and discoidal; armed with 2 pairs of anchors and 7 pairs of hooks. Anchors ventral, approximately uniform in size and shape; bases of each pair connected by a dissimilar transverse bar; bars articulate with each other.

Ref. Sproston (1946, p. 217); Mizelle and Hughes (1938, p. 350), key to spp.; Mizelle *et al.* (1956).

Actinocleidus articularis (Mizelle, 1936) Mueller, 1937 (Syn. *Cleidodiscus a.*): on *Lepomis megalotis,* Ill.; Mizelle (1940): same, Tenn.
A. bakeri Mizelle and Cronin, 1943: on *Lepomis microlophus,* Tenn.
A. bifidus Mizelle and Cronin, 1943: on *Lepomis microlophus,* Tenn.
A. bifurcatus (see *Clavunculus b.*).
A. brevicirrus Mizelle and Jaskoski, 1942: on *Lepomis miniatus,* Tenn.
A. bursatus (see *Clavunculus b.*).
A. crescentis Mizelle and Cronin, 1943: on *Lepomis microlophus,* Tenn.
A. fergusoni Mizelle, 1938: on *Lepomis macrochirus,* Ill., Okla.; Mizelle and Brennan (1942): Tenn., Fla.; Mizelle and Regensberger (1945): Wis.; Hargis (1952*b*): *L. macrochirus, Chaenobryttus coronarius,* Va.; Seamster (1938): *L. macrochirus, L. humilis,* Okla.

A. flagellatus Mizelle and Seamster, 1939: on *Chaenobryttus coro-narius*; Fla.; Mizelle and Cronin (1943): same, Tenn.; Hargis (1953*b*): same, Va.

A. fusiformis (Mueller, 1934) Mueller, 1937 (Syn. *Ancyrocephalus cruciatus, Cleidodiscus f.*): on *Micropterus dolomieui, M. punctulatus, M. salmoides*, Calif., Fla., La., N. Y., Okla., Wis., Va.; cf. Mizelle *et al.* (1956) for refs.; Haderlie (1953): *M. salmoides*, Calif.; dorsal bar a thin, square plate.

A. gibbosus Mizelle and Donahue, 1944: on *Lepomis gibbosus*, Ont.

A. gracilis Mueller, 1937: on *Lepomis macrochirus*, Fla.

A. harquebus Mizelle and Cronin, 1943: on *Lepomis microlophus*, Tenn.

A. incus Mizelle and Donahue, 1944: on *Lepomis gibbosus*, Ont.

A. longus Mizelle, 1938: on *Lepomis cyanellus*, Ill., Okla.; Seamster (1938): Okla.; Summers and Bennet (1938): on *L. microchirus*, La.

A. maculatus Mueller, 1937: on *Lepomis gibbosus*, Fla.; Mizelle (1941*a*): *L. microlophus*, Fla.

A. oculatus (Mueller, 1937) Mueller, 1937 (Syn. *Cleidodiscus o.*): on *Lepomis gibbosus, L. macrochirus*; La., Ont., Va.; cf. Mizelle *et al.* (1956) for refs.

A. okeechobeensis Mizelle and Seamster, 1939: on *Chaenobryttus gulosus, C. coronarius*, Fla.; Hargis (1952*b*): same, Va.

A. recurvatus Mizelle and Donahue, 1944: on *Lepomis gibbosus*, Ont.; Hargis (1952*b*): Va.

A. scapularis Mizelle and Donahue, 1944: on *Lepomis gibbosus*, Ont.

A. sigmoideus Mizelle and Donahue, 1944: on *Lepomis gibbosus*, Ont.; Hargis (1952*b*): Va.

A. subtriangularis Mizelle and Jaskoski, 1942: on *Lepomis miniatus*, Tenn.

A. triangularis (Syn. for *Anchoradiscus t.*).

A. unguis (Syn. for *Clavunculus u.*).

Genus **Ancyrocephalus** Creplin, 1839

Synonyms:

Ancyrocephalus aculeatus (see *Cleidodiscus a.*).
A. angularis (see *Urocleidus a.*).
A. bursatus (see *Clavunculus b.*).

Genus **Haplocleidus** (Mueller, 1936)
Mizelle and Hughes, 1938

Synonyms:

Haplocleidus affinis (see *Urocleidus a.*).
H. dispar (see *Urocleidus d.*).
H. furcatus (see *Urocleidus f.*).
H. parvicirrus (see *Urocleidus p.*).

Genus **Leptocleidus** Mueller, 1936

Synonyms:

Leptocleidus megalonchus (Mueller, 1936) Mizelle and Hughes, 1938 (see *Cleidodiscus m.*).

Genus **Onchocleidus** Mueller, 1936

Synonyms:

Onchocleidus acuminatus (see *Urocleidus a.*).
O. contortus (see *Urocleidus principalis*).
O. cyanellus (see *Urocleidus c.*).
O. dispar (see *Urocleidus d.*).
O. distinctus (see *Urocleidus d.*).
O. ferox (see *Urocleidus d.*).
O. helicis (see *Urocleidus h.*).
O. interruptus (see *Urocleidus i.*).
O. mimus (see *Urocleidus m.*).
O. perdix (see *Urocleidus p.*).
O. principales (see *Urocleidus p.*).
O. similis (see *Urocleidus s.*).
O. spiralis (see *Urocleidus s.*).

Genus **Pterocleidus** Mueller, 1937

Synonyms:

Pterocleidus acer (see *Urocleidus a.*).
P. acuminatus (see *Urocleidus a.*).
P. biramosus (see *Urocleidus b.*).
P. grandis (see *Urocleidus g.*).
P. miniatus (see *Urocleidus m.*).

Genus **Tetracleidus** Mueller, 1936

Synonyms:

Tetracleidus banghami (see *Cleidodiscus b.*).
T. chautauquensis (see *Urocleidus c.*).
T. megalonchus (see *Cleidodiscus m.*).

Suborder POLYOPISTHOCOTYLEA

Monogenea: The functional haptor of the adult is developed immediately anterior to the larval haptor, or, if the latter is retained, it develops 6 (sometimes 2) muscular suckers upon it. The larval anchors are often retained on the end of the body, but the larval hooklets very seldom. The haptor consists of 6 to many separate adhesive units which may be supported by cuticular sclerites. Cephalic glands seldom well developed, but an oral sucker may be present or a pair of cuticular bothria within the mouth. Genitointestinal canal present.

Ref. Sproston (1946, p. 336); Yamaguti (1963, p. 166).

KEY

TO THE GENERA OF POLYOPISTHOCOTYLEA

1. Numerous clamps on haptor .... Family MICROTYLIDAE   2
1. 8 or fewer clamps or anchors on haptor   3
2. Approximately equal number of clamps on either side; clamps of equal size . . . . (fig. 78) *Microcotyle*
2. Unequal number of clamps on the two sides . . (fig. 79) *Lintaxine*
3. Haptor with apical sucker and 3 pairs of anchors; *D. hamulatum* on gills of *Polyodon* . . . . (fig. 80) *Diclybothrium*
3. Haptor with 8 clamps   4
4. Wide club-shaped worm; 2 pairs of apical anchors and 8 clamps . . . . (fig. 81) *Mazocraeoides*
4. Worm more elongate; no apical anchors but 8 clamps present   5
5. Genital sucker present; vagina absent; testis lobed . . . .
   . . . . (fig. 82) *Octomacrum*
5. Genital sucker absent; Y-shaped vagina present; testes follicular; *D. salmonis* on gills of salmonids; also 3 European species . . . . (fig. 83) *Discocotyle*

Genus **Microcotyle** van Beneden and Hesse, 1863
(Fig. 78)

Symmetrical haptor a triangular extension of body; larva bears 2 pairs of anchors; clamps all same size. Genital atrium usually armed. Ovary usually looped in ?-shape. 74 species, mostly marine.

Ref. Sproston (1946, p. 424); Yamaguti (1963, p. 239).

Microcotyle eriensis Bangham and Hunter, 1936: on gills, *Aplodinotus grunniens*, L. Erie, 1.2 to 2.7 mm long, 21 to 28 clamps each side; Meserve (1938): valid sp.

M. macroura MacCallum and MacCallum, 1913: on *Roccus lineatus*, Atlantic coast; Meserve (1938): valid sp.

M. spinicirrus MacCallum, 1918: on gills, *Aplodinotus grunniens*, N. Y. Aquarium, length 2.5 mm, 80 to 90 clamps total, 2 to 3 mm long; Bangham and Hunter (1936): L. Erie; Bangham and Venard (1942): Tenn.; Linton (1940): Iowa; Remley (1942): life cycle, Iowa; Simer (1929): Miss.

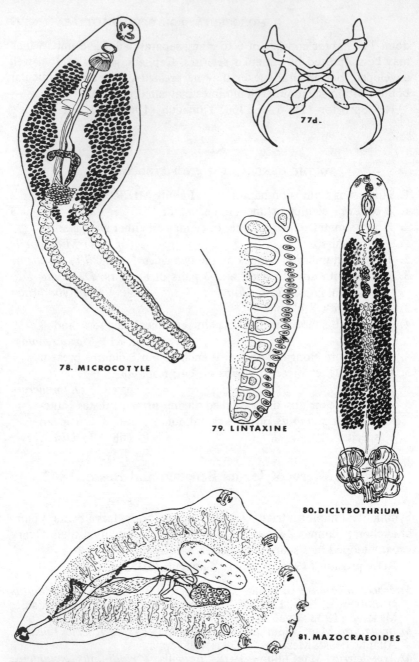

77d.

78. MICROCOTYLE

79. LINTAXINE

80. DICLYBOTHRIUM

81. MAZOCRAEOIDES

Figs. 77d–81. Fig. 77d. *Actinocleidus brevicirrus*, anchors and bars (from Mizelle and Jaskoski, 1942). Fig. 78. *Microcotyle spinicirrus* (from Bangham and Hunter, 1936). Fig. 79. *Lintaxine cokeri*, haptor (from Linton, in Sproston, 1946). Fig. 80. *Diclybothrium hamulatum* (from Simer, 1929). Fig. 81. *Mazocraeoides olentagiensis* (from Srouffe, 1958).

Genus **Lintaxine** Sproston, 1946
(Fig. 79)

Generally similar to *Microcotyle* except that sides of haptor have "grown together," forming two closely parallel rows of clamps: in one row, clamps are three times the size, but only about a third as numerous as in other. One species.

Ref. Sproston (1946, p. 460); Yamaguti (1963, p. 264).

Lintaxine cokeri (*Heteraxine c.* Linton, 1940) Sproston, 1946: on gills, *Aplodinotus grunniens*, Iowa, 5 mm long, 10 and 30 clamps; Manter and Prince (1953): redescription of Linton's material; Price (1961): redescription.

Genus **Diclybothrium** Leuckart, 1835
(Fig. 80)

No circum-oral sucker, but 2 glandular, bothrium-like areas occupy anterolateral region, with a simple mouth between them. Haptor with 6 sclerite-bearing suckers; 3 pairs of anchors at posterior of haptor. Eyes present. Parasites of Acipenseridae and Polyodontidae. Three species.

Ref. Sproston (1946, p. 375); Yamaguti (1963, p. 299).

Diclybothrium armatum Leuckart, 1835: on gills, *Acipenser* spp., Europe, Can. Worms 3.25 to 5.88 mm long, 3 pairs of anchors. Price (1942): redescription.
D. hamulatum (Simer, 1929) Price, 1942: on gills, *Polydon spathula*, Tenn., half as many testes as *D. armatum*; Bangham and Venard (1942): same, Tenn.; Bangham (1955): *Acipenser fulvescens*, Mich.

Genus **Mazocraeoides** Price, 1936
(Fig. 81)

Body broad and flattened. Haptor not differentiated as separate organ, but clamps on margin of body proper; anterior pair may be anterior to middle; short posterior lappet bears 3 pairs of anchors. Intestine much branched and accompanied by vitellaria. Genital pore armed. Eggs large, without polar filaments. On gills of Clupeidae. Five species.

Ref. Sproston (1946, p. 395); Yamaguti (1963, p. 220).

Mazocraeoides megalocotyle Price, 1958 (placed in *Pseudomazocraeoides* by Price, 1961, because clamps are more than twice as large as in related forms): on *Dorosoma cepedianum*, Tenn.
M. olentangiensis Sroufe, 1958 (Syn. *M. similis* Price, 1959): on gills, *Dorosoma cepedianum*, Ohio, Tenn.
M. tennesseensis Price, 1961: on *D. cepedianum*, Tenn.

Genus **Octomacrum** Mueller, 1934
(Fig. 82)

Vagina absent. Unarmed genital sucker present. Testis a single lobed mass in posterior half of body proper. Vitellaria confined to lateral fields, do not enter haptor. Bent crook-shaped anchors. Two species.

Ref. Sproston (1946, p. 403); Yamaguti (1963, p. 198).

Octomacrum lanceatum Mueller, 1934 (Syn. *Octobothrium sagittatum* Wright, 1879): on gills, *Catostomus teres, C. commersoni, Erimyzon sucetta*, Can., N. Y.; Anthony (1963) and Bangham (1944, 1955): on *C. commersoni*, Wis.; Bangham and Adams (1954): *C. commersoni, C. macrocheilus, Mylocheilus caurinus*, Brit. Col.; Fischthal (1945): *Notropis cornutus*, Wis.; Fischthal (1950, 1956): *C. commersoni*, N. Y., Wis.; Hargis (1952b): *Notropis* spp., Va.; Hoffman (unpubl.): *N. cornutus*, W. Va.; *Meyer* (1954): *C. commersoni*, Me.; Sindermann (1953): *E. sucetta*, Mass.; Fritts (1959): *C. macrocheilus*, Idaho.
O. microconfibula Hargis, 1952: on gills, *Notemigonus crysoleucas*, Va.
Octomacrum sp. Bangham and Adams (1954): on *Couesius plumbeus, Richardsonius balteatus*, Brit. Col.

Genus **Discocotyle** Diesing, 1850
(Fig. 83)

Y-shaped vagina opening in 2 marginal pores near level of genital atrium. Cirrus uncuticularized, without hooks or spines. Testes: numerous follicles in posterior of body, ovary near their anterior border. Eggs large, without filaments. Vitellaria massive, lateral. Three species.

Ref. Sproston (1946, p. 400); Yamaguti (1963, p. 189).

Discocotyle sagittata (Leuckart, 1842) Diesing, 1850: on salmonids, identical, or nearly so, with North American form below, Europe.
D. salmonis Schaffer, 1916: on gills, *Salmo gairdneri*, N. Y.; Bangham and Adams (1954): *Prosopium williamsii, Salvelinus malma*, Brit. Col.; Bangham (1955): *Coregonus artedi, C. hoyi, Prosopium cylindraceum*, Mich.; Brinkman (1952): synonymizes with *D. sagittata*; Davis (1946): U. S.; Laird and Embody (1931): treatment; Price (1943): considers this identical with *D. sagittata*, on *S. gairdneri, S. salar, S. trutta, S. fontinalis, C. ontariensis*, U. S.; Van Cleave and Mueller (1934): N. Y.

Order ASPIDOCOTYLEA Monticelli, 1892
(Syn.: *Aspidogastrea* Faust and Tang, 1936)

Endoparasitic trematodes with a discoid, ventral opisthohaptor well set off from body and divided into areoli or loculi with marginal bodies instead of anchors and hooklets.

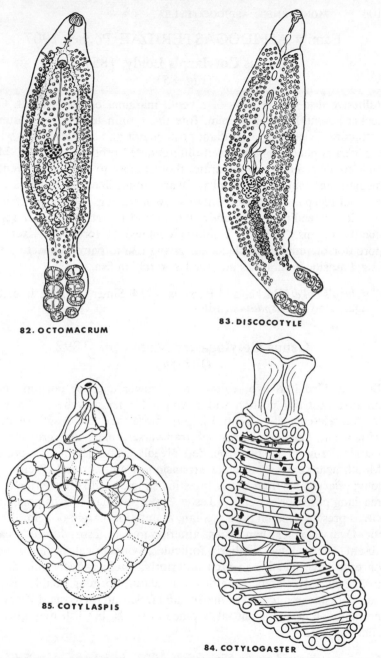

82. OCTOMACRUM

83. DISCOCOTYLE

85. COTYLASPIS

84. COTYLOGASTER

Figs. 82–85. Fig. 82. *Octomacrum lanceatum* (from Price, 1943). Fig. 83. *Discocotyle salmonis* (from Price, 1943). Fig. 84. *Cotylogaster michaelis* (from Monticelli, 1892, in Yamaguti, 1963). Fig. 85. *Cotylaspis sinensis* (from Faust and Tang, 1936, in Yamaguti, 1963).

Family ASPIDOGASTERIDAE Poche, 1907

Genus **Cotylaspis** Leidy, 1857

(Fig. 85)

Adhesive disc with 3 rows of alveoli; marginal organs present. Oral sucker absent. Pharynx present. Intestine terminates short of posterior extremity. Testes single, median, postequatorial. Seminal vesicle winding. Cirrus pouch contains well-differentiated prostate complex. Male and female genital pores separate, though close to each other, ventro-median, near anterior extremity. Ovary submedian, pre-testicular. No seminal receptacle. Uterus contains few large eggs. Vitellaria of large follicles, extending in hindbody, lateral and posterior to other reproductive organs. Terminal genitalia confined to forebody. Excretory pore dorsoterminal; vesicle small, giving rise to paired collecting vessels. Parasitic in molluscs and turtles, rarely in fish.

Cotylaspis cokeri Barker and Parsons, 1914. Simer (1929): in esophagus, *Polyodon spathula*, Miss.

Genus **Cotylogaster** Monticelli, 1892

(Fig. 84)

Body divided into two regions, a narrower anterior portion and a broader posterior portion within which former may be partially retracted telescopically. Ventral haptor more than half length of body, with single longitudinal row of transversely elongated grooves and marginal row of smaller rounded alveoli. Marginal organs present. Mouth near center of disclike expansion of anterior end. Prepharynx long; pharynx medium-sized. Intestine simple, saclike posteriorly, not reaching posterior extremity. Testes 2, post-ovarian. Cirrus pouch and cirrus present. Genital pore median, at anterior margin of ventral haptor. Ovary smaller than testes, anterior to fore testis. Laurer's canal absent. Uterus long. Vitellaria follicular, symmetrical, in linear series on each side, meeting in arch anteriorly; vitelline ducts paired, one from each side. Parenchymatous musculature forming double visceral sac. Parasitic in intestine of marine and freshwater teleosts. Life cycle: normally all development takes place in snails, but fish may also become infected.

Cotylogaster occidentalis Nickerson, 1902: in intestine, *Aplodinotus grunniens*; Bangham and Venard (1942): Tenn.; Dickerman (1948): life cycle; Simer (1929): Miss.; Sogandares (1955).

Adult Digenetic Trematodes
(Platyhelminthes: Trematoda: Digenea)

Contents

Key to the families of adult digenetic tremadodes from North American freshwater fishes 107
Family SANGUINICOLI-DAE 109
Genus *Sanguinicola* 110
Family DIDYMOZOIDAE . 110
Genus *Nematobothrium* .. 110
Family PARAMPOSTO-MIDAE 111
Genus *Pisciamphistoma* (*Paramphistomum*) 111
Family BUCEPHALIDAE . 111
Key to the genera of BUCEPHALIDAE 113
Genus *Bucephaloides* 113
Genus *Bucephalus* 114
Genus *Rhipidocotyle* 114
Genus *Paurorhynchus* 115
Family CRYPTOGONI-MIDAE 115
Key to the genera of CRYPTOGONIMIDAE .. 116
Genus *Allacanthochasmus* . 116
Genus *Neochasmus* 117
Genus *Cryptogonimus* 119
Genus *Acetodextra* 119
Genus *Centrovarium* 120
Genus *Caecincola* 120
Family MONORCHIIDAE . 121
Genus *Asymphlodora* 121
Family HAPLOPORIDAE . 121
Genus *Carassotrema* 123
Family LISSORCHIIDAE . 123
Genus *Lissorchis* 124
Genus *Triganodistomum* .. 124
Family DEROPRISTIIDAE 125

Genus *Pristicola* 125
Genus *Deropristis* 125
Genus *Skrjabinopsolus* ... 126
Genus *Cestrahelmins* 126
Family SYNCOELIDAE .. 127
Genus *Syncoelium* 127
Family GORGODERIDAE. 127
Genus *Phyllodistomum* ... 127
Family AZYGIIDAE 129
Key to the genera of AZYGIIDAE 129
Genus *Leuceruthrus* 130
Genus *Proterometra* 130
Genus *Azygia* 131
Family MICROPHAL-LIDAE 134
Genus *Microphallus* 134
Family HEMIURIDAE 134
Key to the genera of HEMIURIDAE 135
Genus *Lecithaster* 135
Genus *Hemiurus* 136
Genus *Brachyphallus* 136
Genus *Aponurus* 137
Genus *Genolinea* 137
Genus *Derogenes* 139
Genus *Deropegus* 139
Genus *Halipegus* 140
Family PLAGIORCHIIDAE 140
Key to the genera of PLAGIORCHIIDAE 141
Genus *Protenteron* 141
Genus *Glossidium* 141
Genus *Alloglossidium* 142
Family ALLOCREA-DIIDAE 142
Key to the genera of ALLOCREADIIDAE 143

Genus *Creptotrema* 143
Genus *Crepidostomum* ... 144
Genus *Bunodera* 147
Genus *Bunoderina* 148
Genus *Plagiocirrus* 149
Genus *Vietosoma* 149
Genus *Allocreadium* 149
Family MACRODEROI-
DIAE 151
Key to the genera of
MACRODEROIDIDAE .. 151
Genus *Paramacroderoides* . 151
Genus *Macroderoides* 152
Family LEPOCREADIIDAE 154
Subfamily HOMALO-
METRINAE 154

Key to the genera of
HOMALOMETRINAE ... 154
Genus *Homalometron* 155
Genus *Microcreadium* 155
Genus *Barbulostomum* 156
Family OPECOELIDAE .. 156
Key to the genera of
OPECOELIDAE 157
Genus *Podocotyle* 157
Genus *Plagioporus* 157
Family PTYCHOGONI-
MIDAE 159
Genus *Ptychogonimus* 159
Accidental in fish?
Pleurogenes 159
Adult tissue and bladder in-
habiting digenetic trematodes 159

Digenetic trematodes of fish belong to two important biological categories: (1) those that live in fish as adults, producing eggs which pass to the outside environment to continue the life cycle and (2) those that penetrate the skin and become metacercariae, usually encysted in the tissue, where they remain until eaten by the definitive host. To aid in the identification of the trematodes the two categories will be discussed separately.

Adult digenetic trematodes are usually found in the intestine and stomach of fish. Their pathogenicity has not been adequately studied. One species, *Acetodextra ameiuri* is found in the swim bladder and ovary of ictalurids. *Nematobothrium* and *Paurorhynchus* are found in the body cavity. Two genera, *Sanguinicola* and *Cardicola*, live in blood vessels where their eggs are deposited: the eggs become lodged in arteriole termini and cause considerable damage to the gills where they hatch and the miracidia emerge. Blood fluke disease is serious in some western trout hatcheries; no control method is known except to eliminate the responsible snails in the water supply.

KEY

TO THE FAMILIES OF ADULT DIGENETIC TREMATODES
FROM NORTH AMERICAN FRESHWATER FISH

1. Parasites of blood; no suckers; intestine X-shaped . . . .
. . . . (fig. 86) SANGUINICOLIDAE
1. Living in pairs in tissue, usually encysted, occasionally free; oral sucker present . . . . DIDYMOZOIDAE
1. Parasites of digestive tract, urogenital tract, occasionally air bladder, gall bladder, and ovary, exceptionally in body cavity 2
2. Ventral sucker ventral-terminal . . . .
. . . . (fig. 87) PARAMPHISTOMATIDAE
2. Ventral sucker ventral; mouth anterior (in oral sucker) . . . 3
2. Mouth ventral, leads to saccular single gut; sometimes with hoodlike anterior rhynchus with or without papillae . . . .
. . . . BUCEPHALIDAE

107

3. Gonotyl present or absent, and may be inconspicuous; ventral sucker often encircled by fold of body wall or embedded in body parenchyma; vitellaria not extensive; both metacercariae and adults in fish .... CRYPTOGONIMIDAE
(p. 115)

3. Gonotyl absent; ventral sucker not encircled by fold or embedded; vitellaria extensive or not 4

4. Usually single testis 5

4. Testes 2; cirrus spined 6

4. Testes 2 or more; cirrus not spined 7

5. Cirrus spined; metraterm usually spined ... MONORCHIIDAE
(p. 121)

5. Cirrus absent; metraterm not spined ..... HAPLOPORIDAE
(p. 121)

6. Body not spined; vitellaria limited in extent; genital pore marginal .... LISSORCHIIDAE
(p. 123)

6. Body spined; vitellaria follicular, usually in hindbody; genital pore pre-acetabular, not marginal ..... DEROPRISTIIDAE
(p. 125)

7. Testes 2 8

7. Testes numerous, follicular; in branchial cavity of anadromous fish .... SYNCOELIDAE
(p. 127)

8. Worm usually with narrowed anterior and wide, flat, posterior body; in urinary bladder, occasionally in intestine .... GORGODERIDAE in part, *Phyllodistomum*
(p. 127)

8. Worm without wide, flat, posterior body; in intestine, gall bladder, air bladder, or ovary 9

9. Body usually ribbon-like; prepharynx short or absent; vitellaria restricted to a short region behind ventral sucker ...
.... AZYGIIDAE
(p. 129)

9. Body not ribbon-like; vitellaria extensive or not 10

10. Vitellaria grouped into symmetrical bunches; ceca short; ovary pre-testicular; excretory vesicle V-shaped ....
.... MICROPHALIDAE
(p. 134)

10. Vitellaria of compact lobes; ceca long; ventral sucker large; ovary post-testicular; worm usually with tail-like portion;

usually in esophagus and stomach (sometimes in anadromous fish) .... HEMIURIDAE
(p. 134)

10. Vitellaria follicular, usually in two lateral fields; ovary pre-
testicular 11

11. Body covered with small spines; vitellaria not extensive;
uterus with ascending and descending limbs; excretory
vesicle Y-shaped with long stem ...... PLAGIORCHIDAE
(p. 140)

11. Body not covered with spines 12

12. Uterus with ascending limb only, but with pre-testicular coils;
with or without oral papillae; prepharynx usually long;
vitellaria usually extensive (remaining families may be
difficult to identify) 13

12. Uterus with ascending and descending limbs; without oral
papillae; prepharynx short; vitellaria not extensive; suckers very large; usually in stomach of sharks, rare in freshwater .... PTYCHOGONIMIDAE
(p. 159)

13. Excretory bladder I-shaped; uterus passes posterior to testes;
cirrus sac long, slender .... MACRODEROIDIDAE
(p. 151)

13. Excretory bladder tubular or saccular; uterus does not pass
posterior to testes; cirrus sac stout or lacking 14

14. Genital pore to left of midline .... OPECOELIDAE
(p. 156)

14. Genital pore median or nearly so 15

15. Cirrus sac lacking; body spined or not; cercaria without stylet .... LEPOCREADIIDAE: HOMALOMETRINAE
(p. 154)

15. Cirrus sac usually large with stout cirrus; body not spined;
cercaria with stylet .... ALLOCREADIDAE
(p. 142)

? Accidental in fish .... *Pleurogenes*
(p. 159)

Family SANGUINICOLIDAE Graff, 1907

Key characteristics: Digenea without suckers; usually found in circulatory system. Esophagus narrow, long; ceca X- or H-shaped. Eggs
nonoperculate, thin-shelled.

Ref. Yamaguti (1953, p. 258).

Genus **Sanguinicola** Plehn, 1905
(Fig. 86)

Sanguinicolidae: Body lanceolate with fine marginal striations or denticulations except for two extremities, of which anterior may protrude in form of proboscis. Esophagus long, may form fusiform swelling near anterior end. Intestine X-shaped, occasionally divided into 5 branches. Testes in 2 rows in median field between ovary and intestine. Cirrus pouch present. Male genital pore dorsal, median or submedian, close to uterine pore near posterior extremity. Ovary divided into symmetrical wings in posterior half of body. Uterus very short, containing only one egg, opening beside or anterior to male pore; egg usually contains miracidium. Vitellaria lateral to esophagus, intestine, and testes, sometimes also lateral and posterior to ovary; vitelline duct joining oviduct just before ootype is formed. Parasitic in circulatory system of freshwater fishes. Life cycle: adult in fish blood vessels; longevity at least 4 months; miracidia pass out gills; snail intermediate host; cercariae penetrate fish.

Ref. Erickson and Wallace (1959), key to spp.

Sanguinicola davisi Wales, 1958: in *Salmo gairdneri, S.g. kamloops*; snail host is *Oxytrema* (*Goniobasis*); causes heavy mortality, Calif.; Uzmann (1963); *S. gairdneri*, Wash.

S. huronis Fischthal, 1949: in *Micropterus salmoides, M. dolomieui*, Wis.

S. klamathensis Wales, 1958: in *Salmo clarki*; snail host presumably *Oxytrema*; causes mortality, Calif.

S. lophophora Erickson and Wallace, 1959: experimentally in *Notropis hudsonius*; snail host is *Valvata tricarinata*; Minn.; Larson (1961): Minn.

S. occidentalis Van Cleave and Mueller, 1932: in *Stizostedion vitreum*; N. Y.; Fischthal (1947): in *S. vitreum, Perca flavescens*, Wis.

Sanguinicola sp., Fischthal (1947): in *Catostomus commersoni, Notropis cornutus*, Wis.

Family DIDYMOZOIDAE Poche, 1907

Key characteristics: Usually long, slender Digenea, in pairs or singly, encysted or not, in tissues or body cavity of marine fishes, rarely in freshwater fishes.

Ref. Yamaguti (1953, p. 221).

Genus **Nematobothrium** van Beneden, 1858
(No fig.)

Generic diagnosis, Didymozoidae: Hermaphrodite with tendency toward sexual dimorphism. Enclosed in pairs in rounded to long tubular cyst. Body narrow, long, winding, of nearly uniform breadth throughout; may be divided in two directly continuous regions (slender forebody and thicker subcylindrical hindbody). Pharynx present. Ceca long, may be atrophied to some extent. Acetabulum present occasionally. Testes paired, elongated near anterior extremity. Genital pore ventral to oral sucker or pharynx. Ovary and vitellaria single, tubular, long, winding; former extend between testes and shell gland complex, latter between shell gland complex and posterior extremity. Uterus ascends first to posterior end of testes, turns back on itself, and descends to posterior extremity to take final ascending course. Parasitic in marine fishes, rarely in freshwater fishes. Life cycle unknown.

Nematobothrium texomensis McIntosh and Self, 1955: in ovaries, *Ictiobus bubalus*, 8-9 ft. long, Okla.; Self and Campbell (1956): *I. bubalus, I. cyprinella*; Self and Peters (1963): miracidium.

Family PARAMPHOSTOMATIDAE Fischoeder, 1901

Key characteristics: Amphistomate Digenea with very thick body. Ventral sucker terminal or subterminal. Oral sucker terminal.
Ref. Yamaguti (1953, p. 247).

Genus **Pisciamphistoma** Yamaguti, 1953
(Fig. 87)

Body elongate, without ventral pouch. Oral sucker (or pharynx) large, without diverticles. Esophagus slender, with muscular bulb at posterior end. Ceca simple, terminating near posterior extremity. Acetabulum large, very prominent, ventroterminal. Testes small, diagonal, intercecal. Cirrus pouch(?). Genital pore postbifurcal. Ovary median, postequatorial. Uterus passing between two testes, with few coils. Viteline follicles in extracecal fields between level of esophageal bulb and acetabulum. Intestinal parasites of fishes. Life cycle unknown.

Pisciamphistoma reynoldsi Bogitsh and Cheng, 1959: in intestine, *Lepomis macrochirus*, Tenn.
P. stunkardi (Holl, 1929), (Syn. *Paramphistomum s.*): in intestine, *Lepomis gibbosus, Chaenobryttus gulosus*, N. C.; Yamaguti (1953) proposed that the new genus *Pisciamphistoma* be erected for this worm; Harms (1959): *Ictalurus natalis*, Kans.

Family BUCEPHALIDAE Poche, 1907

Gasterostomata: Mouth ventral, without oral sucker. Adhesive organ (rhynchus) at anterior extremity. Pharynx and esophagus present. Intestine simple, saccular or tubular. Testes 2, usually in hindbody. Cirrus pouch in posterior of body, containing seminal vesicle and

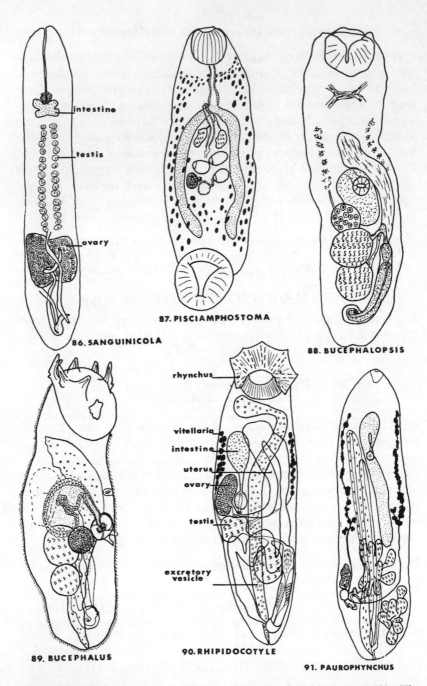

Figs. 86–91. Fig. 86. *Sanguinicola huronis* (from Fischthal, 1949). Fig. 87. *Pisciamphostoma stunkardi* (from Bogitsh and Cheng, 1959). Fig. 88. *Bucephalopsis pusillus* (from Woodhead, 1930). Fig. 89. *Bucephalus*

prostate gland; cirrus projects into genital atrium in form of crooked lobe. Genital pore ventral, terminal or subterminal. Ovary anterior or between testes. No seminal receptacle. Laurer's canal present. Vitellaria follicular, usually in forebody. Uterus convoluted with many small eggs. Excretory bladder tubular, with terminal opening.

Ref. Yamaguti (1953, p. 3).

KEY

TO THE GENERA OF BUCEPHALIDAE

Rhynchus (anterior adhesive organ) without tentacular appendages; ovary pre-testicular . . . . (fig. 88) *Bucephaloides*
Rhynchus sucker-like but with 7 tentacular appendages; ovary pre-testicular . . . . (fig. 89) *Bucephalus*
Rhynchus not sucker-like; but with pentagonal caplike expansion; ovary pre-testicular . . . . (fig. 90) *Rhipidocotyle*
Rhynchus weakly developed; ovary opposite anterior testis; in body cavity . . . . (fig. 91) *Paurorhynchus*

Genus **Bucephaloides** Hopkins, 1954
(Fig. 88)
[Syn.: *Bucephalopsis* (Diesing, 1855) Nicoll, 1914]

Bucephalidae: Rhynchus sucker-like, without tentacular appendages. Mouth opening usually in middle third of body. Intestine short. Testes tandem or oblique, usually post-equatorial. Ovary anterior or opposite to testes. Vitelline follicles in paired pre-ovarian groups. Uterus may ascend up to or near rhynchus. Life cycle: adult in intestine of fish; cercaria emerge from pelycepod mollusc; metacercaria in fish.

Ref. Doss (1963, p. 147); Yamaguti (1953, p. 5).

Bucephaloides ozakii (Nagaty, 1937). Bangham and Adams (1954): in *Salvelinus alpinus*, Brit. Col.
B. pusillus (Stafford, 1904): in *Stizostedion vitreum*; Can.; Cooper (1915): *Micropterus dolomieui, Etheostoma nigrum*, Can.; Fischthal (1947, 1950, 1952): Wis.; Bangham (1944, 1955): Wis., L. Huron; Lyster (1939): Can.; Meyer, F. (1958): *Lepomis cyanellus, L. humilis, Pimephales promelas, Esox lucius, Ictalurus melas,*

---

elegans (from Woodhead, 1930). Fig. 90. *Rhipidocotyle papillosum* (from Kniskern, 1952). Fig. 91. *Paurorhynchus hiodontis* (from Dickerman, 1954).

Perca flavescens; Iowa; Van Cleave and Mueller (1934): N. Y.; Woodhead (1930): Mich.
Bucephaloides sp. Hoffman (1953): progenetic, in musculature, *Notropis cornutus,* N. D.

Genus **Bucephalus** Baer, 1926
(Fig. 89)

Bucephalidae: Rhynchus sucker-like, with 7 tentacular appendages. Mouth opening in middle third of body. Intestine short. Testes tandem or oblique, post-equatorial. Genital lobe present. Ovary pre-testicular. Vitelline follicles divided in front of ovary into two distinct groups. Excretory bladder variable in length. Life cycle: as *Bucephaloides* (see p. 183 for metacercaria).

Ref. Doss (1963, p. 149); Srivastava (1963); Yamaguti (1953, p. 4).

Bucephalus elegans Woodhead, 1930: in ceca, *Ambloplites rupestris,* mollusc host *Eurynia iris,* Mich.; Bangham (1941, 1955): *P. flavescens,* L. Huron; Fischthal (1947): *A. rupestris, Perca flavescens,* Wis.; Fischthal (1956): *A. rupestris,* N. Y.; McDaniel (1963): *Lepomis macrochirus,* Okla.; Van Cleave and Mueller (1934): *P. flavescens, Roccus chrysops, Esox niger, Ictalurus nebulosus,* N. Y.
B. papillosum (see *Rhipidocotyle papilosum*).
Bucephalus sp. Bangham (1942, 1955): in *Roccus chrysops, Notropis hudsonius,* Mich., Tenn.; Fischthal (1950): *Cottus bairdi,* Wis.

Genus **Rhipidocotyle** Diesing, 1858
(Fig. 90)

Bucephalidae: Rhynchus with pentagonal cap or hoodlike expansion and suctorial pit ventroposteriorly. Mouth opening usually in middle third. Intestine short. Testes tandem or oblique, in posterior half of body. Cirrus pouch and genital pore as in *Bucephalopsis.* Ovary pretesticular. Vitellaria usually divided into two pre-ovarian groups. Uterus not extending as far forward as in *Bucephalus* and *Bucephalopsis.* Excretory bladder very long. Life cycle: adult in intestine and ceca of fish; cercaria in clams; metacercaria in fish.

Ref. Yamaguti (1953, p. 7); Van Cleave and Mueller (1934, p. 188).

Rhipidocotyle lepisostei Hopkins, 1954: in *Lepisosteus spathula,* metacercaria in *Mugil* spp., La.
R. papillosum (Woodhead, 1929): in ceca, *Micropterus dolomieui,*

M. salmoides, and stomach and intestine of *Esox lucius*, mollusc host *Elliptis dilatatus*, Mich.; Bangham (1955): L. Huron; Fischthal (1947): *Lepomis macrochirus*, Wis.; Fischthal (1956): *M. dolomieui*, N. Y.; Van Cleave and Mueller (1934): N. Y.

R. septpapillata Krull, 1934: *Lepomis gibbosus*, metacercaria in *Fundulus diaphanus*, *L. gibbosus*, Va. Kniskern (1950): *Micropterus dolomieui*, *L. gibbosus*, mollusc host *Lampsilis*.

Genus **Paurorhynchus** Dickerman, 1954
(Fig. 91)

Bucephalidae: Rhynchus weakly developed. Pharynx about one-third of body length from anterior end; intestine long, turning backward to region of anterior testis. Testes large, strongly lobed, oblique, in posterior two-fifths of body; cirrus pouch half as long as posterior testis, moderately thick-walled, containing seminal vesicle and pars prostatica. Genital pore treminal. Ovary small, opposite anterior testis, ootype post-ovarian. Laurer's canal opening at level of ootype. Vitellaria a linear series of irregular acini, extending on each side of body from level of cecal flexure to level of anterior testis. Gravid uterus occupying all available space between level of anterior end of vitellaria and posterior extremity; eggs small. Excretory bladder long, tubular, reaching to near pharynx. Life cycle unknown.

Ref. Yamaguti (1958, p. 14).

Paurorhynchus hiodontis Dickerman, 1954: in body cavity, *Hiodon tergisus*, L. Erie; Margolis (1964): in *H. alosoides*, Sask. Very fragile worm.

Family CRYPTOGONIMIDAE Ciurea, 1933

Family diagnosis: Body small, oval to elongate. Eyespots present or absent. Circumoral coronet of spines present or absent. Oral sucker terminal or subterminal. Pharynx present. Esophagus short. Ceca usually long, sometimes short. Acetabulum usually small and embedded in parenchyma or enclosed in genital atrium, occasionally reduced or lacking. Testes usually double, exceptionally single or numerous inter- or extracecal in hindbody. Seminal vesicle well developed. Cirrus pouch absent. Gonotyl may be present. Genital pore usually preacetabular. Ovary lobed or not, pre-testicular, occasionally intertesticular, exceptionally double. Seminal receptacle present. Vitellaria follicular, grouped in bunches, exceptionally compact, mostly in lateral fields of hindbody, occasionally in forebody. Uterus usually extending into post-testicular region of body. Eggs with or without polar filament. Excretory vesicle V- or Y-shaped; arms usually wide and very long, reaching to near pharynx. Parasites of marine and freshwater teleosts, exceptionally of reptiles.

Ref. Dunagan (1960), key to cercariae.

KEY

TO THE GENERA OF CRYPTOGONIMIDAE

(Larval forms not included; see p. 178 for key to metacercariae)

1. Circumoral spines present   2
1. Circumoral spines absent . . . . :. . .   3
2. Vitellaria mostly pre-acetabular; ventral sucker in middle third
    of body; ovary with few lobes . . . . (fig. 92) *Allacanthochasmus*
2. Vitellaria post-acetabular; ventral sucker near intestinal fork
    in first third of body; ovary a transverse band of follicles
    . . . . (fig. 93) *Neochasmus*
3. Body elongate . . . . (fig. 94) *Cryptogonimus*
3. Body oval to elliptical   4
4. Ceca reach to testes   5
4. Ceca terminate at or near posterior extremity; in gonads,
    visceral cavity, and air bladder . . . . (fig. 95) *Acetodextra*
5. Testes symmetrical or somewhat diagonal, at about middle
    of hindbody; vitellaria extend in dorsolateral fields between
    level of intestinal bifurcation and testes; uterus occupies
    post-testicular region . . . . (fig. 96) *Centrovarium*
5. Testes diagonal, some distance behind ventral sucker; vitel-
    laria in neck or shoulder region; uterus extends back of
    testes . . . . (fig. 97) *Caecincola*

Genus **Allacanthochasmus** Van Cleave, 1938
(Fig. 92)

Cryptogonimidae: Body small, more or less elongate, spined. Oral sucker nearly terminal, with single complete circle of spines. Pre-pharynx present, pharynx prominent, esophagus short; ceca do not quite reach posterior. Ventral sucker smaller than oral sucker, in middle third of body. Testes symmetrical, diagonal, in posterior half of body. Seminal vesicle anterior or dorsal to ventral sucker. Genital atrium opening immediately in front of ventral sucker. Ovary lobed, extending transversely between ventral sucker and testes. Seminal receptacle between ventral sucker and ovary. Vitellaria extend along ceca in lateral fields from intestinal bifurcation to ovary or testes. Uterine coils occupy most of post-testicular zone. Excretory bladder Y-shaped; arms reach pharynx. In intestine of freshwater fishes. Life cycle: metacercariae in fish (see p. 193).

Ref. Doss *et al.* (1963, p. 26); Yamaguti (1958, p. 226).

Allacanthochasmus artus Mueller and Van Cleave, 1932: in *Roccus chrysops*, N. Y.; Bangham (1942): *Pylodictis olivaris, R. chrysops, R. interrupta,* Tenn.; Bangham (1955): L. Huron; Bangham and Venard (1942): *R. chrysops, R. interrupta, P. olivaris;* Miller (1958); Mueller (1934); Nigrelli (1940): *R. interrupta,* Ill., N. Y.
A. varius Van Cleave, 1922: in *Roccus chrysops,* Mississippi R.; Bangham (1933): *Micropterus dolomieui;* Bangham (1942): *R. chrysops, R. interrupta,* Tenn.; Bangham (1955): L. Huron; Bangham and Hunter (1939): *R. chrysops, Labidesthes sicculus,* L. Erie; Bangham and Venard (1942): Tenn.; Hugghins (1959): *R. chrysops,* S. Dak.; Nigrelli (1940): N. Y.; Pearse (1924): Wis.; Van Cleave and Mueller (1934): N. Y.

Genus **Neochasmus** Van Cleave and Mueller, 1932
(Fig. 93)

Cryptogonimidae: Body plump, with eyespots. Oral sucker terminal, large, with single crown of spines. Prepharynx present. Pharynx usually small; esophagus short; ceca not reaching posterior. Ventral sucker deeply submerged. Testes symmetrical, near ceca ends. Seminal vesicle well developed, extending posteriad of ventral sucker. Genital pore just in front of ventral sucker. Gonotyl rudimentary. Ovary divided into numerous follicles extending transversely between ventral sucker and testes and overreaching ceca laterally. Seminal receptacle pre-ovarian. Vitellaria in lateral and posterior fields in middle of body. Uterine coils mostly post-testicular. Excretory bladder not studied. In intestine of fishes and reptiles. Life cycle unknown.

Ref. Yamaguti (1958, p. 226).

Neochasmus ictaluri Sogandares-Bernal, 1955: in *Ictalurus furcatus,* La.
N. umbelus Van Cleave and Mueller, 1932: in *Etheostoma nigrum, Micropterus dolomieui, M. salmoides, Necturus maculosus* (salamander), *Roccus interrupta,* N. Y.; Bangham (1942): *Microperca proeliaris,* Tenn.
Neochasmus sp. Hoffman, 1957 (unpubl.): in *Fundulus diaphanus,* N. D.
Neochasmus sp. Peters, 1959 (pers. comm.): progenetic in flesh of fish.

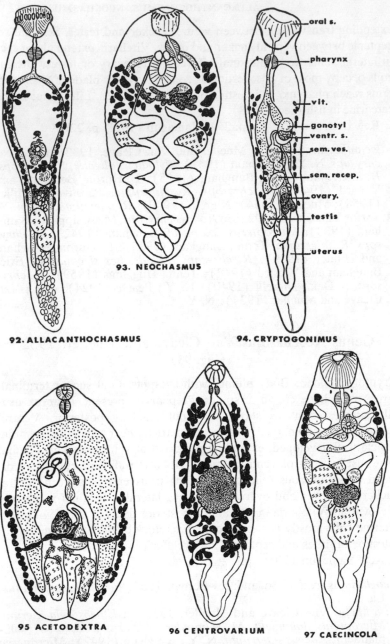

92. ALLACANTHOCHASMUS

93. NEOCHASMUS

94. CRYPTOGONIMUS

oral s.

pharynx

vit.

gonotyl
ventr. s.
sem. ves.

sem. recep.

ovary.

testis

uterus

95 ACETODEXTRA

96 CENTROVARIUM

97 CAECINCOLA

Figs. 92–97. Fig. 92. *Allacanthochasmus artus* (from Mueller and Van Cleave, 1932. Fig. 93. *Neochasmus ictaluri* (from Sogandares, 1955). Fig. 94. *Cryptogonimus chyli* (from Mueller and Van Cleave, 1932). Fig. 95. *Acetodextra amiuri* (from Mueller and Van Cleave, 1932). Fig. 96.

Genus **Cryptogonimus** Osborn, 1910
(Fig. 94)

Cryptogonimidae: Body elongate with tendency toward uniform breadth, spined, with eyespots. Oral sucker large; prepharynx and pharynx average; esophagus short; ceca extend to posterior third. Ventral sucker median, double—one directly behind other, with genital pore between, enclosed in circular muscular fold of body wall. Testes diagonal, in posterior half of body. Seminal vesicle elongated saccular, extending posteriad of ventral sucker. Prostatic complex present. No cirrus or cirrus pouch. Genital atrium tubular, passing between two ventral suckers. Ovary submedian, pre-testicular. Seminal receptacle pre-ovarian. Vitellaria laterad of ceca. Uterus convoluted in post-testicular region. Excretory bladder Y-shaped, bifurcating in gonad region; arms reach level of eyespots. Gastrointestinal in freshwater fishes. Life cycle: metacercaria in muscle of fish (see p. 199).

Cryptogonimus chyli Osborn, 1910: in *Micropterus dolomieui*, N. Y. Also recorded from *Ambloplites rupestris*, *Esox masquinongy*, *Lepomis gibbosus*, *L. macrochirus*, *Micropterus dolomieui*, *M. salmoides*, *Perca flavescens*, *Pomoxis annularis*, *P. nigromaculatus*, by Bangham (1941, 1944, 1955): Ohio, Wis., L. Huron; Fantham and Porter (1947): Que.; Fischthal (1947, 1956): N. Y., Wis.; Hunter (1930): N. Y.; Pearse (1924): Wis.; Van Cleave and Mueller (1934): N. Y. For additional hosts see Doss *et al.* (1964, p. 343).

C. diaphanus (see *Protenteron diaphanum*: Plagiorchidae).

Genus **Acetodextra,** Pearse, 1924
(Fig. 95)

Cryptogonimidae: Body flattened, plump. Oral sucker subterminal. Esophagus average; ceca wide, terminating near posterior. Ventral sucker on right of median line behind intestinal fork. Testes elongated, between cecal ends. Seminal vesicle extends posteriad of ventral sucker. Genital pore between ventral sucker and intestinal fork. Ovary median, equatorial or postequatorial. Uterine coils occupy intercecal area. Vitellaria in lateral fields from ventral sucker to cecal ends. Excretory bladder Y-shaped. In ovary and air bladder of freshwater

---

Centrovarium lobotes (from Miller, 1941). Fig. 97. *Caecincola parvulus* (from Van Cleave and Mueller, 1934).

fishes. Life cycle: metacercaria in liver of *Noturus*. Adults probably passed during spawning, and ova "explode" from worm.

Acetodextra amiuri (Stafford, 1904) Pearse, 1924: in *Ictalurus melas*, *I. natalis*, *I. nebulosus*, *metacercaria* in *Noturus gyrinus*, Wis.; Arnold (1933, 1934): *I. nebulosus*, N. Y., Wis.; Bangham (1955): same, Tenn.; Bangham and Hunter (1939): *I. punctatus*, *N. gyrinus*, L. Erie; Cable (1960); Coil (1954): *I. melas*, *I. natalis*, *I. nebulosus*, *I. punctatus*, *Noturus flavus*, *Cottus bairdi*, L. Erie; Corkum *et al.* (1958). *I. melas*, Ill.; Hunter and Hunter (1932): *I. nebulosus*; Lyster (1939): Can.; Mueller and Van Cleave (1932): *I. natalis*, *I. nebulosus*, N. Y.; Perkins (1956): ovary, *I. punctatus*, and "exploding" of ova from worm; Van Cleave and Mueller (1934): N. Y.

Genus **Centrovarium** Stafford, 1904
(Fig. 96)

Cryptogonimidae: Body flattened cigar-shaped, with eye spots. Oral sucker ventroterminal. Prepharynx present. Pharynx small, esophagus short; ceca terminate about midway. Ventral sucker small, about one-third from anterior end. Testes nearly symmetrical, near middle of body. Seminal vesicle elongated, anterodorsal to ventral sucker. Genital pore median, in front of ventral sucker. Ovary large, lobed, median equatorial. Seminal receptacle and Laurer's canal present. Vitellaria in dorsolateral fields between fork and testes. Uterus occupying post-testicular area. Excretory bladder Y-shaped; arms wide, reaching to level of pharynx. In stomach and intestine of freshwater fishes. Life cycle: metacercaria in muscle of fish (see p. 201); snail host unknown.
Ref. Doss *et al.* (1964, p. 178).

Centrovarium lobotes (MacCallum, 1895). Recorded from *Amblo-plites, Anguilla, Esox, Ictalurus, Micropterus, Perca. Stizostedion*, by Bangham (1955); Bangham and Hunter (1939); Cooper (1915); Hoffman (1953); Hoffman (unpubl.): progenetic meta-cercaria; Hunter (1930); Hunter and Hunter (1931); Lyster (1939); Miller (1941): review; Pearse (1924); Stafford (1904); Van Cleave and Mueller (1934).

Genus **Caecincola** Marshall and Gilbert, 1905
(Fig. 97)

Cryptogonimidae: Very small, elliptical, spined. Oral sucker nearly terminal, large. Prepharynx distinct, pharynx large, esophagus average. Ceca short, reaching anterior testis. Ventral sucker small, pre-equatorial. Testes diagonal, in posterior half of body. Seminal vesicle bipartite. No copulatory organ. Genital pore median, at anterior of ventral sucker. Ovary trilobate, in front of posterior testis. Vitelline follicles massed in lateral fields adjacent to ventral sucker. Uterus descends to posterior and then ascends: Excretory bladder Y-shaped, stem long. In stomach, ceca, and intestine of freshwater fishes. Life cycle: pleurolophocercous cercaria in *Amnicola* (*Marstonia*); metacercaria in *Lepomis*.

Ref. Doss *et al.* (1964, p. 159).

Caecincola parvulus Marshal and Gilbert, 1905. Recorded from *Ambloplites rupestris*, *Micropterus dolomieui*, *M. floridana*, *M. salmoides*, *Pomoxis annularis*, by Bangham (1955): L. Huron; Bangham and Venard (1942): Tenn.; Fischthal (1947, 1950): Wis.; Horsefall (1934); Lundahl (1939, 1941): life cycle; Pearse (1924): Wis.; Sparks (1951): Tex.; Van Cleave and Mueller (1934): N. Y.; Venard (1940): Tenn.

Family MONORCHIIDAE Odhner, 1911
(See Yamaguti, 1953, p. 51; 1958, p. 60)
Genus **Asymphylodora** Looss, 1899
(Fig. 98)

Body small, elliptical to fusiform, spined, with well-developed cervical gland. Oral sucker large, subterminal; esophagus variable; ceca reach equator or less. Ventral sucker in anterior half of body. Testis single, in posterior half of body. Cirrus spined, protrusible. Genital pore marginal or submarginal at level of ventral sucker. Ovary median or submedian, pretesticular. Seminal receptacle very small. Vitellaria lateral in ventral sucker-ovarian zone; sometimes extend to testis or farther. Uterus bilateral, post-testicular. Excretory bladder tubular. In intestine of freshwater fishes.

Ref. Stunkard (1959); Yamaguti (1953, p. 51; 1958, p. 60).

Asymphylodora amnicolae Stunkard, 1959: in *Fundulus*, *Lepomis*, *Micropterus*, *Perca*, tailless cercaria in *Amnicola*, metacercaria in same snail, progenetic metacercaria (adult) in same snail (Stunkard, 1959), Mass.; Larson (1961): progenetic in *Amnicola*, Minn.

Family HAPLOPORIDAE Nicoll, 1914

Body small to very small, rather plump, spinulate. Oral sucker subterminal, prepharynx present, pharynx usually well developed, esophagus long in relation to ceca; latter short, occasionally saccular, rarely reaching to near posterior extremity. Acetabulum in middle third of

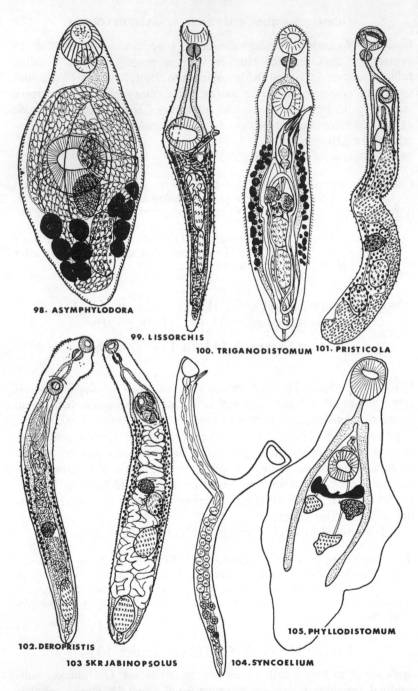

98. ASYMPHYLODORA

99. LISSORCHIS

100. TRIGANODISTOMUM

101. PRISTICOLA

102. DEROPRISTIS

103 SKRJABINOPSOLUS

104. SYNCOELIUM

105. PHYLLODISTOMUM

Figs. 98–105. Fig. 98. *Asymphylodora amnicolae* (from Stunkard, 1959).
Fig. 99. *Lissorchis gullaris* (from Self and Campbell, 1956). Fig. 100.
Triganodistomum hypentelii (from Fischthal, 1942). Fig. 101. *Pristicola*

body. Testis single, submedian, rarely median, post-acetabular. Seminal vesicle bipartite, anterior portion enclosed in hermaphroditic pouch. Hermaphroditic duct well developed. Genital pore median, pre-acetabular. Ovary small, median, usually posterodorsal to acetabulum, in front of testis or level with it. Seminal receptacle present or absent. Laurer's canal present. Vitellaria of paired compact lobes or bunches, behind or beside ovary, or in lateral or dorsal field of hindbody, largely in acetabulo-testicular zone. Uterus winding in hindbody; eggs relatively large, containing miracidia. Excretory vesicle tubular or saccular, occasionally Y-shaped. In marine and freshwater fishes.

Genus **Carassotrema** Park, 1938
(No fig.)

Body small, orange-yellow in life, spined. Oral sucker followed by wide prepharynx; pharynx large, longer than broad; esophagus moderately long. Ceca terminate near posterior extremity. Acetabulum nearly as large as oral sucker or even larger, at about junction of anterior with middle third of body. Testis single, median, in posterior half of body. External seminal vesicle post-acetabular. Hermaphroditic sac anterodorsal to acetabulum. Genital pore pre-acetabular. Ovary median, just in front of testis. Seminal receptacle represented by dilated germiduct or Laurer's canal. Uterus coiled between ovary and hermaphroditic sac. Vitellaria of simple or branched tubular lobes, extending in lateral fields of hindbody as well as in post-testicular area. Excretory vesicle Y-shaped; arms reach acetabular level. Intestine of freshwater fishes.

Carassotrema koreanum Park, 1938: in *Carassius auratus*, Korea; Yamaguti (1958): *C. auratus, Cyprinus carpio*, Japan; Long and Lee (1958): in several freshwater fish, China.
C. mugilicola Shireman, 1964: in *Mugil cephalus*, brackish water, La.

Family LISSORCHIIDAE Poche, 1926

Body elongate. Suckers and pharynx well developed. Ceca not reaching posterior. Testes tandem, in hindbody. No cirrus pouch. Seminal vesicle and cirrus well developed; cirrus spined. Genital pore lateral,

---

sturionis (from Cable, 1952). Fig. 102. *Deropristis inflata* (from Cable, 1952). Fig. 103. *Skrjabinopsolus manteri* (from Cable, 1952). Fig. 104. *Syncoelium filiferum* (from Lloyd and Guberlet, 1936). Fig. 105. *Phyllodistomum etheostomae* (from Fischthal, 1943).

level with ventral sucker. Ovary pre-testicular and post-acetabular. Uterus reaches farther back than testes. Vitellaria lateral, in hindbody, not extensive. Excretory bladder Y-shaped. In intestine of freshwater fishes.

Genus **Lissorchis** Magath, 1917
(Fig. 99)

Description as for Lissorchiidae; ovary many-lobed.

Lissorchis fairporti Magath, 1917: in *Ictiobus cyprinella, I. bubalus*; Iowa; Magath (1918): xiphidiocercaria in *Helisoma*, metacercaria in *Chironomus*; Cort *et al.* (1950): believe that Magath was mistaken in larval forms; Simer (1929): Miss.
L. gullaris Self and Campbell, 1956: in *Ictiobus bubalus, I. cyprinella, I. niger*, Okla.

Genus **Triganodistomum** Simer, 1929
(Fig. 100)

Description as for Lissorchiidae: ovary three-lobed. Life cycle: metacercaria in oligochaetes and planaria.
  Yamaguti, 1953, p. 51, places this genus in Monorchiidae.
  Ref. Fischthal (1942), key; Yamaguti (1953, p. 51; 1958, p. 66).

Triganodistomum attenuatum Mueller and Van Cleave, 1932: in *Catostomus commersoni*; Anthony (1963): same. Wis.; Bangham (1944, 1951, 1955): *C. commersoni, Gila straria*, Wis., Wyo., Mich.; Bangham and Adams (1954): *C. catostomus. C. macrocheilus.* Brit. Col.: Fischthal (1947, 1950): Wis.; Fischthal (1956): *C. commersoni*, N. Y.; Fried *et al.* (1964); Krueger (1954): Ohio; Meyer, M. (1954): Me.; Meyer, F. (1958): Iowa.
T. crassicrurum Haderlie, 1953: in *Catostomus rimiculus.* Calif.
T. garricki (Syn. *Alloplagiorchis garricki* Simer, 1929): in *Carpiodes difformis*, Miss.
T. hypentelii Fischthal, 1942: in *Hypentelium nigricans*, Mich.; Hoffman (unpubl.): *Moxostoma aureolum*, W. Va.
T. mutabile (Cort, 1919): in *Erimyzon sucetta*, Mich., Minn.; Bangham (1942): Tenn.; Wallace (1941): cercaria in *Helisoma*, metacercaria in oligochaets and planaria.
T. polyobatum Haderlie, 1950: in *Catostomus occidentalis*, Calif.
T. simeri Mueller and Van Cleave, 1932: in *Catostomus commersoni*, N. Y.; Bangham (1955): L. Huron; Krueger (1954): Ohio.
T. translucens Simer, 1929: in *Ictiobus bubalus*, Miss.
Triganodistomum spp. Bangham (1944, 1951, 1955): in *Moxostoma rubreques, Catostomus fecundus, Notropis hudsonius*, Wis., Wyo., L. Huron; Hugghins (1959): *C. commersoni*, S. Dak.

Family DEROPRISTIIDAE (Skriabin, 1958) Peters, 1961
KEY

TO THE GENERA OF DEROPRISTIIDAE

1. Circumoral spines present .... (fig. 101) *Pristicola*
1. Circumoral spines absent 2
2. Neck swollen; prepharynx distinct; in marine fish 3
.... (fig. 102) *Deropristis*
2. Neck not swollen 3
3. Esophagus short; ceca reach posterior end of body ....
.... (fig. 103) (*Pristotrema*) *Skrjabinopsolus*
3. Esophagus long; ceca do not reach posterior end of body ....
.... (fig. 135) *Cestrahelmins*

Genus **Pristicola** Cable, 1952
(Fig. 101)

Deropristiidae: Body subcylindrical with blunt ends, spined. Oral sucker terminal, large, with double, ventrally interrupted crown of spines. Prepharynx distinct; pharynx rounded; esophagus short; ceca terminate at posterior. Ventral sucker small, near anterior. Testes tandem, near posterior. Cirrus pouch behind ventral sucker, enclosing bipartite seminal vesicle, prostatic complex, and cirrus. Hermaphroditic duct runs over ventral sucker and opens in front of it. Ovary pretesticular, submedian. Uterine coils reach posterior; metraterm well developed, behind ventral sucker; eggs numerous. Vitellaria extend in lateral and partly dorsal field from seminal vesicle to just posttesticular. In intestine of sturgeons.

Pristicola sturionis (Little, 1930) Cable, 1952 (Syn. *Dihemistephanus s.* Little, 1930): in *Acipenser sturio*, Ind.

Genus **Deropristis** Odhner, 1902
(Fig. 102)

Deropristiidae: Body elongate, swollen at neck, covered with slender spines. Oral sucker subterminal; prepharynx distinct; esophagus short; ceca terminate at posterior. Ventral sucker small, behind swollen neck. Testes close tandem, near posterior. Cirrus pouch long, extending far back of ventral sucker, enclosing bipartite seminal vesicle, prostatic complex, and spined cirrus. Hermaphroditic duct present. Genital pore immediately in front of ventral sucker. Ovary median or submedian, post-equatorial, separated from anterior testis by uterus. Seminal receptacle large. Laurer's canal present. Uterine coils descend as far as anterior testis and then ascend; metraterm distinct, spined, behind ventral sucker; eggs small, numerous. Vitelline follicles small, extending along each side of body between metraterm and anterior testis. Excretory bladder Y-shaped with short stem and long arms. In intestine of fishes.

Deropristis inflata (Molin, 1859) Odhner, 1902: in *Anguilla, Ictalurus, Rana* (frog); Cable and Hunninen (1942): trichocercous cercaria in *Bittium*, encysts in *Nereis* (annelid), adult in *Anguilla rostrata*.

Genus **Skrjabinopsolus** Ivanov, 1935
(Fig. 103)
(Syn.: *Pristotrema* Cable, 1955)

Deropristiidae: Body elongate, subcylindrical, spined, without neck swelling. Oral sucker subterminal; pharynx large; esophagus short; ceca terminate at posterior. Ventral sucker small, about one-third from anterior end. Testes tandem to oblique, in posterior of body. Cirrus pouch large, claviform, extending posteriad of ventral sucker, enclosing bipartite seminal vesicle, pars prostatica, and eversible spined cirrus. Genital pore immediately in front of ventral sucker. Ovary submedian, post-equatorial, separated from anterior testis by uterine coils. Seminal receptacle present. Uterine coils extend back as far as posterior testis; metraterm well developed, spined; eggs medium, numerous. Vitelline follicles extend along each side of body between cirrus pouch and anterior testis. Excretory bladder Y-shaped, stem reaching to posterior testis. In intestine of sturgeons.

Skrjabinopsolus manteri (Cable, 1952) Cable, 1955: in *Acipenser rubicundus, Scaphirhynchus platyorhynchus*, Ind.; Peters (1960): cercaria.

Genus **Cestrahelmins** Fischthal, 1957
(Fig. 135)

Body elongate, spinose, swollen at neck. Oral sucker subterminal; prepharynx extremely short, almost nonexistent; esophagus very long; ceca short, extending half the length of hindbody. Acetabulum pre-equatorial. Testes 2, oblique, post-ovarian in hindbody. Cirrus pouch large, extending far back of acetabulum; contains large bipartite seminal vesicle, prostate glands, and spined, eversible cirrus. Genital atrium small, opening on left, close to anterolateral margin of acetabulum. Ovary on right, pre-testicular. Seminal receptacle large. Laurer's canal present. Vitelline follicles post-acetabular and pre-testicular, in lateral fields meeting dorsally. Uterine coils occupy post-testicular and testicular regions of hindbody; metraterm large, saccular, covered inside with acicular spines. Excretory bladder saccular. Parasite of freshwater fish.

Cestrahelmins laruei Fischthal, 1957: in intestine, *Esox masquinongy*, Wis.

Family SYNCOELIIDAE Odhner, 1927
Genus **Syncoelium** Looss, 1899
(Fig. 104)

Body smooth or papillate, moderately large; forebody narrow; hindbody cylindrical or flattened elliptical. Oral sucker subterminal, followed by prominent pharynx. Esophagus short; ceca simple, united near posterior extremity. Acetabulum with stalk of varying length, toward midbody (usually pre-equatorial). Testes follicular, numerous, mostly in two longitudinal rows between base of acetabular peduncle and ovary. Seminal vesicle tubular, long, more or less winding. Prostatic complex may or may not be distinctly developed just behind intestinal bifurcation. Hermaphroditic duct posterior to oral sucker. Ovary divided into several rounded or somewhat elongate lobes immediately behind testes. Vitellaria of several small compact lobes massed together between ovary and posterior cecal arch. Uterus descends to posterior end of body, then ascends in close transverse coils, overreaching ceca laterally, median and almost straight in forebody. Excretory vesicle short; arms united over pharynx. In branchial cavity of marine fishes. Life cycle unknown.
Ref. Yamaguti (1953, p. 213; 1958, p. 313).

Syncoelium filiferum (Sars, 1885). Lloyd and Guberlet (1936): in branchial cavity, *Oncorhynchus gorbuscha, O. nerka,* Pacific coast.

Family GORGODERIDAE Looss, 1901
Dawes (1946, p. 93); Yamaguti (1953, p. 55; 1958, p. 378).

Genus **Phyllodistomum** Braun, 1899
(Fig. 105)
(Syn.: *Catoptroides* Odhner, 1902)

Forebody tapered anteriorly; hindbody foliate, with more or less crenulated margin. Oral sucker terminal, mouth ventroterminal. No pharynx.

Ceca somewhat sinuous, terminating near posterior extremity. Ventral sucker pre-equatorial, small to medium-sized. Testes intercecal, diagonal or symmetrical, in broadest part of hindbody. Seminal vesicle saccular. Genital pore post-bifurcal. Ovary submedian pre-testicular. No seminal receptacle. Vitellaria compact or lobed, paired behind ventral sucker. Uterus occupies most of hindbody intruding into extracecal fields; eggs embryonated. Excretory bladder tubular. In urinary bladder of freshwater and marine fishes and amphibia. Life cycle: macrocercous cercaria in *Musculium*; encyst in sporocysts or in arthropods.

Ref. Dawes (1946, p. 93); Jaiswal (1957), review; Thomas (1956), life cycle; Yamaguti (1953, p. 55; 1958, p. 85).

Phyllodistomum americanum Osborn, 1903: in salamanders, also *Ictalurus nebulosus, Esox lucius, Perca flavescens.* This record in fish may be an error.

P. brevicecum Steen, 1938: in *Cottus bairdi, Umbra limi,* Ind.; Fischthal (1947, 1950), Wis.; Peckham and Dineen (1957): Ind.

P. carolini Holl, 1929: in *Ictalurus natalis,* N. C.; Harms (1959): *I. melas,* Kans.; Wu (1938): syn. of *P. staffordi.*

P. coregoni Dechtiar, 1966: in *Coregonus clupeaformis,* Ont.

P. caudatum Steelman, 1938: in *Ictalurus melas,* Okla.; Beilfuss (1954): macrocercous cercariae in *Musculium,* may encyst in sporocyst, adult in *I. melas*; Harms (1959): *I. melas, I. natalis,* Kans.

P. etheostomae Fischthal, 1943: in *Etheostoma blennioides, E. maculatum, Percina caprodes,* Mich.; Anthony (1963): *P. caprodes,* Wis.; Fischthal (1947, 1950): *Catostomus commersoni, E. flabellare, E. nigrum, Cottus bairdi, P. caprodes,* Wis.; Fischthal (1956): *E. nigrum,* N. Y.; Hoffman (1952): N. D.

P. fausti Pearse, 1924: in *Aplodinotus grunniens,* Wis.; Simer (1929): Miss.

P. folium (Olfers, 1816) Braun, 1899: Europe; Stafford (1904): in *Esox lucius,* Can.

P. hunteri (Arnold, 1934): in *Ictalurus nebulosus,* N. Y.; Bangham (1942): Tenn.

P. lachancei Choquette, 1947: in *Salvelinus fontinalis,* Que.; Fischthal (1947): *Salmo gairdneri,* Wis.

P. lacustri (Loewen, 1929): in *Ictaluris lacustris*; Minn.; Bangham (1942): *I. anguilla, I. punctatus, I. melas, I. natalis, Noturus olivaris,* Tenn.; Coil (1954): growth and variation in *I. lacustris*; Harms (1959): *I. punctatus,* Kans.

P. lohrenzi (Loewen, 1935): in *Lepomis cyanellus,* Kans.; Venard (1940): Tenn.; Bangham (1942): *L. macrochirus, L. microlophus, Micropterus salmoides,* Tenn.; Sparks (1951): Tex.; Beilfuss (1954): macrocercous cercaria in *Musculium,* metacercaria in sporocysts or free cercaria eaten by caddis fly nymph, then to final host; Byrd *et al.* (1940): flame cell formula of cercaria; Steelman (1939): *Cercaria coelocerca.*

P. lysteri Miller, 1940: in *Catostomus commersoni,* Can.; Fischthal (1950): redescription, Wis.

P. nocomis Fischthal, 1942: in *Nocomis biguttatus,* Mich.; Fischthal (1947): Wis.

P. notropidus Fischthal, 1942: in *Notropis cornutus,* Mich.; Fischthal (1947, 1950): *N. cornutus,* Wis.

P. pearsii Holl, 1929: in *Enneacanthus glorius.* Also reported from *Micropterus salmoides, Aphredoderus sayanus, Centrarchus macropterus, Lepomis gibbosus,* by Bangham (1942): Tenn., Fischthal (1947): Wis.; Van Cleave and Mueller (1934): N. Y.

P. semotili Fischthal, 1942: in *Semotilus atromaculatus,* Mich.

P. staffordi Pearse, 1924: in *Ictalurus nebulosus,* Wis. Also reported from *I. melas, I. natalis, Esox masquinongy,* by Anthony (1963): Wis.; Bangham (1941, 1942, 1944, 1955): Ohio, Tenn., Wis.; Bangham and Adams (1954): Brit. Col.; DeRoth (1953): Me.; Fischthal (1947, 1950): Wis.; Fischthal (1956): *Noturus insignis,* N. Y.; Haderlie (1953): Calif.; Hugghins (1959): *I. melas,* S. Dak.; Lyster (1939): *Esox masquinongy,* Can.; Meyer (1954): Me.; Sindermann (1953): Mass.; Van Cleave and Mueller (1934): N. Y.

P. superbum Stafford, 1904: Fantham and Porter (1947): in *Salvelinus fontinalis, Esox lucius, E. niger, Catostomus commersoni, Micropterus dolomieui,* Que.; Lyster (1939): illus., *E. lucius;* Pearse (1924): *Perca flavescens,* Wis.; Sindermann (1953): *P. flavescens, Lepomis gibbosus,* Mass.; Sogandares-Bernal (1955): brackish water, *Micropogon undulatus;* Van Cleave and Mueller (1934): *P. flavescens, Stizostedion vitreum,* N. Y.

P. undulans Steen, 1938: in *Cottus bairdi, Umbra limi,* Ind.; Fischthal (1947): Wis.

Phyllodistomum sp. Bangham (1942): Tenn.; Bangham (1944): in *Ambloplites rupestris, Salmo gairdneri;* Fischthal (1947): *Esox lucius, Lepomis gibbosus, Salmo gairdneri;* Fischthal (1950): *E. masquinongy;* Fischthal (1956): *Catostomus commersoni,* N. Y.; Meyer (1958): hepatic bile duct, *C. commersoni.*

Family AZYGIIDAE Odhner, 1911

Key characteristics: Usually elongate worms; ventral sucker in anterior of body; genital sucker and circumoral ring absent.

Van Cleave and Mueller (1934, p. 254); Yamaguti (1953, p. 127; 1958, p. 193).

KEY

TO THE GENERA OF AZYGIIDAE

1. Ovary post-testicular; thick worm . . . . . . (fig. 106) *Leuceruthrus*
1. Ovary pre-testicular 2

2. Uterus and vitellaria extend into forebody; excretory arms
   united anteriorly ....(fig. 107) *Proterometra*
2. Uterus and vitellaria do not extend into forebody, excretory
   arms not united anteriorly; usually slender worm........
   ....(fig. 108) *Azygia*

Genus **Leuceruthrus** Marshall and Gilbert, 1905
(Fig. 106)

Azygiidae: Body tongue-shaped, smooth. Oral sucker subterminal, large, surmounted by pre-oral lobe. Pharynx well developed; esophagus very short; ceca terminate near posterior. Ventral sucker smaller than oral sucker, pre-equatorial; testes diagonal, post-acetabular. Copulatory apparatus present. Genital pore a little in front of ventral sucker. Ovary median, about midway between testes and posterior. Vitellaria occupy greater part of lateral fields posterior to ventral sucker. Uterus closely coiled between 2 testes. Excretory bladder reaches ovary. In stomach of freshwater fishes. Life cycle unknown.

Ref. Yamaguti (1953, p. 131; 1958, p. 197).

Leuceruthrus micropteri Marshall and Gilbert, 1905: in stomach, *Micropterus salmoides*, Wis.; Bangham (1939): in *Amia calva, M. dolomieui, M. salmoides*, L. Erie; Pearse (1924): in *M. dolomieui, Roccus chrysops*, mouth and stomach, Wis.; Hoffman (1957): in *A. calva*, Minn.

Genus **Proterometra** Horsefall, 1933
(Fig. 107)

Azygiidae: Body plump, rather elliptical. Oral sucker subterminal, very large; pharynx small; esophagus very small; ceca terminate at posterior. Ventral sucker much smaller than oral sucker, post-equatorial. Testes nearly side by side, symmetrical, close to posterior. Cirrus pouch rounded, pre-acetabular; contains winding seminal vesicle and prostatic complex. Genital pore pre-acetabular. Ovary median, between and in front of testes. Uterus forms a loop between ovary and ventral sucker and another between the 2 suckers. Vitellaria fill lateral fields between oral sucker and near posterior. Excretory bladder Y-shaped; arms united anteriorly over oral sucker. In intestine and esophagus of freshwater fishes. Life cycle: cysticercous cercaria in snails, no second intermediate host.

Ref. Van Cleave and Mueller (1934, p. 350); Yamaguti (1953, p. 130; 1958, p. 197).

Proterometra catenaria Smith, 1935: in *Lepomis cyanellus*, Fla.

P. *dickermani* Anderson, 1962: *L. gibbosus*, *L. macrochirus*, Mich.; Anderson and Anderson (1963): cercaria, metacercaria, and progenetic adult in *Goniobasis*.

P. *hodgesiana* Smith, 1936: in *L. cyanellus*; cysticercous cercaria in *Goniobasis*, Fla.

P. *macrostoma* Horsefall, 1933: in esophagus, *L. humilis*, cysticercous cercaria in *Goniobasis* and *Pleurocera* (Syn. *Cercaria melanophora* Smith, 1932; *Cercaria fusca* Pratt, 1919); Bangham (1955): *L. gibbosus*, *Ambloplites rupestris*, L. Huron; Dickerman (1934, 1936, 1945): *Pomoxis annularis, L. macrochirus, L. gibbosus*; Hussey (1943): miracidium.

P. *sagittaria* Dickerman, 1946: in esophagus, *L. gibbosus*, experimental, cysticercous cercaria in *Goniobasis* and *Pleurocera*.

Genus **Azygia** Looss, 1899
(Fig. 108)
(Syn.: *Megadistomum* Stafford, 1904; *Mimodistomum* Stafford, 1904; *Hassallius* Goldberger, 1911)

Azygiidae: Body medium to large, much elongated, strongly muscular. Oral sucker subterminal, medium-sized. Pharynx long or globular, esophagus very short; ceca terminate near posterior. Ventral sucker medium-sized, in anterior of middle third of body. Testes tandem or diagonal, usually in posterior third of body. Cirrus pouch subglobular to piriform, immediately in front of ventral sucker or overlapping it. Genital pore pre-acetabular. Ovary immediately pre-testicular. Uterus closely coiled in intercecal field between ovary and ventral sucker; numerous eggs. Vitellaria in lateral fields posterior to ventral sucker. Excretory bladder bifurcating behind gonads; arms long but not united anteriorly. In stomach or intestine of freshwater and marine fishes. Life cycle: cysticercous cercaria in snails, eaten by host fish; small fish may act as "carriers."

Ref. Petrushevskaya (1962), taxonomy; Stunkard (1956), review; Van Cleave and Mueller (1934); Wootton (1957), key to spp.; Yamaguti (1953, p. 128; 1958, p. 194).

Azygia acuminata Goldberger, 1911: in stomach, *Amia calva*, Ind.; Britt (1947): chromosomes; Cooper (1915): *A. calva*; Manter (1926): Iowa, Ill.; Pearse (1924): *A calva, Roccus chrysops, Stizostedion vitreum*, Wis.; Sillman (1962): Mich.; Van Cleave and Mueller (1934): as syn. of *A. longa*; Wootton (1957): *Ictalurus*

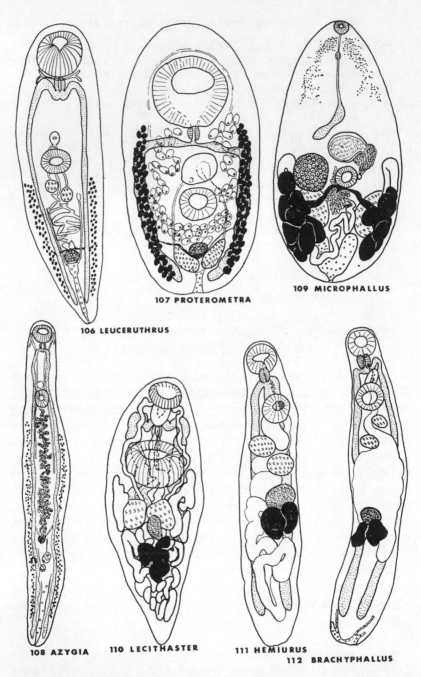

106 LEUCERUTHRUS

107 PROTEROMETRA

109 MICROPHALLUS

108 AZYGIA

110 LECITHASTER

111 HEMIURUS

112 BRACHYPHALLUS

Figs. 106–112. Fig. 106. *Leuceruthrus micropteri* (from Marshall and Gilbert, 1905, in Yamaguti, 1953). Fig. 107. *Proterometra sagittaria* (from Dickerman, 1946). Fig. 108. *Azygia acuminata* (from Sillman, 1962).

nebulosus, Lepomis macrochirus, Esox niger, experimentally in *Perca flavescens,* paratenic in *Eucalia, Anguilla,* snail host *Campeloma,* Mass.

A. angusticauda (Stafford, 1904) Manter, 1926 (Syn. *Mimodistomum a.* Stafford; *Azygia loossi* Marshall and Gilbert, 1905). Reported from *Ambloplites, Amia, Esox, Ictalurus, Lepomis, Lota, Micropterus, Perca, Pomoxis, Roccus, Salvelinus, Stizostedion,* by Anthony (1963): Wis.; Bangham (1941, 1944, 1955): Wis., L. Huron; Bangham and Hunter (1939): Ohio; Bangham and Venard (1946); Bhalerao (1943): Bengal; Choquette (1951): Que.; Cross (1938): Wis.; DeRoth (1953): Me.; Fantham and Porter (1947): Que.; Fischthal (1947, 1950, 1952): Wis.; Foote and Blake (1945); Hunninen (1936); Hunter (1942): Conn.; Hunter and Rankin (1940): Conn.; Lyster (1939); MacLulich (1943) Ont.; Meyer (1953): Me.; Miller (1940, 1941): Ont.; Odlaug *et al.* (1962): Minn.; Stafford (1904); Szidat (1932); Van Cleave and Mueller (1934): N. Y.; Wooton (1957): Mass. Life cycle unknown.

A. bulbosa Goldberger, 1911 (Syn. of *A. longa* Manter, 1926).

A. longa (Leidy, 1851) Manter, 1926. Reported in stomach and ceca of *Ambloplites, Amia, Anguilla, Esox, Etheostoma, Ictalurus, Lepomis, Lota, Lucioperca* (Russia), *Micropterus, Osmerus, Perca, Percina, Salmo, Salvelinus, Trichiurus,* by Bangham (1933): Ohio; Bangham and Venard (1946): Ont.; Choquette (1951); DeRoth (1953): Me.; Dickerman (1934); Fantham and Porter (1947): Que.; Hunninen (1936); Linton (1940): Mass.; Lyster (1939): Que.; Miller (1941): review; Nordlie (1960); Sillman (1962): life cycle, cysticercous cercaria in *Amnicola,* Mich.; Sindermann (1953): Mass.; Stafford (1904); Van Cleave and Mueller (1934): N. Y., cercaria eaten by host fish or "carrier" fish; (Sillman, 1962; Stunkard, 1955).

A. loossii Marshall and Gilbert, 1905 (Syn. of *A. angusticauda; A. micropteri* (MacCallum, 1921). Syn. *Eurostomum m. M.* is probably a syn. of *A. longa* (Velasquez, 1958).

A. sebago Ward, 1910: in *Salmo sebago, Perca flavescens, Anguilla rostrata, Esox reticulatus,* Me.; Heitz (1917): *A. chrysopa, Osmerus mordax, Esox reticulatus, S. sebago,* Me., and *E. lucius, Lota maculosa, Ictalurus nigricans,* Can.; Hunter (1930); Manter (1926): as syn. of *A. longa;* Stunkard (1955): life cycle, *Amnicola, A. rostrata,* guppies, *Lepomis macrochirus,* Mass.; Van Cleave and Mueller (1934); Wootton (1957): life cycle.

A. tereticollis—Distomum tereticolle Leidy, 1851. Syn. of *A. longa* (Yamaguti, 1958).

Fig. 109. *Microphallus ovatus* (from Van Cleave and Mueller, 1934). Fig. 110. *Lecithaster confusus* (from Odhner, 1905, in Yamaguti, 1953). Fig. 111. *Hemiurus levinseni* (from Miller, 1941). Fig. 112. *Brachyphallus crenatus* (from Miller, 1941).

Family MICROPHALLIDAE
(See below)

Genus **Microphallus** Ward, 1901
(Fig. 109)

Body piriform, unspined. Oral sucker subterminal, small; prepharynx distinct, pharynx small, esophagus variable; ceca very short. Ventral sucker small, in middle third of body. Testes symmetrical, posterior to ventral sucker. Seminal vesicle in front of ventral sucker. No cirrus sac. Genital pore to left of ventral sucker. Ovary on right of ventral sucker. Vitellaria lobed, symmetrical, position variable in lateral fields of posterior half of worm. Eggs small, numerous. Excretory bladder Y-shaped. In stomach and intestine of freshwater fishes. Life cycle: metacercaria in crayfish.

Ref. Baer (1944, p. 70), key; Yamaguti (1958, p. 242).

Microphallus medius Van Cleave and Mueller, 1932: in *Perca flavescens*, N. Y. Yamaguti (1953) states that this is *Maritreminoides*, a bird trematode, accidentally liberated in fish.

M. obstipes (Van Cleave and Mueller, 1932). Removed to *Maritrema obstipum* by Mueller (1934). In *Ambloplites rupestris*, N. Y. Yamaguti (1953) states that this is *Maritreminoides*, a bird trematode, accidentally liberated in fish.

M. opacus (Ward, 1894) Ward, 1901 (Syn. according to Yamaguti (1953): *M. opacus ovatus* Strandine, 1943). Recorded from *Ambloplites, Amia, Anguilla, Esox, Ictalurus, Labidesthes, Micropterus, Perca*, by Bangham (1939, 1955): Mich., L. Huron; Dolley (1933): Ind.; Fantham and Porter (1947): Que.; Hare (1943): Ohio.

Experimentally in *Plethodon, Amblystoma, Rana, Crysemys, Emys, Thamnophis, Natrix, Didelphys, Procyon* (Rausch, 1947) and mice (Sogandares, 1965).

M. ovatus Osborn, 1919. [Syn. of *M. opacus* according to Rausch (1947)]. Reported in stomach of *Micropterus dolomieui*, intestine of *Ambloplites rupestris, Ictalurus nebulosus, Esox niger*, by Bangham (1941): Ohio; Van Cleave and Mueller (1934): N. Y. Metacercaria in crayfish.

Family HEMIURIDAE Lühe, 1901

Small to medium distomes, some with tail-like portion. Suckers and pharynx well developed; ventral sucker usually prominent. Esophagus short. Ceca of some united posteriorly. Testes tandem, diagonal or symmetrical, usually in hindbody. Seminal vesicle free in parenchyma. Cirrus pouch may be present. Genital pore median, usually near oral sucker, pharynx, or intestinal bifurcation. Ovary usually post-testicular. Seminal canal and Laurer's canal usually present. Vitellaria compact, lobed or tubular, usually post-ovarian. Uterus descends, then ascends, sometimes ascending only. Eggs numerous, usually nonfilamented. Excretory bladder Y-shaped; arms united anteriorly. In esophagus and stomach, rarely intestine or gall bladder or outside intestinal tract of fishes, rarely of frogs.

Ref. Yamaguti (1953, p. 178; 1958, p. 260).

KEY

TO THE GENERA OF HEMIURIDAE

1. Excretory arms united anteriorly 2
1. Excretory arms not united anteriorly . . . .
. . . . (fig. 110) (*Dichadena*) *Lecithaster*
2. Short tail present 3
2. No tail 4
3. Pars prostatica very long . . . . (fig. 111) *Hemiurus*
3. Pars prostatica short; cirrus lacking . . . . (fig. 112) *Brachyphallus*
4. Vitellaria double, compact or lobed 5
4. Vitellaria divided into 7 rounded lobes . . . . . (fig. 113) *Aponurus*
5. Ventral sucker slightly pre-equatorial, equatorial, or post-
   equatorial 6
5. Ventral sucker near anterior extremity; hermaphroditic duct
   enclosed in sac . . . . (fig. 114) *Genolinea*
6. Genital cone or papilla present, enclosed in sac 7
6. Genital cone or papilla and hermaphroditic duct absent . . . .
   . . . . (fig. 116) *Halipegus*
7. Ventral sucker pre-equatorial; in freshwater hosts . . . . *Deropegus*
7. Ventral sucker usually equatorial; in marine fish . . . .
   . . . . (fig. 115) *Derogenes*

Genus **Lecithaster** Lühe, 1901
(Fig. 110)
(Syn.: *Dichadena* Linton, 1910)

Hemiuridae: Body small or fusiform, smooth, without tail. Oral sucker subterminal; pharynx well developed, esophagus short, ceca not quite reaching posterior. Ventral sucker larger than oral sucker, about one-third of body length from anterior. Testes nearly symmetrical, behind ventral sucker. Seminal vesicle saccular, dorsal or posterodorsal to ventral sucker. Hermaphroditic duct enclosed in oval to elliptical pouch. Genital pore ventral to pharynx or intestinal bifurcation. Ovary 4- or 5-lobed, usually in posterior half of body. Seminal receptacle often conspicuous. Vitellaria of 7 rosette-shaped or claviform lobes. Uterus may reach posterior; eggs small. Excretory arms not united anteriorly but reach to pharynx. In intestine of marine fishes.

Lecithaster gibbosus (Rud., 1802) Lühe, 1901. Nicoll (1907): North American hosts.
L. salmonis Yamaguti, 1934. Bangham and Adams (1954): *Oncorhynchus nerka, Salvelinus alpinus*, Brit. Col.; Shaw (1947): *Oncorhynchus tschawytscha*, Ore.

Genus **Hemiurus** Rudolphi, 1809
(Fig. 111)
(Syn. *Apoblema* (Dujardin, 1845); *Eurycoelum* Brock, 1886, after Chaudoir, 1848)

Hemiuridae: Body small to medium, with "tail." Cuticular denticulations present. Oral sucker nearly terminal. Pharynx normal. Ceca extending into "tail." Ventral sucker variable in size, near anterior. Testes diagonal, short distance behind ventral sucker. Seminal vesicle pre-testicular, constricted into 2 portions, anterior portion with muscular coat. Pars prostatica long, winding. Hermaphroditic duct slender. Genital pore ventral to oral sucker or pharynx. Ovary post-testicular. Vitellaria immediately post-ovarian, with 2 compact lobes. Uterus winding into "tail" or not; eggs small, numerous. Excretory arms united anteriorly. In stomach of marine fishes.

Hemiurus appendiculatus (Rud., 1802) Looss, 1899. Reported from *Salmo salar, Osmerus mordax, Anguilla anguilla*, Can.
H. levinseni Odhner, 1905. Margolis (1956): *Oncorhynchus gorbuscha*, Brit. Col.

Genus **Brachyphallus** Odhner, 1905
(Fig. 112)

Hemiuridae: Body elongated with "tail," often with ventral pit between ventral sucker and genital pore. Oral sucker subterminal, large; pharynx globular; esophagus short; ceca reach to near tail end. Ventral sucker nearly as large as oral sucker, near anterior. Testes diagonal, behind ventral sucker. Seminal vesicle constricted into 2 parts, anterior or anterodorsal to ventral sucker. Pars prostatica short. Hermaphroditic duct short; genital pore ventral to pharynx or esophagus. Ovary in middle third of body. Vitellaria of 2 indented or lobed masses. Utcrus not extending into "tail." Excretory arms united dorsal to pharynx. In freshwater and marine fishes.

Brachyphallus crenatus (Rudolphi, 1802) Odhner, 1905. Recorded from *Osmerus mordax, Oncorhynchus tschawytscha, Salmo salar* (see Miller, M. J., 1941, review), by Bangham and Adams (1954): *Salvelinus alpinus*, Brit. Col.; Shaw (1947): *O. tschawytscha*; Ore.

Genus **Aponurus** Looss, 1907
(Fig. 113)

Hemiuridae: Body bulb- or spindle-shaped, no "tail." Oral sucker subterminal, may or may not be surmounted by pre-oral lobe. Pharynx round, esophagus short; ceca terminate near posterior. Ventral sucker larger than oral sucker, one-third of body length or more from anterior. Testes diagonal, in middle third of body. Seminal vesicle globular to elongate, anterior to ventral sucker or partly overlapping it. Pars prostatica tubular, enclosed in elongated, well-developed hermaphroditic pouch. Genital pore ventral to oral sucker, pharynx, or intestinal bifurcation. Ovary median or submedian, post-testicular. Vitellaria of 7 rounded lobes, post-ovarian. Seminal receptacle sometimes large. Uterus usually reaching posterior. Excretory arms united dorsal to pharynx. In stomach of marine fishes.

Aponurus sp. Shaw (1947): in *Salvelinus malma, Salmo gairdneri,* Ore.

Genus **Genolinea** Manter, 1925
(Fig. 114)
(Syn.: *Parasterrhurus* Manter, 1934)

Hemiuridae: Small to medium, nearly cylindrical, no "taii." Oral sucker subterminal, followed by globular pharynx. Esophagus short; ceca wide with sinuous wall, terminating near posterior. Ventral sucker larger than oral sucker, about one-third body length from anterior. Testes obliquely tandem, posterior to ventral sucker. Seminal vesicle tubular, winding, anterior, anterodorsal or dorsal to ventral sucker. Pars prostatica short. Hermaphroditic duct short or long, winding, enclosed in hermaphroditic pouch. Genital atrium opening at level of intestinal bifurcation. Ovary in posterior half of body. Seminal re-

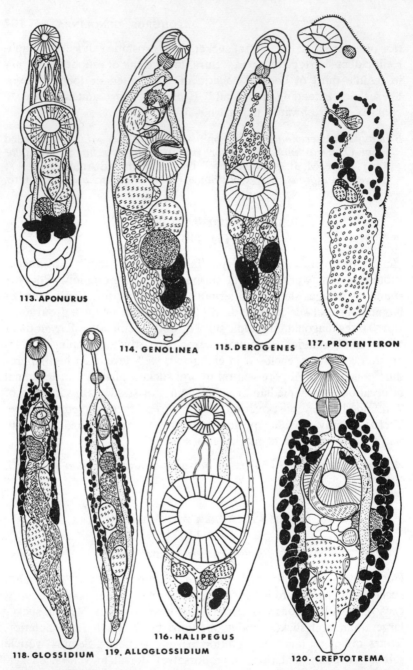

113. APONURUS

114. GENOLINEA   115. DEROGENES   117. PROTENTERON

116. HALIPEGUS

118. GLOSSIDIUM   119. ALLOGLOSSIDIUM

120. CREPTOTREMA

Figs. 113–120. Fig. 113. *Aponurus laguncula* (from Looss, 1908, in Yamaguti, 1953). Fig. 114. *Genolinea oncorhynchi* (from Margolis and Adams, 1956). Fig. 115. *Derogenes varicus* (from Miller, 1941). Fig. 116. *Halipegus perplexus* (from Simer, 1929). Fig. 117. *Protenteron diaphanus*

ceptacle often large. Vitellaria of 2 compact lobes, obliquely tandem, post-ovarian. Uterus extends beyond vitellaria to near posterior. Excretory arms united dorsal to pharynx. In stomach of marine fishes.

Genolinea oncorhynchi Margolis and Adams, 1956: in *Oncorhynchus gorbuscha*; also list of spp. of *Genolinea*, Brit. Col.

Genus **Derogenes** Lühe, 1900
(Fig. 115)

Hemiuridae: Body small, no "tail." Oral sucker subterminal with pre-oral lobe. Pharynx globular. Esophagus short. Ceca reach to posterior. Ventral sucker larger than oral sucker, toward midbody. Testes nearly symmetrical, posterior to ventral sucker. Seminal vesicle simple, saccular, anterior to ventral sucker. Pars prostatica usually long. Hermaphroditic duct enclosed in pouch projecting into genital atrium in conical papilla. Genital pore near pharynx. Ovary post-testicular. Seminal receptacle present. No Laurer's canal. Vitellaria of 2 symmetrical compact masses, immediately post-ovarian. Uterus reaches to posterior. Excretory arms united dorsal to pharynx. In esophagus, stomach, and gall bladder of marine fishes.

Derogenes varicus (Mueller, 1784) Looss, 1901. Miller (1941): in *Salmo salar, Osmerus mordax* and also review.
Derogenes sp. Haderlie (1953): in *O. kisutch*, Calif.; Shaw (1947): in *Salmo clarki*, Ore. These are probably *Deropegus aspina* according to McCauley and Pratt (1961).

Genus **Deropegus** McCauley and Pratt, 1961
(No fig.)
(Syn.: *Halipegus aspina* Ingles, 1936; *Derogenes* sp.
of Haderlie, 1953)

Hemiuridae: Body elongate lanceolate, often slightly constricted in region of ventral sucker. Oral sucker subterminal; ventral sucker in anterior half of body. Prepharynx and esophagus absent or very short. Intestinal ceca inflated; turn anteriorly from pharynx before passing to posterior end of body. Testes spherical, posterior to ventral sucker and separated from it by coils of uterus; lie obliquely with left testis

---

(from Miller, 1941). Fig. 118. *Glossidium corti* (from Van Cleave and Mueller, 1934). Fig. 119. *Alloglossidium geminus* (from Van Cleave and Mueller, 1934). Fig. 120. *Creptotrema funduli* (from Mueller, 1934).

usually somewhat anterior to right. Genital pore immediately post-bifurcal, submedian. Seminal vesicle short and broad; prostate gland subspherical; hermaphroditic duct anterior to prostate gland, opens into genital atrium through conical papilla. Ovary submedian behind testes and separated from them by loops of uterus. Vitellaria of 2 compact, unlobed masses behind ovary. Laurer's canal present, inflated proximally to form seminal receptacle. Seminal receptacle-uterus contains few eggs, present in descending loop of uterus; egg-filled uterine loops lie transversely between ovary and ventral sucker; metraterm weakly developed, unites with male duct to form hermaphroditic duct. Excretory pore terminal; excretory vesicle Y-shaped, arms join dorsal to oral sucker. Parasites of freshwater fishes and amphibians.

Deropegus aspina (Ingles) McCauley and Pratt, 1961: in stomach of *Salmo clarki, S. gairdneri, Oncorhynchus kisutch, O. tshawytscha,* and frogs, Ore.

Genus **Halipegus** Looss, 1899
(Fig. 116)
(Syn.: *Genarchella* Travassos, Artigas and Pereira, 1928)

Hemiuridae: Body muscular. Oral sucker subterminal, surmounted by pre-oral lobe. Pharynx muscular. Esophagus short. Ceca reach to posterior. Ventral sucker well developed, in middle third of body. Testes nearly symmetrical, behind ventral sucker. No cirrus pouch or hermaphroditic pouch. Seminal vesicle and prostatic complex free in parenchyma. Genital pore ventral or posterior to pharynx. Ovary median, submedian, or lateral, near posterior. Seminal receptacle formed by dilatation of Laurer's canal. Vitellaria of 2 compact lobes or 2 groups of 4 or 5 lobes each, post-ovarian. Uterus winding mostly in intercecal field; eggs elongate, with polar filament. Excretory arms united dorsal to oral sucker or pharynx. In buccal cavity, esophagus, stomach, rarely ears of frogs, occasionally in fishes.

Halipegus perplexus Simer, 1929: in intestine, *Polyodon spathula,* Miss.

Family PLAGIORCHIIDAE Lühe, 1901
(Syn.: *Lepodermatidae* Odhner, 1910)

Ref. Yamaguti (1953, p. 164; 1958, p. 244).

KEY

TO THE GENERA OF PLAGIORCHIIDAE

1. Ceca short, do not reach ventral sucker or scarcely pass it . .
   . . . . (fig. 117) *Protenteron*
1. Ceca terminate at or near posterior extremity   2
2. Vitellaria in ventral sucker testicular zone . . . .
   . . . . (fig. 118) *Glossidium*
2. Vitellaria intruding into forebody . . . . . (fig. 119) *Alloglossidium*

Genus **Protenteron** Stafford, 1904
(Fig. 117)

Plagiorchiidae: Body spined, broadest at middle. Eyespots lateral to pharynx. Oral sucker terminal; prepharynx longer than either pharynx or esophagus; ceca divergent, terminating in region of ventral sucker. Ventral sucker small, about one-third from anterior. Testes obliquely side by side, in midbody. Cirrus pouch reaching to ovary. Genital pore just in front of ventral sucker. Ovary just in front of testes. Uterus reaches to posterior; eggs small. Vitellaria extend in lateral fields from level of intestinal bifurcation to ovary. In intestine of freshwater fishes.

Protenteron diaphanum Stafford, 1904: in *Ambloplites rupestris*, Can.; Miller (1941): transfers to *Cryptogonimus*.

Genus **Glossidium** Looss, 1899
(Fig. 118)

Plagiorchiidae: Body elongate, tapered toward extremities, spined. Oral sucker subterminal; prepharynx distinct, pharynx large, esophagus short; ceca wide, reaching to posterior. Ventral sucker small, in anterior half of body. Testes obliquely tandem, behind ventral sucker, separated from each other by uterus. Cirrus pouch claviform, enclosing bipartite seminal vesicle, prostatic complex, and cirrus. Genital pore submedian, just in front of ventral sucker. Ovary behind ventral sucker, submedian. Seminal receptacle present. Uterus passing between testes and reaching posterior. Vitellaria extend in lateral fields in ovario-testicular zone. Excretory bladder Y-shaped. In intestine of fishes. Life cycle unknown.

Glossidium geminum (Mueller, 1930), (Syn. *Plagiorchis g.* Mueller). Van Cleave and Mueller (1934) transferred this species to *Alloglossidium*. Yamaguti (1953) transferred it to *Glossidium* presumably because the vitellaria do not extend as far forward. Reported in *Ictalurus melas, I. natalis, I. nebulosus,* by Anthony (1963): Wis.; Bangham (1937, 1941, 1944, 1955): Ohio, Wis., L. Huron; Bangham and Venard (1946); Fischthal (1947, 1950, 1952): Wis.; Hunter (1942); Meyer (1954): Me.; Miller (1940); Olsen (1937); Sindermann (1953); Mass.

Genus **Alloglossidium** Looss, 1899
(Fig. 119)

Very similar to *Glossidium* but anterior follicles of vitellaria extend forward to intestinal bifurcation or pharynx.

Ref. Doss *et al.* (1963, p. 35); Olsen (1937), key; Van Cleave and Mueller (1934), key; Yamaguti (1958, p. 245).

Alloglossidium corti (Lamont, 1921), (Syn. *Plagiorchis c.* Lamont; *P. ameiurensis* McCoy, 1928). Recorded from *Ambloplites, Chaenobryttus, Ictalurus, Noturus,* by Allison, R. (1957): chemotherapy; Anthony (1963); Bangham (1939, 1941, 1942, 1944); Bangham and Adams (1954): Brit. Col.; Fischthal (1947, 1950) Wis.; Fischthal (1956): in *Noturus insignis,* N. Y.; Fritts (1959): Idaho; Haderlie (1953): Calif.; Meyer (1958): Iowa; Van Cleave and Mueller (1934): N. Y. See Doss *et al.* (1963) for complete list.

Life cycle: xiphidocercaria in *Helisoma*; metacercaria in dragonfly nymphs (Crawford, 1937; McMullen, 1935).

A. kenti Simer, 1929: in *Ictalurus punctatus,* Tenn.; Van Cleave and Mueller (1934): syn. of *A. corti.*

Family ALLOCREADIIDAE Stossich, 1903

Body small to medium-sized; may be divided into two distinct regions. Accessory suckers sometimes present. Anterior extremity simple or with projection. Oral sucker subterminal or terminal, simple or with sphincter or appendages. Prepharynx, pharynx, and esophagus present. Intestine bifurcate, without anterior limbs; anus or cloaca present or absent. Testes double, rarely single, sometimes 9 or 10, in intercecal field of body. Cirrus pouch usually present, sometimes very poorly developed. Seminal vesicle present or absent. Genital pore usually preacetabular, median or submedian, rarely marginal or dorsal, sometimes near pharynx or oral sucker. Ovary pre-testicular, occasionally inter-testicular, post-acetabular median or submedian. Seminal vesicle and Laurer's canal present or absent. Vitellaria more or less extensive, divided into numerous follicles extending chiefly in lateral fields, usually reaching to posterior extremity. Uterus pre-testicular, sometimes between, ventral, or posterior to testes. Excretory bladder tubular or saccular.

Ref. Yamaguti (1953, p. 63).

KEY

TO THE GENERA OF ALLOCREADIIDAE

1. Oral sucker surmounted by papillae or lobes   2
1. Oral sucker not surmounted by papillae or lobes   5
2. Head papillae 2; testes 2 .... (fig. 120) *Creptotrema*
2. Head papillae 2; testes 4, in 2 pairs ....
   .... (fig. 121) (*Megalogonia*) *Crepidostomum*
2. Head papillae 4 to 6   3
3. Uterus not reaching to posterior extremity; many species;
   sometimes in gall bladder .... (fig. 121) *Crepidostomum*
3. Uterus reaching to posterior extremity   4
4. Vitellaria extend length of body except for two extremities;
   ceca long .... (fig. 122) *Bonodera*
   (p. 147)
4. Vitellaria mostly pre-equatorial; ceca half as long ....
   .... (fig. 123) *Bunoderina*
   (p. 148)
5. Vitellaria extensive   6
5. Vitellaria limited to rather large, sparse follicles between ventral sucker and ovary .... (fig. 124) *Plagiocirrus*
   (p. 149)
6. Uterus reaching to or near posterior extremity; vitellaria in
   bifurcotesticular zone; testes symmetrical ....
   .... (fig. 125) *Vietosoma*
   (p. 149)
6. Uterus not reaching to or near posterior extremity   7
7. Vitellaria in hindbody only; cirrus pouch large, anterior or
   dorsal to ventral sucker, sometimes farther posterior .....
   .... (fig. 130) *Allocreadium*
   (p. 149)
7. Vitellaria intruding into forebody; cirrus pouch short, stout,
   pre-acetabular .... *A. ictaluri* (*Lepidauchen i.*)
   (p. 150)

Genus **Creptotrema** Travassos, Artigas, and Peravia, 1928
(Fig. 120)

Allocreadiidae: Body somewhat fusiform. Oral sucker large, surmounted dorsally by pair of head lobes. Pharynx well developed.
Esophagus short. Ceca terminate near posterior extremity. Ventral
sucker larger than oral sucker, nearer to anterior extremity than to
posterior. Testes diagonal or tandem, toward middle of hindbody. No

external seminal vesicle. Cirrus pouch elongated claviform, overlapping ventral sucker. Genital pore median, near intestinal bifurcation. Ovary on right or left, post-acetabular. Vitellaria of rather large follicles, extending along ceca between pharynx and posterior extremity. Uterus with few coils between testes and ventral suckers; eggs few, large. Excretory bladder tubular, reaching to testicular level. Life cycle unknown.

Ref. Manter (1962); Van Cleave and Mueller (1934, p. 359); Yamaguti (1953, p. 110).

Creptotrema funduli Mueller, 1934 (Syn. *Allocreadium commune* Cooper, 1915): in intestine, *Fundulus diaphanus, Labidesthes sicculis*, N. Y.; Bangham (1937): in *Eucalia inconstans*; Bangham (1941, 1955): *Fundulus chrysotus, Labidesthes sicculus, Umbra limi*, Fla., L. Huron; Manter (1962): this should probably be *Plagioporus*; Meyer, M. (1954): *Fundulus diaphanus*; Me. *Creptotrema* sp. Bangham (1951): metacercaria in *Cottus semiscaber*, Wyo.

Genus **Crepidostomum** Braun, 1900
(Fig. 121)

Allocreadiidae: Body elongated-oval to subcylindrical. Oral sucker terminal, surmounted anterodorsally by a half-crown of 2 to 6 head papillae. Mouth aperture ventroterminal or ventral. Prepharynx present, pharynx well developed. Esophagus short or moderate. Ceca terminate near posterior extremity. Ventral sucker in anterior half of body. Testes tandem (sometimes appear as 4), in posterior half of body. Cirrus pouch more or less elongated claviform; overreaches ventral sucker; contains seminal vesicle, prostatic complex, and ductus ejaculatorius. Genital pore median, pre-acetabular. Ovary submedian, between ventral sucker and anterior testis. Seminal receptacles and Laurer's canal present. Uterus coiled between anterior testes and ventral sucker. Vitellaria circumcecal, extending from forebody to posterior extremity. Excretory bladder reaches to or beyond anterior testis. Life cycle: adult in fish; oculate xiphidiocercaria in sphaerid clams; metacercaria in aquatic insects, usually mayflies, and crustacea.

Ref. Dollfus (1949), review; Doss *et al.* (1964, p. 331); Hopkins (1931), key to spp.; Hopkins (1934), monograph; Lacey (1965), histology; Van Cleave and Mueller (1934, p. 200); Yamaguti (1953, p. 117; 1958, p. 130).

Crepidostomum ambloplitis Hopkins, 1931 (See *C. cooperi*).
C. brevivitellatum Hopkins, 1934: in *Anguilla rostrata* and *Lota maculosa*. Me.; Fantham and Porter (1947): *A. rostrata*, Que.; Lyster (1939): same.

C. canadense Hopkins, 1931: in *Perca flavescens,* Ont.; Hopkins (1934): in *Boleosoma nigrum,* Ont.; Lyster (1940): Syn. of *C. isostomum.*

C. cooperi Hopkins, 1931 (Syn. *C. ambloplitis, C. fausti, C. solidum*). Recorded from *Ambloplites, Catostomus, Chaenobryttus, Coregonus, Cottus, Cyprinus, Esox, Etheostoma, Fundulus, Ictalurus, Lepomis, Micropterus, Notemigonius, Noturus, Perca, Pomoxis, Prosopium, Roccus, Salmo, Salvelinus;* also from *Emyda* and *Chelydra* (turtles) and *Necturus* (salamander).

    References and localities.—Anthony (1963): Wis.; Bangham (1942, 1947, 1955): L. Huron, Tenn., Wis.; Choquette (1948, 1954): Que., life cycle; Fantham and Porter (1947): Que.; Fischthal (1947, 1956): Wis., N. Y.; Hopkins (1934): monograph; Ill., Miss., N. Y., Okla.; Lyster (1939): Can.; Meyer, M. (1954): Me.; Miller (1941): Can.; Shaw (1947): Ore.; Sindermann (1953): Mass.; Van Cleave and Mueller (1934): N. Y.; Venard (1941): Tenn.; Welker (1962): gall bladder of common grackle.

C. cornutum (Osborn, 1903) Stafford, 1904. Recorded from *Ambloplites, Amia, Anguilla, Carassius* (exper.), *Chaenobryttus, Ictalurus, Lepomis, Micropterus, Morone, Notemigonius, Pomoxis, Salmo, Salvelinus;* also salamander and frog.

    References and localities.—Abernathy (1937): experimental; Ameel (1937): life cycle; Anthony (1963): Wis.; Bangham (1939, 1941, 1942, 1944, 1955): N. Y., Ohio, Tenn., Wis.; Cheng (1957): metacercaria; DeRoth (1953): Me.; Fantham and Porter (1947): Que.; Fischthal (1947, 1950): Wis.; Henderson (1938): metacercaria; Hopkins (1934): monograph; Hunter (1930): N. Y.; Hussey (1941): life cycle; Krueger (1954): Ohio; Lyster (1939): Can.; Meyer, M. (1954): Me.; Parker (1941); Pearse (1924): Wis.; Simer (1929): Miss.; Sindermann (1953): Mass.; Sogandares (1965): La.; Venard (1941): Tenn. *Life cycle:* cercaria in *Sphaerium, Musculium;* metacercaria in crayfish.

C. farionis (O. F. Müller, 1784) Nicoll, 1909 (Syn. *C. laureatum*). Recorded from *Coregonus, Cristovomer, Etheostoma, Lepomis, Leucichthys, Lota, Notropis, Oncorhynchus, Perca, Prosopium, Salmo, Salvelinus, Thymallus.*

    References and localities.—Alexander (1960): Ore.; Bangham (1951): Wyo., L. Huron; Bangham and Adams (1954): Brit. Col.; Bergman (1956): Iceland; Choquette (1948): Can.; Crawford (1943): life cycle; Dyk (1958): life cycle, Czech.; Fischthal (1947): Wis.; Fox (1962): Mont.; Fritts (1959): Ill.; Haderlie (1953): Calif.; Hopkins (1934): monograph; Jones and Hammond (1960): Utah; Linton (1940); MacLulich (1943): Can.; Meyer, M. (1954): Me.; Pearse (1924): Wis.; Warren (1952): Minn.; also recorded from Europe.

    Life cycle: cercaria in *Pisidium, Sphaerium;* metacercaria in mayfly nymphs and *Gammarus.*

C. fausti Hunninen and Hunter, 1933 (See *C. cooperi*).

C. hiodontos Hunter and Bangham, 1932 (See *C. illinoiense*).

C. ictaluri (Surber, 1928), (Syn. *Megalogonia i.*). Recorded from *Ictalurus punctatus, I. melas, I. nebulosus, I. natalis, Micropterus salmoides, Noturus gyrinus, N. insignis, N. miuris.*

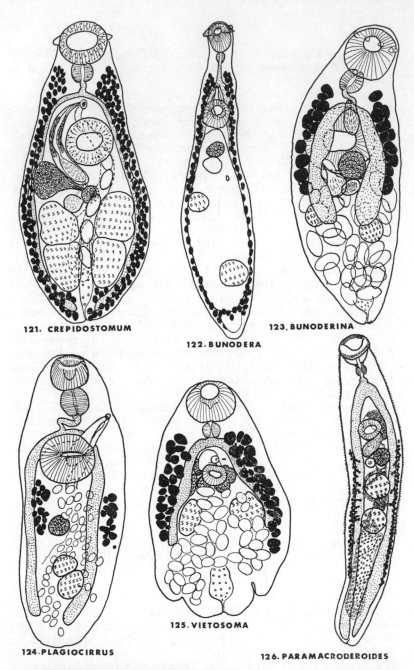

121. CREPIDOSTOMUM

122. BUNODERA

123. BUNODERINA

124. PLAGIOCIRRUS

125. VIETOSOMA

126. PARAMACRODEROIDES

Figs. 121–126. Fig. 121. *Crepidostomum ictaluri* (from Surber, 1928). Fig. 122. *Bunodera luciopercae* (from Hopkins, 1934). Fig. 123. *Bunoderina eucaliae* (from Miller, 1936). Fig. 124. *Plagiocirrus primus* (from Van Cleave and Mueller, 1932). Fig. 125. *Vietosoma parvum* (from Van

References and localities.—Bangham (1942, 1955): Tenn., L. Huron; Fischthal (1956): N. Y.; Harms (1959): Kans.; Hopkins (1934): monograph; Lyster (1939): Can.; Sindermann (1953): Mass.; Van Cleave and Mueller (1934): N. Y.

Life cycle: mollusc host unknown; metacercaria in mayfly nymphs.

C. illinoiense Faust, 1919. (Syn. *C. hiodontos*), Hunter and Bangham, 1932: in intestine, *Etheostoma nigrum, Hiodon tergisus, Pomoxis annularis.*

References and localities—Hopkins (1934): monograph; Peters (1963): miracidium; Self (1954): Okla.

C. isostomum Hopkins, 1931. Recorded from *Aphredoderus sayanus, Cottus asper, Etheostoma nigrum, Percina caprodes, Percopsis omiscomaycus.*

References and localities.—Bangham (1942, 1955): Tenn., L. Huron; Bangham and Adams (1954): Brit. Col.; Fischthal (1947): Wis.; Hopkins (1934): life cycle, monograph; Van Cleave and Mueller (1934): N. Y.

Life cycle: oculate xiphidocercaria in *Sphaerium*, metacercaria in mayflies, but not completed experimentally.

C. laureatum Cooper, 1915 (see *C. farionis*): in *Etheostoma, Lepomis, Perca*; Shaw (1947): *Salmo gairdneri*, Ore.

C. lintoni (Pratt, in Linton, 1901). (Syn. *C. petalosum*). Reported in *Acipenser fulvescens, A. rubicundus, Ambloplites rupestris, Scaphirhynchus platorhynchus.*

References and localities.—Bangham (1939, 1955): L. Erie, L. Huron; Faust (1918); Fischthal (1952): Wis.; Hopkins (1934): monograph; Lyster (1939): Can.

C. petalosum Lander (see *C. lintoni*).

C. solidum Van Cleave and Mueller, 1932 (see *C. cooperi*).

C. transmarinum Nicoll, 1909 (see *C. farionis*).

Crepidostomum sp. Reported from *Amia, Amphiodon, Aphredoderus, Cottus, Cyprinus, Gila, Ictalurus, Lepomis, Lota, Oncorhynchus, Salmo, Salvelinus, Stizostedion, Thymallus*, by Bangham (1939, 1942, 1944, 1951): Fla., Tenn., Wis., Wyo.; Bangham and Adams (1954): Brit. Col.; Fritts (1959): Idaho; Neiland (1952): Ore.; Pearse (1924): Wis.; Shaw (1947): Ore.; Wales (1958): cause of mortality, Calif.

Genus **Bunodera** Railliet, 1896
(Fig. 122)

Allocreadiidae: Body elongate, broadened in uterine region, attenuated at neck, with pair of eye spots, one on each side of pharynx. Oral sucker terminal, transverse row of 4 muscular head lobes along an-

Cleave and Mueller, 1932). Fig. 126. *Paramacroderoides echinus* (from Venard, 1941).

terior dorsal margin. Prepharynx present. Esophagus long; ceca terminate near posterior extremity. Ventral sucker pre-equatorial. Testes diagonal, in posterior part of body. Cirrus pouch plump, overlapping ventral sucker, enclosing bipartite seminal vesicle, prostatic complex, and cirrus. Genital pore median, pre-acetabular. Seminal receptacle and Laurer's canal present. Uterus passing between 2 testes, or overreaching them ventrally. Vitellaria extend in lateral fields of hindbody and intrude into neck region. Excretory bladder saccular or tubular. Life cycle: adult in intestine and ceca of fish; xiphidiocercaria in sphaerid clams; metacercaria in copepods and crayfish.

Ref. Hopkins (1934), monograph; Van Cleave and Mueller (1932, p. 41; 1934, p. 210); Wisniewski (1958), life cycle; Yamaguti (1953, p. 117; 1958, p. 128).

Bunodera armatum (MacCallum, 1895), (see *Homalometron armatum*).

B. cornuta (Osborn, 1903), (see *Crepidostomum cornutum*).

B. luciopercae (O. F. Müller, 1776). (Syn. *B. nodulosa* (Froelich, 1791)]: in *Perca flavescens*; Fischthal (1947): same, Wis.; Linton (1940): same, Mass.; Meyer (1954): *Micropterus salmoides*, Wis.; Pearse, (1924): Wis.; Van Cleave and Mueller (1934): N. Y. Also Europe.

B. nodulosa (Froelich, 1791) Railliet (1896), (see *B. luciopercae*).

B. sacculata Van Cleave and Mueller, 1932 (Yamaguti, 1958, places this worm in genus *Bunoderina*): in *Micropterus salmoides*, *Perca flavescens*, N. Y.; Anthony (1963): Wis.; Bangham (1944, 1955): *Lepomis gibbosus*, L. Huron; Bangham and Venard (1946): *P. flavescens*, *Notropis cornutus*; Fischthal (1947): *Stizostedion vitreum*, Wis.; Lyster (1939): Can.; Sindermann (1953): *P. flavescens*, *Roccus americana*, Mass.

Genus **Bunoderina** Miller, 1936
(Fig. 123)

Very similar to *Bunodera* but vitellaria in lateral fields between pharynx and cecal ends. Manter (1962) removes this genus to Callodistomatidae.

Ref. Doss (1963, p. 157); Yamaguti (1953, p. 120).

Bunoderina eucaliae Miller, 1936 (Miller, 1940, reduces this to *Bunodera*): in *Eucalia inconstans*, Can.; Anthony (1963): *Umbra limi*, Wis.; Bangham (1944, 1955): *E. inconstans*, *U. limi*, Wis., L. Huron; Bangham and Adams (1954): *E. inconstans*, *Gasterosteus aculeatus*, Brit. Col.; Fischthal (1950): same. Life cycle: cercaria probably in *Pisidium* (Hoffman, 1955).

B. sacculata (see *Bunodera s.*).

Genus **Plagiocirrus** Van Cleave and Mueller, 1932
(Fig. 124)

Allocreadiidae: Small, plump worm. Oral sucker subterminal, pharynx large; esophagus moderately long; ceca terminate near posterior extremity. Ventral sucker larger than oral sucker, in anterior half of body. Testes tandem, close together, intercecal, in posterior of body. Cirrus pouch claviform with posterior overlapping ventral sucker. Genital pore near left margin of body at level of pharynx. Ovary on right of median line, post-equatorial. Vitelline follicles form longitudinally elongated lateral groups between ventral sucker and posterior end of ovary. Uterus occupies most of intercecal field of hindbody; eggs relatively large. Excretory vesicle tubular.

Ref. Yamaguti (1958, p. 180).

Plagiocirrus primus Van Cleave and Mueller, 1932: in intestine, *Notemigonus crysoleucas*, N. Y.; Anthony (1963): same host, Wis.; Fischthal (1947): *Etheostoma flabellare*, Wis.; Fritts (1959): *Catostomus macrocheilus*, Idaho.

P. testeus Fritts, 1959: in *Catostomus macrocheilus*, Idaho.

Plagiocirrus sp. Bangham and Adams, 1954: in *Catostomus commersoni*, Brit. Col.

Genus **Vietosoma** Van Cleave and Mueller, 1932
(Fig. 125)

May not belong in Allocreadiidae; original describers place it in Plagiorchidae: Body comparatively broad, unarmed. Oral sucker large, prepharynx present, esophagus very short, ceca not reaching to posterior extremity. Ventral sucker small, pre-equatorial. Testes symmetrical, medial to posterior portion of ceca. Cirrus pouch short, preacetabular. Genital pore median, just in front of ventral sucker. Ovary lobed, pre-equatorial, overlapping ventral sucker. Uterus passing between 2 testes and reaching posterior extremity; eggs large. Vitellaria extend in lateral fields along entire length of intestinal limbs. Excretory bladder saccular, with terminal pore.

Ref. Van Cleave and Mueller (1932, 1934); Yamaguti (1953, p. 86; 1958, p. 185).

Vietosoma parvum Van Cleave and Mueller, 1932: in intestine and stomach, *Ictalurus punctatus*, N. Y.; Lyster (1939): in *I. punctatus*.

Genus **Allocreadium**, Looss, 1900
(Fig. 130)

Body more or less elongate, with blunt-pointed extremities, unarmed. Suckers and pharynx well developed. Ceca reach to near posterior extremity. Ventral sucker in anterior half of body. Testes tandem, toward middle of hindbody. Cirrus pouch well developed, containing winding seminal vesicle, prostatic complex, and cirrus, anterior or dorsal to ventral sucker, sometimes extending farther posterior. Genital pore close to intestinal bifurcation. Ovary between ventral sucker and anterior testis, submedian. Seminal receptacle and Laurer's canal present. Vitellaria in hindbody or extending into forebody; may or may not be continuous behind posterior testis. Uterus between anterior testis and ventral sucker. Excretory bladder may reach to posterior testis. In freshwater and marine fishes. Life cycle: adult in fish, sometimes progenetic in arthropod host; opthalmocercaria in sphaerid clams and limpets; metacercaria in arthropods and clams.

Ref. Hopkins (1933, 1934); Mehra (1962), review; Peters (1957), review; (1960), cercaria; Van Cleave and Mueller (1932, p. 86; 1934, p. 199); Yamaguti (1953, p. 69; 1958, p. 101).

Allocreadium armatum (see *Homalometron armatum*).
A. boleosomi Pearse, 1924 (Syn. *Homalometron b.*): in *Etheostoma nigrum, Percina caprodes*, Wis.; Bangham and Hunninen (1939): L. Erie; Peters (1957): as synonym of *Plagioporus b.*
A. colligatum Wallin, 1909: in stomach, *Semotilus bullaria*, Me.
A. commune (Olsson, 1876): Scandinavia; Cooper (1915): in intestine, *Fundulus diaphana*; gall bladder, *Notropis cornutus*, Can; Dobrovolny (1939): as synonym of *Plagioporus sinitsini*.
A. halli Mueller and Van Cleave, 1932 (see *A. ictaluri*).
*A. ictaluri*** Pearse, 1924: in *Ictalurus punctatus*, Wis. [Syn. *Lepidauchen i.* (Pearse, 1924) Yamaguti, 1958; *Maculifer chandleri* Harwood, 1935; *A. halli* Mueller and Van Cleave, 1932]. Reported from *Catostomus commersoni, Ictalurus furcatus, I. melas, I. nebulosus, I. punctatus*, by Arnold (1933); Bangham and Venard (1942): Tenn.; Fischthal (1952); Hopkins (1934): metacercaria in Unionidae clams; Krueger (1954): Ohio; Meyer, F. (1958): Iowa; Peters and Self (1963): cercaria in limpet, *Laevapex fusca*, metacercaria in mantle of limpets and *Lampsilis*, Okla.; Seitner (1951): cercaria in *Pleurocerca*, metacercaria in Unionidae; Worley and Bangham (1951): Que.
A. lobatum Wallin, 1909: in stomach, *Semotilus bullaris*, Me.; Anthony (1963): in *S. atromaculatus, N. cornutus*, Wis.; Bangham (1941, 1951, 1955): in intestine, *Campostoma anomalum, Catostomus fecundus, Gila straria, Notropis cornutus, N. deliciosus, Prosopium williamsi, Rhinichthys cataractae, Richardsonius baleatus, R. carringtoni, Salmo clarki, Salvelinus fontinalis, Semotilus atromaculatus, S. corporalis, Nocomis biguttatus, Notropis cornutus, N. hudsonius*, Ohio, Wyo., L. Huron; Bangham and Venard (1946): DeGiusti (1962): ophthalmoxiphidocercaria in *Pisidium*, meta-

---

* S. H. Hopkins (pers. comm., 1953) places this species in *Lepidauchen*. Yamaguti (1958) apparently places it in *Lepidauchen* because the vitellaria intrude into the forebody. Cable (pers. comm., 1966) retains it in *Allocreadium* mainly because it has no spines as do all other *Lepidauchen* (Lepocreadiidae).

cercaria in *Gammarus*, adult in *S. atromaculatus*, but may become progenetic adult in *Gammarus*; Fischthal (1947, 1950, 1956): *Catostomus commersoni, Exoglossum maxilingua, Notropis cornutus, Semotilus atromaculatus*, N. Y., Wis.; Hoffman (1953): N. Dak.; Hoffman (1959): *Semotilus corporalis*, W. Va.; Hunninen (1936): *S. corporalis*; Meyer (1954): Me.; Pearse (1924): Wis.; Peters (1957): cercaria probably in sphaerid clams.

A. shawi (see *Podocotyle shawi*).

? *Lepidauchen* sp. Holloway and Bogitsh, 1964, from *Ictalurus nebulosus*, Va., may be related to *Allocreadium ictaluri* ?

Family MACRODEROIDIDAE McMullen, 1937

Family diagnosis, Plagiorchioidea: Excretory bladder in larval and adult forms I-shaped, variable in length. Cercarial stage characterized by bladder form and small number of stylet glands. Body of adults elongate and spined. Intestinal ceca extend into posterior end. Genital pore anterior to acetabulum and near median line. Cirrus well developed. Testes oblique and posterior to ovary. Ovary just posterior to acetabulum and lateral to median line. Vitellaria are massed follicles. Uterus passes between testes into posterior end. In digestive tract of fish.

KEY

TO THE GENERA OF MACRODEROIDIDAE

Seminal receptacle present; cirrus pouch does not overlap ventral
  sucker . . . . (fig. 126) *Paramacroderoides*
Seminal receptacle absent; cirrus pouch overlaps ventral sucker
  . . . . (fig. 127) *Macroderoides*

Genus **Paramacroderoides** Venard, 1941
(Fig. 126)

Allocreadiidae: Body small, flattened subcylindrical, coarsely spined. Oral sucker large, with spines larger than body spines. Prepharynx present. Esophagus short, bifurcating halfway between two suckers. Ceca terminate at posterior extremity. Ventral sucker comparatively small, about one-third of body length from anterior extremity. Testes median, tandem, in middle third of body. Cirrus pouch long, slender, extending back of ventral sucker, containing unequally divided seminal vesicle, prostate, and cirrus. Genital pore median, just pre-acetabular. Ovary slightly to left, adjacent to posterior end of cirrus pouch. Seminal receptacle on right of ovary. Uterus descends to near posterior extremity. Vitelline follicles irregular in shape; extend in lateral fields from behind ventral sucker to middle of post-testicular region. Excretory bladder tubular, about one-fifth as long as body.

Ref. Yamaguti (1953, p. 88; 1958, p. 184).

Paramacroderoides echinus Venard, 1941: in intestine of *Lepisosteus platostomus*, Tenn.; Bangham (1942), Tenn.; Holliman and Leigh (1953): life cycle, snail host *Helisoma*; metacercaria in *Gambusia, Jordanella, Heterandria, Mollienesia, Fundulus*; experimental adult in *L. platyrhincus*, Fla.

Paramacroderoides sp. Leigh, 1957: in *Lepisosteus platyrhincus*, snail host *Helisoma*; metacercaria in *Gambusia*.

Genus **Macroderoides** Pearse, 1924
(Fig. 127)
(Syn.: *Plesiocreadium* Winfield, 1929)

Macroderoididae: Body elongate, heavily spined throughout. Oral sucker medium-sized. Pharynx and prepharynx present. Esophagus bifurcating about midway between suckers; ceca terminate at posterior extremity. Ventral sucker comparatively small, pre-equatorial. Testes obliquely or exactly tandem, in posterior half of body. Cirrus pouch overlaps ventral sucker; contains seminal vesicle, weakly developed prostatic complex, and long thick cirrus. Genital pore median, or slightly to right or left, immediately pre-acetabular. Ovary behind ventral sucker, somewhat to one side. No seminal receptacle. Vitellaria confined to hindbody; may or may not reach posterior extremity. Uterus descends to posterior extremity and then ascends, passing between 2 testes or dorsal and ventral to them; eggs small. Excretory bladder elongate saccular, with terminal pore. Life cycle: adult in intestine of fish; lophocercous xiphidiocercaria in snails (*Helisoma*); metacercaria in fish and tadpoles.

Ref. Van Cleave and Mueller (1932, p. 49; 1934, p. 212), key to spp.; Yamaguti (1953, p. 87; 1958, p. 182).

Macroderoides flavus Van Cleave and Mueller, 1932: in intestine and rectum, *Esox niger. E. americanus*, N. Y.; Bangham (1941, 1942): Ohio, Tenn.; Fantham and Porter (1947): Que.; Fischthal (1947): *Esox lucius*, Wis.; Hunter (1939): Conn.; Meyer, M. (1954): Me.; Sindermann (1953): *E. niger*, Mass.

M. parva (Hunter, 1932): in *Lepisosteus osseus, Amia calva*; Anthony (1963): *A. calva*, Wis.; Fischthal (1947, 1950): *A. calva*, Wis.

M. spinifera Pearse, 1924: in *Lepisosteus platystomus. Ictalurus natalis, I. nebulosus*, Wis.; Bangham (1939, 1941, 1942): L. Erie, Tenn.; Fischthal (1950): *Esox masquinongy*, Wis.; Leigh (1958): cercaria in *Helisoma*; metacercaria in muscle of fish and tadpoles;

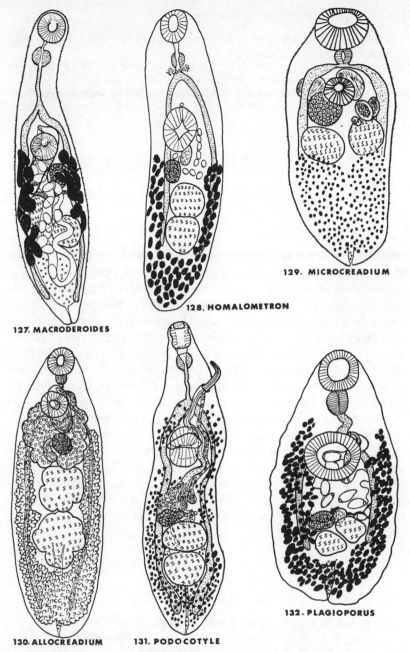

129. MICROCREADIUM

128. HOMALOMETRON

127. MACRODEROIDES

130. ALLOCREADIUM   **131. PODOCOTYLE**

132. PLAGIOPORUS

Figs. 127–132. Fig. 127. *Macroderoides flavus* (from Van Cleave and Mueller, 1932). Fig. 128. *Homalometron armatum* (from Miller, 1959). Fig. 129. *Microcreadium parvum* (from Simer, 1929). Fig. 130. *Allocreadium lobatum* (from Van Cleave and Mueller, 1934). Fig. 131. *Podocotyle shawi* (from McIntosh, 1939). Fig. 132. *Plagioporus cooperi* (from Hunter and Bangham, 1932).

adult in *L. platyrhincus*, Fla.; Meyer (1954): *Esox niger*, Me.; Simer (1929): *L. osseus*, Miss.

M. typica (Winfield, 1929) Van Cleave and Mueller, 1932: in *Amia calva*, Mich.; Anthony (1963): Wis.; Bangham (1944, 1955): Wis., L. Huron; Cort (1939); McMullen (1935): cercaria in *Helisoma*; metacercaria in fish and tadpoles; Sogandares-Bernal (1965): La.

Family LEPOCREADIIDAE Nicoll, 1934

Subfamily HOMALOMETRINAE Cable and Hunninen, 1942

(Syn.: Anallocreadiinae, Hunter and Bangham, 1932)

Medium-sized distomes, body not elongate, with or without spines. Mouth subterminal. Prepharynx, pharynx, and esophagus present. Intestinal ceca almost always long, rarely joining excretory vesicle or with separate anal openings. Excretory vesicle tubular, variable in length, frequently reaching pharyngeal level; rarely Y-shaped and then with short arms not receiving main tubules; excretory pore posterior, terminal or subterminal. Main excretory tubules never reach cephalic region but divide near ventral sucker to form anterior and posterior collecting tubules. Flame cells numerous. Genital pore median, immediately in front of or rarely posterior to ventral sucker; cirrus sac lacking. Testes in posterior half of body, occasionally multiple, usually 2, median and tandem, sometimes diagonal or opposite in short-bodied forms. Ovary pre-testicular, rarely behind level of anterior testis. True seminal receptacle and Laurer's canal present. Vitellaria well developed, rarely restricted to post-testicular regions. Cercariae trichocercous, stylet lacking, with conspicuous eye spot, remnants of which may persist in adults. Simple rediae in gastropods; encyst in invertebrates.

KEY

TO THE GENERA OF HOMALOMETRINAE

1. Oral papillae present; oral sucker cupuliform (in brackish water) . . . . (no fig.) *Barbulostomum*
1. Oral papillae absent; oral sucker not cupuliform   2
2. Cirrus pouch absent; vitellaria in hindbody, extracecal and post-testicular . . . . (fig. 128) *Homalometron*
2. Cirrus pouch short, anterior to ventral sucker; vitellaria in entire post-testicular area . . . . (fig. 129) *Microcreadium*

Genus **Homalometron** Stafford, 1904
(Fig. 128)
(Syn.: *Anallocreadium* Simer, 1939)

Lepocreadiidae: Body elongate, spinulate. Oral sucker and prepharynx present. Esophagus short; ceca terminate at posterior extremity. Ventral sucker in anterior half of body. Testes tandem in post-equatorial intercecal field. Seminal vesicle free. No cirrus pouch. Genital pore immediately pre-acetabular. Ovary submedian, in front of anterior testis. Seminal receptacle and Laurer's canal present. Vitellaria in hindbody, extracecal and post-testicular. Uterus winding between anterior testis and genital pore; eggs few, comparatively large. Excretory bladder reaching to posterior testis. Life cycle: adult in intestine of fishes; metacercaria in clams.

Ref. Miller (1959), review; Yamaguti (1953, p. 71; 1958, p. 137).

Homalometron armatum (MacCallum, 1895), (Syn. *Distomum isoporum* var. *armatum* MacCallum, 1895): in *Aplodinotus grunniens*, *Lepomis gibbosus*; Hopkins (1937): cercaria in *Amnicola*, metacercaria in clams; Miller (1959): redescription from *A. grunniens*, *L. humilis, L. microlophus*; Ramsey (1965): La.
H. boleosomi (see *Allocreadium boleosomi*).
H. pallidum Stafford, 1904. In *Fundulus heteroclitus, Roccus americana*, and other brackish-water fish. Fantham and Porter (1947): Que.; Larson (1964): *F. diaphanus*, Minn.; Manter (1926): illus.; Stunkard (1964): life cycle, cercaria in *Hydrobia*, encysts in *Hydrobia* and *Gemma*.
H. pearsei (Hunter and Bangham, 1932) Manter, 1947. [Syn. *Anallocreadium p.* Hunter and Bangham; *A. armatum* Miller (1940)]: in *Aplodinotus grunniens*, L. Erie; Sogandares-Bernal (1955): La.

Genus **Microcreadium** Simer, 1929
(Fig. 129)

Allocreadiidae: Body small, rather plump. Oral sucker large, esophagus very short. Ceca reach to near posterior extremity. Ventral sucker small, in anterior half of body. Testes side by side in equatorial intercecal zone. Cirrus pouch short, anterior to ventral sucker. Genital pore just in front of ventral sucker. Ovary submedian, between right testis and ventral sucker. Seminal receptacle present. Vitellaria occupy entire post-testicular area. Uterus winding between testes and ventral sucker; eggs few. Excretory bladder apparently reaching to behind testes.

Ref. Yamaguti (1953, p. 71; 1958, p. 113).

Microcreadium parvum Simer, 1929: in intestine, *Aplodinotus grunniens*; Hopkins (1937): cercaria in *Amnicola*.

Genus **Barbulostomum** Ramsey, 1965
(no fig.)

Lepocreadiidae: Homalometrinae: With the characteristics of the subfamily. Differs from *Homalometron* in cupuliform shape of oral sucker and presence of oral papillae.

Barbulostomum cupuloris Ramsey, 1965: in intestine, *Lepomis microlophus, L. punctatus miniatus*, from slightly brackish water, La.

Family OPECOELIDAE Ozaki, 1925

Suckers well developed; ventral sucker well removed from posterior end, often embedded in a protruding fleshy lobe, stalk, or disc. Prepharynx short, if present; pharynx well developed; ceca simple, extending well toward posterior end of body, sometimes with separate anal openings posteriorly, uniting and having a single such opening, or joining excretory vesicle and thus forming uroproct with excretory pore serving as anus. Cuticle always unarmed but often thick and wrinkled; cercarial eye-spot pigment absent. Genital pore anterior to ventral sucker, often far so and to left of median line, rarely submarginal or dorsal. Cirrus sac present or absent, sometimes with accessory sucker near genital pore. Seminal vesicle unipartite, usually long and coiled within cirrus sac; external seminal vesicle absent. Testes 2 or rarely multiple, intercecal, tandem, diagonal, or even side by side in short-bodied forms. Ovary pre-testicular; seminal receptacle if present an enlargement of Laurer's canal. Vitellaria well developed and extensive, with follicles usually along most of body length. Uterus pre-testicular, metraterm independent of cirrus sac and not especially thickened or otherwise modified; eggs large, operculate, occasionally with antopercular knobs or long filaments. Excretory vesicle sac-shaped or tubular, variable in length, sometimes extending almost to intestinal bifurcation but usually not anterior to ovary. Main excretory tubule on each side extends anteriorly a variable distance but never to pharyngeal level and becoming recurrent before receiving two secondary tubules, one extending anteriorly and one posteriorly. Each secondary tubule serves two groups of flame cells: excretory formula is $2 [(n + n) + (n + n)]$ often $n = 2$ in cercaria and remains unchanged in adult. Cercaria: distome and cotylomicrocercous, i.e., tail short and consisting of a cup with glands in wall or a glandular core and used for attachment to substratum. Eye spots absent; develop in sausage-shaped sporocysts in prosobranch gastropods; metacercariae in invertebrates or rarely fishes.

Ref. Cable (1956); Manter (1947).

KEY

TO THE GENERA OF OPECOELIDAE

Vitellaria in hindbody; ovary 3-lobed; cirrus pouch extends behind ventral sucker....................(fig. 131) *Podocotyle*
Vitellaria intrudes into forebody; ovary rounded; cirrus pouch does not extend behind middle of ventral sucker..........
.................................(fig. 132) *Plagioporus*

Genus **Podocotyle** (Dujardin, 1845)
(Fig. 131)
(Syn.: *Sinistroporus* Stafford, 1904)

Opecoelidae: Body more or less elongated, without spines. Ventral sucker in anterior half of body; may be surmounted by puckered margin of sucker peduncle. Oral sucker and pharynx well developed, esophagus variable, ceca terminate near posterior end. Testes tandem, toward middle of hindbody or further behind. Cirrus pouch long, slender or claviform; may extend backward beyond ventral sucker. Genital pore to left of median line at level of esophagus or intestinal bifurcation. Ovary pre-testicular. Seminal receptacle and Laurer's canal present. Vitellaria usually in hindbody, circumcecal. Uterus winding forward in front of ovary. Excretory bladder reaches ovary or more anteriorly. In intestine of freshwater and marine fish. Life cycle unknown.

Ref. Park (1937), revision; Yamaguti (1953, p. 72; 1958, p. 119).

Podocotyle shawi McIntosh, 1939 (Yamaguti, 1958, considers this to be *Cainocreadium*): in *Onchorhynchus kisutch*, Ore.; Fritts (1959): *O. nerka, Prosopium williamsi*, Idaho; Griffith (1953): *O. nerka, Salmo gairdneri*, Wash.; Shaw (1947): Ore.
P. simplex (Rudolphi, 1809) of Stafford: in *Salmo salar* and marine fish, Can.
P. virens (Sinitsin, 1931), (see *Plagioporus v.*).
Podocotyle sp. Shaw (1947): in *S. clarki, S. gairdneri*, Ore.

Genus **Plagioporus** Stafford, 1904
(Fig. 132)
(Syn.: *Lebouria* Nicoll, 1909)

Opecoelidae: Body flattened, elliptical, fusiform or oval. Oral sucker, pharynx, and acetabulum well developed; latter pre-equatorial, sometimes equatorial. Ceca terminate at or near posterior extremity. Testes tandem or oblique, toward midbody in subgenus *Plagioporus*, near posterior extremity in subgenus *Caudotestis*. Cirrus pouch more or less claviform; contains winding seminal vesicle, rather indistinct pars prostatica surrounded by prostate cells and eversible ductus ejaculatorius. Genital pore submedian (usually on left), level with esophagus or intestinal bifurcation, or occasionally with pharynx. Ovary pretesticular, submedian, sometimes median. Seminal receptacle and Laurer's canal present. Vitellaria extend into forebody, not reaching to posterior extremity in subgenus *Caudotestis*. Uterus winding between ovary or anterior testis and acetabulum. Excretory vesicle tubular, reaching to ovary, occasionally to anterior end of acetabulum (*Paraplagioporus*). Flame cell formula: $2[(2+2) + (2+2)] = 16$ in *P.* (*Plagioporus*) *ira* and *P.* (*Paraplagioporus*) *isagi*. Parasites of marine and freshwater fishes. Divided into three subgenera, *Plagioporus*, *Caudotestis*, and *Paraplagioporus*, according to difference in position of testes or in anterior extent of excretory vesicle.

Ref. Dobrovolny (1939a, 1939b), key to species, life cycle; Van Cleave and Mueller (1934, p. 359); Yamaguti (1953, p. 75; 1958, p. 116).

Plagioporus angusticolis (Hausmann, 1896) Dobrovolny, 1939: in *Cottus gobio*, Europe; Haderlie (1953): *Salmo gairdneri*, Calif.; Mathias (1936): cercaria in *Neritina*, metacercaria in *Asellus*, *Gammarus*, adult in *Anguilla*, *Cottus*.

P. boleosomi (see *Allocreadium b.*).

P. cooperi (Hunter and Bangham, 1932): in *Amocrypta*, *Notropis*, *Rheocrypta*; Bangham (1951, 1955): *Cottus*, *Etheostoma*, *Gila*, *Notropis*, *Richardsonius*, Wyo., L. Huron.

P. lepomis Dobrovolny, 1939: in intestine of *Lepomis megalotis peltastes* and other centrarchids; cercaria in *Goniobasis*; metacercaria in *Hyalella*; Yamaguti (1958): transfers to *Podocotyle*, Mich.

P. macrouterinus Haderlie, 1953: in *Ptychocheilus grandis*, Calif.

P. serotinus Stafford, 1904: in *Moxostoma macrolepidotum*, Can.; Haderlie (1953): *Archoplites interruptus*, Calif.; Manter (1954): questions this specific identification; Miller (1940, 1941): *Moxostoma* sp., Can.

P. serratus Miller, 1940: in *Hiodon tergisus*, Que.

P. siliculus Sinitsin, 1931: in *Salmo clarki*, cotylocercous xiphidiocercaria in *Goniobasis*, metacercaria in muscles of crayfish, *Polamobius*, Ore.

P. sinitsini Mueller, 1934: in gall bladder, *Catostomus commersoni*, *Notropis cornutus*; Dobrovolny (1939): *Allocreadium commune* as synonym and experimental in *Nocomis biguttatus*, *Notropis cornutus*, cercaria in *Goniobasis*, metacercaria in sporocysts in same snail; Bangham (1955): *C. commersoni*, *Nocomis biguttatus*; Fischthal (1947, 1950), Wis.; Fischthal (1956): *C. commersoni*, *Semotilus corporalis*, *S. atromaculatus*, N. Y.

P. truncatus Linton, 1940: in *Roccus americana* and other marine fish, Mass.

P. virens Sinitsin, 1931: in *Cottus* (freshwater); cercaria in *Flumini-cola*, metacercariae in same snail, Ore.
Plagioporus sp. Haderlie, 1953: in intestine, *Salmo gairdneri*, Calif.; Alexander (1960): Ore.; Harms (1959): *Ictalurus melas*, Kans.

Family PTYCHOGONIMIDAE Dollfus, 1937
[See Yamaguti (1953, p. 173; 1958, p. 255).]

Genus Ptychogonimus Lühe, 1900
(Fig. 133)

Body large, plump, covered with thick, transversely wrinkled cuticle. Oral sucker very large, subterminal. Pharynx well developed. Prepharynx and esophagus very short. Ceca wide, nearly reaching posterior. Ventral sucker slightly anteriad. Testes near posterior. Vas deferens coiled, seminal vesicle small. Genital sucker with 3 circular concentric muscular folds between pharynx or oral sucker and ventral sucker. Ovary submedian between ventral sucker and testes. Laurer's canal opening dorsally to ovary. Vitellaria occupy most of lateral fields of hindbody. Uterus descends then ascends on each side of ovary and testes; eggs medium. Excretory pore terminal; arms parallel to each other, forming conspicuous circumoral ring. In stomach of sharks, rare in freshwater fishes. Life cycle: metacercaria in crayfish.

Ptychogonimus fontanus Lyster, 1939: in *Perca flavescens*, Que.; Fantham and Porter (1947): Que.; Lyster (1940): *Salvelinus fontinalis*.

ACCIDENTAL IN FISH?

Pleurogenes sp. (Lecithodendriidae), (fig. 134). Usually present in frogs but reported once from *Salvelinus fontinalis* (Fantham and Porter, 1947).

ADULT TISSUE AND BLADDER-INHABITING DIGENETIC TREMATODES

Most adult trematodes live in the gastrointestinal tract, but a few are found elsewhere. This list is included as an aid to identification.

Family SANGUINICOLIDAE: Parasites of blood vessels; long slender trematodes; eggs and miracidia may plug gill vessels—*Sanguinicola davisi, S. huronis, S. klamathensis, S. occidentalis.*

Family DIDYMOZOIDAE: Large worms living in pairs in tissue—*Nematobothrium texomensis*; up to 9 ft. long; in ovaries of *Ictiobus.*

Family BUCEPHALIDAE: Mouth in ventral sucker; central saccular gut—*Paurorhynchus hiodontis*; in body cavity of *Hiodon.*

160

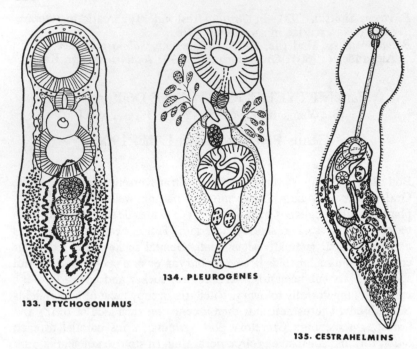

134. PLEUROGENES

133. PTYCHOGONIMUS

135. CESTRAHELMINS

Figs. 133–135. Fig. 133. *Ptychogonimus megastomus* (from Jacoby, 1899, in Yamaguti, 1953). Fig. 134. *Pleurogenes* sp. (from Fantham and Porter, 1947). Fig. 135. *Cestrahelmins laruei* (from Fischthal, 1957).

Family HETEROPHYIDAE: Gonotyl (genital sucker) present— *Acetodextra* in catfish ovary, body cavity, and air bladder.
Family ALLOCREADIIDAE and OPECOELIDAE: *Crepidostomum* and *Plagioporus* in gall bladder occasionally.
Family GORGODERIDAE: Large flat posterior body—*Phyllodistomum* in urinary bladder and ureters.

Metacercarial Trematodes

Contents

Key to the groups of cercariae which infect fish 164
Order STRIGEATOIDEA .. 165
Suborder STRIGEATA 166
  Key to the furcocercous cercariae of fish trematodes of North America 166
Superfamily STRIGEOIDEA 169
  Key to the larval genera of Strigeoidea of fish 169
  Larval genus Diplostomulum ... 169
  Key to the known species of Diplostomulum of North American freshwater fishes 170
  Larval genus Tetracotyle. 171
  Key to the known species of Tetracotyle of North American fishes 173
  Larval genus Neascus ... 174
  Key to known species of Neascus of North American freshwater fishes .. 175
  Larval genus Prohemistomulum 177
Key to the metacercariae, exclusive of strigeoids, of North American fishes 178
Superfamily CLINOSTOMATOIDEA 181
  Genus Clinostomum 182
Superfamily BUCEPHALOIDEA 182
  Genus Bucephaloides 183
  Genus Bucephalus 183
  Genus Rhipidocotyle . . 183
Order ECHINOSTOMIDA . 183

Family ECHINOSTOMATIDAE 185
  Genus Petasiger 185
  Genus Echinochasmus ... 185
  Genus Stephanoprora ... 186
Family CATHAEMASIIDAE 186
  Genus Cathaemasia 186
  Genus Ribeiroia 186
Superorder EPITHELIOCYSTIDIA 187
Order PLAGIORCHIIDA . 187
Family ALLOCREADIDAE 187
  Genus Macroderoides ... 187
  Genus Paramacroderoides 188
Family TROGLOTREMATIDAE 188
  Genus Nanophyetus 188
  Genus Sellacotyle 189
Order OPISTHORCHIIDA. 189
Family OPISTHORCHIIDAE 189
  Genus Opisthorchis 190
  Genus Metorchis 190
  Genus Amphimerus 190
Family HETEROPHYIDAE 191
  Genus Apophallus 191
  Genus Allacanthochasmus 193
  Genus Ascocotyle 193
  Subgenus Leighia 194
  Subgenus Phagicola 195
  Genus Pygidiopsoides ... 195
  Genus Cryptocotyle 197
  Genus Euhaplorchis 197
  Genus Parastictodora ... 197
  Genus Scaphanocephalus. 198

Genus *Galactosomum* ... 198
Family CRYPTOGONI-
    MIDAE 199
Genus *Cryptogonimus* ... 199

Genus *Acetodextra* 199
Genus *Centrovarium* 201
Genus *Caecinocola* 201
Genus *Neochasmus* 201

Metacercariae may be found in all tissues of fish. Usually they can be recognized by their suckers. In some young forms the suckers may not be apparent. They might be confused with small larval tapeworms, but the latter possess typical calcareous concretions. Most metacercariae become encysted; a few do not.

In some instances descriptions of metacercariae from fish were not available. Drawings of the adults are included in those cases, for the metacercaria often has some of the same characteristics as the adult; this should help in identifying unknown metacercaria.

The life cycles involve the fish's being eaten by the proper final host: fish-eating bird, mammal, other fish, and possibly reptile.

The first intermediate hosts of all Digenea are molluscs (snails and clams) wherein the trematodes develop and multiply, finally producing many free cercariae which, in the case of fish metacercariae, actively penetrate and migrate into the fish tissues. Discussions of the developmental stages may be found in general or medical parasitology texts. However, because the fisheries parasitologist may want to acquaint himself with the cercariae for life-cycle studies or for trematode control, a brief key to the various kinds of cercariae is given here. There is no available synopsis of American cercariae, but Chandler (1952) distributed a mimeographed key to the furcocercous cercariae which has been very helpful to many researchers.

When a cercaria penetrates and migrates into the tissues of a fish, it causes obvious mechanical damage and hemorrhage. It is not known whether there is also toxic damage. The damage done by one cercaria is negligible, but if enough are present the fish will be killed (Hoffman 1956, 1958; Hoffman and Hoyme, 1958; Hoffman and Hundley, 1957). After the cercaria has localized and transformed into a metacercaria, not much further damage will be done unless so many cercariae accumulate that their mass interferes with the fish's metabolism. Although control of snails is difficult, it is the logical method to use in conjunction with fish culture. It may not be feasible, except in localized areas, to control snails in larger waters. However, molluscicides are being studied extensively by public health workers because of human

163

schistosomiasis, and some good molluscicides which are not toxic to fish will probably be found.

A key is included as an aid to those who might be concerned with the cercariae of fish parasites. For definitive identification, consult the original descriptions of the cercarial species.

KEY

TO THE GROUPS OF CERCARIAE WHICH INFECT FISH*

1.		Furcocercous (forked tail)	2
1.		Not furcocercous	9
2	(1).	Furcocystocercous or gasterostomate	8
2	(1).	Not furcocystocercous or gasterostomate	3
3	(2).	Longifurcate (furcae nearly as long as or longer than tail stem)	4
3	(2).	Brevifurcate (furcae shorter than tail stem)	6
4	(3).	Excretory pores at ends of furcae; 6 flame cells in tail stem; ventral sucker absent or weak.............. .... (*Prohemistomulum*) .... Superfamily CYATHOCOTYLOIDEA	
4	(3).	Excretory pore lateral or posterior on furcae; ventral sucker usually well developed .... Superfamily STRIGEOIDEA	5
5	(4).	Attached part of excretory bladder with 3 ciliary tufts of individual cilia*........................ (*Neascus, Diplostomulum*) Family DIPLOSTOMATIDAE	
5	(4).	Attached part of excretory bladder without ciliary tufts or individual cilia*. .(*Tetracotyle*) Family STRIGEIDAE	
6	(3).	Pharynx present ....(*Clinostomum*) Superfamily CLINOSTOMATOIDEA	
6	(3).	Pharynx absent; excretory pores at end of furcae	7
7	(6).	Ventral sucker present; no undulating membrane; 2 flame cells in tail stem (not in fish; included to aid identification) ....Family SPIRORCHIDAE and .... Superfamily SCHISTOSOMATOIDEA	
7	(6).	Ventral sucker absent; furcae with dorsoventral undulating membrane; no flame cells in tail stem; body with long undulating membrane (lophocercous).. ....Family SANGUINICOLIDAE	
8	(2).	Furcocystocercous; distomate; pharyngeate........ ....Family AZYGIDAE	

* This key characteristic may not fit American species; it was used by Odening (1962) for European forms and is included here for comparison.

8   (2). Gasterostomate; tail stem short and bulbous; furcae
         very long and active; development in branched sporo-
         cysts in lamellibranchs . . . . . . Family BUCEPHALIDAE
9   (1). Echinostomate (oral collar, but not spines, may be
         present) . . . . Superfamily ECHINOSTOMATOIDEA
9   (1). Not echinostomate   10
10 (9). Amphistomate; excretory bladder consists of a mem-
         brane rather than definite cells . . . .
         . . . . Family PARAMPHISTOMATIDAE
10 (9). Not amphistomate; excretory bladder consists of thick-
         walled layer of epithelium   11
11 (10). Without caudal excretory vessels at any stage of de-
         velopment; stylet present or absent   12
11 (10). With caudal excretory vessels during development;
         stylet always absent   13
12 (11). Distomate; pharyngeate; stylet horizontal; encystment
         in invertebrates . . . . (Plagiorchiidae, Lissorchiidae)
         . . . . Superfamily PLAGIORCHIOIDEA
12 (11). Distomate; stylet usually not horizontal if present; en-
         cystment in invertebrates and vertebrates (Troglotre-
         matidae cercariae are microcercous) (Acanthocolpi-
         dae, Allocreadiidae, Monorchiidae, Gorgoderidae,
         Troglotrematidae) . . Superfamily ALLOCREADIOIDEA
13 (11). Primary excretory pores on margins of tail near body-
         tail furrow; tails pleuro- or parapleurolophocercous;
         encyst in lower vertebrates . . . .
         . . . . . (Opisthorchiidae, Heterophyidae, Cryptogonimidae)
         . . . . Family OPISTHORCHIIDAE
13 (11). Primary excretory pores on tail distant from body-tail
         furrow; cercariae cystophorous or similar; encyst in
         copepods . . . . (Ptychogonimidae, Hemiuridae)
         . . . . Superfamily HEMIUROIDEA

Of the above, the following use fish for the second intermediate host:
Cyathocotyloidea, Diplostomatidae, Strigeidae, *Clinostomum*, Bucephalidae, Echinostomatoidea, Plagiorchiidae, Allocreadiidae, Trog-
lotrematodae, Opisthorchiidae, Heterophyidae, and Cryptogonimidae.
The remainder are adults in fish except that some Allocreadiidae and
Heterophyidae use the fish for the second intermediate host as well
as final host.

Order STRIGEATOIDEA

Cercariae usually fork-tailed; miracidia with 1 or 2 pairs of flame cells.
Ref. LaRue (1957).

Suborder STRIGEATA

Cercariae fork-tailed; usually distomate; excretory bladder V-shaped; penetration glands present.

KEY

(Some non-fish parasites are included to avoid confusion.)

1.		Longifurcous	2
1.		Brevifurcous	26
2	(1).	Pharyngeal	3
2	(1).	Apharyngeal	23
3	(2).	Distomate	4
3	(2).	No ventral sucker; 3 pairs of penetration glands in posterior of body; short nonbifurcating gut (Rhabdocaeca group)	17
4	(3).	Penetration glands anterior, 2 pairs	5
4	(3).	Penetration glands not anterior, or none described	14
5	(4).	No caudal bodies	6
5	(4).	Caudal bodies present .... (no fish parasites yet demonstrated)	
6	(5).	No commissure in excretory system	7
6	(5).	Commissure in excretory system	12
7	(6).	Ceca extend at least to ventral sucker	8
7	(6).	Ceca do not extend to ventral sucker .... (not fish parasite)	
8	(7).	Ceca extend to ventral sucker or slightly beyond.... .... (not fish parasite)	
8	(7).	Ceca extend to posterior of body	9
9	(8).	Ceca inconspicuous .... (*Tetracotyle* in leeches) .... *Cercaria dubia*	
9	(8).	Ceca well developed	10
10	(9).	1 pair of flame cells in tail .... *Diplostomula* of frogs	
10	(9).	2 or 3 pairs of flame cels in tail	11
11	(10).	Penetration glands lateral to ventral sucker ....*Diplostomula* of frogs	
11	(10).	Penetration glands anterior to middle of ventral sucker ....Probably not fish parasites	
12	(6).	Commissure anterior to ventral sucker	13
12	(6).	Commissure posterior to ventral sucker .... (*Diplostomula* in frogs) *Cercaria ranae*	

13 (12). Unpigmented eye spots; penetration glands in separate groups on each side of ventral sucker (*Tetracotyle* in snail) .... *Cotylurus flabelliformis*

13 (12). No unpigmented eye spots; penetration glands in median field anterior to ventral sucker and extend to fork of ceca (*Tetracotyle* in snail) .... *Cotylurus cornutus*

14   (4). 2 pairs of penetration glands; without postventral sucker constriction when hanging in water........   15

14   (4). More than 2 pairs of penetration glands; with postventral sucker constriction when hanging in water ...   16

15 (14). No caudal bodies................ (not fish parasites)

15 (14). Caudal bodies present; body spines in rows; anterior end with forward-pointing spines............. ..*Diplostomum spathaceum* and related *Diplostomum* spp.

16 (14). 4 pairs of posterior penetration glands............ .... (*Tetracotyle* in leech) *Apatemon* spp.

16 (14). 3 pairs of posterior penetration glands; no posterior commissure; body spines in rows (*Diplostomulum* in perch eyes exper.)................ *Cercaria marginatae*

17   (3). Unpigmented eye spots present (may be difficult to see); no caudal bodies........... *Uvulifer ambloplitis*

17   (3). No eye spots.............................   18

18 (17). No caudal bodies present (may be *Uvulifer ambloplitis?*) .... *Cercaria flexicorpa*

18 (17). Caudal bodies present; penetration glands numerous, anterior to bifurcation of ceca, or mostly so; excretory system with 4 main trunks; ceca very broad........ .... *Cyathocotylids*   19

19 (18). Long furcae, with fin folds present at least at tips of furcae; flame cells 30 to 36, arranged in 5 groups, always 3 pairs in tail; ventral sucker rudimentary (*Vivax* group)   20

19 (18). Long slender furcae without fin folds; no ventral sucker   21

20 (19). Fin folds extend full length of furcae; 6(?) or 8 gland cells in oral sucker................ *Prohemistomatinae* C. *vivax* Sons., 1892 = *Prohemistomum vivax* (exper. in *Gambusia* and *Tilapia*), Europe, Egypt. Others in marine snails.

20 (19). Fin folds on distal half of furcae; 8 gland cells in oral sucker; develop in Melaniidae....... *Prohemistomatinae* *Szidatia joyeuxi*, encysts in frogs and freshwater fish, Tunis.

Neogogatea kentuckiensis, encysts in freshwater fish, America.

Cercaria tatei Johnston and Angel, 1940.

20 (19). Fin folds at extremities of furcae only (freshwater fish, China) .... *Prosostephanus industrius*

21 (19). Flame cells 15 + 3, latter scattered in length of tail . .
.... *Paracoenogonimus*
P. ovatus in freshwater fish (Europe).
P. szidati in minnows (Indiana).

21 (19). Flame cells not over 24 in all.................... 22

22 (21). Flame cells 9 + 3 (*Novena* group of Dubois, 1941)
.... *Cyathocotyle*

22 (21). Flame cells 5 + 2; no ventral sucker...........
.... (*Testis* group of Sewell, 1922)

22 (21). Flame cells total 12, none in tail (China) ....
.... *Cercaria tauiana*

23 (2). Distome................ (possibly not fish parasites)

23 (2). Monostome 24

24 (23). 3 pairs posterior penetration glands; pigmented eye spots; no fin folds; no pharynx or ceca...........
.... *Multicellulata* group   25

24 (23). Numerous (14 or more) penetration glands in cluster in middle of body.......... (possibly not fish parasites)

25 (24). Caudal bodies present (*Neascus*) . . *Cercaria multicellulata*
C. paramulticellulata
C. posthodiplostomum minimum

25 (24). No caudal bodies; fin folds on furcae...... *C. louisianae*
C. isomi

BREVIFURCOUS

26 (1). Pharynx present .... *Clinostomum*

26 (1). Pharynx absent 27

27 (26). Ventral sucker present; no undulating membrane; 2 flame cells and 1 excretory canal in tail stem (not fish parasites) .... *Schistosomatoidea*
Spirorchidae

27 (26). Ventral sucker absent; body with long undulating membrane (lophocercous); furcae with dorsoventral undulating membrane; no flame cells in tail stem ...
.... *Sanguinicolidae*

Superfamily STRIGEOIDEA

Cercaria usually longifurcate, tail stem usually slender; oral sucker well developed; ventral sucker usually present. Metacercaria with oral sucker, usually ventral sucker, holdfast organ posterior to ventral sucker, concretions in reserve excretory system, sometimes lateral pseudosuckers, and body is usually divided into definite forebody and hindbody.

Illustrations of most species can be found in Hoffman (1960).

KEY

TO THE LARVAL GENERA OF STRIGEOIDEA IN FISH

1. Metacercaria enclosed in hyaline cyst of parasite origin. . . .   2
1. Metacercaria not enclosed in cyst of parasite origin (except *Bolbophorus confusus*), but may be enclosed in connective tissue cyst of host origin; lateral pseudosuckers usually prominent; hindbody may or may not be well defined; termini of excretory tubules contain round or oval calcareous concretions . . . . (fig. 136) *Diplostomulum*
2. Metacercaria with lateral pseudosuckers; reserve excretory system consists of large continuous space in dorsal and lateral regions of forebody, sometimes accordian-like . . . .
. . . . (fig. 137) *Tetracotyle*
2. Metacercaria without lateral pseudosuckers; reserve excretory system not like (2) above   3
3. Hindbody usually well developed, but that of *Ornithodiplostomum* not well set off; reserve excretory system well developed, with 3 main branches in forebody giving it a striated appearance . . . . (fig. 138) *Neascus*
3. Hindbody not apparent; reserve excretory system well developed, with 3 main branches united posteriorly, giving it a "W" appearance . . . . (fig. 139) *Prohemistomulum*
(p. 177)

Larval Genus **Diplostomulum** Hughes, 1929
(Fig. 136)

All known species of this group belong to the Diplostomatidae.

Characteristics of larval group: (1) forebody foliaceous, concave ventrally; (2) hindbody a small conical prominence on posterodorsal part of forebody; (3) reserve system (bladder) of more or less definitely arranged tubules with calcareous corpuscles, round or ellipsoidal, disposed in vesicles at termini of small branches; (4) usually a pair of lateral organs (so-called lateral suckers) on anterolateral edges beside oral sucker; (5) no true cyst of parasite origin, except in *Bolbophorus confusus*.

Ref. Hoffman (1960), synopsis.

KEY

TO THE KNOWN SPECIES OF DIPLOSTOMULUM OF
NORTH AMERICAN FRESHWATER FISHES

1. Found in eyes.................................... 2
1. Found in musculature or brain..................... 4
2. Found in lens; hindbody usually distinct.........*D. spathaceum*
2. Found in vitreous chamber........................ 3
3. Worm 3 times as long as broad; hindbody small ...*D. scheuringi*
3. Worm less than 3 times as long as broad; hindbody usually
    distinct ....*D. huronense*
4. Found in brain of *Eucalia inconstans*.........*D. baeri eucaliae*
4. Found in musculature............................ 5
5. Ventral sucker nearly as large as oral sucker; parasite freed
    easily.................(*D. corti*) *Hysteromorpha triloba*
5. Ventral sucker about half as large as oral sucker; parasite dif-
    ficult to free from host cyst....................... 6
6. Cyst not pigmented.............................*D. ictaluri*
6. Cyst usually pigmented; 5 reserve excretory channels instead
    of many termini with calcareous concretions; cyst of para-
    site origin present...................*Bolbophorus confusus*

Diplostomulum of *Bolbophorus confusus* (Kraus, 1914). Dubois, 1935: common in some parts of Europe; Fox (1961, 1965a): large black-spot in musculature of *Salmo trutta, S. gairdneri,* [Mont., young cyst not pigmented; Fox and Olson (1965): life cycle; in pelican, *Helisoma*, and several species of fish; Fox (1965b): cyst of parasite origin also present; Hugghins (1956): adult in pelican, S. Dak.; Olson (1966).

Diplostomulum corti (see *Hysteromorpha triloba*).

Diplostomulum of *Diplostomum baeri eucaliae* Hoffman and Hundley, 1957: unencysted in brain of *Eucalia inconstans*, snail host *Stagnicola* spp., adult in wild mallard duck, exper. in screech owl and unfed chick, N. Dak.

Diplostomulum emarginatae Oliver, 1942. Probably *D. spathaceum*.

Diplostomulum flexicaudum (see *D. spathaceum*).

Diplostomulum of *Diplostomum huronense* (LaRue, 1927) Hughes and Hall, 1929 (a synonym of *D. spathaceum* according to Dubois and Mahon, 1959): in lens and vitreous chamber of *Perca flavescens, Percopsis omiscomaycus*, snail host unknown, adult in gulls, Mich.; Bangham (1944): *P. flavescens,* Wis.; Meyer, F. (1958): *Lepomis humilis,* Iowa.

Diplostomulum ictaluri Haderlie, 1953: in flesh of *Ictalurus catus*, Calif.

Diplostomulum scheuringi Hughes, 1929. Recorded from vitreous chamber of Centrarchidae, Cyprinidae, Esocidae, Etheostomidae, Gadidae, Percidae, Percopsidae, Salmonidae, Siluridae. by Anthony (1963): Wis.; Bangham (1944): Wis.; Chandler (1951): Minn.; Etges (1961): snail host *Helisoma*, metacercaria in eyes and brain of newts and fish, also encysts in mice. Fischthal (1947, 1950, 1956): N. Y., Wis.; Haderlie (1953): free in coelom, Calif.; Hughes (1929): Mich.; Hunter (1942): Conn.; Mueller and Van Cleave (1931): N. Y.; Palmer (1939): Mass.; Sindermann (1953): Mass.; Adult unknown.

Diplostomulum of *Diplostomum spathaceum* (Rudolphi, 1819) Braun, 1893 (Syn. probably *D. emarginatae, D. flexicaudum, D. indistinctum, D. gigas*). Dubois (1953) considers American form, known as *D. flexicaudum*, to be identical with European *D. spathaceum*. Bangham (1951, 1955): in lens of *Catostomus* spp., Wyo., Mich.; Caballero and Winter (1954): *Micropterus dolomieui*, Mex.; Cort (1937, 1941): development in *Lymnaea* snails, Mich.; Cort and Ameel (1951): same; Cort *et al.* (1957): snail hosts *Fossaria, Stagnicola, Lymnaea*; Davis (1936): penetration studies; Ferguson (1943): experimental in lens of eye, *Lepomis macrochirus, Pimephales promelas, Salmo gairdneri*, tadpoles, frogs, turtles, chicks, ducklings, mice, rats, guinea pigs, rabbits, N. J.; Ferguson and Hayford (1941): adult in gulls, serious blinding of *S. gairdneri*, N. J.; Hoffman (unpubl.); blinding of *Ictalurus punctatus*, Ohio, longevity of metacercariae at least 1½ years; Hughes and Berkhout (1929): described as *D. gigas*; Hussey (1941): embryology; Larson (1965): life cycle of related form which causes tumor-like structure on lens of eye of bullheads, also migration studies on cercariae in fish; Mueller (1940): in *Pomolobus pseudoharengus*; Odlaug (1954): adult in gulls, Minn.; Olivier and Cort (1942): cercaria erroneously described as *Cotylurus communis*; Van Haitsma (1931): life cycle = metacercaria in *Catostomus*; adult in gulls, cercaria from *Lymnaea*.

Diplostomulum of *Hysteromorpha triloba* (Rudolphi, 1819) Lutz, 1931 (Syn. *Diplostomulum corti* Hughes, 1929). Rather large metacercaria in musculature of *Ictalurus* spp., *Hyborhynchus notatus, Catostomus* spp., *Notemigonus crysoleucas*; snail host *Gyraulus hirsutus*; adult in herons, cormorants, and unfed chick; Calif., Ill., Iowa, Manitoba, Mich., Minn., N. Dak., S. Dak., Wis. (Hugghins, 1954, 1959; Haderlie, 1953; Meyer, F., 1958; Hoffman, 1960).

Larval Genus **Tetracotyle**
(Fig. 137)

All known species of this group belong to the Strigeidae.

Characteristics of larval group: (1) forebody oval or ovate-oblong in contour and relatively thick, concave ventrally or cup-shaped; (2) hindbody a short, rounded prominence at posterior end of forebody,

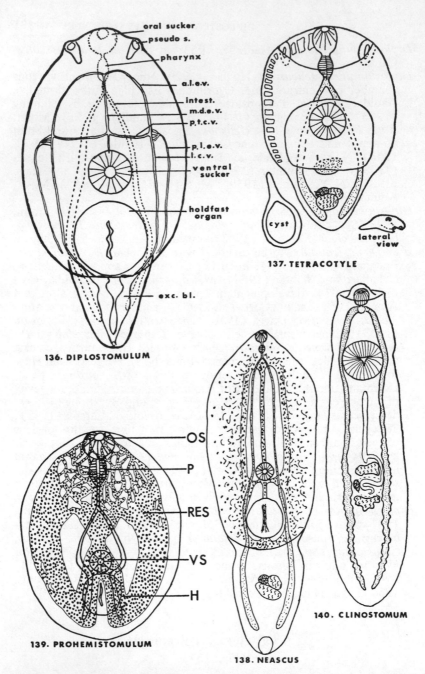

oral sucker
pseudo s.
pharynx
a.l.e.v.
intest.
m.d.e.v.
p.t.c.v.
p.l.e.v.
l.c.v.
ventral sucker
holdfast organ
exc. bl.

136. DIPLOSTOMULUM

137. TETRACOTYLE

cyst
lateral view

OS
P
RES
VS
H

139. PROHEMISTOMULUM

138. NEASCUS

140. CLINOSTOMUM

Figs. 136–140. Fig. 136. *Diplostomulum* of *Diplostomum baeri eucaliae.*
a.l.e.v. = anterior lateral excretory vessel; l.c.v. = lateral collecting vessel;
m.d.e.v. = median dorsal excretory vessel; p.l.e.v. = primary lateral excretory vessel; p.t.c.v. = posterior transverse commissural excretory vessel

sometimes inconspicuous; (3) reserve bladder a large continuous space occupying dorsal and lateral regions of forebody, with sheetlike extension into ventral lip of anterior suctorial pocket, with small spherical calcareous concretions in reserve excretory vessels and mostly in anterior part of worm; (4) a pair of lateral pseudosuckers (cotylae) on anterolateral edges beside oral sucker; (5) a true cyst of parasite origin.

This larval group has been described by Faust (1918) and Hughes (1928). The last synopsis of the *Tetracotyle* group was by Hoffman (1960).

KEY

TO THE KNOWN SPECIES OF TETRACOTYLE OF
NORTH AMERICAN FISHES

(Undoubtedly there are many undescribed species of *Tetracotyle*)

1. Cyst with prominent tail-like projection; found in musculature of *Eucalia inconstans* . . . . *Apatemon pellucidus*
1. Cyst without prominent tail-like projection; not found in musculature but found in viscera, particularly pericardium   2
2. Hindbody well defined; "suctorial pocket" well developed . . . . *T. lepomensis*
2. Hindbody lacking or not well defined; "suctorial pocket" lacking   3
3. Lateral cotylae as large as oral sucker . . . . *T. tahoensis*
3. Lateral cotylae decidedly smaller than oral sucker   4
4. Diameter of holdfast organ about one-half to one-third total length of parasite; difficult to remove parasite from cyst . . . . *T. diminuta*
4. Diameter of holdfast organ less than one-third total length of parasite   5
5. Intact cyst about 1000µ in diameter; parasite about 700µ long;

(from Hoffman, 1960). Fig. 137. *Tetracotyle* of *Apatemon gracilis pellucidus* (from Hoffman, 1959). Fig. 138. *Neascus* of *Posthodiplostomum minimum* showing three main reserve excretory vessels which give it "strigeid" (fluted) appearance (orig.). Fig. 139. *Prohemistomulum* (cyathocotylid) of *Neogogatea kentuckiensis*. OS = oral sucker; P = pharynx; RES = reserve excretory system; VS = ventral sucker; H = holdfast (from Hoffman and Dunbar, 1963). Fig. 140. *Clinostomum marginatum* (from Van Cleave and Mueller, 1934).

pharynx observed with difficulty; parasite easily freed from
cyst . . . . *Cotylurus communis*
5. Intact cyst about 600μ in diameter; parasite about 450μ long;
pharynx easily seen . . . . *T. intermedia*

Tetracotyle of *Apatemon gracilis pellucidus* (Yamaguti, 1933) Dubois,
  1953. Encysted in muscle, *Eucalia inconstans*, snail host unknown,
  adult reared in unfed chicks, N. Dak. (Hoffman, 1959).
Tetracotyle of *Cotylurus communis* (Hughes, 1928; LaRue, 1932):
  encysted in pericardial cavity, *Catostomus commersoni*, *Percopsis
  omiscomaycus*, *Stizostedion canadense*, *S. vitreum*. Snail host unknown. Adult in *Larus argentatus*, Mich.; Hoffman (unpubl.): in
  S. canadense, Minn., Wis.
Tetracotyle diminuta Hughes, 1928: encysted in pericardial cavity and
  adipose tissue behind eyes, *Perca flavescens*, *Percopsis omisco-
  maysus*, Mich.
Tetracotyle intermedia Hughes, 1928: encysted in pericardium, *Prosopium quadrilaterale*, *Leucichthys artedi*, Mich. Also reported in
  Russia.
Tetracotyle lepomensis Bogitsh, 1958: in mesenteries, *Lepomis macro-
  chirus*, Va.
Tetracotyle parvulum (*Diplostomum parvulum* Stafford, 1904). May
  be *Cotyluris communis*; cf. Hughes (1929).
Tetracotyle tahoensis Haderlie, 1953: encysted in pericardium, *Ca-
  tostomus tahoensis*, Calif.
Tetracotyle spp. Reported also by Bangham (1944, 1951, 1955):
  Mich., Wis., Wyo.; Bangham and Adams (1954): Brit. Col.; Bang-
  ham and Venard (1942): Tenn.; Fischthal (1947, 1950, 1956):
  N. Y., Wis.; Hunter (1942): Conn.; Sindermann (1953): Mass.;
  Van Cleave and Mueller (1934): N. Y.; Yamaguti (1942): Japan.

Larval Genus **Neascus** Hughes, 1927
(Fig. 138)

All known species of this group belong to the Diplostomatidae.

Characteristics of larval group: (1) forebody much like *Diplostomulum*; (2) hindbody more extensively developed than in *Diplostomulum*; (3) reserve bladder more extensively developed than in
Diplostomulum and with calcareous granules not confined to termini
of small branches which do not end blindly but constitute anastomoses;
(4) no lateral pseudosuckers or earlike processes; (5) generally encysted with true cyst of parasite origin.

The last synopsis of the group is that of Hoffman (1960).

KEY

(Key extracted and slightly modified from Hoffman, 1960. Undoubtedly there are many other undescribed species of *Neascus*.)

1. In cranial cavity of *Notropis cornutus* and *Pimephales promelas*; small oval cyst (also see no. 2) . . . .
. . . . *Ornithodiplostomum ptychocheilus*
1. In viscera, mesenteries, peritoneum   2
1. In musculature and integument   4
2. Hindbody relatively short; constriction slight; relatively small (cyst 750μ); in mesenteries of *Cyprinidae* . . . .
. . . . *Ornithodiplostomum ptychocheilus*
2. Hindbody relatively large; constriction pronounced; much larger (metacercaria more than 1 mm) . . . .
. . . . *Neascus* of *Posthodiplostomum* spp.   3
3. In liver, kidneys, and on heart of centrarchids . . . .
. . . . *P. minimum centrarchi*
3. In mesenteries of cyprinids . . . . *P. m. minimum*
3. In mesenteries of *Umbra limi*; forebody very large; ventral sucker in center of forebody; adult unknown . . . .
. . . . *Neascus* (*Posthodiplostomum?*) *grandis*
4. Black pigment surrounding cyst*; resembles *P. minimum* somewhat   5
4. No black pigment surrounding cyst; parasite cyst 570 x 340μ; parasite 600 to 700μ long . . . . *Neascus ellipticus*
5. Metacercaria nearly fills parasite cyst   6
5. Metacercaria does not fill parasite cyst; quite large (parasite cyst 450μ long; metacercaria 870μ long); metacercaria with finger-like anterior papilla; in perch; adult unknown . . . . . .
. . . . *Neascus longicollis*
6. Parasite cyst pear-shaped or egg-shaped   7
6. Parasite cyst oval or round   8
7. Parasite cyst about 330 by 200μ; egg-shaped and flattened dorsoventrally; in many fish . . . . *Uvulifer ambloplitis*
7. Parasite cyst smaller (about 270 by 160μ), abruptly narrowed at one end; in perch; adult not demonstrated, possibly is *Uvulifer semicircumcisus* in kingfisher . . . . . *Neascus pyriformis*
8. No ventral sucker; reserve excretory system similar to that of

_____

* Sometimes the pigment does not appear until two to three weeks after infection.

N. ambloplitis but forebody greatly cup-shaped . . . .
. . . . *Crassiphiala bulboglossa*
8. Ventral sucker present; reserve excretory system similar to that
of *N. ambloplitis* although simpler; cyst nearly round; in
dace; adult unknown . . . . *Neascus rhinichthysi*

Neascus of *Crassiphiala bulboglossa* Van Haitsma, 1925. *Metacercaria*
described by Hughes (1928): Black-spot skin cysts reported from
Cyprinidae, Cyprinodontidae, Esocidae, Etheostomidae, Percidae,
and Umbridae (see Hoffman, 1960, for complete host list). Snail
host is *Helisoma* (Hoffman, 1956; Larson, 1961); final host is kingfisher (Hoffman, 1956); widespread.
Neascus of *Mesoophorodiplostomum pricei* (Krull, 1934) Dubois,
1936 (Syn. *Neodiplostomum p.*): in *Fundulus* spp.; snail host unknown; adult in gulls (exper.), U. S.
Neascus ellipticus Chandler, 1951: nonpigmented cyst in musculature,
Perca flavescens; other hosts unknown, Minn.
Neascus of *Cercaria flexicorpa* Hobgood, 1938. Probably a synonym
of *Uvulifer ambloplitis*, Okla.
Neascus grandis Mueller and Van Cleave, 1932: in viscera, *Umbra
limi*. Resembles *Posthodiplostomum prosostomum* adult, N. Y.;
Myer, D. (1960): encysted in eyes, *U. limi*, Ill.
Neascus longicollis Chandler, 1951: pigmented cyst in skin, *Perca
flavescens*, Minn.
Neascus of *Ornithodiplostomum ptychocheilus* (Faust, 1917) Dubois,
1936: in viscera of Cyprinidae, in cranium of some, but not all.
Snail host *Physa*. Adult hosts are certain ducks; unfed chicks (Hoffman, 1954, 1958), Ill., Mich., Mont., N. Dak.
Neascus nolfi Hoffman, 1955. I consider this invalid because the description was apparently based on immature specimens in which the
reserve excretory system was not yet apparent.
Neascus pyriformis Chandler, 1951: pigmented cyst in skin, *Perca
flavescens*, Minn.
Neascus of *Posthodiplostomum minimum*. The metacercaria was described by Hughes (1928) as *Neascus vancleavei* and has been reported from many species of fish (Hoffman, 1958). In experimental
infections the cyprinid line will not infect centrarchids and the centrarchid line will not infect cyprinids and other fish families. Therefore, this species complex consists of at least two subspecies (below)
and probably others. Since the review of Hoffman (1958) the metacercaria has been reported by Alexander (1960), Anthony (1963),
Baldauf (1958), Bangham and Adams (1954), Bogitsh (1962,
1963), Colley and Olson (1963), Fischthal (1956), Hugghins
(1959), Kellog and Olson (1963), Lewis and Nickum (1964), Lyster (1939), McDaniel (1963), Meyer, F. (1958), Reed (1955),
and Wilson (1958). The adult has been reported from herons by
Lumsden and Zischke (1963) and from robins and red-winged
blackbirds by Ulmer (1960).
Neascus of *Posthodiplostomum minimum centrarchi* (MacCallum,
1921; Dubois, 1936) Hoffman, 1958: encysted in kidney, liver,

pericardium, and spleen of Centrarchidae; Hoffman (unpubl.): longevity at least 4 years in fish at 12°C. Snail host *Physa* spp. Final host herons, loons, unfed chicks (exper.), U. S., Can., Cuba.

Neascus of *P. m. minimum* (MacCallum, 1921; Dubois, 1936) Hoffman, 1958: encysted in mesenteries of Cyprinidae; life cycle otherwise as above.

Neascus rhinichthysi Hunter, 1933: in skin of *Rhinichthys* spp., N. Y.

Neascus of *Uvulifer ambloplitis* (Hughes, 1927) Dubois, 1938 (Syn. *Neascus* of McCoy, 1928; *Neascus* of Hobgood, 1938; *Neascus wardi*; *U. claviformis*; *U. magnibursiger*; the latter two are synonyms according to Dubois, 1964). This common black-spot has been recorded from Centrarchidae, Cyprinidae, and Esocidae, but Hoffman and Putz (1965) were unable to infect Cyprinidae, Catostomidae, Cottidae, Ictaluridae, Etheostomidae, and Salmonidae. Hoffman (unpubl.): metacercariae live at least 4 years in fish at 12°C. Snail host *Helisoma*; adult host kingfisher, widespread in U. S.

Neascus wardi Hunter, 1928. Probably a synonym of *U. ambloplitis*.

Larval Genus **Prohemistomulum** Ciurea, 1933
(Fig. 139)

All known species of this group belong to the Cyathocotylidae.

Characteristics of larval group: (1) body round or oval, flat, foliaceous, not separated into two parts; (2) no lateral pseudosuckers; (3) holdfast well developed; (4) reserve excretory system (bladder) of 2 main vessels, one lateral and other median, containing small calcareous corpuscles. Peripheral vessels bifurcate anteriorly, giving rise to median which also connects with them at posterior extremity.

The nature of the reserve excretory system is very helpful for separating some *Prohemistomulum* species from some *Neascus* species which have indistinct hindbodies. The author's interpretation of the excretory system of *Neogogatea kentuckiensis* is here offered as an addition to the larval group description: (4) reserve excretory system of 3 broad vessels, continuous and looped to form a crude W, fusing anteriorly in many anastomoses, and containing calcareous corpuscles throughout (fig. 139).

Ref. Hoffman (1960); Erasmus and Bennet (1963): excystation.

Holostephanus ictaluri Vernberg, 1952: adult in intestine, *Ictalurus punctatus*, Ind.

Prohemistomulum of *Linstowiella szidati* (Anderson, 1944) Anderson and Cable, 1950: in musculature of *Notropis cornutus*, cercariae from *Campeloma rufum*, exper. adult in baby chick and *Ardea herodias*, Ind.; Lumsden and Winkler (1962): adult in opossum, La.

Prohemistomulum of *Mesostephanus appendiculatoides* (Price, 1934) Lutz, 1935. Hutton and Sogandares (1960): in brackish fish (*Mugil*

spp.) which move into fresh water, adult in *Pelecanus*, exper. adult
in opossum, raccoon, *Nyctocorax n. hoactli, Larus delawarensis*,
cercaria in *Cerithium*.

Prohemistomulum of *Mesostephanus appendiculatus* (Ciurea, 1916)
Lutz, 1935. Life cycle: *Cerithidea* snail, *Fundulus, Gillichthys*,
baby chicks (Martin, 1961). Adult in dog and cat, U. S. (Kuntz
and Chandler, 1956).

Prohemistomulum of *Neogogatea kentuckiensis* (Cable, 1935) Hoffman and Dunbar, 1963 (Syn. *Mesostephanus k.* (Cable, 1935)
Myer, D., 1960). Metacercaria in musculature of Centrarchidae,
Clupeidae, Cyprinidae, Ictaluridae, Salmonidae, Percidae, *Platy-
poecilus maculatus*, and frogs (*Rana* spp.); *Cottus bairdi* refractory;
snail hosts *Goniobasis livescens, Anaplocamus dilatatus, Mudalia
carinata*; experimental final host was unfed chick, Ohio, W. Va.
(Myer, D., 1960; Hoffman and Dunbar, 1963).

Prohemistomulum of *Prohemistomum chandleri* Vernberg, 1952. Dubois (1953) and Myer, D. (1960): probably a synonym of *Neogogatea kentuckiensis*.

KEY

TO THE METACERCARIAE EXCLUSIVE OF STRIGEOIDS
OF NORTH AMERICAN FISHES

Many metacercariae are difficult to identify; this key is intended to
steer the worker toward the correct identification, but the illustrations
and original descriptions should be consulted often. Five undescribed
larval genera (*Acetodextra, Allacanthochasmus, Neochasmus, Cathaemasia, Opisthorchis*) are included in the hope that it will aid in
their discovery and description.

1. Mouth near middle of body; no true suckers; intestine
    usually a saclike structure containing globules; excre-
    tory bladder S-shaped or simple saccular; cyst oval,
    whitish to yellowish . . . . Family BUCEPHALIDAE   2
1. Mouth at anterior of body in typical oral sucker; ven-
    tral sucker usually present   3
2 (1). Anterior with simple globular sucker-like rhynchus;
    cyst about 100 x 74μ; some spp. progenetic . . . .
    . . . . *Bucephalus*
    . . . . (probably *Bucephaloides* also)
2 (1). Anterior with fan-shaped rhynchus; papillae of adult
    not evident; encysted at base of fins; small spines on
    body; cyst about 400 x 250μ . . . . (fig. 141) *Rhipidocotyle*
3 (1). Typical echinostome crown area at anterior end with
    one row of circumoral spines; excretory vesicle Y-
    shaped . . . . Family ECHINOSTOMATIDAE   5

3 (1). No echinostome crown area.................... 4

4 (3). Internal anatomy similar to echinostomes but head collar and circumoral spines absent; oral sucker may form pentagonal hoodlike expansion; excretory bladder Y-shaped 6

4 (3). Internal anatomy not similar to echinostomes...... 7

5 (3). 22 well-developed crown spines arranged in right and left rows separated dorsally and ventrally by space equal to width of oral sucker .... (fig. 144) *Stephanoprora*

5 (3). 19 or 20 crown spines, not separated by a space....
.........................(fig. 143) *Echinochasmus*
(fig. 142) *Petasiger*

6 (4). Adult in body cavity of kingfisher (metacercaria probably in fish but not described) ....*Cathaemasia*

6 (4). Metacercaria primarily in lateral line of fish; excretory vesicle with short stem and large arms containing conspicuous spherical granules........(fig. 145) *Ribeiroia*

7 (4). Circumoral spines present...................... 8

7 (4). Circumoral spines absent...................... 11

8 (7). Usually 1 row of circumoral spines; body partly or entirely spined; heterophyid gonotyl sucker may be present 9

8 (7). Usually 2 rows of circumoral spines; body entirely spined; heterophyid genital sucker may be present; usually in brackish-water fishes .... (fig. 153) *Ascocotyle* (p. 193) and subgenus *Leighia* (p. 194)

8 (7). 2 rows of circumoral spines preceded by incomplete dorsal row; excretory vesicle saccular; esophagus short; ceca long.........(fig. 147) *Paramacroderoides*
(p. 188)

9 (8). In brackish-water fishes...................... 10

9 (8). In freshwater fishes (metacercariae not completely described) ....*Allacanthochasmus*
(p. 193)
and (fig. 165) *Neochasmus*
(p. 201)

10 (9). Excretory vesicle Y-shaped; anterior half of body spined................(fig. 154) subgenus *Phagicola*
(p. 195)

10 (9). Excretory vesicle simple tubular; body entirely spined; disc-shaped concretions in ceca................
.........................(fig. 155) *Pygidiopsoides*
(p. 195)

11   (7). Heterophyid genital sucker (gonotyl) present......
........................Heterophyidae, in part   12
11   (7). Heterophyid genital sucker (gonotyl) absent (in some
heterophyid metacercariae, genital sucker not yet developed)   16
12 (11). Anterior of worm greatly fan-shaped, giving worm a
T-shape; adult, but not metacercaria, reported from
North America............(fig. 159) *Scaphanocephalus*
(p. 198)
12 (11). Anterior of worm not fan-shaped...............   13
13 (12). Excretory vesicle Y-shaped....................   14
13 (12). Excretory vesicle saccular; ceca short, containing concretions; in brain of *Fundulus* in brackish water.....
........................(fig. 157) *Euhaplorchis*
(p. 197)
13 (12). Excretory vesicle saccular; ceca long; exper. in musculature of *Fundulus* (marine)................
........................(fig. 160) *Galactosomum*
(p. 198)
14 (13). Esophagus long; ceca long; oral sucker larger than
ventral sucker; in *Fundulus* in brackish water......
........................(fig. 158) *Parastictodora*
(p. 197)
14 (13). Esophagus short or moderate; ceca long or short....   15
15 (14). Esophagus moderate; ceca long; small heterophyid
with prominent anterior excretory tubules; brackish
and freshwater........(fig. 156) *Cryptocotyle concavum*
(p. 197)
15 (14). Probably—esophagus short; ceca long; oral sucker
smaller than ventral sucker (metacercaria not described yet) ....*Acetodextra*
(p. 199)
15 (14). Esophagus short; ceca moderate; oral sucker larger
than ventral sucker..........(fig. 161) *Cryptogonimus*
(p. 199)
16 (11). Large metacercaria; usually in yellowish cyst; oral
sucker surrounded by collar-like fold when retracted;
short esophagus swollen posteriorly without forming a
typical pharynx; large ventral sucker near anterior end
........................(fig. 140) *Clinostomum*
16 (11). Smaller metacercariae; not in yellowish cysts.......   17
17 (16). Excretory vesicle Y-shaped...................   18
17 (16). Excretory vesicle saccular   20

18 (17). Esophagus and ceca short.................... 19
18 (17). Esophagus moderate, ceca long; Y-shaped excretory
vesicle with saccular or long stem; elongated worms
with prominent anterior excretory tubules........
.... (fig. 151) *Amphimerus, Metorchis*
(p. 190)
19 (18). Oral sucker larger than ventral sucker; small metacercaria....................... (fig. 164) *Caecincola*
(p. 201)
19 (18). Oral sucker about same size as ventral sucker; large
metacercaria; gonads well developed............
.... (fig. 163) *Centrovarium*
(p. 201)
20 (17). Esophagus long; excretory vesicle saccular, S-shaped;
elongated worms; one species causes black-spot cysts
.... (fig. 152) *Apophallus*
(p. 191)
20 (17). Esophagus moderate or short................. 21
21 (20). Esophagus moderate, ceca long; oral sucker larger than
ventral sucker.............. (fig. 146) *Macroderoides*
(p. 187)
21 (20). Esophagus short, ceca long or short; oral sucker smaller than, or about same size as, ventral sucker; body
small, piriform 22
22 (21). Ceca short .... (fig. 149) *Sellacotyle*
(p. 189)
22 (21). Ceca long 23
23 (22). Small piriform metacercaria..... (fig. 148) *Nanophyetus*
(p. 188)
23 (22). Probably small elongate metacercaria similar to *Am-phimerus* except for saccular excretory vesicle (metacercaria not described yet in North America)......
.... (fig. 150) *Opisthorchis*
(p. 190)

Superfamily CLINOSTOMATOIDEA
(Fig. 140)

Metacercaria: Body stout, linguiform, convex dorsally, concave ventrally. Oral sucker surrounded by collar-like fold when retracted.
Esophagus swollen bulbously at posterior end without forming typical
pharynx. Large ventral sucker in anterior third of body. Two immature testes at about middle of hindbody or near posterior extremity.

Ovary to right between testes. Immature uterus anterior to testes. Vitellaria not formed. Excretory vesicle small, V-shaped, with dorso-terminal pore. In fish, frogs, salamanders, and land snail (*Subulina*). Adults develop in mouth and esophagus of birds. Cercaria: brevifurcous, pharyngeate.

Ref. Yamaguti (1958, p. 686); Agarwal (1959), key to spp.

Clinostomum marginatum (Rud., 1819). This common, large yellow grub was first reported in North America by Leidy (1856). It has since been reported from so many fish that it is safe to assume that it is capable of infecting any species of our freshwater fish. Reported by Anthony (1963), Bangham (1926, 1927, 1933, 1939, 1941, 1944, 1951, 1955), Bangham and Hunter (1938), Bangham and Venard (1942, 1946), Chandler (1951), Cooper (1915), DeRoth (1953), Edney (1940), longevity 3 years 9 months, Elliot and Russert (1949), Fischthal (1945, 1950), Haderlie (1953), Harms (1959), Hoffman (unpubl.): longevity at least 2 years, Hollis and Coker (1948), Hopkins (1933), Hugghins (1959), Hunninen (1936), Hunter (1942), Hunter and Dalton (1939), Krueger (1954), Linton (1940), Lyster (1939), Meyer, F. (1958), Meyer, M. (1954), Nigrelli (1935, 1936), Osborn (1911, 1912), Pearse (1924), Pratt (1923), Schwartz (1956), Simer (1929), Sindermann (1953), Van Cleave and Mueller (1934), Worley and Bangham (1952). Snail host *Helisoma* (Krull, 1934); final host heron (Osborn, 1912; Hunter and Hunter, 1934). Cercaria (Krull, 1934; Larson, 1961). Incidence and possible control in Kansas (Klass, 1963).

The yellow grub is very unsightly to fishermen and, if numerous, does considerable damage to fish in hatcheries and in nature. A related *Clinostomum* species has been found to infect the upper respiratory tract of man and other mammals in Asia, although it is usually found in the mouth of herons.

Superfamily BUCEPHALOIDEA

Cercariae furcocercous and gasterostomate; tail stem short and bulbous; furcae very long and active; excretory vesicle cylindrical; development in branched sporocysts in lamellibranchs of fresh and brackish waters.

Family BUCEPHALIDAE

Gasterostomate metacercaria: Mouth ventral, without oral sucker. Rhynchus (adhesive organ) at anterior extremity. Pharynx and esophagus present near middle of body. Intestine saccular or tubular.

Testes long, (if present) usually in hindbody. Ovary in front of or between testes. Excretory vesicle usually tubular with terminal open-vesicle very long. Life cycle: see adult (p. 112).

Genus **Bucephaloides** Hopkins, 1954
(Metacercarial description not available)

Bucephaloides clara Komiya, 1943: in musculature, *Carassius auratus*, China.

B. strongylurae Hopkins, 1954: metacercaria free in body cavity of *Menidia beryllina*, which sometime reaches fresh water, Tex.

Genus **Bucephalus** Baer, 1826
(See adult)

Metacercaria: Body elongate. Rhynchus sucker-like with 7 tentacular appendages. Mouth opening in middle third of body, intestine short. Excretory vesicle tubular, variable in length. Life cycle: see adult (p. 114).
Ref. Woodhead (1930).

Bucephalus elegans Woodhead, 1930: in fins and muscle of fish, Mich.; Fischthal (1947): in *Catostomus commersoni*, *Hyborhynchus notatus*, *Hypentelium elegans*, *Nocomis biguttatus*, *Notropis* spp., Wis.

Genus **Rhipidocotyle** Diesing, 1858
(Fig. 141)

Metacercaria: Like *Bucephalus* except rhynchus with a pentagonal cap or hooklike expansion and suctorial pit ventroposteriorly. Excretory vesicle very long. Life cycle: see adult (p. 114).

Rhipidocotyle papillosa (Woodhead, 1929). Eckmann (1932): encysts at base of fin rays, *Ambloplites rupestris*, Mich.; Fischthal (1947, 1956): in *Micropterus dolomieui*, Wis., in *Cottus bairdii*, N. Y.; adult in predatory fish.

R. septpapillata Krull, 1934: in *Fundulus diaphanus*, *Lepomis gibbosus*, Va.; Kniskern (1950, 1952): in *L. gibbosus*, *Lebistes* sp., *Micropterus salmoides*, *Semotilus atromaculatus*.

Order ECHINOSTOMIDA LaRue, 1957
Suborder ECHINOSTOMATA Szidat, 1939

Cercariae with large bodies and strong tails; collar and collar spines present. Cystogenous glands numerous. Development in collared rediae with stumpy appendages.

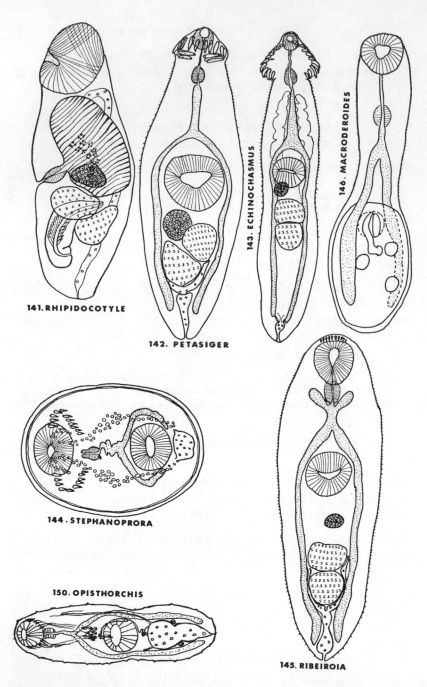

Figs. 141–146. Fig. 141. *Rhipidocotyle papillosum* (from Kniskern, 1952).
Fig. 142. *Petasiger nitidus*, adult, modified from Beaver (1939). Fig. 143.
Echinochasmus donaldsoni, adult, modified from Beaver (1941). Fig.
144. *Stephanoprora* sp., encysted (from Lee and Seo, 1959). Fig. 145.

Family ECHINOSTOMATIDAE

Metacercaria: More or less elongated distomes with head collar usually provided with single or double crown of spines. Cuticle usually with spines or scales. Oral sucker, pharynx, and esophagus present. Ceca reaching to posterior extremity. Ventral sucker well developed. Excretory vesicle Y-shaped. Adults usually in intestine of birds and mammals.

Genus **Petasiger** Dietz, 1909
(Fig. 142)

Metacercaria: Body small, plump, broadly fusiform with maximum breadth at middle; neck region more or less constricted. Head collar reniform, large, with double, dorsally uninterrupted row of spines. Ventral sucker equatorial. Cercaria: echinostome.
Ref. Yamaguti (1958, p. 637).

Petasiger nitidus Linton, 1928: adult in *Colymbus auritus*; Beaver (1939): metacercaria encysted in esophagus of *Ambloplites rupestris, Ictalurus nebulosus, Lebistes reticulatus, Lepomis macrochirus, Notropis hudsonius, Perca flavescens, Umbra limi*; cercaria from *Helisoma* spp. must be eaten by fish; exper. adult in canaries, Mich.

Genus **Echinochasmus** Dietz, 1909
(Fig. 143)

Metacercaria: Unpigmented host cyst. Body doubled in parasite cyst, which is 68 to 76µ by 40 to 44µ. Head collar strongly developed, reniform, with single, dorsally interrupted row of spines 3 to 7µ long. Ventral sucker well separated from oral sucker. Excretory bladder Y-shaped, extending to oral sucker, enlarged and containing large (4µ) granules. Cercaria: echinostome.
Ref. Yamaguti (1958, p. 643).

Echinochasmus donaldsoni Beaver, 1941: metacercaria in gills, *Amia calva, Eucalia inconstans*; exper. in *Ictalurus nebulosus, Lebistes reticulatus, Lepomis macrochirus, Mollienesia latipinia, Notropis* spp., *Perca flavescens, Umbra limi*; cercaria from *Amnicola* spp.; adult in *Podilymbus podiceps* (grebe), exper. in pigeon, Mich.
E. schwartzi (Price, 1931): in gills, *Fundulus heteroclitus* (Lillis and Nigrelli, 1965).

Ribeiroia ondatrae, adult, modified from Beaver (1939). Fig. 146. *Macroderoides spiniferus* (from Leigh, 1958).

Genus **Euparyphium** Dietz, 1909

Euparyphium melis (Schrank, 1788). In nares and cloaca of *Ambloplites rupestris, Ictalurus nebulosus, Lepomis macrochirus, Perca flavescens, Percina caprodes, Umbra limi* (normal intermed. host apparently frog tadpoles), cercaria in *Stagnicola emarginata*, adult in *Mustela vison*, Mich. (Beaver, 1941); Lumsden (1961) = *E. beaveri* Yamaguti, 1958.

Genus **Stephanoprora** Odhner, 1902
(Fig. 144)

Metacercaria (Lee and Seo, 1959): Cysts ovoid, unpigmented; worm about 152 by 108μ; 22 well-developed crown spines of nearly equal size in right and left rows separated dorsally and ventrally by space about equal to width of oral sucker. Y-shaped excretory bladder extending to oral sucker and filled with spherical granules. Cercaria: echinostome.

Ref. Yamaguti (1958, pp. 468, 648).

Stephanoprora denticulata (Rud., 1802). Reported from gills of *Fundulus heteroclitus* but not described (Stunkard and Uzmann, 1962).
S. polycestus Dietz, 1909. Metacercaria in fish fed to crows which became infected (Beaver, 1936). Adult found in grebe (Pratt and Matthias, 1962).
Stephanaprora sp. Lee and Seo, 1959: experimentally infected gulls, *Eucalia inconstans, Lebistes reticulatus*, Mich.

Family CATHAEMASIIDAE

Elongate distomes with internal anatomy similar to that of Echinostomatidae. Head collar and circumoral spines absent, but oral sucker may form pentagonal hoodlike expansion. Excretory system Y-shaped. Adults in birds.

Genus **Cathaemasia** Looss, 1899
(No fig.)

Cathaemasia pulchrosoma (Travassos, 1917) Manter, 1949: adult in body cavity of *Ceryle alcyon* (kingfisher); metacercaria probably in fish. This may be *Pulchrosomum* sp. (Stunkard, 1965).
Ref. Yamaguti (1958, p. 618).

Genus **Ribeiroia** Travassos, 1939
(Fig. 145)

Metacercaria: Body flat, wide, spined. Very little development. Oral sucker large, terminal. Ventral sucker large, protrusible, in posterior third. Prepharynx present; long esophagus with lateral excretory "siphons" containing conspicuous spherical granules. Cercaria: gymnocephalous. Adult: in mammals and proventriculus of birds.

Ref. Yamaguti (1958, p. 621).

Ribeiroia ondatrae (McMullen, 1938; Beaver, 1939) Yamaguti, 1958 (Syn. *Psilostomum ondatrae* Price, 1931, of Beaver, 1939). Beaver (1939): small (320μ) cysts primarily in lateral line of *Ambloplites rupestris, Ictalurus* sp., *Lepomis gibbosus, L. macrochirus, Micropterus dolomieui, Perca flavescens*; cercaria from *Helisoma antrosum*; exper. adult from proventriculus of chicken, duck, canary, Mich.; Kuntz (1951): metacercaria in *A. rupestris, L. gibbosus*; cercaria from *H. antrosum*; exper. adult in muskrat (*Ondatra zibethica*), Mich.; Lumsden and Zischke (1963): adult in *Hydranassa tricolor*, Pa., discussion of synonomy; Price (1931): from muskrat, gull, Ont., Ore.; Newsom and Stout (1933): from chickens, Colo.; Beaver (1939): Osprey, Cooper's hawk, Mich.

Superorder EPITHELIOCYSTIDIA LaRue, 1957

Primitive excretory bladder of cercaria surrounded by, and then replaced by, layer of cells derived from mesoderm; hence definitive bladder thick-walled and epithelial. Cercarial tail single, reduced in size, or lacking; caudal excretory vessels present or lacking; miracidium with one pair of flame cells.

Order PLAGIORCHIIDA LaRue, 1957

Cercariae completely lacking caudal excretory vessels at any stage of development; stylet present or lacking.

Family MACRODEROIDIDAE McMullen, 1937

(See section on adult trematodes for family diagnosis.)

Genus **Macroderoides** Pearse, 1924
(Fig. 146)

Metacercaria: Body pear-shaped to elongate, spined. Oral sucker large, prepharynx long. Esophagus moderate, branching in middle third of body; ceca terminates at posterior extremity. Ventral sucker smaller than oral sucker, in middle third of body. Reproductive system represented by three small gonads. Excretory vesical saccular, filled with refractile granules. Cercaria: xiphidiocercaria without fin folds. Adult: in intestine of freshwater fish.

Macroderoides spinifera Pearse, 1924. Leigh (1958): metacercaria encysted in muscle, *Fundulus* sp., *Gambusia affinis, Heterandria*

formosa, Mollienesia latipinna; also in frog tadpoles (*Rana* spp.);
cercaria in *Helisoma duryi*; adults in *Lepisosteus platyrhincus*, Fla.
M. typica (Winfield). McMullen (1935): metacercaria exper. in bullheads, but normal host is frog tadpole, Mich.

Genus **Paramacroderoides** Venard, 1941
(Fig. 147)

Metacercaria: Body small, pear-shaped to elongate, spined. Oral
sucker large, terminal with one incomplete circle of dorsal spines
followed by two complete circles. Prepharynx short, esophagus short;
ceca extend nearly to posterior extremity. Immature gonads present.
Excretory bladder saccular, filled with refractile granules. Cercaria:
xiphidiocercaria with fin folds. Adult: in intestine of gars (*Lepisosteus*).

Paramacroderoides echinus Venard, 1941. Leigh and Holliman
   (1956): metacercaria encysted in muscle, *Fundulus* sp., *Gambusia
   affinis, Heterandria formosa, Jordanella floridae, Mollienesia latipinna*; cercaria in *Helisoma duryi*; adult in *Lepisosteus platyrhincus*,
   Fla.

Family TROGLOTREMATIDAE (Odhner, 1914)
Wallace, 1935

Thick monostomes or distomes 2 to 13 mm long, living in cystlike
cavities in birds or mammals, or smaller piriform distomes living in
intestines of mammals. Body covered with spines. Digestive tract with
pharynx, esophagus, and two intestinal ceca of varying length. Excretory vesicle Y-shaped or saclike. Genital pore with no external modi-
fication, shortly before or behind ventral sucker. Seminal vesicle large,
cirrus pouch absent except in *Troglotrema*. Testes equal and side by
side, anterior or posterior to ventral sucker. Ovary anterior to testes
except in *Nephrotrema,* in which latter are anterior to ventral sucker.
Seminal receptacle and Laurer's canal present. Uterus long and loosely
coiled, eggs small, 17μ to 25μ long; or uterus short, eggs larger, 60μ
to 85μ long. Cercariae of microcercous type.

Genus **Nanophyetus** Chapin, 1927
(Fig. 148)

Metacercaria: Body small, piriform, slightly flattened. Oral sucker
large, subterminal, no prepharynx. Esophagus short, ceca long. Ventral sucker nearly equal to oral sucker, nearly equatorial. Excretory

vesicle saccular. Cercaria: microcercous xiphidiocercaria. Adult: intestinal parasites of Carnivora and man.

Nanophyetus salmincola Chapin, 1926 (Syn. *Troglotrema s.*; *Distomulum oregonensis* Ward and Mueller, 1926). Small encysted metacercaria in kidney, musculature, eye orbit, etc. of *Carassius, Cottus, Gila, Oncorhynchus, Salmo, Salvelinus*. Cercaria from *Oxytrema silicula*. Adults in intestine of canids, cat, bear, mink, hog, guinea pig, raccoon, hamster, and man. Bennington and Pratt (1960): latest report on life cycle; Wood and Yasutake (1956): histopath.; Farrell *et al.* (1964), and Pratt *et al.* (1964): longevity 2 to 4 years. Gebhardt, Millemann, Knapp and Nyberg (1966): naturally in *Cottus perplexus, Lampetra* spp., *Richardsonius balteatus* and the giant salamander, *Dicamptodon ensatus*; experimentally in *S. gairdneri, S. salar, S. trutta, Salvelinus fontinalis, S. namaycush, Lampetra richardsoni, C. perplexus, Carassius auratus, R. bolteatus, Catostomus macrocheilus, Lepomis macrochirus, Gasterosteus aculeatus* and *Gambusia affinis*; Ore. This parasite causes serious disease of salmonids in hatcheries in Oregon and Washington. It also carries a rickettsia which causes a serious disease of canids.

Genus **Sellacotyle** Wallace, 1935
(Fig. 149)

Metacercaria: Body small, piriform. Oral sucker large, prepharynx very short, pharynx subglobular, esophagus short; ceca short, not passing ventral sucker. Ventral sucker large, equatorial. Juvenile testes anterolateral to excretory vesicle. Excretory vesicle saccular, filled with highly refractive droplets. Cercaria: microcercous distome. Adult: in intestine of mammals.

Sellacotyle mustelae Wallace, 1935: small cyst in flesh and mesenteries of *Ictalurus melas, Schilbeodes gyrinus, Moxostoma aureolum, Pimephales promelas, Semotilus* sp., *Notropis* sp.; cercariae in *Campeloma rufum*; adult embedded in intestinal mucosa of mink, dog, cat, rat, ferret, fox, raccoon, and skunk, Minn.; Erickson (1946): in mink, Minn.; Harkema and Miller (1961): in raccoon, S. C.

Order OPISTHORCHIIDA LaRue, 1957

Cercariae with caudal excretory vessels during development; stylet always lacking. Dunagan (1960), key to cercariae.

Family OPISTHORCHIIDAE Braun, 1901

Metacercaria: Body small, sometimes elongated. Suckers well developed in metacercariae but become weak in adults. Short prepharynx, esophagus moderate; ceca usually reach posterior extremity. Gonads very small or not present. No sucker-like structures (gonotyls) in genital atrium. Excretory vesicle Y-shaped but with saccular stem.

Genus **Opisthorchis** Blanchard, 1895
(Fig. 150)

Metacercaria: Body flattened, elongated moderately, spined or not. Ventral sucker large (small in adult), near equator. Oral sucker well developed; short prepharynx. Esophagus short; ceca terminate near posterior extremity. Excretory stem saccular, sigmoid. Anterior excretory tubules prominent. Cercaria: lophocercous. Adult: in bile ducts, gall bladder, and pancreas of mammals.

Ref. Yamaguti (1958, pp. 484, 492, 855).

Opisthorchis tonkae Wallace and Penner, 1939: encysted in *Notropis deliciosus*; exper. adult in muskrat, Minn.; Sillman (1953): metacercaria in *Lepomis gibbosus, Hyborhynchus notatus*; exper. adult in mouse, muskrat, cat, rat; cercaria in *Amnicola limosa*, Mich.

Genus **Metorchis** Looss, 1899
(No fig.)

Metacercaria: Body small; cuticle spined or not. Oral sucker well developed, pharynx usually small. Esophagus short; ceca terminate at posterior extremity. Ventral sucker about equal to oral sucker, situated near equator. Reproductive system not developed. Excretory vesicle Y-shaped but with large stem. Cercaria: oculate with dorsal and ventral fin folds. Adult: in mammals.

Ref. Yamaguti (1958, p. 859).

Metorchis conjunctus (Cobbold, 1860) Looss, 1899 (Syn. *Parametorchis canadense* Price, 1929). Cameron (1944, 1945): encysted (0.4 mm cyst) in flesh of *Catostomus commersoni*; cercaria in *Amnicola limosa*; adult in bile ducts of liver of mammals including man, Can.; Erickson (1946): adult in mink, Minn.; Jordan and Ashby (1957): adult in dog, S. C.; Rawson (1960): adults killed 10 sled dogs, Sask.

Genus **Amphimerus** Barker, 1911
(Fig. 151)

Metacercaria: Body flattened, much elongated, spined. Ventral sucker large (small in adult). Prepharynx distinct, esophagus moderate length; ceca reach to posterior extremity. Excretory vesicle Y-shaped but branches small and saccular stem prominent. Anterior excretory tubules very prominent. Cercaria: oculate, lophocercous, with pigment cells. Adult: in reptiles, birds, mammals.

Ref. Yamaguti (1958, pp. 484, 692, 853).

Amphimerus elongatus Gower, 1938. Wallace (1939, 1940): small cyst in flesh of *Notropis deliciosus*; cercaria in *Amnicola limosa*; exper. adult in chicks, ducklings, Minn.; Gower (1938): adult in ducks and swans, metacercaria in fish, Mich.; Hoffman (1965): in *Notropis cornutus*, W. Va.

A. pseudofelineus (Ward, 1901) Barker, 1911. Evans (1963): metacercaria in *Catostomus commersoni*, Manitoba.

Family HETEROPHYIDAE

Metacercaria: Distomes small, covered with scalelike spines. Oral sucker and pharynx present. Ceca long or short. Ventral sucker well developed or not, median or submedian. Gonotyl (genital sucker) may or may not be developed yet. Excretory vesicle variable, F-, Y-, or T-shaped, occasionally saccular.

Ref. Yamaguti (1958, pp. 699, 865); Van Cleave and Mueller (1934, p. 216); see also section on adult trematodes.

Genus **Apophallus** Lühe, 1909
(Fig. 152)

Metacercaria: Body small, flattened, elongated, spined. Oral sucker small, prepharynx present, esophagus long; ceca terminate near posterior extremity. Ventral sucker near middle of body. Gonads may be present. Gonotyl not yet developed in species described. Excretory vesicle saccular, sigmoid (becomes Y-shaped in adult). Cercaria: pleurolophocercous; ventral sucker rudimentary. Adult: in intestine of birds and mammals.

Ref. Cameron (1945).

Apophallus brevis Ransom, 1920. Miller (1941, 1942, 1946): metacercaria enclosed in black cyst in *Salvelinus fontinalis*, *Salmo trutta*, but not *S. gairdneri* or *S. namaycush*; cercaria develops in *Amnicola limosa*, Que.; Ransom (1920): adult in gulls and loons; Cameron (1945): exper. adult in cats, pigeons natural in loons; Penner (1940): see Odlaug (1956): adult in muskrat, Minn.

A. donicus (Skrj. and Lindtrop, 1919) Price, 1931 (Syn. *Rossicotrema d.*, *Cotylophallus venustus*, *C. similis*). Metacercaria not reported from North America yet, but adult of this European parasite

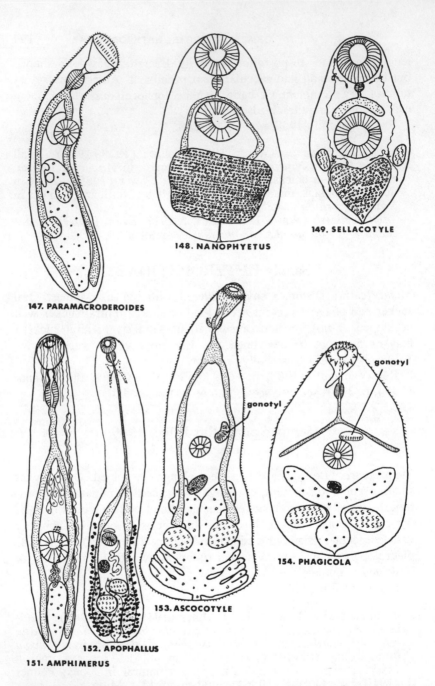

147. PARAMACRODEROIDES

148. NANOPHYETUS

149. SELLACOTYLE

151. AMPHIMERUS

152. APOPHALLUS

153. ASCOCOTYLE

gonotyl

gonotyl

154. PHAGICOLA

Figs. 147–154. Fig. 147. *Paramacroderoides echinus* (from Leigh and Holliman, 1956). Fig. 148. *Nanophyetus salmincola* (from Bennington and Pratt, 1960). Fig. 149. *Sellacotyle mustelae* (from Wallace, 1935). Fig. 150. *Opisthorchis*, European (from Vogel, 1934) (find on preceding plate). Fig. 151. *Amphimerus elongatus* (from Wallace, 1939). Fig. 152.

recorded in North America by Price (1931); Cameron (1945): Que.; Shaw (1947): Ore.; Douglas (1951): from dog, cat, rabbit, mouse, rat, fox, mink, merganser, heron, hawk, owl, gull, tern, Calif.

A. *imperator* Lyster, 1940. Cameron (1945): probably a synonym of *A. brevis*.

A. *itascensis* Warren, 1953: pigmented cyst, *Perca flavescens*, *Notropis* sp., Minn. Cyst shaped like balloon tire. Adult unknown.

A. *venustus* (Ransom, 1920), (Syn. *Cotylophallus v.*) Cameron (1945): metacercaria in unpigmented oval cysts, *Amia calva*, *Catostomus commersoni*, *Cyprinus carpio*, *Esox lucius*, *Ictalurus nebulosus*, *Lepisosteus osseus*, *Lepomis gibbosus*, *Micropterus dolomieui*, *Moxostoma aureolum*, *Notropis cornutus*, *Perca flavescens*, *Stizostedion vitreum*; adult in cats, dogs, seals, raccoons and great blue heron; cercaria in *Goniobasis liviscens*, Que.; Babero and Shepperson (1958): in raccoon, Ga.

Apophallus sp. Chandler, 1951: in *Perca flavescens*, Minn.

Apophallus sp. Bangham, 1951: in *Salmo clarki*, Wyo.

Genus **Allacanthochasmus** Van Cleave, 1933
(See adult)

Metacercaria: Not described, but probably—body small, elongate, spined. Oral sucker nearly terminal with single complete circle of spines. Prepharynx present, pharynx well developed, esophagus short, ceca long. Ventral sucker smaller than oral sucker, in middle third of body. Genital atrium (gonotyl) probably not well developed yet. Excretory vesicle Y-shaped; arms reach to pharynx.

Ref. Yamaguti (1958, p. 226); Van Cleave and Mueller (1934, p. 224).

Allacanthochasmus varius Van Cleave, 1922. Van Cleave and Mueller (1934): encysted in various fishes, especially minnows.

Genus **Ascocotyle** Looss, 1899
(Fig. 153)

Metacercaria: Body small, usually pear-shaped, spined. Eye spots present. Oral sucker with sensory anterior lip and posterior solid muscular appendage of varying lengths surrounded by membrane embedded in parenchyma of forebody; with or without spines. Pre-

---

Apophallus americanus (from Van Cleave and Mueller, 1932). Fig. 153.
Ascocotyle chandleri adult, modified from Lumsden (1963). Fig. 154.
Phagicola sp. from *Gambusia affinis* (orig.).

pharynx, pharynx, and esophagus present. Ceca extend to level of ventral sucker or sometimes to posterior of body. Genital sac contains variously shaped, spined, or unspined gonotyl which may not be well developed in metacercaria. Gonads already present in some species, testes side by side in posterior half of body; ovary usually roundish and anterior to testes. Coarse vitelline follicles may already be present lateral to testes. Excretory vesicle Y-shaped but posterior stem may be enlarged or branched. Cercaria: parapleurolophocercous, pleurolophocercous, or ophthalmogymnocephalous. Cercariae of subgenus *Ascocotyle* parapleurolophocercous, body narrow, more or less uniform in diameter, oral sucker not recessed into a chamber. Sogandares and Lumsden (1963) propose subgenera *Ascococotyle, Phagicola*, and *Leighia* for *Ascocotyle* "complex." Adult: in birds and mammals. Vitellaria extend to ventral sucker, with two rows of oral spines; uterus mainly confined to postventral sucker area.

Ref. Yamaguti (1958, p. 704); Sogandares-Bernal and Lumsden (1963); Burton (1958), key to spp.

Subgenus **Ascocotyle** Looss, 1899
(Fig. 153)

Metacercaria: Two rows of oral spines; body entirely spined.

A. diminuta Stunkard and Haviland, 1924 (Syn. of *Phagicola angrense*).

A. leighi Burton, 1956: encysted beneath endothelium in conus arteriosus, *Mollienesia latipinna*; exper. adult in baby chick, Fla.; Sogandares and Bridgman (1960): *M. latipinna*, La.; Sogandares and Lumsden (1964): *Belonesox belizanus, Cyprinodon variegatus, M. latipinna, M. sphenops*, Fla., La., Tex.; adult in *Hydranassa tricolor, Cosmerodius albus*.

A. tenuicollis Price, 1935. Leigh (1956): encysted in *Gambusia, Mollienesia, Chaenobryttus*; exper. adult in chick, natural in herons, Fla.; Burton (1956): in conus, *Gambusia, Mollienesia*, Fla.; Price (1935): in heron, Tex.

Subgenus **Leighia** Sogandares and Lumsden, 1963

Metacercaria: Probably very similar to *Ascocotyle*. Cercaria: Ophthalmogymnocephalus, body subspherical, oral sucker recessed into chamber. Adult: vitellaria extend to ventral sucker, with two rows of oral spines; uterus extends to level of pharynx.

Ascocotyle (Leighia) chandleri Lumsden, 1963: progenetic adult encysted in liver, *Cyprinodon variegatus, Mollienesia latipinna*; exper. adult in chick, Tex.

A. (Leighia) mcintoshi Price, 1936. Leigh (1956): encysted in mesenteries and viscera, *Gambusia affinis*, *Mollienesia latipinna*; exper. adult in chicks, natural in *Guara alba*, Fla.; Price (1936): adult in *G. alba*, Fla.

Subgenus **Phagicola** Faust, 1920
(Fig. 154)

Metacercaria: Probably very similar to *Ascocotyle* (A.) but with one, rarely two, or no rows of oral spines; when one row is present there may be a few dorsal accessory crown spines; anterior half of body spined. Cercaria: pleurolophocercous; body usually piriform when at rest; oral sucker not recessed into a chamber.

Ascocotyle (*Phagicola*) *angrense* (Travassos, 1916) Travassos, 1930 (Syn. *Ascocotyle rana*, *A. diminuta*, *Phagicola lageniformis*). Metacercaria reported by Sogandares and Bridgman (1960), Stunkard and Uzmann (1955), Hutton and Sogandares (1958, 1959), Sogandares and Lumsden (1963), from gills, *Belonesox belizanus*, *Cyprinodon variegatus*, *Fundulus chrysotus*, *F. grandis*, *F. heteroclitus*, *F. jenkinsi*, *F. majalis*, *F. similis*, *Lucania parva*, *Mollienesia latipinna*, *M. sphaenops*, Conn., Fla., La., N. Y., Md., Mex. Adult in many birds (ardeiform) and sometimes mammals.

A. (*Phagicola*) *longa* (Ransom, 1920) Witenberg, 1929. Hutton (1958): encysted in mullet heart; one row of oral spines; exper. adult in raccoon, Fla.; Hutton and Sogandares (1958, 1959, 1960); Harkema and Miller (1962): adult in pelicans, egret, chicks, herons, cormorant, gull, opossum, rat, hamster, mink, raccoon, Fla., S. C.

A. (*P.*) *molienesicola* (Sogandares and Bridgman, 1960), (Syn. *Pseudascocotyle m.*): encysted in intestinal wall, muscle, and gills, *Mollienesia latipinna*; no oral spines, La.

Genus **Pygidiopsoides** Martin, 1951
(Fig. 155)

Metacercaria: Small encysted worms. Mouth surrounded by a single row of large spines (6μ); remnants of eye spots present. Oral sucker oval in lateral view, not tapering posteriorly into a cone. Prepharynx, pharynx, and esophagus well developed; ceca terminate near posterior border of ventral sucker and filled with disc-shaped concretions. Ventral sucker near midbody level, well developed. Excretory vesicle simple, spindle-shaped to tubular, containing granular concretions.

Pygidiopsoides spindalis Martin, 1951. Martin (1964): small oval cysts in gills, *Fundulus parvipinnis*; simple-tailed monostome cercariae from *Cerithidea californica*; exper. adults in chicks and cats; Calif.

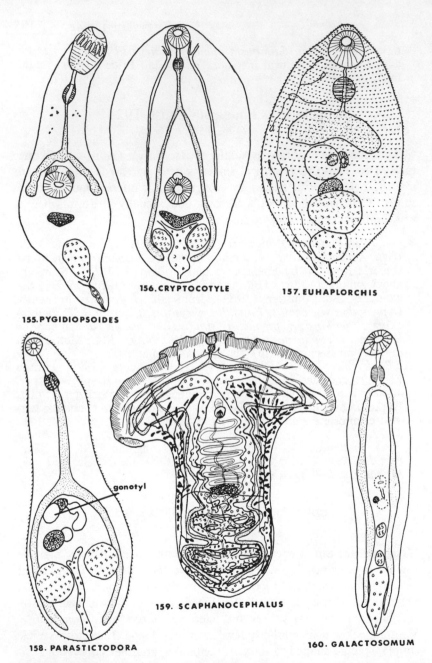

155. PYGIDIOPSOIDES

156. CRYPTOCOTYLE

157. EUHAPLORCHIS

gonotyl

158. PARASTICTODORA

159. SCAPHANOCEPHALUS

160. GALACTOSOMUM

Figs. 155–160. Fig. 155. *Pygidiopsoides spindalis*, adult, modified from Martin (1951). Fig. 156. *Cryptocotyle concava* from *Catostomus commersoni* (orig.). Fig. 157. *Euhaplorchis californiensis*, adult, modified from Martin (1950). Fig. 158. *Parastictodora hancocki*, adult, modified from Martin (1950). Fig. 159. *Scaphanocephalus expansus* (from Yama-

Genus **Cryptocotyle** Lühe, 1899
(Fig. 156)

Metacercaria: Body small to very small, round to tongue-shaped, spined. Oral sucker subterminal, prepharynx present, esophagus short; ceca terminate near posterior extremity. Ventral sucker embedded in parenchyma. Muscular gonotyl present. Excretory vesicle Y- or T-shaped; arms run sharply toward sides in posterior third before continuing anteriorly. Cercaria: monostomatous, lophocercous, or pleurolophocercous, oculate. Adult: in intestine of birds and mammals.

Cryptocotyle concavum (Crepl., 1825) Fischoeder, 1903. Hoffman (1957): cysts in flesh of *Catostomus commersoni*; adult in baby chicks, N. Dak.; Wooten (1957): pigmented cysts beneath skin and in flesh of *Gasterosteus aculeatus* but not *Cyprinus carpio* or *Gambusia affinis*; cercariae in *Amnicola longinqua*; exper. adult in young chicks and ducklings, Calif. This small trematode has probably been overlooked by many people.

C. lingua (Crepl., 1825) Fischoeder, 1903. Very common black-spot in marine fish of northeast coast, more lately on west coast (see Sindermann and Farrin, 1962); adult in gulls and terns, exper. in mammals.

Genus **Euhaplorchis** Martin, 1950
(Fig. 157)

Metacercaria: Body small, spined. Pigment of cercarial eye spots still present. Oral sucker subterminal, large; prepharynx distinct; pharynx well developed; esophagus short; ceca very short, voluminous, containing wafer-shaped concretions, terminating at level of ventral sucker. Ventral sucker and gonotyl (present in older specimens) near midbody. Excretory vesicle saccular, becoming filled with concretions. Cercaria: oculate, parapleurolophocercous. Adult: probably in gulls.

Euhaplorchis californiensis Martin, 1950: encysted on brain, *Fundulus parvipinnis*; cercaria in *Cerithidea californica*; exper. adult in chicks, Calif.

Genus **Parastictodora** Martin, 1950
(Fig. 158)

(Tentatively considered a subgenus of *Stictodora* by Yamaguti, 1958.) Metacercaria: Body small, flattened, becoming club-shaped, spined.

---

guti, 1942). Fig. 160. *Galactosomum spinetum* (from Sogandares-Bernal and Hutton, 1960).

Oral sucker subterminal, prepharynx conspicuous; esophagus very long, reaching to ventral sucker; ceca reach to near posterior extremity. Ventral sucker embedded in parenchyma, modified into nonsuctorial organ with numerous spines projecting into genital atrium, gonotyl immediately adjacent and present in older specimens. Excretory vesicle Y-shaped with long stem. Cercaria: oculate with lateral fin fold on anterior half of its long tail and ventral fin fold on posterior half.

Parastictodora hancocki Martin, 1950: cysts in lower jaw, around eyes, tissue around bases of fins, *Fundulus parvipinnis, Gillichthys mirabilis*; cercariae in *Cerithidea californica*; exper. adults in chicks, Calif.

Genus **Scaphanocephalus** Jägerskiöld, 1903
(Fig. 159)

Metacercaria: Forebody widened laterally, winglike, with anterior border becoming crenulated and striated; hindbody subcylindrical. Cuticle spined. Oral sucker at marginal notch, pharynx present, esophagus short; ceca form incomplete loops into each wing of forebody, serpentine posteriorly, opening at posterior extremity into excretory vesicle by short narrow passage. Ventral sucker embedded in body parenchyma, opening into genital atrium lying immediately behind. Testes present, ramified, tandem, in posterior of body. Genital atrium with gonotyl projecting from dorsal wall. Ovary multilobate, medium, pre-testicular. Uterus winding between ovary and genital atrium. Vitellaria present in lateral fields. Excretory vesicle Y-shaped with short stem; bifurcations unite between two testes and in front of anterior testis, approach each other in front of ventral sucker, and turn outward abruptly. Cercaria: unknown. Adult: in birds.
    Ref. Yamaguti (1942, 1958, p. 709).

Scaphanocephalus expansus (Crepl., 1842) Jägerskiöld, 1904. Yamaguti (1942): metacercaria under scales and in fins of marine fish in Japan; Hoffman (1953): adult from osprey, Iowa. The adult of this parasite has been reported from Europe, Egypt, Asia, and North America.

Genus **Galactosomum** Looss, 1899
(Fig. 160)

Heterophyidae: Metacercaria: Body spined. Oral sucker subterminal; prepharynx often long; esophagus short; ceca long. Ventral sucker embedded in body parenchyma and opening into genital atrium. Testes tandem or oblique, in posterior of body. Excretory vesicle extends to posterior testis. Life cycle: adult in birds; cercaria in marine snails; metacercaria in fish.

Galactosomum spinetum (Braum, 1901). Sogandares and IIutton (1960): in *Fundulus similis* (exper.), Fla.

Family CRYPTOGONIMIDAE Ciurea, 1933

Metacercaria: Body small, oval to elongate. Eye spots present. Oral sucker terminal or subterminal. Prepharynx short, esophagus short, ceca long or short. Ventral sucker moderate size, but may be reduced in adults. Gonotyl may be present. Gonads and sometimes vitellaria already developed. Excretory vesicle V- or Y-shaped, arms usually long. Cercaria: not described. Adult: in fish.

Genus **Cryptogonimus** Osborn, 1910
(Fig. 161)

Metacercaria: Unpigmented cysts in flesh, 300 to 350μ by 200 to 225μ. Parasite folded double in cyst. Freed parasite about 500 by 160μ. Anterior cuticle spined. Large oral sucker. Prepharynx relatively long. Pharynx large. One pair of large black eye spots. Ventral sucker and gonotyl present. Y-shaped excretory vesicle large and filled with excretory granules; stem about one-third length of body. Immature testes present. Ovary inconspicuous.

Ref. Yamaguti (1958, p. 217); Chandler (1951).

Cryptogonimus chyli Osborn, 1910: metacercaria in flesh of *Ambloplites rupestris*, N. Y.; Chandler (1951): sparse in flesh of *Perca flavescens*, Minn.; Cooper (1915): flesh of *Micropterus dolomieui*, *A. rupestris*, *Notropis hudsonius*; Bangham (1926): in skin and muscle of *M. dolomieui*; Fischthal (1947): in *Etheostoma nigrum*, *E. caeruleum*, *E. flabellare*, *P. flavescens*, Wis.; Fischthal (1952, 1956): in *A. rupestris*, Wis., N. Y.

Genus **Acetodextra** Pearse, 1924
(Fig. 162)

Metacercaria: No description given. See section on adults.
Ref. Yamaguti (1958, p. 220); Perkins (1956).

Acetodextra ameiuri (Stafford, 1904) Pearse, 1924: young specimens (no description) encysted in liver peritoneum, *Schilbeoides gyrinus*, Wis.

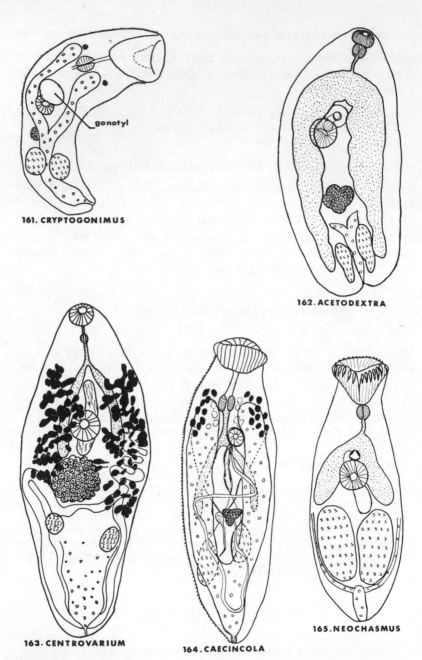

161. CRYPTOGONIMUS

gonotyl

162. ACETODEXTRA

163. CENTROVARIUM

164. CAECINCOLA

165. NEOCHASMUS

Figs. 161–165. Fig. 161. *Cryptogonimus chyli* (from Chandler, 1951). Fig. 162. *Acetodextra amiuri*, adult, modified from Van Cleave and Mueller (1934). Fig. 163. *Centrovarium lobotes*, progenetic, from *Pimephales promelas* (orig.). Fig. 164. *Caecincola parvulus* (from Lundahl, 1941). Fig. 165. *Neochasmus* sp., Israel (from Paperna, 1964).

Genus **Centrovarium** Stafford, 1904
(Fig. 163)

Metacercaria: Body flattened, cigar-shaped, oculate. Oral sucker ventroterminal, followed by short prepharynx. Pharynx small, esophagus short; ceca terminate at about midbody. Ventral sucker moderately small, about one-third from anterior. Reproductive system usually advanced; testes nearly symmetrical in posterior third. Genital pore immediately pre-acetabular. Ovary large, lobed, median, at midbody. Vitellaria may already be present in dorsolateral fields between intestinal bifurcation and testes. Uterus occupies post-testicular area. Excretory vesicle Y-shaped; arms wide, reaching to level of pharynx.

Ref. Yamaguti (1958, p. 218); Van Cleave and Mueller (1934, p. 222).

Centrovarium lobotes (MacCallum, 1895). Hunter (1930): unpigmented cysts in flesh of minnows, N. Y.; Van Cleave and Mueller 1934): in flesh of *Percopsis omiscomaycus*, N. Y.; Pearse (1924): *Notropis hudsonius*, Wis.; Hunter and Hunter (1931): *P. omiscomaycus, Hybornchus notatus, N. deliciosus, N. cornutus, N. bifrenatus, N. heterolepis*, N. Y.; Bangham (1955): *N. volucellus, Pimephales promelas*, Wis.; Hoffman (1965): egg-producing adults in metacercarial cysts (progenesis), *P. promelas*, N. Dak.

Genus **Caecincola** Marshall and Gilbert, 1905
(Fig. 164)
(See section on adult trematodes)

Metacercaria: Body very small, elliptical, spinulate. Large oral sucker nearly terminal. Prepharynx distinct, pharynx well developed, esophagus short, ceca short. Ventral sucker small, pre-equatorial, on ventral surface. Excretory vesicle Y-shaped, stem about one-third of body length. Adults: in stomach and intestine of fish.

Caecincola parvulus Marshall and Gilbert, 1905. Lundahl (1939, 1941): metacercaria encysted beneath skin of *Lepomis gibbosus* exper., Mich.

Genus **Neochasmus** Van Cleave and Mueller, 1932
(Fig. 165)

Cryptogonimidae: Metacercaria: Oral sucker terminal, large, with single crown of spines. Ceca short. Ventral sucker submerged. Testes symmetrical at or near ends of ceca. Genital pore and gonotyl just in front of ventral sucker. Life cycle: adult in intestine of fishes and reptiles; snail host unknown; metacercaria probably in fish muscle.

Neochasmus sp. Peters (1959, pers. comm.): progenetic adult in flesh of fish; Paperna (1964): metacercaria from fish muscle, Israel.

Cestodes

(Platyhelminthes: Cestoda)

Contents

Key to the classes of Cestodes 205
Class CESTODARIA 205
  Order AMPHILINIDEA .. 206
    Genus *Amphilina* 206
Class CESTODA 206
  Key to the orders of adult Cestoda of North American freshwater fishes 206
  Order CARYOPHYLLIDEA 207
    Family CARYOPHYL-LAEIDAE 207
    Key to the genera of Caryophyllaeidae 207
    Genus *Spartoides* 208
    Genus *Biacetabulum* 208
    Genus *Archigetes* 209
    Genus *Bialovarium* 209
    Genus *Khawia* 211
    Genus *Pliovitellaria* 211
    Genus *Capingens* 211
    Genus *Hypocaryophyl-laeus* 213
    Genus *Monobothrium* ... 213
    Genus *Glaridacris* 215
    Genus *Pseudolytocestus* . 216
    Genus *Caryophyllaeus* ... 217
    Genus *Hunterella* 217
    Genus *Atractolytocestus* . 218
  Order PROTEO-CEPHALIDEA 218
    Family PROTEO-CEPHALIDAE 218

Key to the genera of Proteocephalidae of North American freshwater fishes 218
  **Genus *Corallobotrium* . 219
  Genus *Ophiotaenia* 220
  **Genus *Proteocephalus* . 220
Order PSEUDOPHYL-LIDEA 224
Key to the adults of the families of Pseudophyllidea of North American fishes 224
Key to the larval pseudophyllideans of North American freshwater fishes 225
  Family Amphicotylidae ... 225
    Genus *Eubothrium* 226
    Genus *Marsipometra* 227
  Family BOTHRIO-CEPHALIDAE 227
    Genus *Bothriocephalus* .. 229
  Family DIPHYLLOBO-THRIIDAE 230
    *Genus *Diphyllobothrium* .231
    *Genus *Ligula* 232
    *Genus *Schistocephalus* . 232
  Family HAPLO-BOTHRIIDAE 233
    **Genus *Haplobothrium* . 233
  Family TRIAENO-PHORIDAE 234
    **Genus *Triaenophorus* . 234

* Larva only in fish.
** Larva and adult in fish.

Order SPATHEBO-
THRIIDEA 235
Family CYATHO-
CEPHALIDAE 235
**Genus *Cyathocephalus* 235
Family DIPLO-
COTYLIDAE 236
Key to the genera 236

Genus *Bothriomonas* 237
Genus *Diplocotyle* 237
Order TETRA-
PHYLLIDEA 239
**Genus *Pelichnibothrium*
(*Phyllobothrium*) 239
Order CYCLOPHYLLIDEA 240
Genus *Hymenolepis* 240

Cestodes are very common in fish in nature and sometimes in fish culture. Two life-cycle stages are represented in fish: adults inhabit the intestine and pyloric ceca, and plerocercoids of the same or different species are found in the visceral organs and musculature. The first-stage larvae, procercoids, are found in the hemocoele of aquatic copepods, amphipods, and isopods.

Usually no detectable damage results from the intestinal form, although there is some indication that heavy infections of tapeworms, notably *Corallobothrium* of catfish, do retard the growth of the fish. The plerocercoids migrating in the visceral cavity, however, produce adhesions which are very damaging to fish. This is particularly true of *Proteocephalus ambloplitis* in bass, and *Diphyllobothrium* spp. in salmonids. In small fish the wandering plerocercoids can cause death when vital organs are severely damaged. In adult fish the adhesions can cause impaired metabolism, decreased egg production, and sexual sterility.

KEY

TO THE CLASSES OF CESTODES

Strobila monozoic (one set of reproductive organs); embryo with
    10 hooklets .... CESTODARIA
Strobila polyzoic (many proglottides, each containing a set of
    reproductive organs, except Spathebothriidea) or monozoic
    (Caryophyllidea); embryo (onchosphere) with 6 hooklets
    .... CESTODA

Class CESTODARIA Monticelli, 1892

Unsegmented tapeworms, up to 30 cm by 2 cm, containing a single set of genitalia. Parasitic in intestine or body cavity of fishes. Embryo with 10 hooklets.

Order AMPHILINIDEA
Genus **Amphilina** Wagener, 1858
(No fig.)

Body flat, usually more or less elongate; calcareous corpuscles present. Anterior extremity with small sucker-like depression, followed by muscle ending in large anchor cells. Testes scattered, not forming such long strips as in Schizochoerinae and Gephyrolininae; some medial, others lateral, to ascending uterine limbs; propulsion apparatus short, fusiform; cirrus very long, slender, with 10 hooks. Ovary multilobed, irregular in shape, in posterior third of body. No accessory seminal receptacle. Vagina crossing vas deferens, opening on posterior margin of body. Vitellaria strongly developed, branched; unpaired vitelline duct ascending. Descending uterine limb largely on same side as first ascending limb; eggs oval, thick-shelled. In body cavity of sturgeons.

Amphilina bipunctata Riser, 1948: in sturgeon, Ore.

Class CESTODA Southwell, 1930

Polyzoic or monozoic cestodes with scolex of varying structure; neck present or absent; strobila usually with distinct external segmentation (not so in Caryophyllidea); inner longitudinal muscle forming boundary between cortical and medullary parenchyma; reproductive organs usually in medulla, only occasionally in cortex. Excretory system with paired lateral stems, sometimes a network of longitudinal vessels. Embryo (onchosphere) with 6 hooks. Parasitic in intestine of vertebrates.

KEY

TO THE ORDERS OF ADULT CESTODA OF NORTH AMERICAN FRESHWATER FISHES

1. Nonsegmented 2
1. Segmentation usually distinct 3
2. Scolex without true bothria; no external segmentation, but gonads multiple .... SPATHEBOTHRIIDEA
(p. 235)
2. Scolex with suctorial grooves; only one set of reproductive organs .... CARYOPHYLLIDEA
3. Scolex with 2 bothria .... PSEUDOPHYLLIDEA
(p. 224)
3. Scolex with 4 bothridia .... TETRAPHYLLIDEA
(p. 239)
3. Scolex with 4 suckers .... PROTEOCEPHALIDEA
(p. 218)

Order CARYOPHYLLIDEA Ben. *in* Olsson, 1893
Family CARYOPHYLLAEIDAE Leuckhart, 1878

Body elongate, sometimes filiform, variable in length; scolex with suctorial grooves of different shapes, marked off from body or not. Testes medullary, anterior to ovary and uterus. Cirrus pouch pre-uterine, post-testicular. Male genital pore in anterior or posterior half of body. Ovary symmetrically lobed, medullary or partly cortical, post-uterine, usually near posterior extremity. Vitellaria medullary, cortical, or partly cortical and partly medullary, in two lateral fields; may be partly post-ovarian. Excretory system of communicating longitudinal vessels, pore terminal.

Ref. Hunter (1930); Kulakovskaya (1964), life cycles, Czech.

KEY

TO THE GENERA OF CARYOPHYLLAEIDAE FROM NORTH AMERICAN FRESHWATER FISHES*

1. Cirrus opens into utero-vaginal duct 2
1. Cirrus opens separately and ahead of utero-vaginal duct .. 7
2. Uterine coils extend anterior to cirrus pouch 3
2. Uterine coils do not extend anterior to cirrus pouch 5
3. Ovary U- or V-shaped .... (fig. 166) *Spartoides*
3. Ovary H-shaped 4
4. Scolex with 2 saucer-like depressions; neck long, narrow ..
   .... (fig. 167) *Biacetabulum*
4. Scolex with 3 pairs of loculi; neck short, broad ....
   ....(fig. 168) *Archigetes*
5. Ovary V- or U-shaped .... (fig. 169) *Bialovarium*
5. Ovary H-shaped 6
6. Scolex fanlike, frilled at anterior end; loculi absent ....
   .... (fig. 170) *Khawia*
6. Scolex rounded at anterior end; 2 loculi present ....
   .... (fig. 171) *Pliovitellaria*
7. Uterine coils extend anterior to cirrus pouch 8
7. Uterine coils do not extend anterior to cirrus pouch 9
8. Scolex large, expanded anteriorly, with 2 large, deep bothria .
   .... (fig. 172) *Capingens*
8. Scolex small, with 3 pairs of poorly defined loculi ....
   .... (fig. 173) *Hypocaryophyllaeus*
9. Scolex with loculi, terminal introvert, or both 10

* Key prepared by Dr. John Mackiewicz, State University of New York at Albany.

9. Scolex without loculi or bothria 11
10. Scolex truncate with terminal introvert; loculi present or absent; post-ovarian vitellaria usually absent ....
.... (fig. 174) *Monobothrium*
10. Scolex fan-shaped or with raised flat disc; 3 pairs of loculi; post-ovarian vitellaria present .... (fig. 175) *Glaridacris*
(p. 215)
11. Post-ovarian vitellaria present 12
11. Post-ovarian vitellaria absent ..... (fig. 176) *Pseudolytocestus*
(p. 216)
12. Scolex flattened, sometimes crenate and fanlike ....
.... (fig. 177) *Caryophyllaeus*
(p. 217)
12. Scolex conical, may have terminal introvert-like structure .. 13
13. Scolex large, wider than body; narrow neck absent; short, stout forms .... (fig. 178) *Hunterella*
(p. 217)
13. Scolex small; narrow neck present; long, thin forms ......
.... (fig. 179) *Atractolytocestus*
(p. 218)

Genus **Spartoides** Hunter, 1929
(Fig. 166)

Scolex distinct, with 3 pairs of loculi. Male and female pores open separately on ventral surface near posterior tip of last fourth of body. Cirrus sac opens within confines of ovarian wings. One row of main longitudinal muscles; vitellaria and ovarian follicles in part cortical to inner longitudinal muscles; ovarian commissure entirely medullary. Ovary U-shaped; uterine coils extend anteriorly to cirrus sac more than twice length of ovarian wings. Postovarian vitellaria absent. Parasitic in Catostomidae. Life cycle unknown.

Spartoides wardi Hunter, 1929: in *Carpiodes carpio*, *C. thompsoni*, *Ictiobus cyprinellus*, Iowa, Ill., Minn.; Mackiewicz (1964): host not given, Okla.

Genus **Biacetabulum** Hunter, 1927
(Fig. 167)

Scolex well defined, varying but little in shape, bearing one pair of well-defined acetabulum-like suckers, with or without additional loculi. Cirrus opens into utero-vaginal canal before it reaches surficial atrium (like *Caryophyllaeides*). Ovary H-shaped and entirely medullary. Uterine coils extend anteriorly to cirrus sac, with maximum length one-fourth that of testicular field, usually less. Terminal excretory bladder and external seminal vesicle present. Vitellaria mostly in testicular zone, but post-ovarian vitellaria present. Life cycle unknown.

Biacetabulum giganteum Hunter, 1929: in *Ictiobus bubalus, I. Cyprinella*, Iowa, Ill., Minn., Miss.

B. infrequens Hunter, 1927: in *Moxostoma anisurum*, Ill.; Bangham (1941): *Catostomus commersoni*, Ohio: Fischthal (1947): *M. rubreques*, Wis.; Mackiewicz (1965): *Hypentelium nigricans*, N. Dak.; Meyer, F. (1958): *C. commersoni*, Iowa; Van Cleave and Mueller (1934): *C. commersoni*, Ill. The last two records are erroneous, according to Mackiewicz (1965).

B. macrocephalum McCrae, 1962: in *C. commersoni*, Colo.

B. meridianum Hunter, 1929: in *Erimyzon sucetta*, N. C.; Bangham and Venard (1942): same, Tenn.; Self and Timmons (1955): *Carpiodes carpio*, Okla.

Biacetabulum sp. Fischthal (1950): in *Pimephales promelas*, Wis.; Fischthal (1956): *C. commersoni*, N. Y.

Genus **Archigetes** Leuckart, 1878
(Fig. 168)

Caryophyllaeinae: Scolex well defined, hexagonal, bearing 2 bothria-like depressions. Cirrus opens into utero-vaginal canal before it reaches surficial atrium. Ovary H-shaped and medullary. Excretory system without terminal vesicle, but with numerous ampullae at posterior end of body. Uterine coils extend anteriorly beyond cirrus sac; vas deferens expands to form external seminal vesicle. Caudal vesicle carrying embryonic hooks present in oligochaete, but not in fish. Parasitic in body cavity of Tubificidae and intestine of fish.

Archigetes iowensis Calentine, 1962: in intestine of *Cyprinus carpio* and seminal vesicle of tubificed oligochaete, Iowa; Calentine (1964): life cycle.

Genus **Bialovarium** (Fischthal, 1953) Fischthal, 1954
(Fig. 169)

Caryophyllaeidae: Scolex poorly defined, bearing pair of shallow loculi. Cirrus opens into utero-vaginal canal before it reaches surficial atrium. Ovary V-shaped and entirely medullary. Uterine coils extend only to lateral margins of cirrus sac, with maximum length one-fourth that of testicular field. Terminal excretory bladder present. Seminal vesicle not

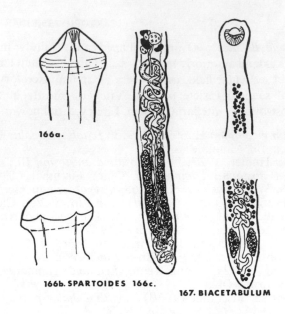

166a.

166b. SPARTOIDES 166c.

167. BIACETABULUM

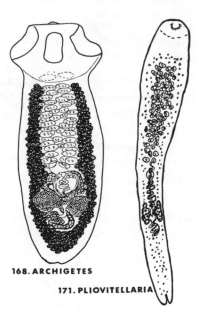

168. ARCHIGETES

171. PLIOVITELLARIA

Figs. 166–168. Fig. 166. *Spartoides wardi. a,* scolex, lateral view; *b,* scolex; *c,* reproductive systems (from Hunter, 1927). Fig. 167. *Biacetabulum infrequens* (from Van Cleave and Mueller, 1934). Fig. 168. *Archigetes iowensis* (from Calentine, 1962).

enclosed in cirrus sac. Post-ovarian vitellaria absent. Parasitic in Cyprinidae. Life cycle unknown.

Bialovarium nocomis Fischthal, 1954: in *Nocomis biguttatus*, Wis.

Genus **Khawia** Hsu, 1935
(Fig. 170)
(Syn.: *Bothrioscolex* Szidat, 1937)

Body slender. Scolex more or less enlarged, frilled or not, without locular depressions. Testes numerous, extending in peripheral medulla from behind scolex to cirrus pouch. No external seminal vesicle. Cirrus pouch well developed; cirrus opens into genital atrium just in front of utero-vaginal aperture. Ovary H-shaped, medullary. Vitellaria extend in lateral fields just outside inner longitudinal muscle sheath between neck and ovary; post-ovarian follicles medullary. Uterine coils between ovary and cirrus pouch, surrounded by layer of accompanying cells. Parasitic in cyprinid fishes. Life cycle unknown.

Khawia iowensis Calentine and Ulmer, 1961: in *Cyprinus carpio*, *Ictiobus cyprinellus*, Iowa; Anthony (1963): *C. carpio*, Wis.

Genus **Pliovitellaria** Fischthal, 1951
(Fig. 171)

Caryophyllaeidae: Scolex poorly defined, varying little in shape, bearing one pair of acetabulum-like bothria without additional loculi. Cirrus opens into utero-vaginal canal before it reaches surficial atrium. Ovary H-shaped and entirely medullary. Uterine coils extend only to anterior margin of cirrus sac, with maximum length to three-fourths to one-half that of testicular field. Terminal excretory bladder present. Seminal vesicle not enclosed in cirrus sac. Vitellaria mostly pre-ovarian, but some post-ovarian. Parasitic in Cyprinidae. Life cycle unknown.

Pliovitellaria wisconsinensis Fischthal, 1951: in *Notemigonus crysoleucas*, *Hyborhynchus notatus*, Wis.

Genus **Capingens** Hunter, 1927
(Fig. 172)

Caryophyllaeidae: Scolex distinct, occupying one-fifth to one-fourth total body length, bearing one pair of well-defined bothria. Scolex does not vary in shape as in other Cestodaria. Vitellaria extend into cortical parenchyma past inner longitudinal muscles having origin within med-

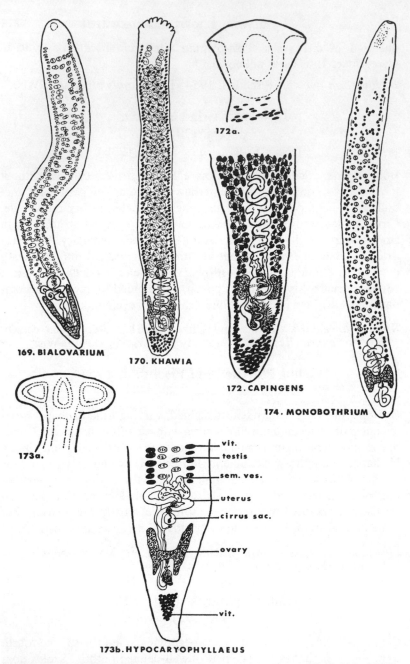

169. BIALOVARIUM

170. KHAWIA

172a.

172. CAPINGENS

174. MONOBOTHRIUM

173a.

vit.
testis
sem. ves.
uterus
cirrus sac.
ovary
vit.

173b. HYPOCARYOPHYLLAEUS

Figs. 169–174. Fig. 169. *Bioalovarium nocomis* (from Fischthal, 1954). Fig. 170. *Khawia iowensis* (from Calentine and Ulmer, 1961). Fig. 171. *Pliovitellaria wisconsinensis* (from Fischthal 1951—find on preceding

ullary parenchyma. These glands form continuous row laterally with post-ovarian vitellaria. Cirrus opens on ventral surface or into shallow genital atrium which is noneversible and anterior to similar atrium for female system. Uterine coils lie anteriorly to cirrus sac, with maximum length one-third or less that of testicular field. Parasitic in *stomach* of Catostomidae. Life cycle unknown.

Capingens singularis Hunter, 1927: in *Carpiodes carpio, Ictiobus urus,* Ill., Minn.; Bangham and Venard (1942): *Carpiodes carpio,* Tenn.; Mackiewicz (1964): *I. bubalus,* Okla.; Self and Campbell (1956): *I. bubalus,* Okla.

Capingens sp. Griffith, 1953: in *Cyprinus carpio,* Wash. This is *Atractolytocestus* according to Mackiewicz (1965).

Genus **Hypocaryophyllaeus** Hunter, 1927
(Fig. 173)

Caryophyllaeidae: Scolex poorly defined, with 3 pairs of loculi. Cirrus opens on ventral surface or into shallow noneversible genital atrium. Ovary H-shaped and entirely medullary. Uterine coils extend anteriorly to cirrus sac, reaching maximum length one-fourth or less that of testicular field. Terminal excretory bladder and external seminal vesicle present. Post-ovarian vitellaria present. Parasitic in intestine of Catostomidae.

Hypocaryophyllaeus gilae Fischthal, 1953: in *Gila atraria,* Wyo.

H. paratarius Hunter, 1927: in *Carpiodes carpio, C. velifer, Ictiobus cyprinella,* Ill., Iowa.

Genus **Monobothrium** Diesing, 1863
(Fig. 174)

Caryophyllaeidae: Scolex round to oval in cross section, bearing 6 shallow longitudinal grooves and terminal funnel-shaped introvert. Cirrus and utero-vaginal canal open together into shallow, eversible, common genital atrium, widely separated by bulky annular pad (male genital papilla?). Ovary H-shaped, entirely medullary. Uterine coils never anterior to cirrus sac, with maximum length one-third that of testicular field, usually less. External seminal vesicle present in North American forms. Terminal excretory bladder present. Post-ovarian

---

plate). Fig. 172. *Capingens singularis* (from Hunter, 1927). Fig. 173. *Hypocaryophyllaeus paratarius* (from Hunter, 1927). Fig. 174. *Monobothrium hunteri* (from Mackiewicz, 1963).

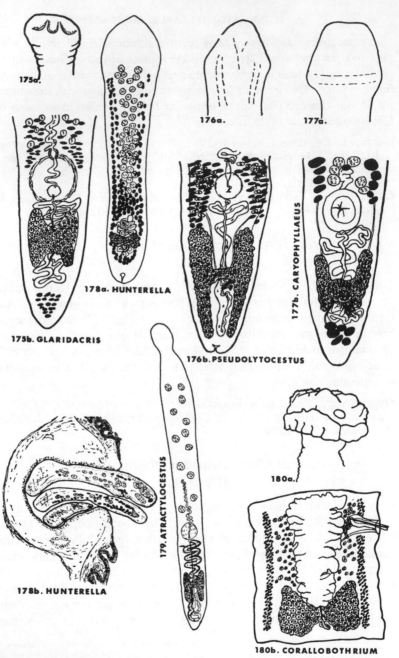

Figs. 175–180. Fig. 175. *Glaridacris catostomi* (from Hunter, 1927). Fig. 176. *Pseudolytocestus differtus* (from Hunter, 1927). Fig. 177. *Caryophyllaeus terebrans* (from Hunter, 1927). Fig. 178. *Hunterella nodulosa*. *a*, entire worm; *b*, section of intestinal pit containing 3 worms

vitellaria may or may not be present. Parasitic in digestive tract of Cyprinidae and Catostomidac. Lifc cyclc unknown.

Monobothrium hunteri Mackiewicz, 1963: in *Catostomus commersoni,* Can., Conn., Mich., N. Y., N. C., W. Va., Wyo., Pa.
M. ingens Hunter, 1927: in *Ictiobus cyprinellus,* Minn.; Bangham (1942): Tenn.; Bangham & Venard (1942): same, Tenn.; Hugghins (1958, 1959): same, S. Dak., is *Hunterella nodulosa;* Krueger (1954): *C. commersoni,* Ohio; Self and Campbell (1956): *I. bubalus,* Okla.; Van Cleave and Mueller (1934): *C. commersoni,* N. Y.

Genus **Glaridacris** Cooper, 1920
(Fig. 175)
(Syn.: *Brachyurus* Szidat, 1938)

Caryophyllaeidae: Scolex well defined, with 3 pairs of loculi or bothria. Scolex may or may not form a definite terminal disc. Cirrus opens on ventral surface or into shallow, noneversible genital atrium. Ovary H-shaped and entirely medullary. Uterine coils never extend anteriorly to cirrus sac; maximum length one-third that of testicular field, usually less. Terminal excretory bladder and external seminal vesicle present. Post-ovarian vitcllaria present. Parasitic in digestive tract of Catostomidae.

Glaridacris catostomi Cooper, 1920: in *Catostomus commersoni,* Mich.; Anthony (1963): *C. commersoni,* Wis.; Bangham (1941): *C. commersoni,* Ohio, Can.; Bangham (1944): *C. commersoni,* Wis.; Bangham (1955): *C. commersoni,* L. Huron; Bangham and Adams (1954): *C. catostomus, C. commersoni, C. macrocheilus,* Brit. Col.; Bangham and Hunter (1939): *C. commersoni,* L. Erie, N. Y.; Cooper (1920): *C. commersoni,* Mich.; DeRoth (1953): *C. commersoni,* Me.; Dolley (1933): Ind.; Fischthal (1947): *Hypentelium nigricans,* Wis.; Fischthal (1950): *C. commersoni, H. nigricans,* Wis.; Fischthal (1952): *Moxostoma rubreques, C. commersoni,* Wis.; Fischthal (1956): *C. commersoni,* N. Y.; Haderlie (1953): *C. occidentalis,* Calif.; Hugghins (1958, 1959): *C. commersoni,* S. Dak.; Hunninen (1935): *C. commersoni,* N. Y.; Hunter (1927): *C. commersoni,* Mich.; Hunter (1942): host not listed, Conn.; Krueger (1954): *C. commersoni,* Ohio; Mackiewicz (1962): review; Linton (1941): *C. commersoni,* Mass.; Meyer,

---

(from Mackiewicz and McCrae, 1962). Fig. 179. *Atractolytocestus huronensis* (from Anthony, 1958). Fig. 180a. *Corallobothrium fimbriatum* (from Essex, 1927, in Wardle and McLeod, 1952). Fig. 180b. *C. giganteum* (from Essex, 1927, in Wardle and McLeod, 1952).

M. (1954): *C. commersoni*, Me.; Meyer, F. (1958): *C. commersoni*, Iowa; Pearse (1924): several spp., Wis.; Sindermann (1953): *C. commersoni, Erimyzon* sp., Mass.; Van Cleave and Mueller (1934): *C. commersoni*, N. Y.; Wardle (1932): *C. catostomus*, Can.

G. *confusus* Hunter, 1927: in *Ictiobus bubalus, Ictiobus* sp., *Dorosoma cepedianum*, Miss., Iowa.; Fischthal (1947, 1956): *C. commersoni*, N. Y., Wis.; Fritts (1959): *C. macrocheilus*, Idaho; Griffith (1953): *C. columbianus, Carpiodes carpio*, Wash.; (this is *Monobothrium* sp., Mackiewicz, 1963); Hoffman (1952): *C. commersoni*, N. Dak.; Hunter (1942): host not listed, Conn.; Meyer, F. (1958): *C. commersoni*, Iowa; Richardson (1941): *C. commersoni*, Can.; Self and Campbell (1956): *I. bubalus, I. niger*, Okla.; Van Cleave and Mueller (1934): *C. commersoni*, N. Y. (last ref. concerns G. *laruei* according to Mackiewicz, 1965).

G. *hexacotyle* (Linton, 1898) Hunter, 1927. (Syn. *Monobothrium h.*) Linton: in *Catostomus* sp., Ariz.

G. *intermedius* Lyster, 1940: in *C. commersoni*, Can.; Fischthal (1947): *C. commersoni*, Wis.; Mackiewicz (1961): considers G. *intermedius* synonymous with G. *laruei*.

G. *laruei* (Lamont, 1921) Hunter, 1927: in *C. commersoni*, Mich., Wis.; Bangham (1951): *Gila atraria*, Wyo. (an error according to Fischthal, 1953; worm later described as *Hypocaryophyllaeus*); Bangham and Venard (1946): same host, Can.; DeRoth (1953): *C. commersoni*, Me.; Mackiewicz (1961): considers this worm a synonym of G. *intermedius*.

G. *oligorchis* Haderlie, 1953: in *Catostomus tahoensis*, Calif.; Meyer, F. (1958): *C. commersoni*, Ia.

Glaridacris sp. Alexander (1960): in *C. catostomus*; Bangham (1951): *Gila atraria*, Wyo.; DeRoth (1953): *C. commersoni*, Me.; Haderlie (1953): *C. occidentalis*, Calif.; Hunter (1942): *Notemigonus crysoleucas*, Conn.; Hunter and Hunter (1931): *C. commersoni*, N. Y.; McPhee (1961): *Lepomis gibbosus*, Wash. (?).

Genus **Pseudolytocestus** Hunter, 1929
(Fig. 176)

Caryophyllaeidae: Scolex has little specialization. Cirrus opens separately on ventral surface or into shallow eversible genital atrium. Ovary H-shaped, almost entirely medullary; only one-third of ovarian follicles extend into cortical parenchyma. Uterine coils never extend anteriorly to cirrus sac, with maximum length one-third that of testicular field. Post-ovarian vitellaria absent. Parasitic in intestine of Catostomidae. Life cycle unknown.

Pseudolytocestus differtus Hunter, 1929: in *Ictiobus bubalus*, Miss.; Bangham and Venard (1942): same, Tenn.; Self and Campbell (1956): same, Okla.

Genus **Caryophyllaeus** Müller, 1787
(Fig. 177)

Caryophyllaeidae: Anterior extremity broadened, folded, or "curled," not specialized into loculi, bothria, or suckers. Cirrus open on ventral surface or into shallow noneversible genital atrium. Medullary ovary H-shaped. Uterine coils never extend anterior to cirrus sac, with maximum length one-third that of testicular field, usually less. No external seminal vesicle; post-ovarian vitellaria and terminal excretory bladder present. Parasitic in digestive tract of Cyprinidae and Catostomidae. Life cycle: larva of procercoid type with a cercomere; develops in body cavity of oligochetes (*Tubifex tubifex, T. barbatus*). If not ingested at this stage by a fish, larval development continues and cercomere is retained. Sexual organs develop only in intestine of fish.

Caryophyllaeus laticeps (Pallas, 1781). Rehder (1959) erroneously reported this worm from *Cyprinus carpio* in Iowa; it proved to be *Khawia iowensis* according to Calentine and Ulmer (1961).

C. terebrans (Linton, 1893) Hunter, 1927: in *Catostomus ardens*, Wyo.; Bangham (1951): *C. fecundus*, Wyo.; Bangham and Adams (1954): *C. catostomus, C. macrocheilus, Cyprinus carpio, Mylocheilus caurinus*, Brit. Col.; Tonn (1955): *C. commersoni* (this is *Hunterella* according to Mackiewicz and McCrae, 1962); Hunter (1927): *Ictiobus bubalus*, Miss.; Linton (1941): *Notropis rubrifrons*, Mass. (erroneous record).

Caryophyllaeus sp. Griffith (1953): Wash. This is *Hunterella* (Mackiewicz and McCrae, 1962).

Caryophyllaeus sp. Wilson (1957): in *Cyprinus carpio*, Kans. This is *Khawia iowensis* (Mackiewicz, 1965).

Genus **Hunterella** Mackiewicz and McCrae, 1962
(Fig. 178)

(Syn.: *Caryophyllaeus terebrans* [partim] (Linton, 1893);
Glaridacris catostomi [partim] Cooper, 1920)

Caryophyllaeidae: Scolex without suckers, loculi, or other organs of attachment. Cirrus opens separately from female gonopore. Ovary H-shaped. Uterine coils not extending anteriorly beyond cirrus pouch. Longitudinal muscles diffuse, not forming definite layers except in posterior region of body. Inner longitudinal muscles loosely arranged in broad band around medullary region. Vitellaria embedded within this medullary region, generally surrounded by fibers of inner longitudinal muscles. Post-ovarian vitellaria and external seminal vesicle present.

Hunterella nodulosa Mackiewicz and McCrae, 1962: in conspicuous pits or nodules in intestine, *Catostomus commersoni*, 14 states; Hugghins (1958, 1959): reported this worm as *Monobothrium ingens*, S. Dak.

Genus **Atractolytocestus** Anthony, 1958
(Fig. 179)

Scolex unspecialized; may possess terminal introvert. Uterus with maximum length one-half that of testicular field. Vitellaria in cortical parenchyma extend at times into medullary parenchyma; vitellaria form continuous row laterally with post-ovarian vitellaria. Uterine coils not passing anterior to cirrus sac. Terminal excretory bladder present. Parasitic in digestive tract of Cyprinidae.

Atractolytocestus huronensis Anthony, 1958: in *Cyprinus carpio*, Mich.; Mackiewicz (1964): same, Okla.; Mackiewicz (1965): this is more widespread than records indicate; see *Capingens* sp.

Order PROTEOCEPHALIDEA Mola, 1928

Scolex with 4 simple suckers; fifth or apical sucker sometimes present. Segmentation usually distinct. Genital pores marginal. Parenchyma usually divided into cortex and medulla by distinct layer of longitudinal muscles or by layer of circular muscle fibers, or by difference in density of texture. Testes, ovary, and vitellaria usually medullary, occasionally cortical. Ovary bilobed, posterior. Vitellaria follicular, in lateral fields or encircling proglottid. Uterus with numerous lateral outgrowths, with or without preformed ventral apertures; eggs usually embryonated. Adults parasitic in fishes, amphibians, and reptiles.
Ref. Frese (1965): review.

Family PROTEOCEPHALIDAE
With characters of the order.

KEY
TO THE GENERA OF PROTEOCEPHALIDAE
OF NORTH AMERICAN FRESHWATER FISHES

1. Scolex enlarged anteriorly with body folds covering suckers
   . . . . (fig. 180) *Corallobothrium*
1. Scolex without body folds covering suckers   2
2. Testes in single continuous field . . . . . (fig. 182) *Proteocephalus*
2. Testes in 2 separate lateral fields . . . . (fig. 181) *Ophiotaenia*

Genus **Corallobothrium** Fritsch, 1886
(Figs. 180, 193)

Proteocephalidae: Scolex unarmed, without rostellum. Suckers on flat anterior face of scolex and covered by body folds and marginal lappets. Neck short, broad; ventral excretory stem with numerous branches opening on ventral surface. Testes in one layer between two excretory stems. Cirrus pouch at anterior half of proglottis. Genital pores irregularly alternating. Ovary intervitellarian, posterior. Vitellaria medullary, in marginal bands medial to excretory stems. Vagina anterior or posterior to cirrus pouch. Adults parasitic in siluroid fishes. Life cycle: adult in silurids; procercoid in copepods; plerocercoid in small fish.

Ref. Van Cleave and Mueller (1934), key to spp.

Jones, Kerley, and Sneed (1956), proposed a new subgenus, *Megathylacoides* for *C. giganteum, C. tva, C. procerum*, and *C. thompsoni.*

Corallobothrium fimbriatum Essex, 1927: in *Ictalurus melas, I. nebulosus, I. olivaris, I. punctatus*; procercoid in *Cyclops*, plerocercoid in *Notropis blennius*, Ill. Reported by Allison, R. (1957), Anthony (1963), Bangham (1941, 1942, 1944, 1955), Bangham and Adams (1954), Bangham and Hunter (1939), Bangham and Venard (1942), DeRoth (1953), Fischthal (1947, 1958), Haderlie (1953), Hare (1943), Harms (1959), Hugghins (1959), Krueger (1954), Meyer, F. (1958), Sindermann (1953), and Wardle (1932); in *Ictalurus melas, I. natalis, I. nebulosus, I. punctatus, Noturus flavis, Pylodictis olivaris, Schilbeodes gyrinus*, from Ala., Can., Iowa, Kans., L. Huron, Mass., N. Y., Ohio, Tenn., Wis. *Life cycle*: adult in catfishes; procercoid in *Cyclops*; plerocercoid in *Notropis.*

C. giganteum Essex, 1927: in *I. punctatus, Pylodictis olivaris*; procercoid in *Cyclops*, Ill.; Bangham (1941): *I. punctatus*, Ohio; Bangham (1942): *I. punctatus, P. olivaris*; Bangham (1955): *I. punctatus*, L. Huron; Haderlie (1953): *I. catus*, Calif.; Harms (1959): *I. punctatus, I. melas, I. natalis*, Kans.; Hugghins (1959); *I. melas*, S. Dak.; Wilson (1957): *I. punctatus*, Kans. *Life cycle*: adult in catfishes; procercoid in *Cyclops*; plerocercoid host unknown.

C. intermedium Fritts, 1959: in *I. nebulosus*, Idaho.

C. minutium Fritts, 1959: in *I. nebulosus*, Idaho.

C. parvum Larsh, 1941: in *I. nebulosus*, procercoid in *Cyclops*; plerocercoids in small fish (*Glaridichthys talcatus*), Mich.; Meyer, M. (1958): *I. nebulosus*, Me.

C. procerum Sneed, 1950: in *I. furcatus*, Okla.; Sneed (1956): anomalous forms.

C. thompsoni Sneed, 1950: in *I. punctatus*, Okla.; Sneed (1956): anomalous forms.

C. tva Jones. Kerley and Sneed, 1956: in *P. olivaris*, Tenn.

Corallobothrium sp. Plerocercoids noted in *Roccus chrysops, Etheostoma nigrum, I. melas, I. punctatus*, by Bangham (1941), Harms (1959), Pearse (1924).

Genus **Ophiotaenia** LaRue, 1911
(Fig. 181)

Proteocephalidae: Scolex globose or somewhat tetragonal, unarmed. No rostellum. Vestigial apical organ may be present. Suckers round, with margins entire. Neck usually long. Gravid proglottides usually longer than wide, but may be broader than long. Inner longitudinal muscle sheath weak. Ventral excretory stems in lateral medulla. Testes in two separated fields anterior to ovary; some follicles may be outside excretory stems. Cirrus pouch may or may not overreach excretory stems. Cirro-vaginal atrium opening indifferently on right or left margin of proglottis in its anterior half. Ovary bilobed, sometimes H- or M-shaped. Vitellaria extend in marginal medulla throughout proglottis length; some follicles may intrude into cortex through space among inner longitudinal muscle bundles. Uterus extends in median field between ovary and anterior end of proglottis, occupying median third or half of proglottis breadth; eggs with three membranes, containing oncospheres. Vagina opening anterior, dorsal, or posterior to cirrus. Parasitic in fishes, amphibians, and reptiles.

Ophiotaenia fragilis (Essex, 1929): in *Ictalurus punctatus*, Ill.

Genus **Proteocephalus** Weinland, 1858
(Figs. 182, 194)

(Syn.: *Ichthyotaenia* Lonnberg, 1894)

Proteocephalidae: Scolex unarmed, with 4 typical suckers; fifth sucker or apical organ present or absent. Unsegmented neck region present. Gravid proglottides wider than long or longer than wide. Inner longitudinal muscle sheath present. Excretory stems slightly medial to outer edge of medulla. Testes in one continuous layer in intervascular medulla dorsal to uterus. Cirrus pouch transverse, at varying levels. Cirro-vaginal atrium opening on right or left margin of proglottis. Vitellaria in lateral fields of medulla outside excretory stems. Uterus extends in median field between ovary and anterior end of proglottis, developing a number of lateral outgrowths; may occupy whole available space of intervascular medulla; eggs globular, embryonated. Vagina opening into genital atrium anterior, dorsal, or posterior to cirrus. Life cycle: procercoid develops in hemocoel of crustacea; plerocercoids in small fish; adults in freshwater fishes, rarely in amphibians and reptiles.

Ref. Van Cleave and Mueller (1934, p. 280), key to species.

Proteocephalus ambloplitis (Leidy, 1887) Benedict, 1900. This species, known as the bass tapeworm, merits extensive review and ex-

perimental infections; only a summary of hosts is given here. Adult
very common in *Micropterus dolomieui* and *M. salmoides*; reported
from Brit. Col., Conn., Kans., La., Mass., Me., Minn., N. H., N. Y.,
S. Dak., Tenn., Wash., W. Va., Wis.; also reported from *Amia calva*,
La., by Sogandares (1955). The plerocercoid has been reported
from the viscera of many species of fish; this should be checked experimentally because the reported worms could have been other species of *Proteocephalus*, or perhaps *Corallobothrium*. Cooper (1915)
hypothesized the life cycle. Bangham (1925) reported the plerocercoid and suggested the copepod as first intermediate host. Hunter
and Hunter (1929) completed the life cycle.

   This parasite has been of much concern in fish culture because
the plerocercoids cause so much fibrosis of the gonads that bass are
sometimes rendered sexually sterile. Life cycle: adult in black bass,
procercoid in copepods; small fish may act as "carriers," plerocercoid in small fish.

P. *arcticus* Cooper, 1921: in *Salvelinus marstoni*; Wardle (1933):
   Oncorhynchus nerka, Salmo clarki, Can.
P. *australis* Chandler, 1935: in *Lepisosteus osseus*, Tex.
P. *cobraeformis* Haderlie, 1953: in *Ptychocheilus grandis*, Calif.
P. *coregoni* Wardle, 1932: in *Coregonus* (atikameg?), Hudson Bay.
P. *elongatus* Chandler, 1935: in *Lepisosteus osseus*, Tex.
P. *exiguus* LaRue, 1911: in *Coregonus* spp., Mich.; Bangham (1955):
   Coregonus spp., Mich.; Bangham and Adams (1954): *Prosopium
   williamsoni, Oncorhynchus nerka*, Brit. Col.; Bauer and Nikolskaya
   (1952): *Coregonus*, Russia; Cross (1934): ciscoes with large num-
   bers of *P. exiguus* had fewer *Acanthocephala* and vice versa, Wis.;
   Guilford (1954): *Petromyzon marinus* (accidental?); Pearse
   (1924): Wis.; Warren (1952): Minn.
P. *filicollis* Rudolphi: in *Gasterosteus* spp., Europe; Hoffman (unpubl.): *Eucalia inconstans*, N. Dak.; Benedict (1900): in *Coregonus nigripinnis, C. prognathus, C. artedi*, Mich.
P. *fluviatilis* Bangham, 1925: in *Micropterus dolomieui*; Bangham
   (1944): *M. dolomieui, M. salmoides*, Wis.; Fischthal (1950): *M.
   salmoides*, Wis.; Hare (1943): same, Ohio; Hugghins (1959):
   same, S. Dak.
P. *laruei* Faust, 1920: in *Coregonus* spp.; Bangham (1940, 1944,
   1951, 1955): *Coregonus* spp., *Petromyzon marinus* (accidental?),
   Mich., Wis., Wyo.; Bangham and Adams (1954): *Oncorhynchus
   nerka*, Canada.
P. *leptosoma* (Leidy, 1888): in *Esox reticulatus*.
P. *luciopercae* Wardle, 1932: in *Stizostedion* spp., Canada.
P. *macrocephalus* (Creplin, 1825): Europe; Bangham (1942): *Pylo-
   dictis olivaris*, Tenn.; Jarecka (1960): life cycle, Poland; Pearse
   (1924): *Anguilla rostrata*, Wis.; Sogandares (1955): same, La.
P. *microcephalus* Haderlie, 1953: in *M. dolomieui*, Calif.
P. *micropteri* (Leidy, 1887): syn of *P. ambloplitis*.
P. *nematosoma* (Leidy, 1890): in *Esox lucius*; Meyer (1954): *E.
   niger*, Me.
P. *osburni* (Bangham, 1925): in *Micropterus dolomieui*, Ohio; de-
   scribed from single immature specimen.

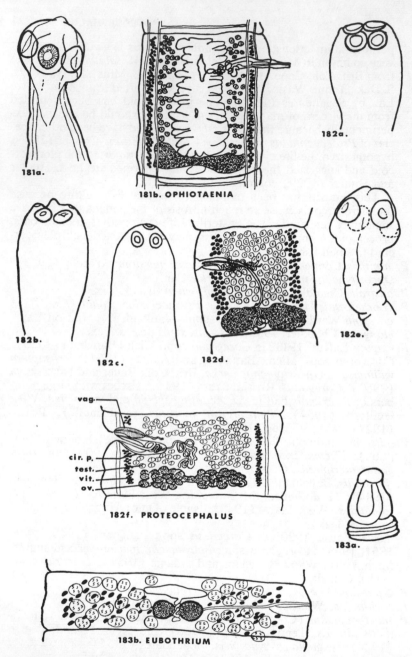

Figs. 181–183. Fig. 181. *Ophiotaenia fragile. a*, scolex; *b*, gravid proglottid (both from Essex, 1929). Fig. 182. *Proteocephalus. a, P. coregoni*, scolex (from Wardle, 1932); *b, P. luciopercae*, scolex (from Wardle, 1932); *c, P. pearsei*, scolex (from LaRue, 1919); *d, P. pearsei*, proglottid (from LaRue, 1919); *e, P. ambloplitis*, scolex; *f, P. ambloplitis*, proglottid (from

P. parallacticus MacLulich, 1943: in *Salvelinus namaycush, S. fontinalis, Salmo trutta,* Can.; Freeman (1964): life cycle, adult in chars; procercoid and plerocercoid in *Cyclops*; Meyer (1954): *S. namaycush,* Me.

P. pearsei LaRue, 1914: adult in *Perca flavescens,* Wis.; Bangham (1944, 1955): same, Wis., Mich.; Pearse (1924): same, Wis.; Van Cleave and Mueller (1934): *P. flavescens, Roccus chrysops,* N. Y. Immature adults reported from many fish by Anthony (1963), Bangham (1928, 1944, 1955), Fischthal (1947, 1950), Meyer, M. (1954), Pearse (1924), Sindermann (1953), Van Cleave and Mueller (1934). Copepod hosts reported by Bangham (1925).

P. perplexus LaRue, 1911: in *Amia calva, Lepisosteus platostomus,* Ill.; Bangham (1942, 1944, 1955): adult in *A. calva*; Fischthal (1950): same; Hunter (1929): same; Van Cleave and Mueller (1934): same, N. Y.; Pearse (1924): adult in *I. nebulosus, A. calva, Esox* sp., *Ambloplites rupestris*; plerocercoids in *Hyborhynchus notatus, Roccus chrysops, Ictalurus,* Wis.

P. pinguis LaRue, 1911: in *E. lucius,* Wis.; Bangham (1941, 1944, 1955): *E. lucius, E. masquinongy, Salmo gairdneri, Salvelinus fontinalis,* Wis., Mich.; Bangham and Adams (1954): *E. lucius,* Brit. Col.; Choquette (1951): *E. masquinongy,* Can.; DeRoth (1953): *E. niger,* Me.; Fischthal (1947, 1950): *E. lucius, E. masquinongy, S. fontinalis, S. trutta,* Wis.; Hugghins (1959): *E. lucius,* S. Dak.; Meyer, M. (1954): *E. niger,* Me.; Meyer F. (1958): *E. lucius,* Iowa; Odlaug *et al.* (1962): *E. lucius,* Minn.; Pearse (1924): *Esox* sp., Wis.; Sindermann (1953): *E. niger, Perca flavescens,* Mass.; Van Cleave and Mueller (1934): *E. lucius, E. niger,* N. Y.; Ward (1910): *E. niger,* Me.; Wardle (1932): *E. lucius*; Worley and Bangham (1952): *E. lucius, E. niger,* Que.; Hunter (1929): life cycle, procercoid in copepods, plerocercoid in fish.

P. primaverus Neiland, 1952: in *Salmo clarki,* procercoid in *Diaptomus,* Wash.; Fritts (1959): *S. clarki,* Idaho.

P. ptychocheilus Faust, 1920: in *Ptychocheilus oregonensis,* Mont.; Bangham (1951): *Gila straria, Richardsonius balteatus,* Wyo.; Bangham and Adams (1954): *Mylocheilus caurinus, P. oregonensis,* Brit. Col.

P. pugitensis Hoff and Hoff, 1929: in *Gasterosteus cataphractus,* Pacific N. Amer.; Bangham and Adams (1954): *G. aculeatus,* Brit. Col.; Guberlet (1929): plerocercoid in *Cyclops,* Wash.

P. pusillus Ward, 1910: in *Salmo sebago,* Me.; MacLulich (1943): Ont.; LaRue (1914): *Salvelinus namaycush,* Me., Ont.; Meyer M. (1954). *S. fontinalis,* Me.

P. salmonidicola Alexander, 1951: in *Salmo gairdneri,* Ore.; Bangham and Adams (1954): *S. clarki, S. gairdneri, Salvelinus alpinus,* Brit. Col.; Haderlie (1953): *S. gairdneri,* Calif.; Jones and Hammond (1960): same, Utah.

Van Cleave and Mueller, 1934). Fig. 183. *Eubothrium. a, E. salvelini,* scolex (from Wardle, 1932); *b, E. crassum,* mature proglottid (from Van Cleave and Mueller, 1934).

P. salvelini Linton, 1897: in *Salvelinus namaycush*, L. Superior.
P. singularis LaRue, 1911: in *Lepisosteus platystomus*, Ill.; Bangham
(1939, 1942): L. Erie, Tenn.; Hare (1943): *L. osseus*, Ohio, and
L. platyrhincus, Tex.; Pearse (1924): *Lepisosteus* spp., Wis.
P. stizostethi Hunter and Bangham (1933): in *Stizostedion glaucum*,
S. canadense, S. vitreum, Micropterus dolomieui, L. Erie and St.
Lawrence R.; Bangham (1944): *S. vitreum, Lepomis microchirus*,
Wis.; Bangham (1955): *S. vitreum*, L. Huron; Connor (1953):
seasonal studies; Fischthal (1947): *S. vitreum, E. lucius*, Que.
P. tumidocollus Wagner, 1953: in *Salmo gairdneri, Salvelinus fon-
tinalis*, procercoids in copepods, Calif.; Wagner (1954): same.
P. wickliffi Hunter and Bangham, 1933: in *Coregonus artedi*, L. Erie;
Meyer, M. (1954): *C. clupeaformis*, Me.
Proteocephalus sp. Reported by many; most of these concern plerocer-
coid forms impossible to identify at this time.

Order PSEUDOPHYLLIDEA

Cestoda: Scolex with two grooves (bothria) or lobes (bothridia),
one dorsal, one ventral; neck conspicuous or not. Strobila with external segmentation well marked, often weak or lacking; proglottides
usually with one set of reproductive organs, sometimes two sets. Genital apertures surficial in some families, marginal or submarginal in
others. Testes follicular, numerous, medullary. Ovary bilobed, medullary. Vitellaria follicular, numerous. Uterine pore always surficial,
ventral or dorsal; eggs commonly but not invariably operculated,
liberating coracidium. Procercoid larval stage in crustaceans; plerocercoid larval stage in fishes. Adults mainly parasitic in fishes, some
in mammals and birds.

KEY

TO THE FAMILIES OF ADULT PSEUDOPHYLLIDEA FROM NORTH
AMERICAN FRESHWATER FISHES

1. Scolex with chitinoid hooks .... TRIAENOPHORIDAE
   (p. 234)
1. Scolex without chitinoid hooks 2
2. Cirro-vaginal pore (genital atrium) marginal; scolex subspherical, bothria distinct although often shallow; eggs not
   operculated but embryonated .... AMPHICOTYLIDAE
2. Cirro-vaginal pore median; scolex club-shaped or elongate;
   eggs operculated but not embryonated 3
3. Pseudoscolex present with shallow surficial bothria; primary
   scolex with 4 protrusible tentacles; eggs probably opercu-

late; small worms . . . . .HAPLOBOTHRIIDAE
(p. 233)
3. Scolex 4-lobed, more or less rectangular, with elongate shallow
bothria; eggs operculate; medium to large worms . . . . . .
. . . . BOTHRIOCEPHALIDAE

KEY

TO THE LARVAL PSEUDOPHYLLIDEANS OF NORTH AMERICAN FRESHWATER FISHES

1. Scolex with chitinoid hooks; large worms in musculature or
   viscera . . . . (fig. 200) *Triaenophorus*
   (p. 234)
1. Scolex without chitinoid hooks 2
2. Scolex with 4 protrusible tentacles . . . . (fig. 199) *Haplobothrium*
   (p. 233)
2. Scolex without tentacles 3
3. Scolex fairly distinct 4
3. Scolex not distinct 5
4. Scolex elongated with elongate bothria; body usually wrinkled,
   giving appearance of segmentation . . . .
   . . . . (fig. 196) *Diphyllobothrium*
   (p. 231)
4. Scolex rectangular to globular . . . . (fig. 195) *Bothriocephalus*
5. Smooth slender worm with dorsoventral groove throughout;
   in body cavity of cyprinids and catostomids . . . (fig. 197) *Ligula*
   (p. 232)
5. Broad worm with obvious evidence of segmentation; anterior
   longitudinal groove (bothrium) on triangular scolex; in
   body cavity of sticklebacks and other fish . . . .
   . . . . (fig. 198) *Schistocephalus*
   (p. 232)

Family AMPHICOTYLIDAE Ariola, 1899

Pseudophyllidea: Scolex unarmed, sometimes replaced by pseudoscolex; bothria distinct, often shallow. Segmentation distinct or ob-
scured by secondary transverse folds. Proglottides serrate or not. Neck
present or absent. Testes numerous, medullary. Genital apertures marginal, irregularly alternating. Ovary in ventral medulla. Vitellaria medullary, cortical or intermuscular. Uterine sac present, its opening
median, ventral, exceptionally dorsal, often not reaching ventral cuticle

until egg formation has commenced. Eggs not operculated, embryonated when laid. Parasitic in fishes.

Genus **Eubothrium** Nybelin, 1922
(Fig. 183)
(Syn.: *Leuckartia* Moniez, 1879, preoccupied)

Amphicotylidae: Scolex subglobular to elongate, often deformed, with simple bothria. Strobila usually distinctly segmented, with dorsal and ventral median furrow. Neck lacking. Inner longitudinal muscle bundles interlaced with transverse muscles. Nerve trunks dorsal to cirrus pouch and vagina. Testes in two lateral fields or almost continuous, medial to nerve trunks. Cirrus pouch not very large or muscular. Genital pores marginal. Ovary kidney-shaped, median or somewhat lateral. Vitellaria cortical, sometimes mainly among inner longitudinal muscle bundles or intruding into peripheral medullary parenchyma, interrupted in median fields. Uterus extending transversely anterior and ventral to ovary, with midventral aperture. Vagina opening anterior to cirrus. Parasitic in teleosts. Life cycle: adult in intestine of fishes; procercoid in copepods; no third host necessary.

Eubothrium crassum (Bloch, 1779): common in Europe; Bangham (1944): in *Lota lota, Petromyzon marinus*, Wis.; *Bangham* (1955): *L. lota, Coregonus artedi, Salmo gairdneri*, Mich.; DeRoth (1953): *S. salar sebago, Salvelinus fontinalis*, Me.; Haderlie (1953): *Oncorhynchus tshawytscha*, Calif.; Pearse (1924): *L. lota, S. namaycush, Micropterus salmoides*, Wis.; Shaw (1947): *S. malma*, Ore.; Sindermann (1953): *Roccus americanus*, Mass.; Van Cleave and Mueller (1934): *L. lota*, N. Y.; Wardle (1932): *Salvelinus alpinus, Myoxocephalus quadricornis*, Can.; Wardle and McCloud (1952): *Salmo salar, S. trutta*; Warren (1952): *Coregonus artedi*, Minn.
E. oncorhynchi Wardle, 1932: in *Oncorhynchus tschawytscha*, Can. Pacific.
E. rogosum (Batsch, 1786): Europe; Bangham and Adams (1954): in *Lota lota*, Brit. Col.; Wardle (1932): same, Can.
E. salvelini (Schrank, 1790). Bangham (1955): in pyloric ceca, *Salvelinus namaycush*, L. Huron; Bangham and Adams (1954): *S. alpinus, S. namaycush, Salmo clarki, S. gairdneri, Mylocheilus caurinus, Ptychocheilus oregonensis*, Brit. Col.; Choquette (1948): *S. fontinalis*, Que.; Dombroski (1955): *Oncorhynchus nerka*, Brit. Col.; Fantham and Porter (1947): *S. fontinalis*, Que.; Fritts (1959): *S. malma*, Idaho; Hoffman (unpubl.): longevity = at least 2 years 7 months; Margolis (1963): adversely affected growth of sockeye salmon; Richardson (1941): *S. fontinalis*, Que.; Vik (1963): life cycle, adult in salmonids; procercoid in copepods; no intermediate fish host; Wardle (1932): *S. fontinalis, S. alpinus, S. namaycush*, Can.; Worley and Bangham (1952): *S. fontinalis*, Que.

Eubothrium sp. Griffith (1953): in *S. gairdneri*, Wash.
Eubothrium sp. Bangham and Adams (1954): in *Oncorhynchus, Prosopium, Catostomus, Couesius, Richardsonius, Cottus*, Brit. Col.
Eubothrium sp. Applegate (1950) and Guilford (1954): in *Petromyzon marinus* (accidental?), Mich., Wis.

Genus **Marsipometra** Cooper, 1917
(Fig. 184)

Amphicotylidae: Scolex approximately pyramidal or arrow-shaped; apical disc narrower than base of scolex; subapical portion may be markedly constricted. Neck with or without zone of proliferation. Anterior proglottides broader than long, posterior ones rectangular. Nerve trunks dorsal to cirrus pouch; vagina close to proglottis margins. Testes in two lateral fields and continuous anterior to uterus; sometimes also posterior to ovary. Prostate gland surrounds distal portion of vas deferens just outside cirrus pouch. Cirrus unarmed. Ovary at posterior end of proglottis or separated from it by testes and vitellaria. Vitellaria in two ventral fields, more or less confluent anteriorly and posteriorly; follicles occurring on dorsal side are joined laterally with ventral follicles. Seminal receptacle sharply set off from seminal duct. Uterine sac occupies greater central portion of medulla, with a number of lateral or radial outgrowths; eggs small, rounded, thin-shelled. Parasitic in Polyodontidae. Life cycle: adult in intestine and ceca of fish; procercoid in *Cyclops*.

Marsipometra confusa Simer, 1930: in ceca, *Polydon spathula*, Miss.; Beaver and Simer (1940): same; Wilson (1956): same.
M. hastata (Linton, 1897): in spiral valve, less often in ceca and stomach, *P. spathula*; Bangham (1942): same, Tenn.; Beaver and Simer (1940): same, Miss.; Hugghins (1959): same, S. Dak.; Meyer, Γ. (1960): *Cyclops* intermediate host, Iowa; Pearse (1924): Wis.; Wilson (1956): *P. spathula*.
M. parva Simer, 1930: in spiral valve, less often in ceca, *P. spathula*, Miss.; Beaver and Simer (1940): same; Wilson (1956): same.
Marsipometra sp. Minckley and Deacon (1959): in *Pylodictis olivaris*, Kans.

Family BOTHRIOCEPHALIDAE Blanchard, 1849

Pseudophyllidea: Small to large forms. Scolex elongated, more or less rectangular, sometimes spherical club- or heart-shaped, usually with apical disc indented on each surficial edge and bearing marginal hooks occasionally. Bothria longitudinally elongated. Neck lacking. Strobila with distinct segmentation, often with secondary segmentation. Proglottides campanulate (bell-shaped), each with more or less distinct median furrow and indistinct marginal groove. Testes medullary, in

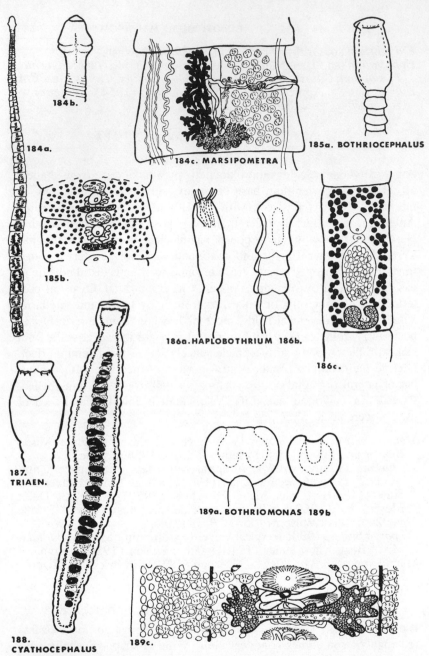

Figs. 184–189. Fig. 184. *Marsipometra confusa. a*, mature specimen; *b*, scolex; *c*, mature proglottid (from Beaver and Simer, 1940). Fig. 185. *Bothriocephalus cuspidatus. a*, scolex; *b*, proglottid (from Van Cleave and Mueller, 1934). Fig. 186. *Haplobothrium globuliforme. a*, primary

two lateral fields continuous from proglottis to proglottis and sometimes across median line. Cirrus pouch round, in median field. Cirrovaginal pore dorsal, median. Ovary bilobed or not, in ventral median medulla. Seminal receptacle present or absent. Vitellaria usually cortical, extending from proglottis to proglottis; may intrude into posterior overhanging borders, occasionally medullary. Uterus has winding uterine duct and roughly spherical median uterine sac; uterine pore midventral, opposite and anterior to cirro-vaginal pore; eggs operculate, not embryonated when laid. Longitudinal excretory stems medullary, in testicular fields or just lateral to them. Parasitic in teleosts.

Genus **Bothriocephalus** Rudolphi, 1808
(Figs. 185, 195)

Bothriocephalidae: Scolex elongate, sometimes spherical or enlarged posteriorly, with apical disc having indented edges; the two indentations may be connected across apex by a groove. Marginal surface of scolex convex or concave, often longitudinally grooved. Bothria longitudinally elongated, of varying width and depth. Neck lacking. Segmentation complete, often with secondary segmentation. Proglottides campanulate, anapolytic; anterior ones bell- or funnel-shaped, posterior ones rectangular. Inner longitudinal musculature well developed. Longitudinal excretory stems 2 or 3 on each side in medulla. Testes in lateral medulla. Cirrus pouch may be surrounded by prostatic cells. Cirro-vaginal pore dorsal, posterior to uterine pore. Ovary compact, transversely elongated, bilobed or not, situated in ventral median medulla. Seminal receptacle absent. Vitellaria entirely cortical, continuous from proglottis to proglottis. Uterine sac and ventral uterine pore median or alternating irregularly from one side of median line to other; eggs thin-shelled, operculated, not embryonated when laid. Parasitic in marine and freshwater teleosts, rarely in amphibians. Life cycle: adult in intestine of fish; procercoid in copepods; small fish sometimes act as "carriers."

---

scolex (from Van Cleave and Mueller, 1934); *b*, scolex and first 3 segments of secondary worm (from Cooper, 1914, in Wardle and McLeod, 1952); *c*, mature proglottid (from Van Cleave and Mueller, 1934). Fig. 187. *Triaenophorus nodulosus*, scolex (from Van Cleave and Mueller, 1934). Fig. 188. *Cyathocephalus truncatus* (from Wisniewski, 1933, in Wardle and McLeod, 1952). Fig. 189. *Bothriomonas fallax. a*, scolex, lateral view; *b*, scolex, ventral view; *c*, gravid proglottid (from Nybelin, 1922, in Wardle and McLeod, 1952).

Bothriocephalus claviceps (Goeze, 1782) Rud., 1810. Bangham (1939, 1942, 1944, 1955): in *Ambloplites rupestris, Anguilla rostrata, Chaenobryttus gulosus, L. gibbosus, L. macrochirus, Micropterus dolomieui, M. salmoides, Percopsis omiscomaycus, Stizostedion canadense,* Fla., Mich., Tenn., Wis.; Cooper (1918): *A. rostrata, L. gibbosus, Gasterosteus bispinosus*; Fischthal (1947, 1950, 1956): L. *gibbosus,* Wis.; Griffith (1953): immature in *L. gibbosus, L. macrochirus,* Wash.; Jarecka (1959): life cycle, Poland; McDaniel (1963): *L. cyanellus,* immature in *L. macrochirus, L. megalotis,* Okla.; Pearse (1924): *A. rostrata, L. gibbosus,* Wis.; Van Cleave and Mueller (1934): *A. rostrata, M. dolomieui,* N. Y.; Venard (1941): Tenn.

B. cuspidatus Cooper, 1917. Anthony (1963): in ceca and intestine, *Perca flavescens, L. macrochirus,* Wis.; Bangham (1944, 1955): *Ictalurus nebulosus, L. gibbosus, L. macrochirus,* Wis., Mich.; Essex (1928): life cycle; Fischthal (1947, 1950): *M. dolomieui, Perca flavescens, Stizostedion canadense, S. vitreum; A. rupestris, Hyborhynchus notatus, I. nebulosus, L. macrochirus,* immature in *I. nebulosus, M. salmoides, P. flavescens,* Wis.; Fritts (1959): *L. gibbosus,* Idaho; Hoffman (1964, unpubl.): *L. macrochirus,* W. Va.; Hugghins (1959): *S. vitreum, S. canadense,* S. Dak.; Meyer, M. (1954): *Lota lota,* Me.; Pearse (1924): *S. vitreum, P. flavescens,* Wis.; Shaw (1947): Ore.; Van Cleave and Mueller (1934): *S. vitreum,* N. Y.; Wardle (1932): *Coregonus* sp., Can.; Worley and Bangham (1952): *S. vitreum,* Que.; *Esox lucius, Hiodon* spp.; *P. flavescens, Stizostedion* spp.

B. formosus Mueller and Van Cleave, 1932: in *Percopsis omiscomaycus, Etheostoma nigrum, Percina caprodes,* N. Y.; Bangham (1941, 1955): *P. omiscomaycus, E. nigrum, Poecilichthys exilis,* Mich., Ohio; Fischthal (1947, 1956): *E. nigrum, Semotilus atromaculatus,* N. Y., Wis.

B. rarus Thomas, 1937: in newts, dev. in copepods; Meyer, M. (1954): in *Fundulus diaphanus,* Me.

B. schilbeodis Cheng and James, 1960: in *Noturus insignis,* Va.

B. speciosus (Leidy, 1858): in *Etheostoma olmstedi.*

B. texomensis Self, 1954: in *Hiodon alosoides,* Okla.

Family DIPHYLLOBOTHRIIDAE Lühe, 1910

Pseudophyllidea: Adult: Scolex usually compressed laterally, with surficial bothria. Neck distinct or indistinct. Segmentation usually distinct. Excretory system forming superficial plexus in cortex. A single, rarely double, set of genital organs per segment. Testes numerous, medullary, may intrude into space among inner longitudinal muscles. Genital apertures surficial and ventral, opening directly and independently, or by way of a common cirro-vaginal pore. External seminal vesicle present. Ovary medullary and ventral, two-winged, posterior. Vitelline follicles cortical, usually meeting in front of genital apertures. Uterus often rosette-shaped, with distal uterine sac opening ventrally behind genital apertures; eggs operculated, not embryonated when laid. Life cycle: adults in fishes, reptiles, birds, and mammals; procercoid in crustaceans; plerocercoid in fishes, amphibians, reptiles, and mammals.

Genus **Diphyllobothrium** Cobbold, 1858
(Fig. 196)

Plerocercoid: Scolex laterally compressed with 2 elongate shallow bothria. Body of worm usually wrinkled, suggesting segmentation. In viscera or musculature. Species difficult to determine.

Migrating larvae in small fish can cause much damage and even mortality (Essex and Hunter, 1926; Hoffman and Dunbar, 1961; Margolis, 1961, pers. comm.). Life cycle: adult in mammals and birds; procercoid in copepods; plerocercoids in fish.

Ref. Kuhlow (1953), determination of European species; Vik (1964), review of genus; M. C. Meyer (1966), review.

Diphyllobothrium cordiceps (Leidy, 1871): in *Salvelinus fontinalis*, Wyo.; Vik (1964): *Salmo mykiss, S. clarki, S. gairdneri, Oncorhynchus kisutch, T. thymallus*, Ore. Adult in pelican, gulls, mergansers; not infective for man.

D. dalliae Rausch, 1956: in *Dallia pectoralis*, Alaska; adult in gulls, dog, *Alopex lagopus*, man; Zhukov (1963): Russia.

D. ditremum (Creplin, 1825): in *S. trutta*, Europe. Adult in fish-eating birds. Hilliard (1960): adult from Alaska.

D. laruei Vergeer, 1942: in *Coregonus* spp., Mich. Adult in cat and dog.

D. latum (Linnaeus, 1758): in *Esox lucius, Perca flavescens, Stizostedion vitreum, S. canadense, Lota lota*. Many dubious records from other fish including trout. Guttowa (1961): copepod hosts, Norway. Adult in bear, dog, man, northern U. S., Can.

D. oblongatum Thomas, 1946: in *Coregonus artedi*. Adult in gulls and tern, Mich. Warren (1952): *C. artedi*, Minn.

D. osmeri (v. Linstow, 1878). Hilliard (1960): in *Osmerus dentex*; exper. in dog, Alaska. Reported from *S. alpinus* and *S. salar* in Europe.

D. salvelini Yeh, 1955: in *S. alpinus*, Greenland.

D. sebago (Ward, 1910) Meyer, M., and Robinson, 1963: in *S. sebago, S. fontinalis, Osmerus mordax*. Adult in gulls, dog, fox, cat, rat, squirrel, but not man, Me.; Meyer, M. and Vik (1963): life cycle.

D. ursi Rausch, 1954: in *O. nerka*; adult in gull and bear, Alaska.

Diphyllobothrium sp. Hoffman and Dunbar, 1961: caused mortality of *S. fontinalis* fingerlings, Que.; this is probably *D. sebago*.

Diphyllobothrium sp. Fox, 1962: in *S. gairdneri*, Mont.

Diphyllobothrium sp. Richardson, 1936: in *S. fontinalis*.

Diphyllobothrium sp. Laird and Meerovitch, 1961: in *S. alpinus*. Adult in man, N. Can.

Diphyllobothrium sp. Reported from *Coregonus* spp., *Gila, Lota, O. nerka, Salmo* spp., *Salvelinus* spp., by Alexander (1960): Ore.; Babero (1953): Alaska; Bangham (1951, 1954, 1955): Ohio, Mich., Brit. Col.; Essex and Hunter (1926): Mont.; Fantham and Porter (1947): Que.; Haderlie (1953): Calif.; Moore (1926): heart worm in *Coregonus*, N. Y., Mueller (1940): N. Y.; Shaw (1947): Ore.; Wardle (1932): W. Can.

Genus **Ligula** Bloch, 1782
(Fig. 197)

Plerocercoid: Large, slender, nonsegmented worm with obvious dorsal and ventral grooves. Usually in body cavity of cyprinids and catostomids. Causes great damage to small fish. Life cycle: adult in fish-eating birds; procercoid in copepods; plerocercoid in body cavity of fish.

Ligula intestinalis (Linnaeus, 1758). Although fish host preference has been shown by Dence (1958) and Lawler (1964), this worm has been reported from *Etheostoma, Catostomus, Coueseus, Hybognathus, Micropterus, Mylocheilus, Notropis, Osmerus, Perca, Pimephales, Prosopium, Ptychocheilus, Richardsonius, Salvelinus,* and *Siphateles,* by Alexander (1960), Bangham (1951, 1955), Bangham and Adams (1954), Choquette (1948), Dence (1943, 1958), Fantham and Porter (1947), Fritts (1959), Haderlie (1953), Hoffman (1954), Hugghins (1959), Lawler (1964), Linton (1893), Meyer, M. (1954), Pearse (1924), Pitt and Grundman (1957), Roussow (1954), Van Cleave and Mueller (1934), and Wardle (1932, 1933), from Brit Col., Calif., Can., Idaho, New Eng., N. Y., N. Dak., Me., Mich., Ore., S. Dak., Que., Wis., Wyo.

Genus **Schistocephalus** Creplin, 1829
(Fig. 198)

Plerocercoid: Scolex triangular with shallow bothrium on dorsal and ventral surface (appears notched). No neck. Segmentation apparent. Worm large and broad. In body cavity of fish. Life cycle: adult in fish-eating bird; procercoid in copepods; plerocercoid in body cavity of fish.

Ref. Thomas (1947), Dubinna (1957), Clark (1954).

Schistocephalus solidus (Müller, 1776). Cooper (1918): in *Gasterosteus* spp., *Uranidea formosa,* N. Bruns.; Haderlie (1953): *G. aculeatus,* Calif.; Hoffman (1956, unpubl.): *Eucalia inconstans,* N. Dak.; Linton (1898): *Cottus bairdi,* Mont.; Meyer, M. (1954): in stomach of *S. fontinalis,* which had eaten it, Me.; Thomas (1947); Clarke (1954): life cycle; Wardle (1933): *Cottus* sp., Can.

S. thomasi Garoian, 1960: plerocercoid in sticklebacks, Mich.

Schistocephalus sp. (prob. *S. solidus*). Bangham and Adams (1954): in *Catostomus, Cottus, Gasterosteus, Oncorhynchus, Prosopium, Richardsonius, Salmo*, Brit. Col.; Cooper (1918): Mont.; Hunter and Hunter (1930): N. Y.; Moore (1926): erroneously identified, should be *Diphyllobothrium*; Wardle (1933): *Cottus, S. fontinalis*, Brit. Col.

Family HAPLOBOTHRIIDAE Meggitt, 1924

Pseudophyllidea: Small slender worms. Pseudoscolex with conical apex, shallow dorsal and ventral depressions and longitudinal marginal grooves continued backward over anterior proglottides; two pairs of auricular appendages posterolaterally; unsegmented neck portion absent. External segmentation confined to anterior region of strobila. Only one set of reproductive organs per segment. Genital pores midventral. Testes and vitellaria in lateral medulla medial to nerve trunks. External seminal vesicle present. Cirrus pouch anterior, median. Cirrus armed with short spines. Ovary in posterior central medulla. Seminal receptacle well marked off from vagina. Uterus with much-coiled proximal duct and large distal sac. Excretory system with large median vessel and two smaller lateral vessels. Parasitic in freshwater teleosts.

Genus **Haplobothrium** Cooper, 1914
(Figs. 186, 199)

Adult: With the characters of the family. Plerocercoid: body divided into two parts; a long, slender, anterior portion (115 by 0.16 mm) comprising scolex and neck, and a posterior portion consisting of a spheroidal bladder-like organ (0.5 by 0.5 mm). Scolex contains 4 protrusible proboscides about 0.42 mm long when evaginated (Essex, 1929). Life cycle: adult in fish; procercoid in copepods; plerocercoid encysted in liver of fish.

Ref. Thomas (1929); Van Cleave and Mueller (1934); Meinkoth (1947).

Haplobothrium globuliforme Cooper, 1914: in intestine, *Amia calva*; Anthony (1963): same, Wis.; Bangham (1942, 1955): same, Mich., Tenn.; Essex (1929): plerocercoid in *Ictalurus nebulosus*, Minn.; Fischthal (1950): *A. calva*, Wis.; Meinkoth (1947): *A. calva, Cyclops*, plerocercoid in *Lebistes reticulatus*; Sogandares (1955): *A. calva, A. rostrata*, plerocercoid in *L. gibbosus*, La.; Van Cleave and Mueller (1934): *A. calva*, N. Y.

Family TRIAENOPHORIDAE Loennberg, 1889
(See description of genus)

Genus **Triaenophorus** Rudolphi, 1793
(Figs. 187, 200)

(Syn.: *Tricuspidaria* Rudolphi, 1793)

Adult: Scolex with dorsal and ventral pair of trident-shaped hooks on apical disc; bothria piriform and shallow. External segmentation absent. Nerve trunks dorsal to cirrus pouch and vagina. Testes medullary, medial to nerve trunks, filling all available space of medulla. Ovary medullary, extending laterally ventral to testes, somewhat approaching pore side. Seminal receptacle large. Vitellaria cortical, diffuse. Uterus with weakly coiled duct and uterine sac; uterine pore opening in front of genital aperture; eggs operculate, thick-shelled. Excretory stems with anastomosing branches and numerous marginal pores. In intestine of teleosts. Plerocercoid: usually enclosed in yellowish connective tissue cyst. Worm about 1 mm in diameter, long (up to 130 mm), coiled, white. Scolex as in adult. Strobila same diameter throughout and lacks sex organs. Caudal filament often present. Life cycle: adult in predatory fish; procercoid in copepods; plerocercoid in forage fish.

Triaenophorus crassus Forel, 1868 (Syn. *T. robustus* Olsson, 1893; *T. tricuspidatus morpha megadentatus* Wardle, 1932). Present in Europe also. Lawler and Scott (1954): review, adult in *Esox* spp.; plerocercoids in *Coregonus* spp., *Lota lota, Oncorhynchus nerka, Percopsis omiscomaycus, Petromyzon marinus, Prosopium* spp., *Salvelinus alpinus, S. namaycush, Stenodus leucichthys, Thymallus signifer* from northern U. S. and Can.; Miller (1952): review, plerocercoid in muscle of *E. lucius, Coregonus* spp., *Prosopium* spp., *S. namaycush, T. signifer.* Later articles include Bangham (1955): Mich.; Guilford (1954): Wis.; Keleher (1952); Lawler (1961): control by fisheries management; Lawler and Watson (1963): sizes of procercoids of *T. crassus* and *T. nodulosus*; McLain (1951): undeveloped in lamprey, Mich.; Margolis (1963): use as a salmon "tag" in ocean; Miller (1954): control through fish management; Odlaug *et al.* (1962): Minn.; Uzmann and Hesselholt (1957): plerocercoid in *O. keta*, Alaska; Warren (1952): Minn.; Watson (1963): lethal temperature for eggs; Watson and Lawler (1963): temperature and hatching; Watson and Price (1960): experimental infection of copepods.

T. nodulosus Pallas, 1760. This may be *T. lucii*; see Yamaguti (1959, p. 57) and Wardle and McCleod (1952, p. 636). Present in Europe also. Lawler and Scott (1954): review, adult in *Esox* spp., *Stizostedion* spp., plerocercoids in viscera of *Catostomus* spp., *Coregonus* spp., *Cottus cognatus, Esox* spp., *Eucalia inconstans, Roccus chrysops, Lepomis* spp., *Micropterus* spp., *Moxostoma* spp., *Notropis* spp., *Perca flavescens, Pomoxis nigromaculatus, Salvelinus fonti-nalis, Thymallus signifer.* Later articles include Bauer (1958): killed small trout, Russia; Chubb (1963): seasonal occurrence, Eng.; Engelbrecht (1961): adults leave starved pike; Lawler and Watson (1963): see *T. crassus*; Lopukhina (1961): effect on small trout, Russia; Matthey (1963): massive mortality of perch, Switz.; Fisch-thal (1950): adult in *Esox* spp., Wis.; Grabiec *et al.* (1962): studies on coracidia; Reichenbach-Klinke (1954): histopathology of cyst, Germ.; Vik (1959): distribution in Norway; Watson (1963), Watson and Lawler (1963), Watson and Price (1963): see *T. crassus.*

T. stizostedionis Miller, 1945: in *Stizostedion vitreum*, procercoid in *Cyclops*, plerocercoid in viscera of *Percopsis omiscomaycus*, Al-berta; Bangham (1955): *S. vitreum*, Mich.; Lawler and Watson (1963): coracidium.

Order SPATHEBOTHRIIDEA Wardle and McLeod, 1952

Cestoda: Scolex without true bothria or suckers. Strobila with in-ternal segmentation but no external segmentation. Testes medullary, in 2 lateral bands. Ovary median. Uterus tubular, coiled, with median pore; eggs operculate. Vitellaria follicular, cortical. Male and vaginal apertures surficial. Adult form apparently neotenic procercoid, sex-ually functional in fishes.

KEY

TO THE FAMILIES

1. Adhesive organ funnel-like .... CYATHOCEPHALIDAE
1. Adhesive organ with 1 or 2 dorsal and ventral hollow spherical
   structures opening apically .... DIPLOCOTYLIDAE

Family CYATHOCEPHALIDAE Nybelin, 1922

With the characters of the order. Scolex with funnel-shaped apical sucking organ. Strobila with series of 20 to 45 sets of reproductive organs, without distinct external segmentation.

Genus **Cyathocephalus** Kessler, 1868
(Fig. 188)

With the characters of the family.

Cyathocephalus americanus Cooper, 1918: syn. of *C. truncatus.*
C. truncatus (Pallus, 1781). Present in Europe also. *American rec-ords.*—Alexander (1960): in *Salmo trutta, S. gairdneri, Salvelinus fontinalis*, Ore.; Bangham (1955): *Coregonus clupeaformis*, Mich.;

Bangham and Adams (1954): *Cottus asper*, Brit. Col.; DeGiusti and Budd (1959): 3-year study in amphipod, Mich.; Nieland (1952): *S. clarki*; Pearse (1924): *Coregonus* sp., Wis.; Wardle (1932): *Coregonus* spp., *Salvelinus alpinus*, and amphipod probably, Can. *European records.*—Bauer (1958, p. 295): weight loss and retardation of ovaries, Russia; Ciric (1963): *Cottus gobio*, Yugoslavia; Senk (1956): *S. trutta*, damage to spawners, Yugoslavia; Vik (1954, 1958): larva in stickleback and *Gammarus*, adult in *S. trutta*, pathology, Norway; Wisniewski (1932): life cycle, Poland.

Family DIPLOCOTYLIDAE Monticelli, 1892

With the characters of the order. Scolex with 1 or 2 spherical hollow structures opening apically. Strobila with many sets of reproductive organs, without distinct external segmentation.

KEY

TO THE GENERA

1. Scolex with cavities completely separated internally; genital apertures on ventral surface . . . . (fig. 190) *Diplocotyle*
1. Scolex with cavities completely fused; genital apertures alternating irregularly from dorsal to ventral . . . .
. . . . (fig. 189) *Bothriomonas*

Genus **Bothriomonas** Duvernoy, 1842
(Fig. 189)

Scolex with cavities completely fused; rudimentary septum at bottom. Strobila with weak inner longitudinal musculature, containing many sets of reproductive organs. Genital apertures alternating irregularly from one flat surface to other. Testes intruding into space between two wings of ovary. Cirrus pouch with strongly developed muscle fibers at base of cirrus. Ovary approximating first toward one body surface and then toward other without reference to position of genital pores; comparatively wide unlobed isthmus, each wing longitudinally elongate and embracing distal uterine coils. Vitellaria lacking in median cortex. Uterine coils occupy about one-fourth breadth of strobila, with its greater part surrounded by gland cells, distal part not particularly wide. Utero-vaginal atrium shallow, with strongly developed sphincter. Eggs with relatively large operculum. Parasitic in sturgeons. Life cycle unknown.

Bothriomonas sturionis Duvernoy, 1842: in *Acipenser oxyrhynchus*, *A. sturio*. Not reported since 1842.

Genus **Diplocotyle** Krabbe, 1874
(Fig. 190)

Diplocotylidae: Scolex with two cavities completely separated internally. Strobila with weakly developed inner longitudinal musculature. All genital pores on ventral surface only. Testes not intruding into space between two ovarian lobes. Cirrus pouch nearly devoid of parenchymatous muscle. Ovarian wing not longitudinally elongated, not embracing distal uterine coils; isthmus unlobed. Uterine coils extend as far outward as ovarian wing, with its small proximal portion surrounded by gland cells and distal portion strongly widened; uterine pore immediately in front of vaginal aperture, with weakly developed sphincter; vagina more or less widened just before opening, its aperture with moderately well-developed sphincter. Eggs with large operculum. Life cycle unknown.

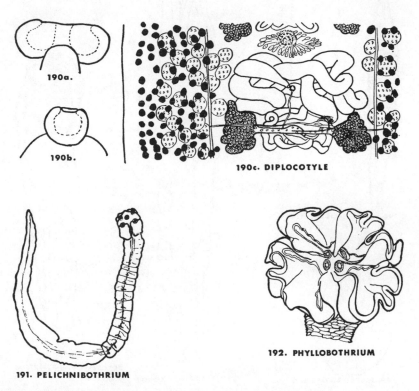

190a.

190b.

190c. DIPLOCOTYLE

191. PELICHNIBOTHRIUM

192. PHYLLOBOTHRIUM

Figs. 190–192. Fig. 190. *Diplocotyle olrikii. a* and *b*, views of scolex; *c*, mature proglottid (from Nybelin, 1922, in Wardle and McLeod, 1952). Fig. 191. *Pelichnibothrium speciosum* (from Yamaguti, 1934, in Wardle and McLeod, 1952). Fig. 192. *Phyllobothrium radioductum*, scolex (from Kay, 1942, in Wardle and McLeod, 1952).

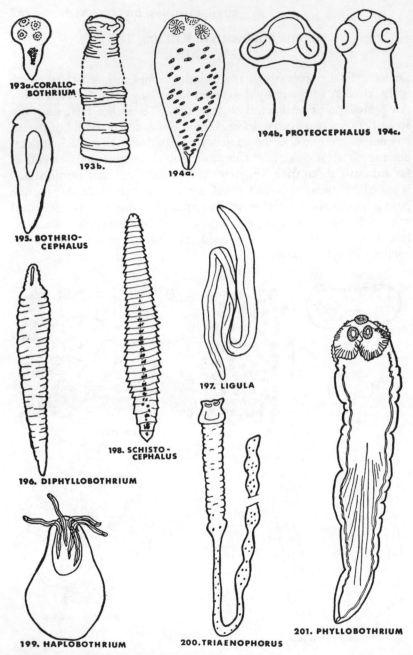

193a. CORALLO-
BOTHRIUM

193b.

194a.

194b. PROTEOCEPHALUS 194c.

195. BOTHRIO-
CEPHALUS

196. DIPHYLLOBOTHRIUM

197. LIGULA

198. SCHISTO-
CEPHALUS

199. HAPLOBOTHRIUM

200. TRIAENOPHORUS

201. PHYLLOBOTHRIUM

Figs. 193–201. Larval cestodes. Fig. 193. *Corallobothrium. a, C. parvum* from small fish (from Larsh, 1941); *b, C. fimbriatum,* plerocercoid (from Wardle, 1932). Fig. 194. *Proteocephalus ambloplitis* plerocercoids. *a,* note calcareous corpuscles (from Van Cleave and Mueller, 1934); *b,* (from Hunter, 1929); *c,* from bluegills (orig.). Fig. 195. *Bothriocephalus cus-*

Diplocotyle olrikii Krabbe, 1874 (Syn. *D. nylandica* (Schneider, 1902); *Bothriomonas intermedius* Cooper, 1918): Wardle (1932): in *Salmo carpio, S. alpinus*, Iceland; *S. alpinus, Artediellus incinatus*, Greenland; *S. alpinus, S. fontinalis, Coregonus* spp., *Myoxocephalus quadriocornis*, N. Can.

Order TETRAPHYLLIDEA Carus, 1863

With 4 sessile or pedunculate bothridia, cup-, leaf-, trumpet-, or ear-shaped. Marine forms; included here because they may be in salmon when they return to fresh water.

Genus **Pelichnibothrium** Monticelli, 1889
(Fig. 191)

Scolex with apical sucker, and accessory sucker in front of anterior border of each bothrium.

See Yamaguti (1959) or Wardle and McLeod (1952) for complete description.

Pelichnibothrium speciosum Monticelli, 1889. Probably same as *Phyllobothrium salmonis* Fujita, 1922. Reported from *Oncorhynchus keta*, N. Pacific; Haderlie (1953): *O. tshawytscha*, Calif.

Genus **Phyllobothrium** Beneden, 1849
(Fig. 201)

Scolex without apical sucker; bothridia sessile or pedunculated, with adherent surfaces smooth, folded or loculated, and borders simple or muscular, curled or frilled; accessory suckers one or two for each bothridium.

See Yamaguti (1959) or Wardle and McLeod (1952) for complete description.

Phyllobothrium ketae Canavan, 1928: adult and larva in *O. ketae*, Alaska.

P. salmonis (see *Pelichnibothrium speciosum*).

---

pidatus, plerocercoid from intestine of *Stizostedion* (from Essex, 1928, in Wardle and McLeod, 1952). Fig. 196. *Diphyllobothrium* sp. (from Wardle, 1932). Fig. 197. *Ligula intestinalis* (from Van Cleave and Mueller, 1934). Fig. 198. *Schistocephalus solidus* (from Vik, 1954). Fig. 199. *Haplobothrium globuliforme* (from Meinkoth, 1947). Fig. 200. *Triaenophorus crassus* (from Miller, 1952). Fig. 201. *Phyllobothrium* plerocercoid (from Dollfus, 1929, in Wardle and McLeod, 1952).

Phyllobothrium sp. Bangham and Adams (1954): in *O. nerka, Salvelinus alpinus, Entosphemus*, Brit. Col; Shaw (1947): *S. clarki*, Ore.

Order CYCLOPHYLLIDEA
Family HYMENOLEPIDIDAE
Genus **Hymenolepis**

Parasites of rodents. Larval forms have been reported from fish, but these may actually be *Proteocephalus*, which *Hymenolepis* resembles.

Hymenolepis sp. larvae. Bangham (1951): in *Gila straria*, Wyo.; Bangham (1955): *Poecilichthys exilis*, Mich.; Bangham and Adams (1954): *Couesius plumbeus*, Brit. Col.; Krueger (1954): *L. cyanellus, L. macrochirus, Notemigonus crysoleucas*, Ohio.

Nematodes

Phylum Nemahelminthes, Class Nematoda

Contents

Key to the genera of adult nematodes from North American freshwater fishes 242
Key to the larval nematodes from North American freshwater fishes 245
Order OXYURIDEA 246
Order TRICHURIDAE 247
   Genus *Capillaria* 247
   Genus *Cystoopsis* 247
   Genus *Hepaticola* 249
Order ASCARIDIDEA 249
   Genus *"Ascaris"* 249
   Family HETERO-
     CHEILIDAE 250
     **Genus *Contracaecum*
     (*Hysterothylacium*) ... 250
   Genus *Raphidascaris* 251
   *Genus *Porrocaecum* 251
   *Genus *Anisakis* 251
Order SPIRURIDEA 251
   Family CAMALLANIDAE 253
   Genus *Camallanus* 253
   Family CUCULLANIDAE 254
   Genus *Bulbodacnitis* 254
   **Genus *Cucullanus*
     (*Dacnitis*) 255

   **Genus *Dacnitoides* 255
   **Genus *Dichelyne* 257
   Family HAPLONE-
     MATIDAE 257
   Genus *Haplonema* 257
   Family HEDRURIDAE .. 258
   **Genus *Hedruris* 258
   Family RHABDO-
     CHONIDAE 258
   Genus *Cystidicola* 259
   Genus *Rhabdochona* 259
   Genus *Spinitectus* 261
   Family SPIRURIDAE 262
   Genus *Metabronema*
     (*Cystidicoloides*) 263
   *Genus *Spiroxys* 263
   Family GNATHO-
     STOMIDAE 265
   *Genus *Gnathostoma* 265
Order FILARIIDEA 267
   Genus *Philonema* 267
   Genus *Philometra*
     (*Ichthyonema*) 268
Order DIOCTO-
   PHYMIDEA 268
   *Genus *Dioctophyma* ... 269
   *Genus *Eustrongylides* .. 269

* Larva only in fish.
** Larva and adult in fish.

Most adult fish nematodes are intestinal-tract residents. Filarid nematodes, however, are found in the body cavity, "cheek galleries," and caudal fin. Larval nematodes of fish may be found in almost every organ, but are common in the mesenteries, liver, and musculature.

The life cycle always involves an invertebrate—copepod, insect nymph, etc.—for the first intermediate host, from where it may return to the fish in the fish's forage. Some, however, use the fish as the second intermediate host and develop to adults in the intestinal tract of piscivorous fish, birds, and mammals.

It is thought that intestinal nematodes of fish produce little pathogenicity. In general, this is probably true, but I have seen fish with large numbers of intestinal *Capillaria* in which the intestinal lining was much inflamed. Some larval nematodes cause considerable damage in the body cavity of fish, notably *Contracaecum* and *Spiroxys*. The large larvae of *Eustrongylides*, *Anisakis*, and *Porrocaecum* are very unsightly, and larval migrans of *Anisakis* can produce an acute abdominal syndrome of man. The filarid, *Philonema*, causes serious visceral adhesions in land-locked salmon and trout. One nematode genus, *Cystidicola*, occurs in the swim bladder of fish. There is no known control of these nematodes.

KEY

TO THE GENERA OF ADULT NEMATODES

(Modified in part from Yorke and Maplestone, 1926)

1.   Esophagus consisting of a narrow tube running through center of row of single cells for most of its length; no lips; in intestine of fish (a European form in liver) .... Order TRICHURIDEA   5

1.   Not so 2

2 (1). Esophagus dilated posteriorly into a bulb .... Order OXYURIDEA

2 (1). Not so 3

3   (2). Head with 3 large lobes or lips (*en face* view); often
esophageal and/or intestinal ceca present; relatively
stout worms . . . . Order ASCARIDEA   6
(p. 249)

3   (2). Not so; no intestinal ceca   4

4   (3). Usually with 2 lateral lips (*Hedruris* has 4); chitinous
buccal cavity or vestibule present; vulva usually in
middle of body or posterior to it; no intestinal ceca
. . . . Order SPIRURIDEA   8
(p. 262)

4   (3). Without lips; vestibule absent or rudimentary; vulva
almost invariably in esophageal region; larvae in utero;
in visceral cavity, or "cheek galleries" and fins . . . . .
. . . . Order FILARIIDEA   17
(p. 267)

Order TRICHURIDEA

5   (1). Esophagus undivided; male with spicule; in intestine;
a European form in liver . . . . (fig. 202) *Capillaria*
(p. 247)

5   (1). Esophagus undivided; male without spicule but with
copulatory sheath; in intestine . . . . (fig. 204) *Hepaticola*
(p. 249)

5   (1). Esophagus divided into anterior muscular portion and
posterior cellular portion; hindbody of female spheri-
cal; in skin blisters . . . . (fig. 203) *Cystoopsis*
(p. 247)

Order ASCARIDEA

6   (3). Alimentary canal simple, no postesophageal ventriculus or esophageal or intestinal diverticula . . . . "*Ascaris*"
(p. 249)

6   (3). Alimentary canal with postesophageal ventriculus and/
or esophageal or intestinal diverticula   7

7   (6). With intestinal cecum (projects anteriorly) and one
esophageal appendix (projects posteriorly); in stomach and intestine . . . . (fig. 205) *Contracaecum*
(p. 250)

7   (6). Without intestinal cecum, but with esophageal appendix . . . . (fig. 207) *Raphidascaris*
(p. 251)

Order SPIRURIDEA

8   (4). With 4 highly specialized lips; males always rolled
about females, posterior end of female invaginated,
forming suckerlike groove from which chitinous hook
projects; in stomach . . . . (fig. 215) *Hedruris*
(p. 258)

8   (4). Without 4 highly specialized lips, etc.   9

9   (8). With large chitinous buccal capsule; in intestine; worm
red; sometimes can be seen protruding from anus . . .
. . . . (fig. 210) *Camallanus*
(p. 253)

9   (8). Without large chitinous buccal capsule; worm not red .   10

10   (9). Head with 2 large lateral lobes; esophagus muscular
throughout, dilated anteriorly to form a pseudobuccal
capsule, and enlarged posteriorly . . . .
. . . . Family CUCULLANIDAE   11
(p. 254)

10   (9). Not so   14

11 (10). With dorsocephalic tubercle; in intestine . . . .
. . . . (fig. 210) *Bulbodacnitis*
(p. 254)

11 (10). Without dorsocephalic tubercle; in intestine   12

12 (11). No intestinal cecum; 2 ovaries . . . . . (fig. 211) *Cucullanus*
(p. 255)

12 (11). With intestinal cecum; 1 or 2 ovaries 13

13 (12). With pre-anal sucker; 1 ovary . . . . (fig. 212) *Dacnitoides*
(p. 255)

13 (12). Without pre-anal sucker; 2 ovaries . . . (fig. 213) *Dichelyne*
(p. 257)

14 (10). Cuticle armed with chitinous hooklike spines in longitudinal rows or in circles along whole, or anterior portion, of body; in intestine . . . . (fig. 218) *Spinitectus*
(p. 261)

14 (10). Cuticle not so armed   15

15 (14). Males with broad caudal alae; usually with 4 pairs
(rarely more) of large pre-anal papillae, almost invari-
ably pedunculated   16

15 (14). Males with or without caudal alae; pre-anal papillae
sessile, usually numerous, in linear row   17

16 (15). Lips large and distinctly trilobed; cuticle of inner surface thickened, tending to interlock with that of opposite lip; *adult in turtles*; larvae in many fish . . . . . .
. . . . (fig. 220) *Spiroxys*
(p. 263)

16 (15). Lips with cuticle of inner surface not thickened, etc.;
spicules unequal; with 4 pairs of pedunculated papillae; eggs with bipolar filaments; gubernaculum present . . . . (fig. 219) (*Cytidicoloides*) *Metabronema*
(p. 263)

17 (15). Spicules equal. . . . .(fig. 214) *Haplonema*
(p. 257)

17 (15). Spicules unequal. 18

18 (17). Males with caudal alae; in swim bladder, heart, intestine. . . . .(fig. 216) *Cystidicola*
(p. 259)

18 (17). Males without caudal alae; females with simple mouth,
long esophagus, embryonated eggs; posterior of body
rather pointed; in intestine. . . . . .(fig. 217) *Rhabdochona*
(p. 259)

Order FILARIIDEA

19 (4). Posterior extremity pointed and curled; female up to 1
foot long; larvae in gravid female. . . .(fig. 222) *Philonema*
(p. 267)

19 (4). Posterior extremity blunt; up to 2 feet long; larvae in
gravid female; larvae also free in fish. . . . .
. . . .(fig. 223) *Philometra*
(p. 268)

KEY

TO THE LARVAL NEMATODES

1. No intestinal cecum (projects anteriorly) or esophageal appendix (projects posteriorly) 2
1. Intestinal cecum present 9
2. Nematode large; in flesh of marine fish, occasionally in anadromous fish . . . . (fig. 208) *Anisakis* and relatives
(p. 251)
2. Nematode usually smaller (except *Eustrongylides*); in mesenteries, visceral organs, occasionally muscle. 3
3. No lips or lateral lobes 4
3. Lips, lateral lobes, or cephalic papillae present. 5
4. Microscopic larvae; not encysted . . . . *Philometra,*
Philonema
(p. 267)
4. Encysted in mesenteries and muscle of fish; cyst about 1 cm in

diameter; *larva blood red*; 12 or 18 small head papillae in 2 circles; recorded from many fish; adult in glands of fore-stomach of aquatic birds . . . . (fig. 225) *Eustrongylides*
(p. 269)

5. Head with no lips, but with circle of 6 papillae; up to 10 mm long; in mesenteries of bullhead, perhaps others . . . .
. . . . (fig. 224) *Dioctophyma renale*
(p. 269)

5. Head with lips   6

6. Head with 2 lateral lips   7

6. Head with 3 or 4 lips   8

7. Head with cuticular head bulb armed with backward-pointing spines . . . . (fig. 221) *Gnathostoma*
(p. 265)

7. Head unarmed . . . . *Dacnitoides, Dichelyne, Cucullanus*
(p. 255)

8. Head with 3 distinct lips; *intestine red*; in mesenteries . . . . . .
. . . . (fig. 220) *Spiroxys*
(p. 263)

8. Head with 4 highly specialized lips . . . . *Hedruris*
(p. 258)

9. Large nematode in flesh of marine fish, occasionally anadro-mous fish . . . . (fig. 209) *Porrocaecum*
(p. 251)

9. Small nematode with 3 lips which may be difficult to see; en-cysted or unencysted in mesenteries and liver; cysts up to 5 mm in diameter . . . . (fig. 206) *Contracaecum*

Class NEMATODA

Body cavity not lined with epithelium; gonads continuous with their ducts. Occasionally posterior portion of digestive tract may atrophy in mature worms. Lateral chords present.

Order OXYURIDEA

Esophagus with posterior bulbar enlargement. Intestine without di-verticula. Caudal extremity of females usually prolonged into a pointed tail.

Oxyurids found in North American fish are probably frog parasites which the fish has eaten. They have been reported by Fischthal (1947): *Salvelinus fontinalis*, Wis.; Pearse (1924): *Lepomis macro-chirus*, Wis.

Order TRICHURIDEA

Anterior region of body filiform; esophagus a narrow tube running through chain of cells. Male: spicule single or apparently absent. Female: one ovary.

Genus **Capillaria** Zeder, 1800
(Fig. 202)

Body capillary; mouth simple, with bacillary bands or not. Esophagus long, gradually increasing in size posteriorly. Male: anus terminal or subterminal; usually small membranous caudal alae or bursa-like structure present; spicule long and slender, with spinose or smooth sheath. Female: vulva near posterior end of esophagus. Oviparous; eggs elliptical, with polar plugs. Parasites of digestive tract, liver, or urinary bladder of vertebrates. Life cycle unknown.

Capillaria catenata Van Cleave and Mueller, 1932: in intestine, *Lepomis gibbosus, Ambloplites rupestris, Micropterus salmoides, Stizostedion vitreum*, N. Y.; Anthony (1963): *A. rupestris*, Wis.; Bangham (1939, 1944, 1951): *A. rupestris, Gila straria, L. gibbosus, L. macrochirus, M. dolomieui, M. salmoides, Pomoxis nigromaculatus, Rhinichthys osculus, Richardsonius balteatus, S. vitreum*, Wis., Wyo.; Fantham and Porter (1947): *Catostomus commersoni*, Que.; Fischthal (1947): *L. gibbosus, L. macrochirus, M. dolomieui, Perca flavescens, S. vitreum*, Wis.; Fritts (1959): *Salmo clarki*, Idaho.
C. catostomi Pearse, 1924: in intestine, *Catostomus commersoni*, Wis.
C. eupomotis Ghittino, 1961: *European*; included because it is common in fish culture in Italy in trout, sunfish, and minnows; in liver.
C. petruschewskii (Shulman, 1948): in *L. gibbosus*, Europe.

Genus **Cystoopsis** Wagner, 1867
(Fig. 203)

Mouth collar set off from rest of head. Buccal capsule with 6 minute denticles. Esophagus proper cylindrical, muscular, followed by capillary tube; anterior portion of tube surrounded by two irregularly alternating rows of granular cells; posterior portion with series of giant nuclei at considerable intervals in addition to outer covering of cells with minute nuclei. Intestine a cylindrical bladder. Male: body cylindrical, with rounded ends, much smaller than female, without cloaca. Intestine biscuit-shaped, post-equatorial. Spicule single, simple, comparatively short. Caudal alae rudimentary. Testis narrow, tubular, turning back on itself posterior to nerve ring; spermatozoa rounded. Female: body divided into filiform forebody and large spherical hindbody forming a cyst. Cyst wall with outer cuticle, layer of flat hypo-

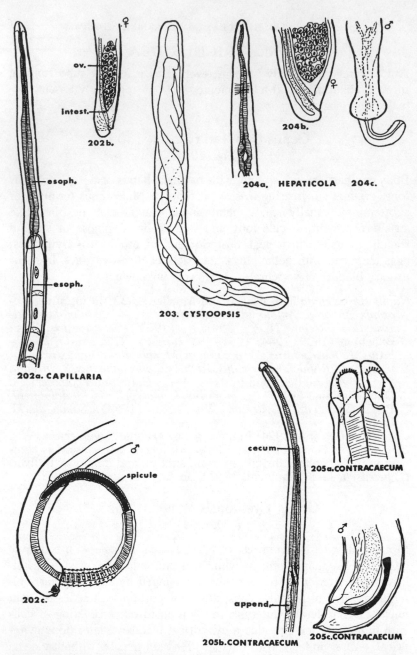

ov.

intest.

202b.

esoph.

esoph.

202a. CAPILLARIA

203. CYSTOOPSIS

204b.

204a. HEPATICOLA 204c.

spicule

202c.

cecum

205a. CONTRACAECUM

append.

205b. CONTRACAECUM

205c. CONTRACAECUM

Figs. 202–205. Fig. 202. *Capillaria catenata. a*, anterior of female showing esophagus and its reduction to a capillary tube in region of paraesophageal cells; *b*, posterior of female showing apex of ovary and intestine; *c*, posterior of male showing everted spicule and sheath, also caudal papillae (from Van Cleave and Mueller, 1932). Fig. 203. *Cystoopsis acipenseri* (from

dermic cells, and inner layer of muscle fibers, most of which run along cell borders of hypodermic cells. Cyst cavity contains transparent colorless fluid, cylindrical intestine, and winding tubular ovary and uterus. Vulva near anterior extremity in region of muscular esophagus. Eggs elongate, with flattened sides, operculate at each pole, embryonated. Parasitic in cutaneous vesicle of fishes. Life cycle unknown.

Cystoopsis acipenseri Wagner, 1867: Europe; Chitwood and McIntosh (1950): in skin blisters, *Acipenser transmontanus*, Ore.

Genus **Hepaticola** Hall, 1916
(Fig. 204)

Body capillary; mouth simple; cuticula without bacillary band. Spicule absent, but membranous eversible sheath present. Female with anterior portion of body half as long as posterior portion. Eggs with polar plugs and striated shell. Life cycle unknown.

Hepaticola bakeri Mueller and Van Cleave, 1932: in intestine, *Catostomus commersoni, Coregonus artedi, Notemigonus crysoleucas*, N. Y.; Bangham (1944, 1951, 1955); Bangham and Adams (1954): *Catostomus, Cottus, Couesius, Lota, Mylocheilus, Oncorhynchus, Prosopium, Ptychocheilus, Rhinichthys, Richardsonius, Salmo, Salvelinus, Umbra*, Brit. Col., Mich., Wyo.; Fischthal (1947, 1950, 1956): *Catostomus, Semotilus, Umbra*, N. Y.; Meyer, M. (1954): *Salvelinus*, Me.

Order ASCARIDEA

Usually large, stout worms. Mouth usually with 3 lips; esophagus frequently more or less enlarged posteriorly, but without definite spherical posterior bulb containing valvular apparatus (except in *Dujardinascaris*, which has small unarmed bulb), with or without posterior ventriculus or ventricular appendix. Intestine with or without anterior diverticula. Spicules equal or unequal. Female not much larger than male. Sometimes intermediate host is required. Parasites of vertebrates.

"*Ascaris*" species reported from North American fish, but undoubtedly belong to other genera. They are:

Saidov, in Yamaguti, 1961). Fig. 204. *Hepaticola bakeri. a*, anterior end; *b*, posterior of female; *c*, posterior of male with spicule sheath everted (from Mueller and Van Cleave, 1932). Fig. 205. *Contracaecum brachyurum. a*, anterior end showing lips; *b*, anterior end showing cecum and appendix; *c*, posterior end of male showing seminal vesicle, intestine, and spicule (black) (from Van Cleave and Mueller, 1934).

"Ascaris" angulata Rudolphi. Pearse (1924): in intestine, *Lepomis gibbosus*, Wis.
"A." labiata Rudolphi. Pearse (1924): in *Ambloplites rupestris*, Wis.
"A." lucii Pearse, 1924: in *Esox lucius*, *A. rupestris*, Wis.
"A." scaphirhynchi Pearse, 1924: in *Scaphirhynchus platorhynchus*, Wis.
"A." tenuissima (Zeder). Linton (1893) quoted by Woodbury (1934): *S. clarki*, Wyo. This is probably *Ichthyobronema t.* (Zeder, 1800).

Family HETEROCHEILIDAE

Ascarididea: Head with 3 large lips. Alimentary canal not simple. Esophagus muscular, with or without ventriculus, from which a posterior cecum or solid glandular appendix may be developed. Intestine with or without anterior cecum lying alongside esophagus.

Genus **Contracaecum** Railliet and Henry, 1912
(Figs. 205, 206)
(Syn.: *Hysterothylacium* Ward and Magath, 1917)

Heterocheilidae: Lips without dentigerous ridges; interlabia present, usually well developed. Ventriculus reduced, with solid posterior appendix. Intestinal cecum present. Male: without definite caudal alae. Postanal papillae up to 7 pairs, partly subventral and partly lateral. Pre-anal papillae numerous. Spicules long, alate, equal or subequal; gubernaculum absent. Female: vulva in anterior region of body. Oviparous. Parasites of fishes, birds, and piscivorous mammals. Life cycle: larval forms infective to fish which eat them. Adults develop in fish-eating birds, fish, and mammals.

Contracaecum brachyurum (Ward and Magath, 1917), (Syn. *Hysterothylacium b.*): in stomach and intestine, *Micropterus*, *Lepomis*, *Lota*, *Esox*; Bangham (1944, 1955): *Ambloplites rupestris*, *E. lucius*, *Micropterus* spp., *Percopsis*, *Poecilichthys*, *Stizostedion*, immature in *Notemigonus*, Mich., Wis.; Bangham and Adams (1954): *L. lota*, Brit. Col.; Odlaug *et al.* (1962): *E. lucius*, *M. dolomieui*, *S. vitreum*, Minn.; Pearse (1924): *M. salmoides*, *P. flavescens*, Wis.; Van Cleave and Mueller (1934): *M. salmoides*, *A. rupestris*, N. Y.; Worley and Bangham (1952): *E. lucius*, Que.
C. collieri Chandler, 1935: in *Cyprinodon variegatus*, marine, Tex.
C. spiculigerum (Rud., 1809): adult cosmopolitan, in cormorants, mergansers, gulls, pelicans. Larva reported from many species of fish; probably no fish host specificity. Reported by Bangham (1939, 1940, 1942, 1951, 1954): Fla., Wis., Tenn.; Haderlie (1953): Calif.; Hugghins (1956, 1959): S. Dak.; McDaniel (1963): Okla.; Meyer, F. (1958): Iowa; Shaw (1947): Ore.; Thomas (1937): life cycle, Ill.; Venard (1940): Tenn.; Wilson (1957): Kans.

Contracaecum spp. Both larvae and immature adults reported from many species of fish, by Anthony (1963): Wis.; Bangham and Adams (1954): Brit. Col.; Fischthal (1947, 1950, 1956): N. Y., Wis.; Fritts (1959): Idaho; Griffith (1953): Wash.; Haderlie (1953): Calif.; Hare (1943): Ohio; Harms (1959): Kans.; Meyer, M. (1954): Me.; Sparks (1951); Worley and Bangham (1952): Que.

Genus **Raphidascaris** Railliet and Henry, 1915
(Fig. 207)

Heterocheilidae, Filocapsulariinae: Lips without dentigerous ridges, with cuticular expansions particularly well developed on subventral lips; interlabia absent. Esophagus followed by small ventriculus, from which springs posterior appendix; intestinal cecum absent. Male: tail conical, slightly curved ventrally; long series of pre-anal and post-anal papillae on each side; spicules equal, alate; gubernaculum absent. Female: tail long, gradually attenuated; vulva in anterior half of body. Oviparous; eggs small, containing few blastomeres when laid. Adults in teleosts. Life cycle: larvae in small fish.

Raphidascaris alius Lyster, 1940: in *Salvelinus fontinalis*, Can.
R. brachyurus (see *Contracaecum brachyurum*).
R. canadense Smedley, 1933: in *Esox lucius*, Can.; Choquette (1951): *E. masquinongy*, Can.; Thomas (1937): adult in *Esox*, larvae encysted in liver and mesenteries of minnows and perch, Ill.
R. cayugensis (Wigdor, 1918): adult in *E. lucius, Ictalurus nebulosus*, larvae probably in minnows, N. Y.
R. laurentianus Richardson, 1937: in *Salvelinus fontinalis*, Can.
Rhaphidascaris spp. Reported by Meyer (1954): *E. niger, S. fontinalis*, Me.; Shaw (1947): *Oncorhynchus tshawytscha*, Ore.

Genus **Porrocaecum** Railliet and Henry, 1912
(Fig. 209)

Large larva of this marine parasite sometimes found in anadromous fish in fresh water. See key, p. 246.

Genus **Anisakis** Duj., 1845
(Fig. 208)

Large larva of this marine parasite sometimes found in anadromous fish in freshwater. See key, p. 245.

Order SPIRURIDEA

Usually more or less filiform worm; mouth usually with 2 lips, but may have 4 or 6 small lips; lips rarely inconspicuous or absent. Behind

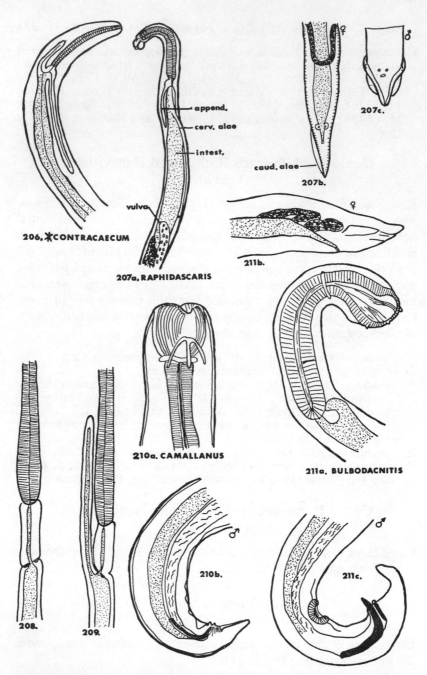

Figs. 206–211. Fig. 206. *Contracaecum* larva from catfish, Alabama (orig.). Fig. 207. *Raphidascaris laurentianus. a*, anterior of female showing appendix, cephalic alae, vulva, vagina, and ovary; *b*, posterior of female showing ovary, intestine, rectal glands, anus, and caudal alae; *c*,

buccal cavity, which is bounded by lips, chitinous capsule frequently present; buccal cavity rarely large and chitinous. Esophagus usually long, cylindrical, divided into two parts; a shorter anterior muscular portion, and a longer glandular posterior portion, rarely undivided; sometimes enlarged anteriorly or posteriorly. Intestine usually simple, without diverticula. Male: spicules usually very unequal, dissimilar. Female: vulva usually near middle of body, sometimes posterior, rarely in esophageal region. Parasites of alimentary canal, respiratory system, orbital, nasal, or oral cavities of vertebrates.

Family CAMALLANIDAE

Key characteristics: mouth elongate dorsoventrally, without lips; with large chitinous buccal capsule.

Genus **Camallanus** Railliet and Henry, 1915.
(Fig. 210)

Camallanidae: Mouth slitlike; buccal capsule with 2 lateral chitinous valves, with longitudinal riblike thickenings internally. From point of junction of valves, dorsally and ventrally, a trident-shaped chitinous process is directed backwards; chitinous ring at junction of valves and esophagus. Esophagus with short anterior muscular portion and long posterior glandular portion enlarged posteriorly. Male: posterior extremity rolled ventrally; small caudal alae present; about 7 pairs of costiform pre-anal papillae, 2 pairs of small ad-anal papillae, numerous post-anal papillae; spicules usually unequal and dissimilar, one feebly chitinized; gubernaculum absent. Female: vulva about middle of body; uteri opposed; posterior ovary lacking; viviparous. Parasites of stomach and intestine of fishes, amphibians, and reptiles. Life cycle: larvae in copepods, possibly other crustacea.

Ref. Van Cleave and Mueller (1934, p. 301), key to spp.

Camallanus ancylodirus Ward and Magath, 1916: in *Carpiodes thompsoni*; Pearse (1924): same, Wis.; Roberts (1957): *Cyprinus carpio*, Okla.; Self and Campbell (1956): *Ictiobus bubalus, I. cyprinella*, Okla.

---

posterior of male (from Richardson, 1937). Fig. 208. *Anisakis* type alimentary canal (from Yorke and Maplestone, 1926). Fig. 209. *Porocaecum* type alimentary system showing intestinal cecum (from Yorke and Maplestone, 1926). Fig. 210. *Camallanus. a, C. oxycephalus*, anterior end showing buccal capsule (from Van Cleave and Mueller, 1934); *b, C. ancylodirus*, posterior end of male (from Ward and Magath, 1916). Fig. 211. *Bulbodacnitis occidentalis. a*, anterior end; *b*, posterior of female; *c*, posterior of male showing blunt spicules (from Smedley, 1933).

C. lacustris Zoega, 1776 (see Yamaguti, 1961, p. 42, for European synonyms). Larva in *Cyclops, Agrion.* Meyer (1954): *Salmo salar sebago,* Me.

C. oxycephalus Ward and Magath, 1916. This red nematode can often be seen hanging out of anus of fish. Reported as adult from *Ambloplites, Ammocrypta, Aplodinotus, Carpiodes, Catonotus, Chaenobryttus, Esox, Etheostoma, Hadropterus, Hiodon, Ictalurus, Labidesthes, Lepomis, Micropterus, Moxostoma, Notropis, Noturus, Perca, Percina, Polyodon, Pomoxis, Pylodictis, Rheocrypta, Rhinichthys, Roccus, Stizostedion*; as immature from *Amia, Catostomus, Erimyzon, Esox, Ictalurus, Lepomis, Perca, Salmo, Semotilus,* by Anthony (1963): Wis.; Bangham (see bibliog.); Fischthal (1947, 1950): Wis.; Harms (1959): Kans.; Hugghins (1959): S. Dak.; Krueger (1954): Ohio; McDaniel (1963): Okla.; Pearse (1924): Wis.; Self and Timmons (1955): Okla.; Sindermann (1953): Mass.; Van Cleave and Mueller (1934): N. Y.; Venard (1941): Tenn.; Wilson (1957): Kans.; Zischke and Vaughn (1962): S. Dak.

C. trispinosus (Leidy, 1851): turtles; McDaniel (1963): immature adults in *Lepomis* spp., Okla.

C. truncatus (Rud., 1914): Europe, possible synonym of *C. lacustris*; Meyer, M. (1954): *Micropterus, Roccus,* Me.

Camallanus spp. Reported by Bangham (1938, 1941, 1942, 1944); Meyer, F. (1958); McLain (1951); Sparks (1951).

Family CUCULLANIDAE

Key characteristics: head with 2 large lateral lobes (lips) each bearing 3 papillae, and bounding a slit-like mouth; esophagus muscular throughout, dilated anteriorly to form a false buccal capsule and enlarged posteriorly.

Ref. Tornquist (1931); Ali (1956).

Genus **Bulbodacnitis** Lane, 1916
(Fig. 211)

Differentiated from *Cucullanus* only by having tubercle on dorsal aspect of head. Parasites of intestine of fishes. Life cycle unknown.

Barreto, 1922, was unable to find any justification for separating this genus from *Cucullanus.* It is included here as an aid in identification.

Bulbodacnitis globosa (Zeder, 1800) Dujardin, 1845 (Syn. *Dacnitis g.*): in *Salmo trutta*; Bangham and Adams (1954): *Prosopium williamsoni, Salmo clarkii, S. gairdneri, Salvelinus fontinalis, S. malma,* Brit. Col.; Woodbury (1934): trout, Wyo.

B. occidentalis Smedley, 1933: in *Salmo gairdneri,* Brit. Col.; Alexander (1960): salmonids, Ore.; Fritts (1959): *Prosopium Williamsoni*, Idaho; Haderlie (1953): *S. gairdneri*, Calif.

B. *scotti* Simon, 1935. Bangham (1951): larva in *Catostomus, Salmo*; adult in *S. trutta, Salvelinus fontinalis, S. namaycush*, Wyo.

Bulbodacnitis sp. Bangham and Adams (1954): larva in *Rhinichthys*; adult in *Mylocheilus caurinus, Rhinichthys cataractae, Richardsonius balteatus*, Brit. Col.

Genus **Cucullanus** Mueller, 1777
(No fig.)
(Syn.: *Bulbodacnitis* Lane, 1916; *Dacnitis* Duj., 1845)

Cucullanidae: Anterior extremity bent dorsally. Lips bounding mouth not chitinized; no chitinous buccal capsule, but pseudocapsule formed by dilation of anterior end of esophagus, which is enlarged posteriorly. Intestine simple. Male: pre-anal sucker without chitinous rim; caudal alae absent; spicules equal; gubernaculum present. Female: vulva near middle of body; vagina directed anteriorly; 2 ovaries. Oviparous, eggs with thin shell. Parasites of intestine of fishes. Life cycle unknown.

Cucullanus chitellarius Ward and Magath, 1916: in *Acipenser rubicundus*; Bangham (1939, 1955): *A. fulvescens*, Mich., Ohio.
C. *globosus* (Syn. of *C. truttae*). Shaw (1946): in *Salmo gairdneri, S. trutta*, Ore.
C. *occidentalis* (Smedley, 1933): in *S. gairdneri*, Brit. Col.; Campana-Rouget (1957): Syn. of *C. truttae*; Yamaguti (1961) *Bulbodacnitis o.* is syn.
C. *scotti* (see *Bulbodacnitis s.*).
C. *truttae* (Fabricius, 1794), (Syn. *C. globosus* Zeder, 1800): in *S. trutta*, Sweden; Iversen (1954): Wash.; Shaw (1947): *S. clarkii, S. gairdneri, Oncorhynchus tshawytscha*, Ore.
Cucullanus sp. Meyer, M. (1954): *Roccus americanus*, Me.

Genus **Dacnitoides** Ward and Magath, 1916
(Fig. 212)

Cucullanidae: Anterior extremity not bent dorsally. Mouth lips not chitinized; no chitinous buccal capsule, but pseudocapsule formed by dilatation of anterior end of esophagus, which is enlarged posteriorly. Intestinal cecum single, dorsal. Male: pre-anal sucker without chitinous rim; tail conical, pointed; no caudal alae; spicules equal; no gubernaculum. Female: tail pointed; vulva just behind middle of body. Only one ovary; posterior uterine branch ends blindly. Parasitic in intestine of fishes: Life cycle unknown.

Dacnitoides cotylophora Ward and Magath, 1916: in *Perca flavescens, Stizostedion vitreum*. Syn. is *Dichelyne c.* Reported from *Amblo-*

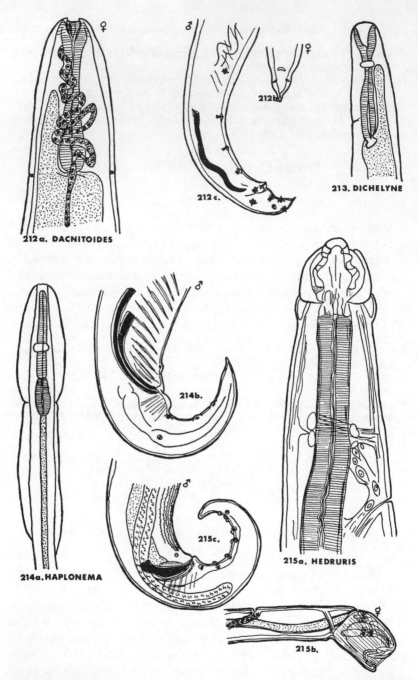

Figs. 212–215. Fig. 212. *Dacnitoides robusta. a*, head of female, dorsal view; *b*, posterior of female; *c*, posterior of male (from Van Cleave and Mueller, 1932). Fig. 213. *Dichelyne fossor* (from Jägerskiöld in Yamaguti, 1961). Fig. 214. *Haplonema* sp. *a*, anterior end; *b*, posterior of male (from

plites, Catostomus, Ictalurus, Lepomis, Micropterus, Perca, Po-moxis, Roccus, Semotilus, Stizostedion, by Bangham (1933, 1939, 1941, 1944, 1955): Fla., Mich.; Fischthal (1947): Wis.; Krueger (1954): Ohio; Meyer, M. (1954): Me.; Odlaug *et al.* (1962): Minn.; Pearse (1924): Wis.; Sindermann (1953): Mass.; Smedley (1934); Worley and Bangham (1952): Que.
D. *robusta* Van Cleave and Mueller, 1932: in *Ictalurus nebulosus,* N. Y.; Bangham and Venard (1942): *Ictalurus anguilla, I. punc-tata;* Campana-Rouget (1957): transferred to *Neocucullanus;* Hunter (1942): *Roccus americanus.*

Genus **Dichelyne** Jägerskiöld, 1902
(Fig. 213)

Cucullanidae: Body of uniform thickness for most part, not enlarged at level of anterior swelling of esophagus. Mouth opening straight forward, sometimes with cuticular transverse ridges or conical structures. Cephalic glands present. Anterior end of esophagus forms false buccal capsule. Intestinal cecum single, dorsal; rarely double, dorsal and ventral. Male: without pre-anal sucker. Tail curved ventrally, conical, sharp-pointed. Pre-anal papillae of 5 pairs, costiform; post-anal papillae few; spicules long, slender; gubernaculum present. Female: tail conical, pointed; vulva behind middle of body. Two ovaries; uterine branches opposed. Oviparous, eggs small. Parasites of teleosts. Life cycle unknown.

Dichelyne cotylophora (see *Dacnitoides cotylophora*).
D. *diplocaecum* Chandler, 1935: in *Ictalurus furcatus,* Tex.
D. *lintoni* (Barreto, 1922): in marine fish including *Fundulus hetero-clitus,* N. C.
Dichelyne sp. Bangham (1938): *Chaenobryttus gulosus, Lepomis microlophus,* Fla.; Bangham (1955): *Etheostoma nigrum,* Mich.; Fischthal (1950): larva in *L. macrochirus,* Wis.

Family HAPLONEMATIDAE
Genus **Haplonema** Ward and Magath, 1916
(Fig. 214)

Haplonematidae: Body rather robust, with anterior end bent or coiled; mouth without lips, head papillae absent; lateral flanges present. No buccal capsule. Esophagus divided into two portions by partition near

---

Van Cleave and Mueller, 1934). Fig. 215. *Hedrurus tiara. a,* anterior of male; *b,* posterior of female, lateral view; *c,* posterior of male (from Van Cleave and Mueller, 1932).

its center, without posterior bulb. Male: tail attenuated, without caudal alae; 2 pairs of pre-anal and 3 pairs of post-anal papillae; spicules equal. Female: posterior extremity straight, with 2 minute papillae; vulva slightly behind middle of body; uteri opposed. Oviparous, eggs with thick smooth shell. Parasites of freshwater fishes. Life cycle unknown.

Ref. Yamaguti (1961, p. 62).

Haplonema aditum Mueller, 1934: in intestine, *Anguilla rostrata*, N. Y.
H. hamulatum Moulton, 1931: in stomach, *Lota lota*, Can.; Bangham (1944, 1955): same, Wis., Mich.
H. immutatum Ward and Magath, 1916: in *Amia calva*, Iowa; Bangham (1942, 1944, 1955): same, Mich., Tenn., Wis.; Bangham and Hunter (1939): same, N. Y.; Pearse (1924): Wis.; Sogandares (1955): La.
Haplonema sp. Bangham (1944): *Esox lucius*, Wis.; Bangham and Adams (1954): larva in *Cottus*; Van Cleave and Mueller (1934): *Anguilla rostrata*, N. Y.

Family HEDRURIDAE
Genus **Hedruris** Nitzsch, 1821
(Fig. 215)

Hedruridae: Body stout posteriorly, slender anteriorly. Lateral lips overlapped by dorsal and ventral lips. Buccal capsule narrow, cylindrical. Esophagus long, slender, apparently undivided, with festooned chitinous ring at anterior extremity. Male: tail spirally twisted and laterally compressed, with 1 pre-anal papilla and 6 post-anal papillae; spicules equal; gubernaculum present or absent. Female: tail invaginated, with clawlike hook for attachment to host. Vulva near anus. Oviparous; eggs elliptical, with polar opercula and containing embryos at deposition. Parasitic in intestine, stomach, and mouth cavity of amphibians, chelonians, and fishes. Life cycle unknown.

Hedruris spinigera Baylis, 1931: in stomach, *Salmo trutta*, N. Zealand.
H. tiara Van Cleave and Mueller, 1932: in stomach, *Esox lucius*, *E. niger*, *Erimyzon sucetta*, N. Y.

Family RHABDOCHONIDAE

Spiruridea: Cuticle with or without ornamentations. Mouth with or without lips. Buccal capsule funnel-shaped or cylindrical, may or may not be provided with longitudinal thickenings or teeth. Esophagus in two parts. Male: posterior extremity rolled up ventrally or spirally coiled. Caudal alae narrow, sometimes with denticulate ridges in pre-cloacal region. Caudal papillae sessile, usually not numerous. Spicules unequal. Female: vulva in anterior or posterior half of body. Oviparous. Parasitic in intestine of fishes and amphibians.

Genus **Cystidicola** Fischer, 1798
(Fig. 216)

(Syn.: *Ancyracanthus* Schneider, 1866)

Rhabdochonidae: Mouth simple or with small lips. Buccal capsule cylindrical, with thick chitinous wall; esophagus very long, divided. Male: posterior extremity spirally coiled, tail rounded at tip; caudal alae narrow; long row of coupled or single pre-anal papillae, a few simple post-anal papillae; spicules unequal, dissimilar. Female: tail straight and blunt; vulva in middle or anterior region. Uterine branches opposed. Oviparous; eggs numerous, thick-shelled; *C. farionis*, possibly others, with polar filaments. Parasitic in swim bladder, air vessels, and rarely esophagus, of freshwater fishes. Life cycle: adult in fish; larvae in *Gammarus*.

Cystidicola canadensis Skinker, 1930: in *Coregonus* spp.
C. cristivomeri White and Cable, 1942: in swim bladder, *Salvelinus namaycush*.
C. lepisostei Hunter and Bangham, 1933: in intestine, *Lepisosteus osseus*, L. Erie; Bangham (1939): same.
C. serratus (Wright, 1879): in heart, *Coregonus albus*; Pearse (1924): in intestine, *Aplodinotus grunniens*, Wis.
C. stigmatura (Leidy, 1886) Skinker, 1931. Found in swim bladder of salmonids primarily in deep lakes. Reported from *Coregonus, Hiodon, Oncorhynchus, Petromyzon, Prosopium, Salmo, Salvelinus*, by Anthony (1963): Wis.; Bangham (1951, 1955): L. Huron, Wyo.; Bangham and Adams (1954): Brit. Col.; Fritts (1959): Idaho; Guilford (1954): L. Mich.; Haderlie (1953): Calif.; MacLulich (1943); Mueller (1940): L. Ont.; Pearse (1924): Wis.; Richardson (1941); Skinker (1931): redescription; Smedley (1933): Ont.; Warren (1952): Minn.
Cystidicola sp. Haderlie (1953): in *Catostomus occidentalis, Hesperoleucus symmetricus, Mylopharodon conocephalus, Ptychocheilus grandis*, Calif.

Genus **Rhabdochona** Railliet, 1916
(Fig. 217)

Rhabdochonidae: Head and body bare, mouth with 2 lips. Buccal capsule funnel-shaped anteriorly, with longitudinal ribs terminating anteriorly in pointed teeth; esophagus of moderate length, composed

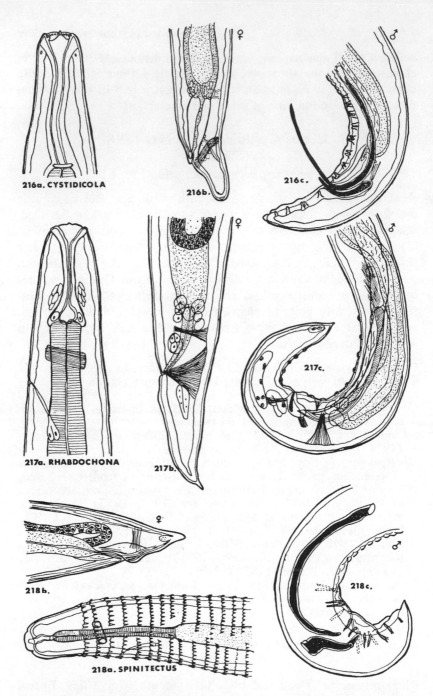

Figs. 216–218. Fig. 216. *Cystidicola cristivomeri. a*, lateral view of anterior; *b*, posterior of female; *c*, posterior of male (from White and Cable, 1942). Fig. 217. *Rhabdochona ovifilamenta. a*, anterior end; *b*, posterior of fe-

of two distinct parts. Male: tail conical, pointed, curved ventrad; caudal alae narrow; numerous, simple, pre-anal papillae and 3 to 6 pairs of post-anal papillae; spicules unequal. Female: tail straight, elongate; vulva in middle region of body; uterine branches opposed. Oviparous, eggs elliptical. Parasitic in intestine of freshwater fishes. Life cycle: *R. cascadilla* larvae develop in mayflies.

Ref. Choquette (1951), review.

Rhabdochona cascadilla Wigdor, 1918. Reported from *Acipenser, Acrocheilus, Ambloplites, Campostoma, Carpiodes, Catostomus, Chrosomus, Cottus, Couesius, Cyprinus, Etheostoma, Eucalia, Exoglossum, Gila, Hiodon, Hyborhynchus, Hypentelium, Micropterus, Mylocheilus, Notropis, Oncorhynchus, Pimephales, Poecilichthys, Ptychocheilus, Rhinichthys, Richardsonius, Salmo, Salvelinus, Semotilus,* by Anthony (1963): Wis.; Bangham (1951, 1954, 1955): L. Huron, L. Pepin, Wyo.; Bangham and Adams (1954): Brit. Col.; Fischthal (1947, 1950, 1956): N. Y., Wis.; Gustafson (1942): life cycle; Meyer, M. (1954): Me. Also reported from gall bladder, *Pimelodella,* Brazil.

R. cotti Gustafson, 1949: in *Cottus cognatus,* eggs with filaments, Wash.; Bangham and Adams (1954): *C. asper, C. cognatus,* Brit. Col.

R. decaturensis Gustafson, 1949: in *Aplodinotus grunniens,* Ill.

R. laurentiana Lyster, 1940: in *Catostomus commersoni, Salvelinus fontinalis,* Can.; Choquette (1951): syn. for *R. cascadilla.*

R. milleri Choquette (1951): in *Moxostoma aureolum,* Can.

R. ovifilamenta Weller, 1938: in *Perca flavescens,* larva probably in *Hyalella,* egg with filament, Mich.

R. pellucida Gustafson, 1949: in *Ptychocheilus oregonensis,* Wash.; Fritts (1959): same, Idaho.

Rhabdochona sp. Reported from *Catostomus, Couesius, Ictalurus, Micropterus, Mylocheilus, Oncorhynchus, Perca, Prosopium, Rhinichthys, Salmo, Salvelinus,* by Bangham (1942, 1951, 1955): L. Huron, Tenn., Wyo.; Bangham and Adams (1954): Brit. Col.; Fritts (1959): Idaho; Griffith (1953): Wash.; Woodberry (1934): Wyo.

Genus **Spinitectus** Fourment, 1883
(Fig. 218)

Cuticle with series of transverse rings, to posterior edge of which are attached backwardly directed spines diminishing in size and number posteriorly. Mouth with indistinct lips; buccal cavity cylindrical or

---

male; *c,* posterior of male (from Weller, 1938). Fig. 218. *Spinitectus gracilis. a,* anterior end; *b,* posterior of female; *c,* posterior of male (from Mueller and Van Cleave, 1932).

funnel-shaped; esophagus consisting of two parts: muscular and glandular. Male: tail spirally coiled; caudal alae narrow, sometimes with denticulate crests in front of cloaca; pre-anal and post-anal papillae present (10 to 15 pairs in all); spicules very unequal, dissimilar. Female: vulva in middle or posterior of body; oviparous; eggs small, ellipsoidal, thick-shelled, sometimes with polar plugs bearing long filaments. Parasitic in stomach and intestine of fishes and frogs. Life cycle: larvae in mayfly larvae.

Ref. Ali (1956), key to species; Schäperclaus (1954, p. 249), cause of mortality.

Spinitectus carolini Holl, 1928. Reported from *Ambloplites, Amia, Aplites, Chaeonbryttus, Esox, Ictalurus, Lepomis, Micropterus, Perca, Pomoxis, Roccus, Stizostedion,* and salamander, *Triturus,* by Anthony (1963): Wis.; Bangham (1939, 1941, 1942, 1944, 1955): Fla., L. Huron, Ohio, Tenn., Wis.; Fischthal (1947, 1950): Wis.; Hoffman (unpubl.): W. Va.; Krueger (1954): Ohio; McDaniel (1963): Okla.; Meyer, M. (1954): Me.; Odlaug *et al.* (1962): Minn.; Sindermann (1953): Mass.

S. gracilis Ward and Magath, 1916: in *Pomoxis nigromaculatus, Aplodinotus grunniens, Roccus chrysops.* Reported from *Ambloplites, Coregonus, Cyprinus, Esox, Eucalia, Ictalurus, Lepomis, Lota, Micropterus, Notropis, Noturus, Osmerus, Perca, Percina, Percopsis, Pomoxis, Prosopium, Roccus, Salmo, Stizostedion,* by Anthony (1963): Wis.; Bangham (1926, 1942, 1944, 1955): L. Huron, Ohio, Tenn., Wis.; Fischthal (1947, 1950): Wis.; Gustafson (1939): life cycle; Harms (1959): Kans.; Krueger (1954): Ohio; Meyer, M. (1954): Me.; Pearse (1924): Wis.; Sindermann (1953): Mass.; Van Cleave and Mueller (1934): N. Y.; Venard (1940): Tenn.

Spinitectus spp. Reported from *Ambloplites, Cyprinus, Ictalurus, Lepomis, Micropterus, Perca, Pomoxis, Roccus, Stizostedion,* by Anthony (1963): Wis.; Bangham (1944, 1955): L. Huron, Wis.; Meyer, M. (1954): Me.

Family SPIRURIDAE Oerley, 1895

Spiruridea: Lateral flanges present or absent. Mouth usually with 2 trilobed lateral lips, occasionally small dorsal and ventral lips also present, or definite lips absent. Behind mouth cavity bounded by lips is usually a more or less cylindrical, chitinized capsule; esophagus long, cylindrical, divided into short, anterior, muscular portion and longer, glandular portion. Male: caudal alae well developed, supported by pedunculate papillae, almost always 4 pre-anal pairs. Spicules usually unequal, dissimilar. Female: vulva usually near middle of body; oviparous. Parasites of digestive tract of vertebrates, including fish.

Genus **Metabronema** Yorke and Maplestone, 1926
(Fig. 219)
(Syn.: *Cystidicoloides* Skinker, 1931)

Spiruridae: Cuticular flanges on both sides of body. Mouth with large lateral lips, and small median lips continuous with lateral by means of cuticular fold; whole head structure strengthened by chitinous support continuous with chitinous wall of buccal capsule. Cervical papillae slightly behind lips; buccal capsule thick-walled, cylindrical; esophagus consisting of two parts. Male: posterior extremity spirally coiled; caudal alae well developed; 4 pairs of pedunculate pre-anal and 4 pairs of pedunculate post-anal papillae, and a pair of large sessile papillae near tip of tail; spicules very unequal; gubernaculum present. Female: tail conical; vulva near junction of anterior and middle third of body, or approximately equatorial or post-equatorial; uteri divergent. Oviparous; eggs thick-shelled, with small button-shaped structure at each end, from which arise 2 very delicate filaments, containing morula when deposited. Parasites of fishes. Life cycle: larvae in mayfly nymphs (Choquette, 1953, 1955).

Metabronema canadense Skinker, 1931 (Syn. *M. amemasu* Fujita, 1939). Van Cleave and Mueller, 1934, consider this a synonym of *M. harwoodi* Chandler, 1931. Choquette (1948) considers it a synonym of *M. salvelini* (Fujita, 1920): in intestine, *Salvelinus fontinalis*, Can.; Fujita (1939): in *S. kundscha*.

M. harwoodi (see *M. salvelini*).

M. prevosti Choquette, 1951: in *Ictalurus nebulosus*, Que.

M. salvelini (Fujita, 1922), (Syn. *Spiroptera s.* Fujita; *M. harwoodi* Chandler, 1931: *M. truttae* Baylis, 1935). Reported from *Cottus asper*, dace, *Esox masquinongy*, *Oncorhynchus kisutch*, *O. nerka*, *Prosopium williamsoni*, *Salmo clarki*, *S. gairdneri*, *S. trutta*, *Salvelinus alpinus*, *S. arcticus*, *S. fontinalis*, *S. kundscha*, *S. namaycush*, by Bangham (1951, 1955), Ohio, Wyo.; Bangham and Adams (1954): Brit. Col.; Choquette (1951): Que.; Fischthal (1947): Wis.; Haderlie (1953): Calif.; Shaw (1947): Ore.; Van Cleave and Mueller (1934): N. Y.

Metabronema sp. Bangham (1951): in *Cottus semiscaber*, Wyo; Bangham and Adams (1954): *Thymallus signifer*, Brit. Col.; Fritts (1959): *Salmo clarki*, Idaho; Meyer, M. (1954) *Roccus americana*, Me.

Genus **Spiroxys** Schneider, 1866
(Fig. 220)

Adult: In stomach of turtles and intestine of amphibians (see Yamaguti, 1961, p. 174). Larvae: mouth with large, distinctly trilobed lips,

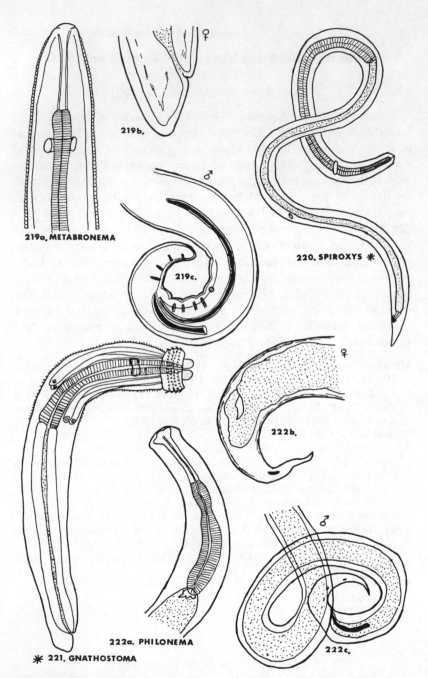

Figs. 219–222. Fig. 219. *Metabronema salvelini. a*, anterior end; *b*, posterior of female; *c*, posterior of male (from Van Cleave and Mueller, 1934). Fig. 220. *Spiroxys contorta* larva from fish (from Hedrick, 1935). Fig. 221. *Gnathostoma procyonis*, third-stage larva from copepod; larva

giving head a triangular appearance. In mesenteries of fish, also amphibia, dragonfly nymphs, and snails (Hedrick, 1935). Life cycle: first intermediate host is *Cyclops* (exper.).

Spiroxys contorta (Rud., 1819). Larvae in *Ictalurus nebulosus, Umbra limi, Rana clamitans* adult and tadpole, *Triturus viridescens,* and Odonata nymphs, adult in turtles (Hedrick, 1935).

Spiroxys sp. Reported from *Ambloplites, Amia, Catostomus, Chrosomus, Cyprinus, Esox, Etheostoma, Eucalia, Ictalurus, Lepomis, Micropterus, Nocomis, Notropis, Noturus, Perca, Pomoxis, Richardsonius, Umbra,* by Anthony (1963): Wis.; Bangham (1941, 1944, 1955): Brit. Col.; Fischthal (1947, 1950, 1956): N. Y., Wis.; Hoffman (unpubl.): W. Va.

Family GNATHOSTOMIDAE

Spiruridea, key characteristics: Mouth with large trilobed lateral lips, behind which is cuticular head bulb with marked transverse striations or with rows of backwardly directed hooks.

Genus **Gnathostoma** Owen, 1836
(Fig. 221)

Head bulb armed with simple hooks, showing no external evidence of presence of ballonets. Body armed all over, or for most part, with cuticular spines; latter scalelike anteriorly, with serrate distal edges, but becoming simpler posteriorly. Male: spicules unequal; 4 pairs of large lateral papillae, 2 pairs of small ventral, caudal papillae. Female: vulva behind middle of body, vagina long, 2 uterine branches. Oviparous, eggs with thin colorless shell markedly thickened at one pole or both. Parasitic normally in walls of stomach or esophagus of carnivorous mammals. Life cycle: larva develops in copepods; encysts in vertebrates as well as in crustaceans.

Gnathostoma procyonis Chandler, 1942: adult in raccoon, Tex.; Ash (1960): larvae in cyclopoid copepods; third-stage larvae in guppies, snakes, turtles, alligators, La.

G. spinigerum Owen, 1836. Adult in many mammals; probably cosmopolitan, including U. S. A. (cf. Yamaguti, 1961, p. 619). First intermediate hosts are copepods; second intermediate hosts are fish, frogs, salamanders, and snakes.

---

in fish probably very similar (from Ash, 1962). Fig. 222. *Philonema oncorhynchi. a,* anterior end; *b,* posterior end of female; *c,* posterior end of male (from Smedley, 1933).

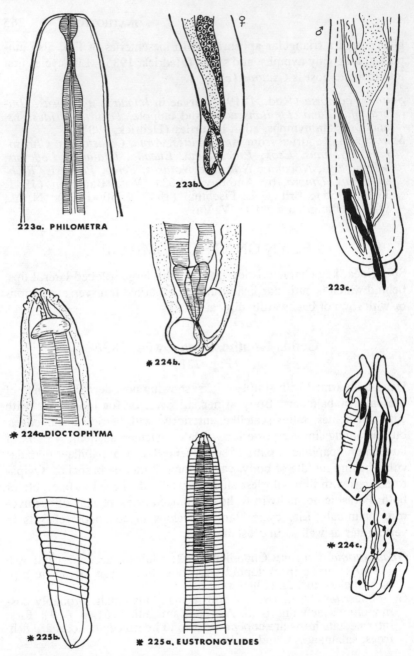

223a. PHILOMETRA

223b.

223c.

※ 224b.

※ 224a.DIOCTOPHYMA

※ 224c.

※ 225b.

※ 225a. EUSTRONGYLIDES

Figs. 223–225. Fig. 223. *Philometra. a, P. cylindracea,* anterior end (from Van Cleave and Mueller, 1934); *b, P. globiceps,* posterior end of female; *c, P. globiceps,* posterior end of male (*b* and *c* from Strassen in Yorke and Maplestone, 1926). Fig. 224. *Dioctophyma renale* larva. *a,* anterior

Order FILARIIDEA

Filariform worms; parasitic in tissues, serous cavities, and blood or lymphatic systems of vertebrates.

Genus **Philonema** Kuitunen-Ekbaum, 1933
(Fig. 222)

Filariidea: Body filiform, anterior extremity rounded, posterior extremity tapering to sharp point in both sexes. Mouth without lips or papillae. Esophagus cylindrical, divided into short anterior portion and longer posterior portion which is distinctly wider than anterior in male. Male: posterior extremity attenuated, spirally coiled, anus well apart (0.35 mm) from tail tip; several pairs of post-anal papillae present. Spicules equal, slender; gubernaculum absent. Female: much larger than male. Anus and vulva atrophied. Uterus occupies almost entire body; ovaries small, one at each end of body. Viviparous. Parasitic in body cavity of fish. Life cycle: larvae in copepods.

"*Philonema* causes adhesions of the viscera of trout and salmon; in severe cases the entire viscera is bound into a solid mass of adhesions preventing normal functions, including reproduction" (Meyer, M., 1960; Richardson, 1937).

Philonema agubernaculum Simon and Simon, 1936: adult in body cavity, *Prosopium williamsoni, Salmo gairdneri, Salvelinus fontinalis*, Wyo.; Bangham (1950): *P. williamsoni, S. trutta, Salvelinus namaycush*, Wyo.; Meyer, M. (1954, 1958, 1960): *S. sebago, S. fontinalis*, larvae in *Cyclops*, Me.; Vik (1964): larger trout acquire infection by eating smelt which have recently ingested infected copepods, Me.

P. oncorhynchi Kuitunen-Ekbaum, 1933 (Syn. *P. salvelini* Richardson, 1936): adult in body cavity, *Oncorhynchus nerka*, Pacific coast, Can.; Baylis (1948): East Greenland; Dombrowski (1955): *O. nerka*, Brit. Col.; Haderlie (1953): *S. gairdneri*, Calif.; Richardson (1937): *S. fontinalis*, Can.; Shaw (1947): *S. gairdneri*, Ore.; Smedley (1933): *O. nerka*, Brit. Col.; Uzmann (1958): control by stocking larger trout, Wash.

P. salvelini Richardson, 1936 (see *P. oncorhynchi*).

Philonema sp. Bangham and Adams (1954): *Prosopium williamsoni, Acrocheilus aleuticus*, Brit. Col; DeRoth (1953): *S. sebago*, Me.; MacLulich (1943): *S. namaycush*, Ont.

---

end; *b*, posterior end (*a* and *b* from Karmanova, 1961); *c*, anterior end (from Woodhead, 1950). Fig. 225. *Eustrongylides* larva from *Lepomis gibbosus*. *a*, anterior end; *b*, posterior end (orig.).

Genus **Philometra** Costa, 1845
(Fig. 223)

(Syn.: *Ichthyonema* Diesing, 1861; *Sanguinofilaria* Yamaguti, 1935)

Filariidea: Female enormously larger than male: body filiform; anterior and posterior extremities rounded; mouth with or without lips. Head and tail papillae present or absent. Esophagus cylindrical, short, bulbous at anterior end; esophageal gland confined to walls of esophagus; rudimentary ventriculus present. Male: posterior extremity rounded; cloaca terminal, bordered by 2 lips; spicules equal, needle-like; gubernaculum present. Female: anus and vulva atrophied; vulva seen at junction of middle and posterior third of body in young worms. Uterus occupies almost entire body; ovaries small, one at each end of body. Parasitic in body cavities or tissues of fishes. Life cycle: adult in tissues of fish; larvae in copepods.

Philometra carassii (Ishii, 1934), (Syn. *Filaria c.*): between caudal fin rays of *Carassius auratus*; Hoffman (unpubl.): same, Ohio.

P. cylindracea Ward and Magath, 1916: in body cavity, *Perca flavescens*, Mich.; Bangham (1944, 1955): *Micropterus salmoides, P. flavescens*, Mich., Wis.; Fischthal (1947): *M. salmoides, Stizostedion vitreum*, Wis.; Hare (1943): *Ambloplites rupestris*, Ohio; Van Cleave and Mueller (1934): *P. flavescens*, N. Y.

P. nodulosa Thomas, 1929: in "cheek galleries" of *Catostomus commersoni*, larvae in *Cyclops*, Ill.; Bangham (1944): *C. commersoni*, Wis.; Fantham and Porter (1947): *C. commersoni*, Que.; Fischthal (1950): *C. commersoni, M. salmoides*, Wis.; Hugghins (1959): *C. commersoni, Ictiobus cyprinellus*, S. Dak.; Van Cleave and Mueller (1934): *C. commersoni*, N. Y.

P. sanguinea (Rud., 1819): European, in tail fin of several fish, including *Carassius auratus*; Ekbaum (1933): Can.; Wierzbicki (1960): life cycle, Poland.

P. translucida Walton, 1927: in *Esox lucius*.

Philometra sp. Bangham (1944, 1955): in *Coregonus clupeaformis, Esox masquinongy, Micropterus dolomieui, Prosopium cylindraceum*, L. Huron, Wis.; Fantham and Porter (1947): *E. niger, P. flavescens, S. fontinalis*, Que.; Fischthal (1947, 1950): *C. commersoni, E. masquinongy, Hypentelium nigricans*, Wis.; Rasheed (1963): revision of genus; Shaw (1947): in abdomen, *S. clarki, S. gairdneri, O. nerka*, Ore.; Travassos (1960): review of genus.

Order DIOCTOPHYMIDEA

Key characteristic: Male with muscular bursa copulatrix not supported by rays. Adult in mammals.

Genus **Dioctophyma** Collet-Meygret, 1802
(Fig. 224)

Adult: in kidney and peritoneal cavity of mammals. First-stage larva: in branchiobdellid oligochaetes. Second-stage larva: encysted in mesenteries of *Ictalurus melas*; head with 6 pointed papillae, pharynx rods and supporting rays of papillae characterize larva.

Dioctophyma renale (Goeze, 1792): cosmopolitan; Woodhead (1950): life cycle, Mich.; Hallberg (1952): exper. infection of ferret; Karmanova (1961): larvae in intestinal wall of *Esox lucius*; Karmanova (1963): in *Lumbriculus*, direct to dogs; fish not obligatory.

Genus **Eustrongylides** Jägerskiöld, 1909
(Fig. 225)

Adult: In glands of proventriculus of fish-eating birds. Larva: red larva encysted in musculature or body cavity of fish; cyst about 1 cm in diameter, flattened; larva up to 10 cm long, about 0.68 mm in diameter, cross-striated; head with 12 papillae, in 2 circles of 6 each.

Eustrongylides sp. Reported by Bangham (1939, 1942, 1951): Fla., Tenn., Wyo.; Bangham and Adams (1954): Brit. Col.; Hoffman (unpubl.): Md., Me.; Hunter (1942); Irving (1954); Meyer, M. (1954): Me.; Mueller (1934): N. Y.; Shaw (1947): Ore.; Sindermann (1953): Mass.

Acanthocephala
(Thorny-headed worms)

Contents

Phylum ACANTHO-CEPHALA 271
Artificial Key to the Genera . . 272
Order NEOECHINO-RHYNCHIDEA 273
Family NEOECHIN-ORHYNCHIDAE 273
Genus *Neoechinorhynchus* 273
Genus *Octospinifer* 275
Genus *Octospiniferoides* . 277
Genus *Paulisentis* 277
Genus *Eocollis* 277
Genus *Gracilisentis* 279
Genus *Tanaorhamphus* . . 279
Order ECHINORHYN-CHIDEA 280

Family ECHINORHYN-CHIDAE 280
Genus *Echinorhynchus* . . 280
Genus *Acanthocephalus* . 281
Genus *Leptorhynchoides* . 283
Family FESSISENTIDAE . 284
Genus *Fessisentis* 284
Family POMPHO-RHYNCHIDAE 284
Genus *Pomphorhynchus* . 284
Family RHADINO-RHYNCHIDAE 285
Genus *Rhadinorhynchus* . 285
SPECIES INQUIRENDA . . 286
LARVAL FORMS IN FISH . 287

There is little difficulty in recognizing an acanthocephalan because of the proboscis which bears chitinoid hooks. The proboscis may become withdrawn while removing the worm from the host, but the body bears little resemblance to other helminths. Usually the retracted proboscis can be everted by allowing the worm to remain in distilled water several minutes to overnight. Most of the key characteristics can be seen in living fish acanthocephala; the number and the shape of the proboscis hooks are the most important.

The life cycle involves the eggs being shed by the adult worm in the fish intestine and being eaten by the first intermediate host: copepod, ostracod, amphipod, or isopod. The first larval stage, acanthor, migrates through the intestinal wall of the crustacean, localizes in the body cavity, and becomes the next larval stage, the acanthella. Usually there is no second intermediate host, and the fish becomes infected by eating the crustacean. Three species at least, however, use fish as a second intermediate host. One of the three species, *Leptorhynchoides thecatum*, will encyst in fish if the larva has been in the crustacean less than 30 days, but needs no second host if eaten by the fish after 30 days in the crustacean.

If acanthocephala are numerous, the damage done to the intestine by the armed proboscis may be serious. The histopathology has been studied by Bauer (1958), Bullock (1963), Nechaeva (1953), Prakish and Adams (1960), and Venard and Warfel (1953). Since the life cycles are relatively simple, it is quite probable that acanthocephala will become more important in fish culture as fisheries work increases.

Phylum ACANTHOCEPHALA Van Cleave, 1948

Key characteristics: Acanthocephala are parasitic worms without alimentary canal, possessing spined retractile proboscis.

Ref. Van Cleave (1948); Petrochenko (1956); Yamaguti (1963).

ARTIFICIAL KEY

TO THE GENERA

1.   Body spined (1 record in Oregon steelhead trout) (= marine?) .... (fig. 238) *Rhadinorhynchus*
(p. 285)

1.   Body not spined 2

2 (1). Prominent globular expansion of neck just posterior to proboscis; more than 3 circles of hooks (in many fish) .... (fig. 237) *Pomphorhynchus*
(p. 284)

2 (1). Anterior end of trunk modified to form a slender false neck and inflated trunk at base of proboscis superficially resembling above; 3 circles of proboscis hooks (in Centrarchidae) .... (fig. 230) *Eocollis*
(p. 277)

2 (1). No expansion of neck or trunk 3

3 (2). Proboscis with 3 circles of hooks 4

3 (2). Proboscis with more than 3 circles of hooks 6

4 (3). Proboscis hooks 6 to a circle (in many fish) ....
.... (fig. 226) *Neoechinorhynchus*

4 (3). Proboscis hooks more than 6 to a circle 5

5 (4). Proboscis hooks 8 to a circle (in Catostomidae) ......
.... (fig. 227) *Octospinifer*

5 (4). Proboscis hooks 8 to 10 to a circle, with prominent root plates (in *Gambusia, Fundulus*) ....
.... (fig. 228) *Octospiniferoides*

5 (4). Proboscis hooks 12 to a circle (in *Dorosoma, Aplodinotus, Ictiobus*) .... (fig. 231) *Gracilisentis*
(p. 279)

6 (3). With 6 diagonal rows of 5 hooks (in *Semotilus atromaculatus*) .... (fig. 229) *Paulisentis*
(p. 277)

6 (3). Hooks not in 6 diagonal rows 7

7 (6). Cement glands syncytial (in *Ictiobus, Dorosoma, Acipenser, Anguilla*) .... (fig. 232) *Tanaorhamphus*
(p. 279)

7 (6). Cement glands not syncytial 8

8 (7). Proboscis hooks with collars (usually in ceca of many fish) .... (fig. 235) *Leptorhynchoides*
(p. 283)

8 (7). Proboscis hooks without collars 9

9 (8). With 4 elongate cement glands; 2 filiform testes (in
    Aplodinotus) .... (fig. 236) *Fessisentis*
                               (p. 284)

9 (8). With 6, sometimes 5 or 7, oval cement glands; 2 oval
    testes 10

10 (9). Ganglion at base of proboscis receptacle (in *Aplodi-
    notus*, Salmonidae, *Catostomus*, *Perca*, *Lepomis*,
    Esox) .... (fig. 234) *Acanthocephalus*
                               (p. 281)

10 (9). Ganglion near middle of proboscis receptacle (in many
    fish) .... (fig. 233) *Echinorhynchus*
                               (p. 280)

Order NEOECHINORHYNCHIDEA Southwell and Macfie, 1925

Body usually small. Proboscis invaginable, usually with comparatively small number of hooks. Trunk spined or not; hypodermal nuclei few, large, amoeboid or fragmented. Proboscis receptacle with single-layered walls. Protonephridial organ absent. Cement gland syncytial, rarely divided into 2 lobes. Eggs elliptical, usually without polar prolongations of middle shell; embryo unarmed. Life cycle: adult in intestine of fish; larva in body cavity of small crustacea.

Family NEOECHINORHYNCHIDAE Van Cleave, 1919

Key characteristics: Trunk not spined; hypodermal nuclei giant and few; lemnisci 2, short or long; cement gland syncytial.

Genus **Neoechinorhynchus** Hamann, 1892
(Fig. 226)

Body usually small, cylindrical, bowed or straight. Lacunar system with median (dorsal and ventral) longitudinal vessels and circular vessels with anastomoses. Giant hypodermal nuclei almost always few (usually 4 or 5 dorsally and 1 or 2 ventrally). Proboscis short, somewhat globular; proboscis hooks in 6 spiral rows of 3 each (usually referred to as 3 circles of hooks); anterior hooks longer and stouter than others. Proboscis receptacle subcylindrical, rather short, single-layered, with ganglion at, or close to, its base. Lemnisci digitiform to filiform, long, with few giant nuclei. Testes contiguous or not, at or near mid-region, sometimes in posterior half of trunk. Cement gland syncytial, with several nuclei; cement reservoir rounded, overlapped by cement gland. Eggs oval to elliptical, with concentric shells. Life cycle: adult in marine and freshwater fish, frogs, and turtles; larva in small crustacea; some species also have second intermediate host; larval form very similar to adult.

Ref. Petrochenko (1956); key to spp.

Neoechinorhynchus australe Van Cleave, 1931: in *Ictiobus* sp., Miss. (see Van Cleave, 1949; see *N. distractum*).

N. crassum Van Cleave, 1919. Recorded from *Campostoma anomala, Catostomus columbianus, C. commersoni, C. macrocheilus, Erimyzon oblongus, Moxostoma erythrurum, M. rubreques, Tinca tinca*, by Anthony (1963): Wis.; Bangham (1941, 1944, 1955): Mich., Ohio, Wis.; Bangham and Adams (1954): Brit. Col.; Fischthal (1947, 1956): N. Y., Wis.; Griffith (1953): Wash.; Lynch (1936); Pearse (1924): Wis.; Van Cleave (1941): hook pattern; Van Cleave (1949): Wis.

N. cristatum Lynch, 1936: in *Catostomus macrocheilus*; Bangham and Adams (1954): *C. catostomus, C. macrocheilus, Mylocheilus caurinus*, Brit. Col.; Fritts (1959): *C. columbianus*, Idaho; Haderlie (1953): *C. humboldtianus, C. tahoensis*, Calif.

N. cylindratum (Van Cleave, 1913) Van Cleave, 1919. Adult reported from *Ambloplites, Amia, Anguilla, Carpiodes, Catostomus, Chaenobryttus, Coregonus, Erimyzon, Esox, Etheostoma, Fundulus, Lepomis, Lota, Micropterus, Notemigonus, Notropis, Perca, Petromyzon, Pomoxis, Roccus, Salvelinus, Stizostedion*; larva reported from *Ambloplites, Catonotus, Centrarchus, Esox, Etheostoma, Gambusia, Lepomis, Menidia, Micropterus, Roccus, Stizostedion*, by Anthony (1963): Wis.; Bangham (1926, 1939, 1941, 1955): Fla., Mich., Ohio; Bangham and Venard (1942): Tenn.; Bogitsh (1961): nature of cyst; Choquette (1951): Can.; Fischthal (1947, 1950, 1956): N. Y., Wis.; Guilford (1954): Mich.; Hare (1943): Ohio; Holloway (1953, 1957): morphology and distribution; Hunter (1930): N. Y.: McDaniel (1963): Okla.; Meyer, M. (1954): Me.; Odlaug *et al.* (1962): Minn.; Pearse (1924): Wis.; Sindermann (1953): Mass.; Sparks (1951): Tex.; Van Cleave (1921, 1923): N. Y.; Van Cleave and Mueller (1934): N. Y.; Venard (1940, 1941): Tenn.; Venard and Warfel (1953): effect on alimentary canal; Ward, Helen (1940): life cycle = ostracod (*Cypria*) first intermediate host, fish as second intermediate host; Warren (1952); Wilson (1957): Kans.; Zischke and Vaughn (1962): S. Dak.

N. distractum Van Cleave, 1949 (Syn. *N. australe*, in part): *Ictiobus* sp., Miss.

N. doryphorum Van Cleave and Bangham, 1949: in *Jordanella floridae*, Fla.

N. longirostris Van Cleave, 1931: in *Dorosoma cepedianum, Ictiobus* sp., Miss.

N. paucihamatum (Leidy, 1890) Petroschenko, 1956 (Syn. *Echinorhynchus p.*): in *Micropterus salmoides, M. nigricans*.

N. prolixoides Bullock, 1963: in *Catostomus commersoni, Erimyzon oblongus,* N. H.

N. prolixum Van Cleave and Timmons, 1952: in *Carpiodes carpio,* Okla.; Self and Timmons (1955): same.

N. rutili (Mueller, 1780), (Syn. *Echinorhynchus tuberosus* Zider, 1803): World-wide circumpolar distribution. Reported from *Ambloplites, Carassius, Catostomus, Cottus, Couesius, Esox, Eucalia, Gasterosteus, Gila, Ictalurus, Lota, Micropterus, Mylocheilus, Notropis, Oncorhynchus, Perca, Prosopium, Ptychocheilus, Pungitius, Richardsonius, Salmo, Salvelinus, Umbra,* by Alexander (1960); Bangham (1951, 1955): Mich., Wyo.; Bangham and Adams (1954): Brit. Col.; Fischthal (1950): Wis.; Hoffman (1953): N. Dak.; Lynch (1936); Merritt and Pratt (1964): life cycle; Plehn (1924 p. 389); Van Cleave and Lynch (1949, 1950): circumpolar distribution; Walkey (1962): life cycle.

N. saginatum Van Cleave and Bangham, 1949: in *Semotilus atromaculatus,* Wis.; Fischthal (1950): Wis.; Meyer, M. (1954): *S. corporalis,* Me.

N. strigosum Van Cleave, 1949: in *Catostomus commersoni, Ictiobus bubalus, Moxostoma aureolum,* Tenn.; Bangham (1955): *C. catostomus, C. commersoni,* Mich.

N. tenellum (Van Cleave, 1913) Van Cleave, 1919: in *Esox lucius, E. niger,* Minn.; Bangham (1944): *E. lucius, E. masquinongy, Stizostedion vitreum,* Wis.; Fischthal (1947, 1950): same.

N. tumidum Van Cleave and Bangham, 1949: in *Coregonus artedi;* Bangham (1955): *Prosopium cylindraceum,* Mich. Also in Russia.

Genus **Octospinifer** Van Cleave, 1919
(Fig. 227)

Neoechinorhynchidae: Body almost cylindrical, tapered posteriorly, with giant hypodermal nuclei mainly in mid-dorsal line. Lacunar system with prominent mid-dorsal longitudinal vessel and reticular anastomoses. Proboscis globular, wider than long, with 8 to 10 spiral rows of 3 hooks each (usually referred to as 8 hooks to each 3 circles of hooks). Proboscis receptacle subcylindrical, short, with ganglion at base. Lemnisci tubular, moderately long, with few giant nuclei. Testes contiguous, in middle third of body. Cement gland syncytial, with several nuclei, well apart from testes. Eggs elliptical, with delicate shell. Parasites of fishes.

Octospinifer macilentus Van Cleave, 1919: in *Catostomus commersoni,* Mich.; Anthony (1963): Wis.; Bangham (1937, 1944, 1955): Ohio, Wis.; Fischthal (1947): Wis.; Harms (1963): life cycle; Meyer, M. C. (1954): Me.; Pearse (1924); Wis.; Sindermann (1954): Mass.; Van Cleave and Mueller (1934): N. Y.

O. torosus Van Cleave and Haderlie, 1950: in *Catostomus occidentalis,* Calif.; Haderlie (1953): same.

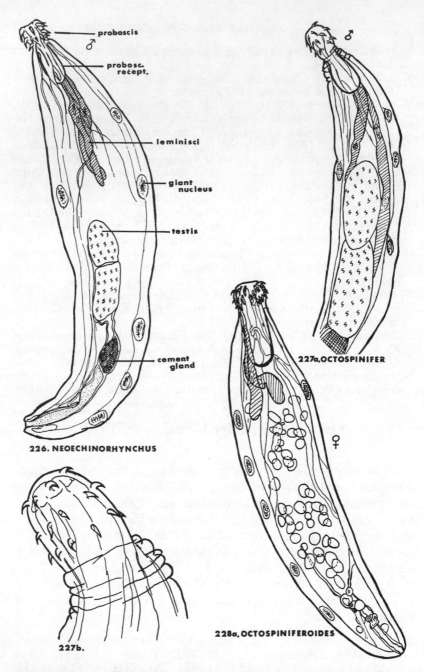

proboscis
probosc. recept.
leminisci
giant nucleus
testis
cement gland

226. NEOECHINORHYNCHUS

227a. OCTOSPINIFER

227b.

228a. OCTOSPINIFEROIDES

Figs. 226–228a. Fig. 226. *Neoechinorhynchus saginatus* (from Van Cleave and Bangham, 1949). Fig. 227. *Octospinifer torosus. a*, anterior end of male; *b*, proboscis (from Van Cleave and Haderlie, 1950). Fig. 228a. *Octospiniferoides chandleri*, entire female (from Bullock, 1957).

Genus **Octospiniferoides** Bullock, 1957
(Fig. 228)

Neoechinorhynchidae: Proboscis globular, with hooks in 3 circles of 8 to 10 hooks each. Hooks small and slender with prominent roots; roots longer than thorn in all 3 circles. Proboscis receptacle thin-walled. Female genital opening nearly terminal. Male unknown. Parasitic in intestine of brackish-water and freshwater fish. Life cycle: intermediate host probably ostracod.

Yamaguti (1963), includes this genus tentatively in *Octospinifer*.

Octospiniferoides chandleri Bullock, 1957: in *Fundulus grandis*, Tex.; Bullock (1964): *Gambusia affinis*, Fla.; Bullock (1965): histochemistry of proboscis.

Genus **Paulisentis** Van Cleave and Bangham, 1949
(Fig. 229)

Neoechinorhynchidae: Size small. Trunk short, broad, thick-walled, with few giant hypodermal nuclei. Proboscis small, short-cylindrical, armed with relatively weak hooks in 6 spiral or diagonal rows of 5 hooks each. Lemnisci moderately long. Testes large, contiguous, with cement gland broadly joined to hind testis. Bursal musculature poorly developed.

Paulisentis fractus Van Cleave and Bangham, 1949: in *Semotilus atromaculatus*.

Genus **Eocollis** Van Cleave, 1947
(Fig. 230)

Neoechinorhynchidae: Trunk with anterior extremity usually inflated into bulb or series of irregular excrescences, followed by narrow, cylindrical, false neck. Giant hypodermal nuclei distinctive of family all restricted to region posterior to trunk bulb and false neck. Posterior to false neck, trunk somewhat swollen, with single giant nucleus dorsally and ventrally. Proboscis short, cylindrical to globular, with 3 circles of 6 hooks each (or 6 spiral rows of 3 each). Neck short. Proboscis receptacle small, single-walled, with ganglion at base. Lemnisci relatively long, narrow, cylindrical, extending through trunk bulb and false neck into cavity of trunk proper, each with median vessel. Testes contiguous, in anterior region of trunk proper. Cement gland large, syncytial, containing 8 giant nuclei; cement reservoir pyriform, small. Life cycle: intermediate hosts unknown.

Eocollis arcanus Van Cleave, 1947: in *Lepomis macrochirus, Pomoxis nigromaculatus*, Ill.

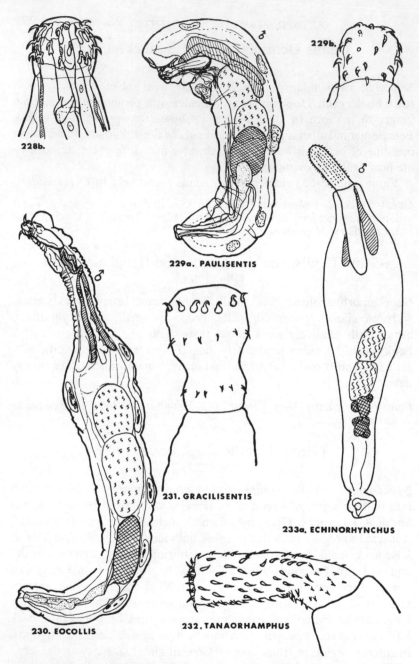

228b.

229b.

229a. PAULISENTIS

231. GRACILISENTIS

233a. ECHINORHYNCHUS

230. EOCOLLIS

232. TANAORHAMPHUS

Figs. 228b–233. Fig. 228b. *Octospiniferoides chandleri*, proboscis (from Bullock, 1957). Fig. 229. *Paulisentis fractus. a*, male with proboscis retracted; *b*, proboscis (from Van Cleave and Bangham, 1949). Fig. 230.

Genus **Gracilisentis** Van Cleave, 1919
(Fig. 231)

Neoechinorhynchidae: Body small. Lacunar system with 2 median longitudinal vessels and transverse anastomoses. Hypodermal giant nuclei mainly in mid-dorsal line. Proboscis short, cylindrical, with slight constriction above basal hook row, with 12 longitudinal rows of 3 hooks each (or 3 transverse rows of 12 each); basal row wider apart from second row than latter is from first, and each hook has narrower root than that of other two rows. Proboscis receptacle subcylindrical, with ganglion near base. Lemnisci slender, longer than proboscis receptacle. Testes contiguous, nearer to posterior extremity than to anterior. Cement gland syncytial, with 12 nuclei. Eggs elliptical, with delicate shell.

Gracilisentis gracilisentis (Van Cleave, 1913) Van Cleave, 1931: in *Aplodinotus grunniens, Dorosoma cepedianum, Ictiobus* sp., Ill., Miss.

Genus **Tanaorhamphus** Ward, 1918
(Fig. 232)

Neoechinorhynchidae: Body small to medium-sized. Trunk subcylindrical, with hypodermal giant nuclei mainly in mid-dorsal line. Proboscis cylindrical, fairly long, with 16 to 20 longitudinal rows of 10 to 16 hooks each. Proboscis receptacle cylindrical, with ganglion at base. Lemnisci long, narrow, with few giant nuclei. Testes contiguous, nearer to posterior extremity than to anterior. Cement gland syncytial, with 16 nuclei. Eggs elongate oval.

Tanaorhamphus ambiguus Van Cleave, 1921 (*Echinorhynchus globulosis* is synonym according to Meyer, 1932). Linton (1893): in *Salmo mykiss*, Wyo.; Meyer (1933): *Acipenser rubicundus*; Van Cleave (1921): *Anguilla chrysypa*.
T. longirostris Van Cleave, 1921: in *Dorosoma cepedianum, Ictiobus* sp.; Bangham (1942): *D. cepedianum, Ictalurus anguilla, I. punctatus*, Tenn.

Eocollis arcanus, male (from Van Cleave, 1947). Fig. 231. *Gracilisentis gracilisentis* (from Van Cleave, 1913, in Ward and Whipple, 1918). Fig. 232. *Tanaorhamphus longirostris* (from Van Cleave, 1913, in Ward and Whipple, 1918). Fig. 233. *Echinorhynchus salmonis* (from Meyer, 1932, in Yamaguti, 1963).

Order ECHINORHYNCHIDEA Southwell
and Macfie, 1925

Body usually small. Proboscis invaginable, with small or large number of hooks; proboscis receptacle with double-layered walls, inserted at base of proboscis. Trunk spinose or not; hypodermal nuclei usually small, numerous. Lacunar system with lateral main vessels. Protonephridial organ absent. Cement gland divided into two or more compact or tubular lobes. Egg usually with polar prolongations of middle shell. Embryo with hooks at each end.

Family ECHINORHYNCHIDAE (Cobbold, 1879)
Yamaguti, 1963

Echinorhynchidae: Trunk aspinose, rarely faintly spinose. Hypodermal nuclei usually small, numerous. Lacunar system with lateral main vessels. Proboscis more or less cylindrical, of moderate length and usually with numerous hooks, or spherical and with few hooks. Proboscis receptacle inserted at base of proboscis, with double-layered walls; ganglion at varying levels. Lemnisci 2, more or less claviform, rather short. Testes oval to elliptical, never cylindrical. Cement glands 4 to 8, usually pyriform, rarely tubular. Eggs elliptical to fusiform.

Ref. Golvan (1950), review; Rodrigo (1960), review; Machado (1959), review.

Genus **Echinorhynchus** Zoega *in* Müller, 1776
(Fig. 233)

Echinorhynchidae: Body small to medium-sized; hypodermal nuclei small, numerous. Lacunar system with lateral main vessels and reticular anastomoses. Proboscis long, cylindrical, directed ventrad, with 9 to 26 longitudinal rows of 5 to 16 hooks each; root of hook simple, becoming smaller toward base of proboscis, where it disappears. Proboscis receptacle cylindrical to claviform, double-walled, with ganglion near middle. Lemnisci usually claviform. Testes oval to elliptical, tandem, contiguous or not, in middle third of trunk. Cement glands 6, more or less compact, one behind another or close together. Eggs much elongated, fusiform, with prominent polar prolongations of middle shell. In freshwater and marine fishes. Life cycle: larva in amphipods; no second intermediate host. Bangham (1955) lists encysted larva of *E. salmonis* in *Osmerus mordax*.

Echinorhynchus clavaeceps Zeder. Van Cleave (1921): in *Anguilla crysypa*, uncertain identification.

E. coregoni (see *E. salmonis*).

E. dirus (see *Acanthocephalus dirus*).

E. globolusus (see *Tanaorhamphus ambiguus*).

E. lateralis (see *Acanthocephalus l.*).

E. leidyi Van Cleave, 1924 (Syn. *E. salvelini* Linkins *in* Ward and Whipple, 1918): *Salvelinus malma, S. namaycush,* Can.; Bangham (1955): *Catostomus commersoni, C. catostomus, Coregonus hoyi, Lota lota, Salmo gairdneri, Salvelinus namaycush,* Mich.; Guilford (1954): *Petromyzon marinus,* Wis.; McLain (1951): *P. marinus,* Mich.; Pearse, (1924): ciscoes, black bass, perch, Wis.; Ward (1937): *L. lota;* Warren (1952): *Coregonus artedi,* Minn.

E. oricola (see *Leptorhynchoides thecatus*).

E. paucihamatus (see *Neoechinorhynchus p.*).

E. proteus (see *Acanthocephalus anguillae*).

E. salmonis Müller, 1784 (Syn. *E. coregoni, E. pachysomus, E. phoenix, E. inflatus, E. maraenae, E. murenae*). Europe and U. S. A. Applegate (1950): *Petromyzon marinus,* Mich.; Bangham (1955): *Acipenser fulvescens, Ambloplites rupestris, Catostomus catostomus, C. commersoni, Coeusius plumbeus, Coregonus* spp., *Lepomis gibbosus, Lota lota, Micropterus dolomieui, M. salmoides, Notropis hudsonius, Osmerus mordax, Perca flavescens, Percopsis omiscomaycus, Petromyzon marinus, Prosopium cylindraceum, Salmo gairdneri, Salvelinus namaycush, Stizostedion canadense, Triglopsis thompsoni,* Mich.; Bangham (1955): encysted larva in *Osmerus mordax;* DeGiusti and Budd (1959): life cycle, larva in amphipod; Griffith (1953): *M. dolomieui,* Wash.; Guilford (1954): *P. marinus,* Wis.; McLain (1951): *P. marinus,* Mich.; Pearse (1924): suckers, whitefish, lake trout, pickerel, ciscoes, bowfin, perch, Wis.; Van Cleave (1921).

E. salvelini (see *E. leidyi*).

E. thecatus (see *Leptorhynchoides t.*).

E. tuberosus (see *Neoechinorhynchus rutili*).

Echinorhynchus sp. Griffith (1953): *Ictalurus nebulosus, Tinca tinca,* Wash.

Genus **Acanthocephalus** Koelreuther, 1771
(Fig. 234)

Echinorhynchidae: Body small to medium-sized. Trunk nearly cylindrical with numerous small hypodermal nuclei. Lacunar system with reticular anastomoses. Neck short. Proboscis fairly long, ovoid or claviform to cylindrical, with 6 to 28 longitudinal rows of 4 to 15 hooks each; hooks increase in size from apex toward middle and thence gradually decrease toward base. Proboscis receptacle saccate to cylindrical, double-walled, with ganglion at base. Lemnisci digitiform or claviform, usually not much longer than proboscis receptacle. Testes oval, tandem, in midregion of body. Cement glands 6, rather compact or pyriform to

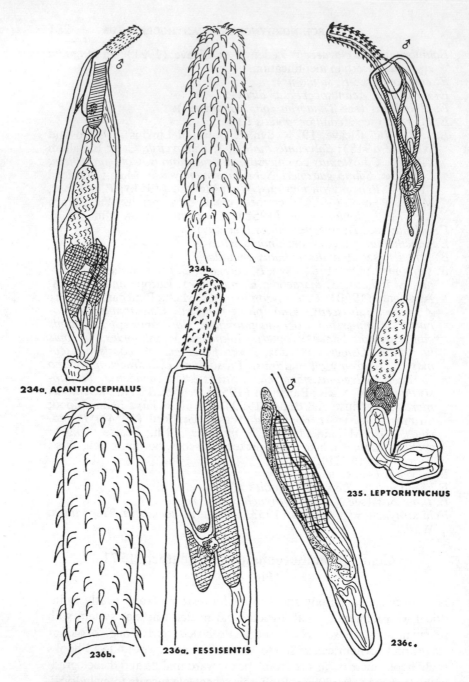

234a. ACANTHOCEPHALUS

234b.

235. LEPTORHYNCHUS

236b.

236a. FESSISENTIS

236c.

Figs. 234–236. Fig. 234. *Acanthocephalus jacksoni. a*, male; *b*, proboscis (from Bullock, 1962). Fig. 235. *Leptorhynchoides thecatus* (from Van Cleave and Mueller, 1934). Fig. 236. *Fessisentis vancleavei. a*, anterior end; *b*, proboscis; *c*, posterior end of male (from Haley and Bullock, 1953).

claviform, close together or in tandem pairs. Eggs greatly elongated, fusiform, with prominent polar prolongations of middle shell. In fishes, amphibians, and reptiles. Life cycle: larva in *Asellus* and *Gammarus*; no second intermediate host.

Ref. Petrochenko (1956), key to spp.; Van Cleave (1952), larva nearly identical with adult.

Acanthocephalus acerbus Van Cleave, 1931: in *Salmo gairdneri*, Japan.
A. aculeatus Van Cleave, 1931: in *Oncorhynchus nerka*, Japan.
A. anguillae (Müller, 1780) Lühe, 1911 (Syn. *E. globulosus, E. linstowi, E. proteus, E. propinquus, E. carpionis*). See Meyer, A. (1933); Nigrelli (1943): mortality of *Roccus lineatus*.
A. dirus (Van Cleave, 1931) Van Cleave and Townsend, 1936 (Syn. *Echinorhynchus d., Pseudoechinorhynchus d.*): in *Aplodinotus grunniens*, Miss., *Ictalurus punctatus, Lepomis macrochirus, Micropterus salmoides*; Bangham and Venard (1942): *A. grunniens*, Tenn.
A. jacksoni Bullock, 1962: in *Salvelinus fontinalis, Salmo gairdneri, Perca flavescens, Catostomus commersoni, Lepomis gibbosus, Esox* sp., immature in *Anguilla, Ictalurus, Microgadus, Notemigonus, Semotilus*, Me.; Bullock (1963): histopathology; West (1964): morphology of acanthor.
A. lateralis (Leidy, 1851) Meyer, 1932 (Syn. *Echinorhynchus globulosus, E. lateralis*). Choquette (1948): in *Salvelinus fontinalis*, Que.; Fantham and Porter (1947): *Ambloplites rupestris, Catostomus commersoni, Perca flavescens, S. fontinalis*, Que.; Golvan (1960): sp. inq.; Haderlie (1953): *Oncorhynchus kisutch, Salmo clarki*, Calif.; Richardson (1936): *S. fontinalis*, Can.

Genus **Leptorhynchoides** Kostylew, 1924
(Fig. 235)

Echinorhynchidae: Trunk cylindrical, aspinose; hydodermal nuclei dendritic. Longitudinal parietal musculature in 4 bands. Proboscis subcylindrical, with 12 to 14 longitudinal rows of 8 to 24 hooks. Proboscis receptacle cylindrical, double-walled; ganglion anterior. Lemnisci tubular to filiform, considerably longer than proboscis receptacle. Testes contiguous, post-equatorial or equatorial. Cement glands compact, 8 in number, massed together immediately behind testes. Eggs elongate, with polar prolongations of middle shell. In freshwater and marine fishes. Life cycle: larva in amphipods; no second intermediate host needed, although larvae less than 30 days old become encysted in mesenteries of fish.

Ref. Golvan (1960), Meyer (1933), Van Cleave and Mueller (1934). Van Cleave and Lincicome (1940) assign this genus to Rhadinorhynchoides; Yamaguti (1963) assigns it to Echinorhynchidae.

Leptorhynchoides thecatum (Linton, 1891) Kostylew, 1924 (Syn. *Echinorhynchus t.* Linton; *E. oricola* Linstow, 1901). Adult reported from the pyloric ceca of *Ambloplites, Amia, Anguilla, Aplodinotus, Carpiodes, Catostomus, Chaenobryttus, Coregonus, Cottus, Cyprinus, Enneacanthus, Esox, Etheostoma, Fundulus, Hiodon, Ictalurus, Ictiobus, Lepisosteus, Lepomis, Lota, Micropterus, Nocomis, Noturus, Osmerus, Perca, Percina, Percopsis, Pomoxis, Pungitius, Roccus, Salmo, Salvelinus, Semotilus, Stizostedion, Umbra*; larva reported from *Etheostoma, Fundulus, Hypentelium, Ictalurus, Moxostoma, Noturus, Percina, Sclerotis, Semotilus, Umbra*.

References and localities—Anthony (1963), Wis.; Bangham (1926, 1933, 1939, 1941, 1944, 1955): Fla., Mich., Ohio, Wis.; Bangham and Venard (1942): Tenn.; Bullock (1962): host specificity; DeGiusti (1939, 1940): life cycle, larva in amphipod; De-Roth (1953): Me.; Haley (1952); Hare (1943): Ohio; Hunter (1930, 1939); Lincicome and Van Cleave (1949): distribution and redescription; MacLulich (1943): Ont.; Meyer, A. (1933); Meyer, F. (1959): Iowa; Meyer, M. (1954): Me.; Odlaug *et al.* (1962): Minn.; Pearse (1924): Wis.; Sindermann (1953): Mass.; Sogandares (1955): La.; Spaeth (1951): effect on amphipod; Van Cleave (1923): Ill.; Van Cleave and Mueller (1934): N. Y.; Venard (1941): Tenn.; Venard and Warfel (1953): effect on fish; Ward (1937); Worley and Bangham (1952): Que.

Family FESSISENTIDAE Van Cleave, 1931
Genus **Fessisentis** Van Cleave, 1931
(Fig. 236)

Trunk cylindrical, of moderate length. Proboscis short, with 12 to 16 longitudinal rows of 6 to 8 hooks each; hooks largest toward middle of proboscis. Proboscis receptacle subcylindrical, double-walled, arising from base of proboscis. Ganglion at base of proboscis receptacle. Lemnisci narrow, much longer than proboscis receptacle. Testes cylindrical, very long. Cement glands 4, elongate pyriform. Eggs much elongated, with polar prolongations of middle shell. In freshwater fishes. Life cycle unknown.

Fessisentis fessus Van Cleave, 1931: in *Aplodinotus grunniens*, Miss. *F. vancleavei* Haley and Bullock, 1953: in *Lepomis gibbosus*, N. H.; Bullock (1962): host specificity.

Family POMPHORHYNCHIDAE Yamaguti, 1939
Genus **Pomphorhynchus** Monticelli, 1905
(Fig. 237)

Body small. Trunk aspinose, hypodermal nuclei small, numerous. Lacunar system with reticular anastomoses. Neck very long, cylindrical, forming globular bulb at anterior end. Proboscis long, almost cylindrical, with 12 to 20 longitudinal rows of 10 to 14 hooks each; posterior hooks much more slender than anterior. Proboscis receptacle double-walled, inserted at posterior end of proboscis and extending backward throughout length of neck. Lemnisci short, claviform. Testes tandem, near middle of trunk. Cement glands 6, rounded to oval. Genital pore of both sexes terminal, not surrounded by spines. Eggs fusiform, with prominent prolongation at each pole of middle shell. In fishes. Life cycle: larva in amphipod; second intermediate host is small fish, but not required by *P. laevis*.

Ref. Petrochenko (1956), key to spp.

Pomphorhynchus bulbocolli (Linkins, 1919) Van Cleave, 1919. Reported from *Ambloplites, Amia, Aplodinotus, Carassius, Carpiodes, Catastomus, Chaenobryttus, Cottus, Cyprinus, Erimyzon, Esox, Etheostoma, Fundulus, Hyborhynchus, Hypentelium, Ictalurus, Ictiobus, Lepomis, Lota, Micropterus, Moxostoma, Mylocheilus, Nocomis, Notemigonus, Notropis, Noturus, Oncorhynchus, Perca, Percina, Percopsis, Pomoxis, Prosopium, Ptychocheilus, Richardsonius, Salmo, Semotilus, Stizostedion, Umbra*, and a snake, *Natrix sipedon*; larva reported from *Ambloplites, Catostomus, Cottus, Etheostoma, Ictalurus, Micropterus, Notropis, Osmerus, Perca, Percina, Percopsis*.

References and localities—Anthony (1963): Wis.; Bangham (1942, 1944, 1951, 1955): Mich., Tenn., Wis., Wyo.; Bangham and Adams (1954): Brit. Col.; Bullock (1962): host specificity; DeRoth (1953): Me.; Dolley (1933): Ind.; Fischthal (1947, 1950): Wis.; Fritts (1959); Griffith (1953): Wash.; Hare (1943): Ohio; Harms (1959): Kans.; Hugghins (1959): S. D.; Jensen (1953): life cycle; Jones and Hammond (1960): Utah; Meyer, A. (1933); Pearse (1924): Wis.; Richardson (1941): Can.; Sindermann (1953): Mass.; Van Cleave (1923, 1931): Ill.; Van Cleave and Mueller (1934): N. Y.; Ward (1940): development and life cycle.

P. tereticolle (Rud., 1809). Marine. Reported from *Roccus lineatus* (Linton, 1889).

Family RHADINORHYNCHIDAE Travassos, 1923
Genus **Rhadinorhynchus** Lühe, 1911
(Fig. 238)

Body cylindrical; hypodermal nuclei small, numerous. Lacunar system with lateral main vessels and reticular anastomoses. Trunk spines usually separated by aspinose area into two groups: anterior group encircles body; posterior confined to ventral side and extends farther backward in female than in male. Proboscis usually very long, clavi-

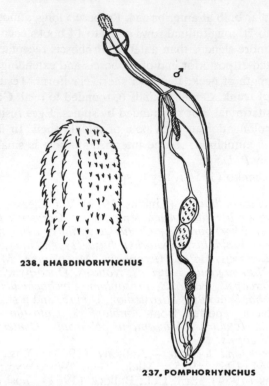

238. RHABDINORHYNCHUS

237. POMPHORHYNCHUS

Figs. 237–238. Fig. 237. *Pomphorhynchus bulbocolli* (from Van Cleave and Mueller, 1934). Fig. 238. *Rhabdinorhynchus peltorhamphi*, proboscis (from Ward, 1951).

form, with 8 to 26 longitudinal rows of 8 to 37 hooks each; ventral hooks usually larger than dorsal; basal hooks project at right angles to proboscis. Proboscis receptacle long, double-walled, with ganglion near middle. Lemnisci usually long, slender. Testes elongate, tandem. Cement glands 2, long, tubular. Genital pore terminal in male, subterminal in female, neither surrounded by spines. Uterus very long. Eggs elongate, with polar prolongations of middle shell. In marine, sometimes freshwater, fishes.

Rhadinorhynchus exilis Van Cleave, 1928: in *Carassius*, China.
R. pristis (Rud., 1802). Reported from *Fundulus majalis* (Meyer, A., 1933).
Rhadinorhynchus sp. Shaw (1947): in *Salmo gairdneri*, Ore.

SPECIES INQUIRENDA

Tetrahynchus sp. Shaw (1947): in *Salmo gairdneri*, Ore.

LARVAL FORMS IN FISH

The following have been found encysted in the mesenteries and liver of small fish: *Neoechinorhynchus cylindratum, Leptorhynchoides thecatum, Pomphorhynchus bulbocolli*, and possibly *Echinorhynchus salmonis*. The marine parasite *Corynosoma hardweni* Van Cleave, 1953, which is adult in seals, has been reported as a larva in *Osmerus mordax* (Meyer, M., 1954).

Leeches
(Annelida: Hirudinea)

Contents

Key to the Species of North American Leeches Reported from Freshwater Fishes 290

Leeches Reported from North American Freshwater Fishes, Phylum Annelida, Class Hirudinea 293

Order PHARYNGOB-
DELLIDA 293
   Genus *Nephelopsis* 293

Order RHYNCHOB-
DELLIDA 293

Family GLOSSIPHO-
NIIDAE 294
   Genus *Placobdella* 294
   Genus *Actinobdella* 295

Family PISCICOLIDAE .. 295
   Genus *Cystobranchus* ... 295
   Genus *Piscicola* 297
   Genus *Piscicolaria* 297
   Genus *Illinobdella* 298

The true fish leeches belong to the family Piscicolidae, but two of the Glossiphoniidae (*Actinobdella triannulata* and *Placobdella pediculata*) show a strong predilection for fish. Other Glossiphoniidae are found on fish occasionally. Leeches attach periodically to fish, take a large blood meal, and leave the fish for varying periods of time. Their life cycles have not been adequately studied. Fish host specificity is apparently lacking (Meyer, 1946), although there may be exceptions. Not enough host records have been made to determine specificity, and there have been no experimental host studies.

The damage done to the fish is proportional to the number of leeches present and the amount of blood they remove. Epizootics have been reviewed by Meyer (1946). Leeches also serve as vectors of *Trypanosoma*, *Cryptobia*, and probably the blood sporozoa *Haemogregarina* and *Dactylosoma*; perhaps also fish bacteria and viruses.

A very unusual case of erratic hirudiniasis was reported by Rupp and Meyer (1954). During warm weather, brook trout had congregated at a spring hole in a Maine lake and were severely attacked by *Haemopsis grandis* and *Macrobdella decora*, which are not considered to be fish leeches.

Some leeches can be identified while alive, but permanent stained slides and serial sections are required for some. Clearing, without permanent staining, is required for determining the ocular arrangement and number.

Leeches are difficult to relax properly before fixation; several methods have been used: (1) Add 70 percent alcohol to the water containing living leeches, gradually increasing the concentration for about 30 minutes until movement ceases. The leeches are then removed, passed between the fingers to straighten them and remove excess mucus, then laid out and kept flat while the fixative is poured on (Mann, 1962). (2) Place specimens in a water-filled petri dish to which a small pinch of smoking tobacco is added. Leave leeches in this until they cease to respond to stimuli. Straighten them on a slide wrapped with tissue paper in a petri dish with just enough water to moisten the paper. Slowly add hot 70 percent alcohol plus a few drops of acetic acid until specimens are covered (Meyer and Penner, 1962). (3) I have used nembutal successfully on some leeches. Confine the leech to 5 ml of water to which a drop of 6 percent nembutal has been added; continue as in (1) or (2).

Slight flattening of leeches is sometimes desirable. Place the live leech in the cover of a Petri dish and put the inverted dish bottom on it. Add the relaxing agent between the two parts. Further flattening may be achieved by placing weight on the inverted dish bottom. Add 70 percent alcohol when the leech is properly relaxed, and let stand overnight with weight in place.

KEY

TO THE SPECIES OF NORTH AMERICAN LEECHES
REPORTED FROM FRESHWATER FISHES

(Modified from Mann, 1962; Meyer, 1946; Moore, 1959)

1.   Mouth a small pore on head sucker through which pharyngeal proboscis may protrude; no jaws; no denticles; blood colorless . . . . Order RHYNCHOBDELLIDA   2

1.   Mouth large, opening from behind into entire sucker cavity; pharynx a fixed crushing tube extending to segment XIII; 3 muscular pharyngeal ridges but no true jaws; eyes 3 or 4 pairs in separate groups; blood red . . . . Order PHARYNGOBDELLIDA, Family ERPOBDELLIDAE (One species, *Nephelopsis obscura*, reported from air bladder of lake trout; probably accidental.)

Order RHYNCHOBDELLIDA

2 (1). Body flat when at rest; not divided into distinct anterior and posterior regions; head narrow with anterior sucker not, or only slightly, distinct from body; usually 3 annuli per segment in midbody; eyes confined to head except *Placobdella hollensis* . . . . Family GLOSSIPHONIIDAE   3

2 (1). Body cylindrical when at rest; usually divided into distinct anterior and posterior regions; head sucker usually set off distinctly; usually more than 3 annuli per segment; eyes may be present on head, neck, and posterior sucker . . . . Family PISCICOLIDAE   8

Family GLOSSIPHONIIDAE

3   (2). Midbody segments with 3 annuli of which first and
            third may be faintly subdivided; body broad and
            flat; salivary glands compact; epididymis a tight
            mass .... (fig. 239) *Placobdella*   4
3   (2). Midbody segments with 2 to 6 annuli; body less broad
            and flat; salivary glands diffuse; epididymis less
            coiled; posterior sucker with marginal circle of
            glands and retractile papillae. (One species on fish:
            3 annuli per segment; about 30 conical papillae on
            posterior sucker; dorsal tubercles prominent, in 5
            longitudinal rows) .. (fig. 240) *Actinobdella triannulata*

Genus *Placobdella*

4   (3). Head distinctly wider than segments just behind; 3
            prominent dorsal ridges bearing tubercles; eyes
            clearly separated by their diameter ....
            .... *Placobdella montifera*
4   (3). Without widened head and dorsal ridges   5
5   (4). Anus of adult between 23rd and 24th segments; post-
            anal somites a slender sucker peduncle ....
            .... *Placobdella pediculata*
5   (4). Anus in or behind 27th segment; postanal somites
            normal   6
6   (5). Minute simple supplementary eyes near mid-dorsal line
            of head; every third annulus deeply pigmented green
            and brown; length 25 to 40 mm .. *Placobdella hollensis*
6   (5). No supplementary eyes; body large and flat   7
7   (6). Dorsal tubercles low, smooth domes, often suppressed;
            opaque and heavily pigmented in bold but variable
            pattern of brown, green, and yellow; venter with
            about 12 bluish or purplish stripes; dorsal and ven-
            tral furrows aligned .... *Placobdella parasitica*
7   (6). Principal dorsal tubercles large, elevated, roughened
            with numerous sensory papillae, many smaller tu-
            bercles; integument translucent and color pattern a
            fine mixture; venter without stripes; dorsal and ven-
            tral furrows not exactly aligned .... *Placobdella ornata*
                                                    (*P. rugosa*)

Family PISICOLIDAE

8 (2). Pulsatile vesicles on margins of body, sometimes not
       visible after preservation   9
8 (2). Pulsatile vesicles absent   12
9 (8). Pulsatile vesicles large, clearly seen after preservation;
       7 annuli per segment; body clearly divided into an-
       terior and posterior regions; one American species
       .... (fig. 241) *Cystobranchus verrilli* *
9 (8). Pusatile vesicles small, difficult to see after preserva-
       tion; 14 annuli per segment; body not divided into
       anterior and posterior regions . (fig. 242) *Piscicola* 10

Genus *Piscicola*

10 (9). Posterior sucker with dark rays and/or oculiform
       spots   11
10 (9). Posterior sucker without such spots or rays ....
       .... *Piscicola punctata*
11 (10). Ocelli 8 to 10, crescent-shaped, on posterior sucker;
       gonopores separated by 2 annuli; sperm duct con-
       voluted .... *Piscicola salmositica*
11 (10). Ocelli 10 to 12, punctiform; no dark rays on posterior
       gonopores; sperm duct looped simply . *Piscicola milneri*
11 (10). Ocelli 12 to 14, punctiform; 14 dark rays on posterior
       sucker; 3 annuli between gonopores; sperm duct
       looped simply .... *Piscicola geometra*

Genus *Piscicolaria*

12 (8). Annuli 3 per segment; no clear division between an-
       terior and posterior of body ....
       .... (fig. 244) *Piscicolaria reducta*
12 (8). Annuli 12 to 14 per segment; contracted specimens re-
       veal distinction between anterior and posterior of
       body .... (fig. 243) *Illinobdella* 13

Genus *Illinobdella*

13 (12). Annuli 12 per segment; ratio of length to width about
       18:1 .... *Illinobdella elongata*
13 (12). Annuli 14 per segment   14

* A new *Cystobranchus* has been described from Virginia (Hoffman, R. L., 1964).

14 (13). Body usually divided into 2 regions; ratio of length to width about 5:1; gonopores separated by 8 annuli (4 annuli faintly subdivided); distinct seminal vesicle . . . . *Illinobdella moorei*

14 (13). Body of uniform width, not divided into 2 regions   15

15 (14). 14 pairs of nephridopores; gonopores separated by 1 annulus; ratio of length to width about 6:1 in preserved specimens . . . . *Illinobdella patzcuarensis*

15 (14). 11 pairs of nephridopores*   16

16 (15). Annuli distinctly defined; ratio of length to width about 15:1 . . . . *Illinobdella richardsoni*

16 (15). Annuli less distinctly defined; ratio of length to width about 9:1; gonopores separated by 4 annuli (2 faintly subdivided); especially characteristic of species are large glands, with ducts entering base of proboscis . . . . *Illinobdella alba*

LEECHES REPORTED FROM NORTH AMERICAN FRESHWATER FISHES

(Phylum ANNELIDA)

Class HIRUDINEA (Leeches)

Ref. Moore (1959); Mann (1962).

Order PHARYNGOBDELLIDA

Mouth large, opening from behind into entire sucker cavity; fixed pharynx a crushing tube extending to somite xiii; eyes 3 or 4 pairs in separate labial and buccal groups; somites 5-annulate but often further divided; 3 muscular pharyngeal ridges but no true jaws or denticles; testisacs very small and numerous, in grape-bunch arrangement; blood red.

Nephelopsis obscura Verrill, 1872. Reported from air bladder of *Salvelinus namaycush*, Wyo. (Meyer and Bangham, 1950).

Order RHYNCHOBDELLIDA

Mouth a small pore on head sucker through which pharyngeal proboscis may be protruded; no jaws; no denticles; blood colorless.

---

* Meyer, M. C. (1965), suspects that re-examination of good serial sections will show that these two species also have 14 pairs of nephridopores.

Family GLOSSIPHONIIDAE

Body not divided externally into distinct anterior and posterior; body at rest usually flattened and much wider than head except some *Actinobdella* and *Placobdella montifera*; head sucker not freely expanded; eyes mostly integrated, confined to head except *Placobdella hollensis*; usually 3 annuli per segment in midbody region.

Genus **Placobdella** Blanchard, 1896
(Fig. 239)

Complete somites 3-annulate, *a*1 and *a*3 may be faintly subdivided; resting form usually very broad and flat; eyes on somite III appear as one pair united in common pigment mass and may be compound; salivary glands compact; epididymis a tight knot.

Placobdella hollensis (Whitman, 1872). Moore (1912): turtles, Minn.; Mullen (1926): on fish, Iowa.
P. montifera Moore, 1912. Bangham (1933): on *Micropterus dolomieui*, Ohio; Harms (1959): on *Ictalurus melas,* Kans.; Mather (1948): various fish, Iowa; Moore (1912): on frogs and toads,

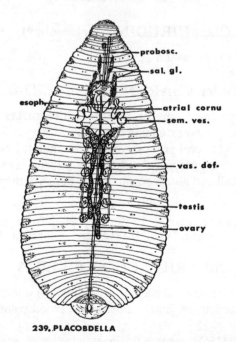

239. PLACOBDELLA

Fig. 239. *Placobdella parasitica* (from Whitman, in Moore, 1959).

Minn.; Pearse (1924*b*): *Cyprinus carpio*, hackleback sturgeon, *M. dolomieui*, Wis.

P. ornata (Verrill, 1872), (Syn. *P. rugosa* Verrill). Mullen (1926): on fish, Iowa.

P. parasitica (Say, 1924). Moore (1901): on turtles, Ill.; Pearse (1924*b*): on *Esox, Fundulus, Lepomis macrochirus*, Wis.

P. pediculata Hemingway, 1912: in gill chamber, *Aplodinotus grunniens*; Branson and Amos (1961): *A. grunniens*, Okla.; DeRoth (1953): *Catostomus commersoni*, Me.; Mullen (1926): *A. grunniens*, Iowa.

Genus **Actinobdella** Moore, 1901
(Fig. 240)

Complete somites with 3 to 6 annuli; resting form not as broad and flat as *Placobdella*; salivary glands diffuse; epididymis loosely coiled; posterior sucker with marginal circle of glands and retractile papillae.

Actinobdella triannulata Moore, 1924. Bangham (1951): on *Catostomus fecundus*, Wyo.; Bangham (1955): *C. catostomus, C. commersoni*, L. Huron; Bangham and Adams (1954): *C. catostomus, C. macrocheilus*, Brit. Col.; Meyer, M. C. (1962): *C. commersoni*, Can., N. Hampshire, Ont.; Meyer and Moore (1954): *C. commersoni*, Ont.

Actinobdella sp. Worley and Bangham (1952): *Perca flavescens*, Que.

Family PISCICOLIDAE

Body at rest cylindrical and usually divided at segment XIII into distinct anterior and posterior regions; head sucker usually distinctly marked off from body; usually more than 3 annuli per segment; simple eyes may be present on head, neck, and posterior sucker.

Ref. Meyer (1940, 1946).

Genus **Cystobranchus** Diesing, 1859
(Fig. 241)

Margins of body with 11 pairs of large pulsatile vesicles, clearly seen after preservation; 7 annuli per segment; body clearly divided into anterior and posterior regions; caudal sucker very large.

Cystobranchus verrilli Meyer, 1940: on *Micropterus dolomieui, Stizostedion vitreum, Pylodictis olivaris*, Ill.; Meyer and Moore (1954): on *Lota lota*; Hoffman (1962): *Lepomis macrochirus*, W. Va.; Mather (1948): completely covering gills of *Ictalurus punctatus*, Iowa.

Cystobranchus virginicus Hoffman, R., 1964; no host datum, Va.

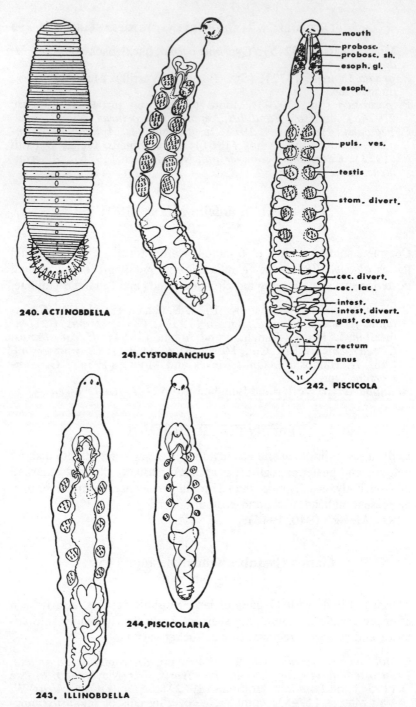

240. ACTINOBDELLA

241. CYSTOBRANCHUS

242. PISCICOLA

- mouth
- probosc.
- probosc. sh.
- esoph. gl.
- esoph.
- puls. ves.
- testis
- stom. divert.
- cec. divert.
- cec. lac.
- intest.
- intest. divert.
- gast. cecum
- rectum
- anus

244. PISCICOLARIA

243. ILLINOBDELLA

Figs. 240–244. Fig. 240. *Actinobdella inequinannulata* (modified from Moore, 1901). Fig. 241. *Cystobranchus verrilli* (from Meyer, 1940).

Genus **Piscicola** Blainville, 1818

Margins of body with 11 pairs of small pulsatile vesicles, difficult to see on preserved specimens; 14 annuli per segment; body not clearly divided into anterior and posterior regions; postceca completely united into one; testisacs 6 pairs.

Piscicola geometra (Linnaeus, 1758). Europe and North America. Meyer (1946): *Cyprinus carpio*, trout, Wis.

P. milneri (Verrill, 1871). Meyer (1946): on *Coregonus clupeaformis*, Mich.; Meyer (1954): *S. salar*, Me.; Meyer and Moore (1954): *Salvelinus namaycush*, Ont.; Pearse (1924b): on cisco and *Lota*, Wis.; Moore (1924): whitefish, Ont.

P. punctata (Verrill, 1871). Reported from *Ictalurus nebulosus, Micropterus dolomieui, M. salmoides, Lepomis cyanellus, L. microlophus*, by Bangham (1933, 1936, 1939) and Venard (1940); Meyer, M. (1940): *Catostomus commersoni, Coregonus clupeaformis, Ictiobus cyprinellus, Moxostoma erythrurum, Osmerus mordax, Salmo gairdneri*, Ill., Mich., N. Y.; Meyer (1946): *Salvelinus fontinalis*; Mullen (1926): *Percina caprodes*, Iowa; Moore (1924): *Stizostedion canadense*, Ont.

P. salmositica Meyer, 1946: on *Salmo gairdneri*, Wash.; Becker and Katz (1965): as vector of *Cryptobia*, Wash.; Cope (1958): *S. clarki*, Wyo.; Earp and Schwab (1954): on fry and eggs, *Oncorhynchus gorbuscha*; Haderlie (1953): *S. gairdneri, O. tshawytscha*, Calif.; Jones and Hammond (1960): *S. gairdneri*, Utah; Meyer (1949): *O. kisutch, O. tshawytscha, O. keta, S. gairdneri*, Wash.

Genus **Piscicolaria** Whitman, 1889
(Fig. 244)

Form clavate, no clear division between anterior and posterior body regions; both suckers much smaller than body diameter; pulsatile vesicles absent; 3 annuli per segment; mouth central in sucker.

Piscicolaria reducta Meyer, 1940: on *Hadropterus phoxocephalus*, Ill.; Harms (1959): *Ictalurus melas, I. punctatus*, Kans.; Meyer (1946): *Lepomis macrochirus*, N. J.; Meyer (1954): *Notemigonus crysoleucas*, Me.

Piscicolaria sp. DeRoth (1953): on *Ictalurus nebulosus, Roccus americanus, Perca flavescens*, Me.; Pearse (1924b): *Ambloplites rupestris, Ictalurus melas, I. nebulosus, I. punctata, Micropterus dolomieui, P. flavescens*, Wis.

Fig. 242. *Piscicola salmositica* (from Meyer, 1946). Fig. 243. *Illinobdella moorei* (*Myzobdella m.* ?) (from Meyer, 1940). Fig. 244. *Piscicolaria reducta* (from Meyer, 1940).

Genus **Illinobdella** Meyer, 1940
(Fig. 243)

Form clavate, but distinction between anterior and posterior body regions more evident in contraction; both suckers much smaller than body diameter; somites 14-annulate; mouth central in sucker; stomach 6-chambered; postceca completely fused.

This genus may be congeneric with *Myzobdella* Leidy according to Meyer, M. (1958) and Raj (1962).

Illinobdella alba Meyer, 1940: on *Aplodinotus grunniens, Ictalurus punctatus, Lepomis macrochirus, Notemigonus crysoleucas, Pomoxis annularis,* Ill.; Bangham (1955): *Perca flavescens,* L. Huron; Bangham and Venard (1942): *L. microlophus,* Tenn.; Meyer (1946): *L. microlophus, Perca flavescens,* Conn., N. Y.; Meyer and Moore (1954): *Micropterus dolomieui, P. flavescens,* Ont.

I. elongata Meyer, 1940: on *Pomoxis annularis,* Ill.; Meyer and Moore (1954): *Micropterus dolomieui,* Ont.

I. moorei (Meyer, 1940) Meyer, 1946: on *Aplodinotus, Chaenobryttus, Esox, Ictalurus, Lepomis, Micropterus, Notemigonus, Perca, Percina, Polyodon, Pomoxis, Stizostedion,* Iowa, Ill., Minn.; Bangham (1944): *Stizostedion,* Wis.; Bangham (1955): *Ambloplites, Perca,* L. Huron; Bangham and Venard (1942): *Chaenobryttus, Micropterus,* Tenn.; Mather (1948): *Ictaluris melas, I. punctata,* Iowa; Meyer and Moore (1954): *Ictalurus, Lepomis, Perca,* Ont.; Raj (1962): 19 spp. fish, Conn.

I. patzcuarensis Caballero, 1940. Meyer (1946): on *Chirostoma grandocule,* Mexico.

I. richardsoni (Meyer, 1940) Meyer, 1946: on *Ictalurus nebulosus, Lepomis microlophus, L. punctatus, Micropterus dolomieui, Pomoxis annularis, P. nigromaculatus,* Ill., N. Y., Ohio; Bangham and Venard (1942): *L. punctatus, P. nigromaculatus,* Tenn.

Illinobdella spp. Reported from *Ambloplites, Amia, Aplodinotus, Catostomus, Chaenobryttus, Esox, Etheostoma, Eucalia, Ictalurus, Lepomis, Micropterus, Perca, Percina, Poecilichthys, Ptychogonimus, Salmo,* by Anthony (1963): Wis.; Bangham (1944, 1951, 1955): L. Huron, Wis., Wyo.; Bangham and Adams (1954): Brit. Col.; Bangham and Venard (1942): Tenn.; Fischthal (1947, 1950): Wis.; Haderlie (1953): Calif.; Hare (1943): Ohio; Worley and Bangham (1952): Que.

Parasitic Copepods

(Arthropoda: Crustacea)

Contents

Artificial Key to the Female
Parasitic Copepods 300
Subclass BRANCHIURA ... 301
  Genus *Argulus* 301
Subclass COPEPODA
  (PARASITIC) 303
  Order CYCLOPIDEA 303

Genus *Ergasilus* 303
Order CALIGIDEA 307
  Genus *Lernaea* 307
Order LERNEOPODIDEA. 310
  Genus *Achtheres* 311
  Genus *Salmincola* 313
  Genus *Cauloxenus* 315

In the Crustacea are two groups commonly referred to as parasitic copepods: the subclass Branchiura (fish lice) and the subclass Copepoda, some of which resemble free-living copepods. The Branchiura is considered an order by some, and placed under the subclass Copepoda, but most recent workers give it subclass rank.

Certain species of parasitic copepods of the genera *Argulus*, *Lernaea*, and *Ergasilus* are very serious pests in fish culture, sometimes in nature, and have become increasingly important in recent years.

Ref. Yamaguti (1963), Bowen (1965), Bowen and Putz (1965), Putz and Bowen (1964).

ARTIFICIAL KEY

TO THE FEMALE PARASITIC COPEPODS

1. On body of fish (except *Salmincola* usually on gills) 2
1. Usually on gills (*Salmincola* sometimes on fins) 4
2. Not fixed in tissue of fish; body flattened dorsoventrally; suck-
      ing discs present . . . . (fig. 245) *Argulus*
2. Attached to body of fish; no sucking discs 3
3. Anterior end of adult female buried in flesh; body cylindrical,
      wormlike; cephalic segment with 2 to 4 soft horns (male
      small, temporary on gills) . . . . anchor worm . . . .
      . . . . (fig. 247) *Lernaea*
                                                                                (p. 307)
3. Anterior appendages buried in base of fins but usually on gills
      (see no. 6) . . . . (fig. 249) *Salmincola*
                                                                                (p. 313)
3. Not buried in flesh but attached near anus of returning salmon;
      no cephalic claspers; genital segment enlarged . . . .
      . . . . salmon louse, *Lepeoptheirus*
4. Body flattened dorsoventrally, pointed cephalic claspers pres-
      ent; genital segment not enlarged; freshwater . . . .
      . . . . (fig. 246) *Ergasilus*
4. Body not flattened dorsoventrally 5

5. Trunk with sigmoid curve; cephalic horns present (resembles *Lernaea*); usually marine, sometimes anadromous . *Lernaeocera*
5. Trunk without sigmoid curve; no cephalic horns, but large cephalic appendages joined distally in a bulla; freshwater . .   6
6. On warm-water fish; abdomen present . . . . . (fig. 248) *Achtheres* (p. 311)
6. On gills, sometimes fins, of salmonids; no abdomen . . . . (fig. 249) *Salmincola* (p. 313)

Subclass BRANCHIURA
Genus **Argulus** Müller, 1785
(Fig. 245)

Argulidae: Two pairs of antennae and preoral sting present. Second maxillae transformed into prehensile discs. Basal plate of maxilliped armed with teeth. Parasites of freshwater and marine fishes. Life cycle: after copulation, females lay eggs in batches on any object in water. If males are absent, it is believed that eggs develop parthenogenetically. Egg hatches in 15 to 55 days, and free-swimming larva must find a fish in 2 to 3 days. After attaching to fish, copepod grows and metamorphoses several times and becomes sexually mature in 30 to 35 days. Copepods remain on fish, surrounded by mucus during winter (Bauer, 1959).

Ref. Bowen and Putz (1965); Bower-Shore (1940), biology; Causey (1957); Hargis (1958), *Argulus* as fortuitous human epizoon; Meehean (1940); Wilson (1944); Yamaguti (1963, p. 320).

Argulus americanus Wilson, 1904. Reported from *Amia calva, Esox masquinongy, Ictalurus nebulosus, Lepomis macrochirus, Sclerotis miniatus, Umbra limi*, from Fla., Ind., Iowa, Mich., N. Y., Tenn., by Bangham and Venard (1942); Meehean (1940); Wilson (1916); Yeatman (1965): redescription; Goin and Ogren (1956): also from amphibia.

A. appendiculosus Wilson, 1907 (Syn. *A. biramosus* according to Meehean, 1940). Reported from *Aplodinotus grunniens, Catostomus* spp., *Dorosoma cepedianum, Ictalurus nebulosus, I. platycephalus, I. punctatus, Ictiobus bubalis, I. cyprinellus, Micropterus salmoides, Perca flavescens, Pomoxis annularis, Roccus chrysops, Stizostedion vitreum*, from Iowa, Ky., Minn., Ohio, Tex., Va., Wis., by Pearse (1924); Schumacher (1952); Wilson (1916).

A. biramosus Bere, 1931 (Syn. of *A. appendiculosus* according to Meehean, 1940): on *Perca flavescens*, Wis.; Allum and Hugghins (1959): epizootic, *Catostomus commersoni, Cyprinus carpio, Ictalurus melas, P. flavescens, Stizostedion vitreum*, S. Dak.; Bere

(1935): *Ambloplites rupestris*, Wis.; Hugghins (1959): *C. commersoni, C. carpio, Esox lucius, I. melas, Ictiobus cyprinellus, P. flavescens, Pomoxis* spp., S. Dak.; Sarig and Lahav (1959): Lindane treatment, Israel; Wilson (1944): possible synonomy with *A. appendiculosus*.

A. canadensis Wilson, 1916: on *Acipenser rubicundus, Coregonus* sp.; Meehean (1940) considers this a synonym of *A. stizostethii*; Davis, W. (1956): on *Alosa sapidissima*; Meyer, M. (1954): *Salvelinus namaycush*, Me.; Wilson (1936, 1944): taxonomy and redescription.

A. catostomi Dana and Herrick, 1837. Reported from *Catostomus catostomus, C. commersoni, Cyprinus carpio, Erimyzon sucetta, Hypentelium nigricans*, by Meehean (1940); Anthony (1963): *C. commersoni*, Wis.; Bangham (1944): same host, Wis.; Bere (1935): same, Wis.; DeRoth (1953): same, Me.; Fischthal (1947, 1950): same, Wis.; Meyer, M. (1954): same, Me.; Mueller (1937): same, N. Y.; Pearse (1924): same, Wis.; Wilson (1932): Woods Hole.

A. coregoni Thorell, 1865. Yamaguti (1963): on *Salmonidae*, Europe and N. America.

A. diversus Wilson, 1944 (Syn. *A. maculosus* Meehean, 1940): from *Erimyzon sucetta*, Fla.; Yeatman (1965): *Ictaluris natalis*, N. C., redescription.

A. flavescens Wilson, 1916: on *Amia calva*. Reported from *A. calva, Chaenobryttus gulosus, Cyprinus carpio, Erymizon sucetta, Fundulus* sp., *Hypentelium nigricans, Ictalurus natalis, I. nebulosus, I. punctatus, Lepomis microlophus, Micropterus floridana, M. punctulatus, M. salmoides, Pylodictis olivaris*, from Fla., Iowa, La., Nova Scotia, Okla., by Bangham (1938); Causey (1957); Meehean (1940); Mueller (1936); Roberts (1957).

A. floridensis Meehean (1940): described from one freeswimming specimen.

A. foliaceus (see *A. japonicus*).

A. funduli Krøyer, 1863: on Atlantic fish including *Fundulus* spp.; Yeatman (1966): *Cyprinodon variegatus*, Fla.

A. ingens Wilson, 1912: on *Lepisosteus tristoechus*, Mass. Must be a mistake because *L. tristoechus* is a Cuban gar.

A. japonicus Thiele, 1900 (Syn. *A. foliaceus* Nettowich; *A. pellucidus* Wagler): on *Carassius auratus, Cyprinus carpio*, cosmopolitan. Hirschmann and Partsch (1953): redescription.

A. laticauda Smith, 1873: on marine fish including *Anguilla rostrata*; Hargis (1958): fortuitous in human eye.

A. lepidostei Kellicott, 1878: on *Lepisosteus osseus*, N. Y.; Bangham (1941): *L. platyrhincus*, Fla.; Bangham and Venard (1942): *L. platostomus*, Tenn.; Wilson (1916): redescription.

A. longicaudatus Wilson, 1944: on *Pomoxis annularis*, Tex.

A. lunatus, Wilson, 1944: on *Carassius auratus*, Va.

A. maculosus Meehean, 1940 (see *A. diversus*).

A. maculosus Wilson, 1902: on *Ambloplites rupestris, Ictalurus natalis, I. nebulosus*, Ind. Also on *Esox masquinongy, E. nobilior*

(Pearse, 1924*b*; Wilson, 1944; Meehean, 1940); Yeatman (1965): redescription.

A. mississippiensis Wilson, 1916: on *Lepisosteus osseus*, La.; Causey (1957): *L. osseus*, La.

A. nobilis Thiele, 1924. Wilson (1924): redescribed from *Lepisosteus*, La.

A. pellucidus Wagler, 1935 (see *A. japonicus*).

A. pugettensis Dana, 1852. Meehean (1940): on marine fish and *Salmo gairdneri*, Pacific coast; Wilson (1908): on *Oncorhynchus kisutch*.

A. stizostethi Kellicott, 1880: on *Stizostedion glaucum* (?), *S. canadense, S. vitreum, Roccus chrysops*. Meehean (1940) considers *A. canadensis* a synonym. Also reported from *Acipenser fulvescens, Coregonus* sp., *Perca flavescens, Roccus chrysops, Salmo salar, Salvelinus fontinalis*, stickleback, suckers, from Iowa, Minn., New Brunswick, Can., Ohio, Wis., by Bangham (1941); Tidd (1931).

A. trilineatus Wilson, 1904: on goldfish. Meehean (1940) considers this a synonym of *A. japonicus*; Wilson (1944): valid species; Cockerell (1926): on *C. auratus*, Colo.; Guberlet (1928): same host.

A. versicolor Wilson, 1902: on *Esox lucius, E. niger, E. reticulatus, Perca flavescens, Pomoxis nigromaculatus, Stizostedion vitreum*, Mass., Md., Ga.; Fischthal (1947): *P. nigromaculatus*, Wis.; Hunter and Rankin (1939): *E. niger*, Conn.; Wilson (1904): redescribed; Sindermann (1953): *E. niger, Lepomis gibbosus*, Mass.; Yeatman (1965): redescription.

Subclass COPEPODA (Parasitic)
Order CYCLOPIDEA Yamaguti, 1963

Key characteristics: Adult female attached to host, but not permanently; legs usually well developed.

Genus **Ergasilus** Nordmann, 1832
(Fig. 246)

Ergasilidae: Body cyclops-like, narrowed posteriorly. Head sometimes fused with, sometimes separated from, first segment. Abdomen 3-segmented in female, 4-segmented in male; caudal rami short. Egg sac long, often cigar-shaped or rather plump; eggs small, numerous. First antenna 6-segmented; second antenna 5-segmented; terminal segment forming stout, clasper-like claw, without accessory claw. Mouth parts removed some distance behind second antennae. Mandible 2-segmented; proximal segment with pectinate palp; distal segment with setose, pectinate, or serrate margin. First maxilla with two setae; second maxilla with shaggy tip. Maxillipeds lacking in female, but strongly

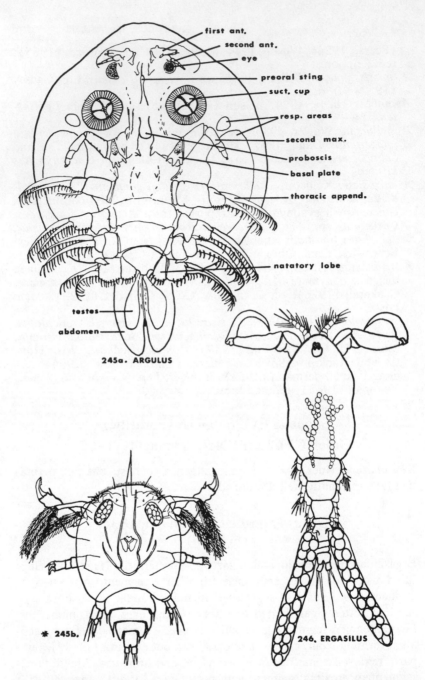

first ant.
second ant.
eye
preoral sting
suct. cup
resp. areas
second max.
proboscis
basal plate
thoracic append.
natatory lobe
testes
abdomen

245a. ARGULUS

＊ 245b.

246. ERGASILUS

Figs. 245–246. Fig. 245a. *Argulus japonicus*, male, ventral view (from Meehean, 1940). Fig. 245b. *Argulus lepidostei*, newly hatched larva (from Wilson, 1916). Fig. 246. *Ergasilus versicolor*, female (from Wilson, 1911).

developed (3-segmented) in male. Rami of first three pairs of legs usually 3-segmented; fourth exopod and endopod 1- and 3-segmented; Fifth leg uniramose, 1-segmented, rarely obsolete. Female parasitic on gills of freshwater and marine teleosts, male free-swimming. Life cycle: eggs laid in egg sacs, where embryonic development takes place and free-swimming nauplius hatches from egg. Larva passes through four copepodid stages, accompanied by molting. Copulation occurs while free-swimming, after which male dies. Female enters gill cavity where she is retained by gillrakers, then creeps to gills and attaches by clasper-like claws.

Ref. Gnadeberg (1948), biology; Markewitsch (1957), biology, development, and keys to spp.; Smith (1949), check list; Yamaguti (1963, p. 28).

Ergasilus auritus Markevich, 1940: Russia; Roberts (1963): *Gasterosteus aculeatus, Onchorhynchus nerka*, Brit. Col., redescription.

E. caeruleus Wilson, 1911 (Syn. *E. confusus* Bere, 1931; and *E. skrjabini* Mueller, 1936, according to Yamaguti, 1963). Reported from *Acrocheilus alutaceus, Ambloplites rupestris, Anguilla rostrata, Aphredoderus sayanus, Catostomus catostomus, C. commersoni, C. macrocheilus, Centrarchus macropterus, Chaenobryttus gulosus, Coregonus artedi, Couesius plumbeus, Erimyzon sucetta, Gila straria, Ictalurus* sp., *Lepomis cyanellus, L. gibbosus, L. humilis, L. macrochirus, Lota lota, Micropterus dolomieui, M. salmoides, Moxostoma rubreques, Mylocheilus caurinus, Noturus gyrinus, Oncorhynchus nerka, Perca flavescens, Percopsis omiscomaycus, Pomoxis* spp., *Ptychocheilus oregonensis, Rhinichthys osculus, Richardsonius balteatus, Stizostedion canadense, S. vitreum*, by Anthony (1963), Bangham (1933, 1940, 1941, 1944, 1951, 1955), Bangham and Adams (1954), Bangham and Venard (1942), Causey (1957), Fischthal (1947, 1950), Moore (1938), Mueller (1940), Pearse (1924b), Smith (1949): epizootic, Tidd (1931), Venard (1940, 1941), Wilson (1914): from Brit. Col., Fla., La., Mich., N. J., N. Y., Ohio, Ont., Tenn., Wis., Wyo. Reported also from epizootic of whitefish, W. Ont.

E. celestis Mueller, 1936 (see *E. versicolor*).

E. centrarchidarum Wright, 1882. Redescribed by Wilson (1932). Reported from *Ambloplites rupestris, Chaenobryttus gulosus, Erimyzon sucetta, Lepomis aureus, L. auritus, L. cyanellus, L. gibbosus, L. macrochirus, Microgadus tomcod, Micropterus dolumieui, M. salmoides, Osmerus mordax, Perca flavescens, Pomoxis annularis, Roccus chrysops, Stizostedion canadense, S. glaucum, S. vitreum*, by Bangham (1932, 39, 1941), Bangham and Venard (1942), Mueller, (1936), Tidd (1931), Wilson (1916), from Fla., Iowa, Ind., Mass., Mich., N. Bruns., N. Y., Ohio, Que., Tenn., Vt.

E. chautauquaensis Fellows, 1887: found free-swimming, L. Champlain; Wilson (1911): free-swimming, Wis.

E. confusus (see *E. caeruleus*).

E. cotti Kellicott, 1879: on *Etheostoma caeruleum*, L. Erie; Tidd (1931): *Cottus bardii*, Ohio.

E. elegans Wilson, 1916: on *Ictalurus melas, Lepisosteus osseus, L. platostomus*, Iowa; Bangham (1944): *Ambloplites rupestris*, Wis.; Bangham and Venard (1942): Tenn.; Mueller (1936, 1942): *Semotilus corporalis*, resembles *E. elegans*, N. Y.

E. elongatus Wilson, 1916: on *Polyodon spathula*, Iowa, Ill.; Causey (1957): same host, La.; Thomsen (1944): marine, Uruguay.

E. fragilis Mueller, 1936. Apparently a synonym of *E. elegans* or *E. megaceros*.

E. funduli Krøyer, 1863: on *Fundulus limbatus, E. heteroclitus*, La.; Wilson (1932): redescribed.

E. labracis Krøyer, 1863: on *Roccus lineatus*, Atlantic; Wilson (1911, 1932): redescription. Serious parasite in marine fisheries research; larvae sometimes so numerous that they nearly cover gills.

E. lanceolatus Wilson, 1916: on *Dorosoma cepedianum*, Ky.

E. lizae Krøyer, 1863: on marine fish including *Fundulus heteroclitus, F. similis*, Atlantic; Allison (1962): became established on *Lepomis macrochirus, L. microlophus, Micropterus salmoides* in fresh water and caused much damage.

E. luciopercarum Henderson, 1927: on *Stizostedion vitreum*, Que.

E. magnicornis Yin, 1949: on *Carassius auratus*, China; Yin (1956): redescribed.

E. manicatus Wilson, 1911: on *Gasterosteus bispinosus, Menidia notata*, Mass.; Bere (1936): *Gambusia holbrooki*, Fla.; Wilson 1932: review.

E. megaceros Wilson, 1916 (Syn. *E. fragilis* Mueller, 1936): in nasal fossae of *Ictalurus* sp., Iowa; Mueller (1936): *Semotilus corporalis*, N. Y. Also *Erimyzon sucetta, Notemigonus crysoleucas*, Fla.

E. nerkae Roberts, 1963: on *Oncorhynchus nerka*, Brit. Col., Wash.

E. nigratus Wilson, 1916: on *Micropterus salmoides*, Iowa.

E. osburni Tidd and Bangham, 1945: on *Lota lota*, Ont.

E. skrjabini Mueller, 1936 (see *E. caeruleus*).

E. turgidus Fraser, 1920: marine, on *Cymatogaster aggregatus*, Brit. Col.; Carl (1937): *Gasterosteus aculeatus*; Cope (1959): G. *aculeatus*; *Oncorhynchus nerka*, Alaska.

E. versicolor Wilson, 1911: on *Ictalurus natalis*, Ind. Reported from *Alosa chrysochloris, Catostomus commersoni, Chaenobryttus coronarius, Erimyzon sucetta, Ictalurus furcatus, I. melas, I. natalis, I. nebulosus, I. punctatus, Lepisosteus spatula, Lepomis cyanellus, L. macrochirus, L. megalotis, Micropterus punctulatus, Notropis cornutus, Pylodictis olivaris, Roccus chrysops, R. mississippiensis*, from Iowa, Ind., Kans., La., Mass., Mich., N. Y., Ohio, Okla., Tenn., Wis., by Bangham (1941, 1944, 1955); Bangham and Hunter (1939), Bangham and Venard (1942), Causey (1957), Harms (1959), McDaniel (1963), Mueller (1940), Sindermann (1953).

Ergasilus sp. Bangham (1944): on *Catostomus commersoni*, Wis.; Bangham (1951): *C. fecundus*, Wyo.; Bangham and Adams (1954): *Cottus asper, Prosopium williamsii*, Brit. Col.; Fischthal (1950): *C. commersoni*, Wis.; Fischthal (1956): *C. commersoni*,

Hypentelium nigricans, Semotilus corporalis, N. Y.; Haderlie (1953): *Lepomis macrochirus, Micropterus salmoides,* Calif.; Marshall and Gilbert (1905): *Perca flavescens,* Wis.; Meyer (1954): *Gasterosteus aculeatus, L. auritus, L. gibbosus, Micropterus dolomieui, Roccus americanus,* Me.

Order CALIGIDEA Stebbing, 1910

Key characteristics: Adult female permanently fixed to host; legs reduced. Body rigid in female only. Male of *Lernaea* and *Lerneocera* does not develop beyond fourth copepodid stage at which time it becomes sexually mature.

Genus **Lernaea** Linnaeus, 1746
(Fig. 247)
(Syn.: in part *Lernaeocera*)

Lernaeidae, Lernaeinae: Female: Head a rounded knob projecting from anterior margin of cephalothorax and placed nearly at right angles to body axis, with deeply buried tripartite eye near center of dorsal surface; one or two pairs of horns, simple or forked, on lateral margins of cephalothorax; sometimes an unpaired horn on center of dorsal margin, or paired horns arising from dorsal side of cephalothorax; all horns conical and soft, branched or not. Neck soft, slender, cylindrical, enlarged gradually into cylindrical trunk; trunk with 2-lobed or double pregenital prominence in front of vulvae; abdomen short, more or less distinctly 3-segmented, bluntly rounded, terminating in pair of small, segmented caudal rami; egg strings elongated conical or ovoid; eggs multiseriate. First antenna nearly cylindrical, 3- or 4-segmented. Second antenna 2- or 3-segmented, tipped with small, stout claw. Proboscis conical, very short. Mandible clawlike, without teeth. First maxilla nodular, tipped with minute conical chitinous projection. Second maxilla terminating in 2 stout claws. Maxilliped 3-segmented; second segment with minute thick-stalked spine and rounded conical protuberance on distal inner margin; terminal segment with 5 claws of different lengths. Four pairs of legs biramose, first pair just behind head, others at increasing distances posteriorly; a fifth pair of 1-segmented stumps just in front of vulvae. Parasitic on outside surface of freshwater fishes, boring into underlying tissues, occasionally on amphibians. Life cycle: eggs hatch in 1 to 3 days, releasing nauplius, a six-legged elliptical larva. Nauplius metamorphoses into first or second copepodid stage in 4 to 16 days. No further development unless proper host is supplied. Larva passes through 5

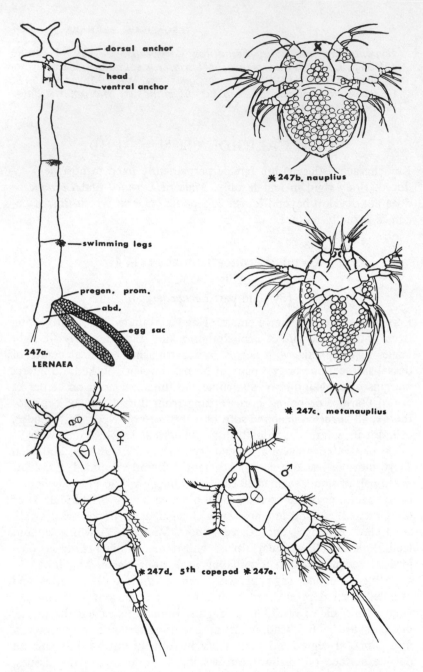

Fig. 247. *Lernaea cyprinacaea. a*, female (composite, from Harding, 1950, and Tidd, 1933); *b*, nauplius (from Grabda, 1963); *c*, metanauplius (from Grabda, 1963); *d, L. cyprinacaea*, 5th copepodid, female and male (from Grabda, 1963).

successive copepodid stages before female attaches. Copulation occurs during fourth copepodid stage and male disappears, presumably dying.

Ref. Gnanamuthu (1951), life cycle; Grabda (1958, 1963), life cycle; Harding (1950), key to species; Nakai (1927), life cycle; Wilson (1917); Yamaguti (1963, p. 176).

Lernaea anomala Wilson, 1917 (Syn. *L. insolens* Wilson, 1917): on *Micropterus salmoides*, N. C., Causey (1957): *M. punctulatus, M. salmoides*, La.

L. carassii Tidd, 1933 (see *L. cyprinacaea*).

L. catostomi (Krøyer, 1863), (Syn. *L. tortua* according to Harding, 1950): on *Moxostoma macrolepidotum*, Mo.; Delco (1962): *Notropis venustus, N. lutrensis, N. volucellus, Gambusia affinis, Lepomis macrochirus, L. punctatus*, Tex.; Dolley (1940): *Erimyzon sucetta, Notropis cornutus*, Miss.; Kellicott (1878): *Ictalurus*; Monod (1932); Tidd (1931): *N. deliciosus*, Ohio; Wilson (1917): *I. furcatus*, Iowa.

L. cruciata (Le Sueur, 1824), (Syn. *Lernaeocera c.*): on *Cichla aenea*, L. Erie; Causey (1957): *Ambloplites rupestris, Micropterus dolomieui, M. salmoides*, La.; Kellicott (1879): *A. rupestris*, Mich.; Monod (1932); Sindermann (1953): *Lepomis gibbosus, M. dolomieui, Roccus americana*, probably *Salmo trutta*, Mass.; Tidd (1931); Wilson (1916): *Moxostoma macrolepidotum, Stizostedion canadense*.

L. cyprinacaea Linnaeus, 1761 (Syn. *L. elegans* Leigh-Sharpe, 1925; *Lernaeocera esocina* Hermann, 1783; *L. carassii* Tidd, 1933; and probably *L. ranae* Stunkard and Cable, 1913). This dangerous parasite apparently so lacks host specificity that it can probably infect all freshwater fish and even frog tadpoles and salamanders. Reported from *Acrocheilus alutaceus, Amia calva, Archoplies interruptus, Carassius auratus, Catostomus* spp., *Cyprinus carpio, Erimyzon sucetta, Gambusia affinis, Hyborhynchus notatus, Ictalurus* spp., *Lavinia exilicauda, Lebidesthes reticulatus, Lepomis* spp., *Micropterus* spp., *Mylopharodon concocephalus, Notemigonus crysoleucas, Notropis* spp., *Oncorhynchus tshawytscha, Orthodon microlepidotus, Pimephales promelas, Pomoxis* spp., *Ptychogonimus oregonensis, Rhinichthys* spp., *Salmo* spp., *Salvelinus fontinalis, Tilapia* sp., *Umbra pygmaea*, also frog tadpoles and salamanders, by Baldauf (1961): Tex.; Camper (1962, pers. comm.): tadpoles, N. C.; Chiriac (1959): E. Europe; Enders and Rifenburg (1928): Ind.; Fryer (1961): Africa; Grabda (1958, 1963): life cycle, Poland; Griffith (1953): Wash.; Guidice (1950): Mo.; Haderlie (1953): Calif.; Haley and Winn (1959): Md.; Hugghins (1959): S. Dak.; Hundley (1957): N. Dak.; Kasahara (1962): biology and control with Dipterex, Japan; Lahav and Sarig (1964): biology; Lahav, Sarig, and Shilo (1964): eradication with Dipterex; McDaniel (1963): Okla.; McNeil (1961): Ariz.; Nakai (1927): life cycle, Japan; Nakai and Kokai (1931): no intermediate host, Japan; Okade (1927): Japan; Osborn (1962, pers. comm.): tadpole hosts, Mo.; Shields and Tidd (1963): tadpole hosts, Ohio;

Shilo, Sarig and Rosenberger (1960): treatment, Israel; Shpol-yanskaya (1953): life cycle, blood picture, Russia; Stolyarov (1936): life cycle and path.; Tidd (1933, 1934, 1962, 1965): exper., Ohio; Wilson (1917): life cycle; Yashouv (1959): host preference, Israel.

L. dolabrodes Wilson, 1917: on *Lepomis macrochirus*, Wis.

L. elegans (see *L. cyprinacaea*).

L. esocina (Burmeister, 1835), (Syn. *Lernaeocera branchialis*): on many fish, Europe; Grabda (1955, 1956): on trout, life cycle, Poland.

L. gasterostei (Bruhl, 1860), (Syn. of *L. esocina* according to Marke-witsch (1957); Grabda (1956): host specificity.

L. insolens Wilson, 1916 (see *L. anomala*).

L. laterobranchialis Fryer, 1959: on *Tilapia*, Africa.

L. lophiara Harding, 1950: on several fish, including *Tilapia*, Africa.

L. pectoralis Kellicott, 1882 (Syn. *Lernaeocera p.*): on *Notropis cornutus*, Mich.

L. pomotidis (Krøyer, 1863), (Syn. *Lernaeocera p.*): on *Lepomis macrochirus*, Iowa; Wilson (1916): same host and copepodid stage on *Ictalurus nebulosus*, La.

L. ranae Stunkard and Cable, 1931: on *Rana clamitans*; probable synonym of *L. cyprinacaea*.

L. tenuis Wilson, 1916: on *Aplodinotus grunniens*, Iowa; Causey (1957): same, La.; Wilson (1917).

L. tilapiae Harding, 1950: on *Tilapia* spp., Africa.

L. tortua Kellicott, 1882 (see *L. catostomi*).

L. variabilis (Wilson, 1916): on *Lepomis macrochirus*; larvae on gills, *Ictalurus melas, Lepisosteus platostomus, Stizostedion cana-dense*, Iowa.

Lernaea sp. Nigrelli (1943): killed fish in aquaria, lowered temper-ature for control; Woodbury (1934): on 50 percent of trout in Yellowstone Lake.

Order LERNEOPODIDEA Yamaguti, 1963

Copepoda: Body of both sexes usually rigid, showing no movable articulation and often no trace of segmentation, exceptionally with-out any distinction of body regions, with or without dorsal, lateral, or posterior processes. Sexual dimorphism universally present. Carapace often present, but never any paired dorsal plates as in Caligidea. Pro-boscis short and blunt, more often lacking; first antenna minute, with only a few segments; second antenna very small, uniramose or bira-mose, sometimes prehensile. First maxillipeds of female often modified into cylindrical arms united throughout their length or at tips to form pedicel of attachment bulla, but sometimes not prehensile or even ab-sent. In some genera, second maxillipeds migrate forward during de-velopment and in adult often lie in front of first maxillipeds. Swimming legs more often lacking in female, but one pair or two usually present in male. Egg sacs claviform, spherical or cylindrical; eggs multiseriate and pressed so tightly together that they are flattened into polygons; nauplius and often metanauplius stages passed within egg, larva emerging in first copepodid stage and attaching itself at once to host by means of frontal filament. Female becomes fixed parasite attached immovably to host; male is dwarfed, clings to female, and can crawl about over her body.

Genus **Achtheres** Nordmann, 1832
(Fig. 248)

Lerneopodidea: Female: Body divided into three regions, cephalothorax, trunk, and abdomen. Cephalothorax with partial dorsal carapace including antennal area; trunk plump, may be more or less distinctly segmented, at least on ventral surface, narrowed into well-defined neck anteriorly, where often one segment is clearly separated; abdomen conical, usually unsegmented, but sometimes segmented (*A. micropteri*). No posterior processes, caudal rami, or genital process. Egg cases short and stout; eggs large, multiseriate. First antenna distinctly 3-segmented, tipped with 3 setae. Second antenna biramose, turned down across frontal margin; endopod 1-segmented, exopod 2-segmented, maxilla with 2 or 3 terminal setae, with or without palp. First maxillipeds separate to very tip, where they are joined to ordinary bulla. Second maxillipeds prehensile, between bases of first maxillipeds. Male: medium-sized (1 mm); cephalothorax in line with body axis, much smaller than trunk and separated from it by a distinct groove; trunk spindle-shaped, clearly segmented. No abdomen or caudal rami. First antenna 3-segmented; second antenna biramose; exopod 2-segmented, tipped with claw. Maxilla with 2 terminal setae, without palp. First maxilliped much longer than second, both projecting strongly from cephalothorax some distance behind other mouth parts. Parasitic on inner surface of gill arches of freshwater teleosts. Life cycle: nauplii and metanauplii develop within egg. Larva hatches into first copepodid stage, which is free-swimming. Second copepodid stage attaches to fish by attachment filament. After mating, male clings to body of female. Distal end of larval filament develops into attachment bulla. Second copepodid stage becomes adult (Wilson, 1911; Zandt, 1935).

Achtheres ambloplitis Kellicott, 1880: on *Ambloplites rupestris*; Bangham (1955): same host, Mich.; Hugghins (1959): *Ictalurus melas*, S. Dak.; Pearse (1924): *Coregonus harengus, Lota maculosa, Micropterus dolomieui*, Wis.; Tidd (1931): Ohio; Wilson

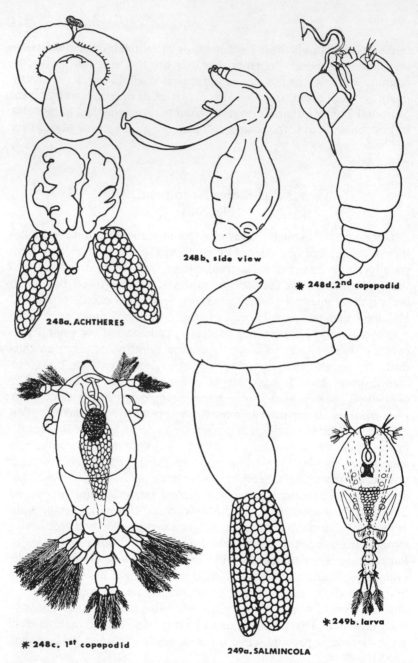

248b. side view

248d. 2nd copepodid

248a. ACHTHERES

※ 248c. 1st copepodid

249a. SALMINCOLA

※ 249b. larva

Figs. 248–249. Fig. 248. *Achtheres ambloplitis. a*, adult, dorsal view (from Wilson, 1915); *b*, adult, side view (from Wilson, 1915); *c*, first copepodid (from Wilson, 1911); *d*, second copepod stage (from Wilson, 1911). Fig. 249. *Salmincola edwardsii. a*, adult (from Wilson, 1915); *b*, free-swimming larva (from Fasten, 1912).

(1911): development; Wilson (1915): redescription; Wilson (1916, 1922, 1924): additional hosts.

A. coregoni (Smith, 1874) Wilson, 1915. Bangham (1951): on *Prosopium williamsi*, Wyo.; Bangham (1955): *P. cylindraceum*, Mich.; Bere (1931, 1935): on cisco, Wis.; Pearse (1924): *Salvelinus namaycush*, Wis.; Tidd (1931): *C. clupeaformis*, Ohio; Warren (1952): cisco, Minn.; Wilson (1908, 1916).

A. corpulentus Kellicott, 1882: on *Coregonus artedi*, *C. clupeaformis*, *C. hoyi*, *C. prognathus*; Bangham (1955): L. Huron; Mueller (1940): L. Ontario; Pearse (1924): *C. johannae*, Wis.; Wilson (1915): redescription.

A. lacae Krøyer, 1863: on *Perca*; Causey (1957): *Ictalurus furcatus*, La.; Wilson (1915): *Roccus sexatilis*, Potomac R.

A. micropteri Wright, 1882. Reported from *Ambloplites rupestris*, *Chaenobryttus gulosus*, *Ictalurus punctatus*, *Lepomis macrochirus*, *Micropterus dolomieui*, *M. salmoides*, by Anthony (1963): Wis.; Bangham (1939, 1941, 1944, 1955): Fla., Ohio, L. Huron, Wis.; Bangham and Hunter (1939): L. Erie; Bangham and Venard (1942): Tenn.; Bere (1931, 1935): Wis.; Causey (1957): La.; Fischthal (1947, 1950): Wis.; Pearse (1924): Wis.; Tidd (1931): L. Erie; Venard (1940): Tenn.; Wilson (1915): redescription, Ind., N. Y.

A. pimelodi Krøyer, 1863. Reported from *Ictalurus anguilla, I. melas, I. natalis, I. nebulosus, I. punctatus, Pylodictis olivaris*, by Bangham and Venard (1942): Tenn.; Harms (1959): Kans.; Hugghins (1959): S. Dak.; Pearse (1924): Wis.; Tidd (1931): L. Erie: Wilson (1915): redescription, Iowa.

Genus **Salmincola** Wilson, 1915
(Fig. 249)
(Syn.; in part—*Lernaeopoda, Basanistes*)

Lerneopodidea: Female: Cephalothorax short, stout, inclined at angle to body axis; separated from trunk by groove, but no definite waist; no distinct dorsal carapace. Trunk short and stout, often flattened dorsoventrally, with no signs of segmentation. No abdomen, caudal rami, or posterior processes. Small transparent genital process present in young female and often in adult. Egg strings usually long and slender; egg small, multiseriate. First antenna 2- or 3-segmented, usually showing no segmentation. Second antenna biramose, both rami 1-segmented, endopod larger than exopod. Maxilla with 3 or 4 spines, with or without a palp. First maxillipeds long or short, joined at tip by button or mushroom-shaped bulla, often joined also at base around back of thorax, forming pair of shoulders. Second maxilliped with stout basal segment and slender terminal claw. Male: small size (0.5 mm); cephalothorax about same length as trunk, the two bent into semicircle; no dorsal carapace. Trunk a little stouter than cephalothorax and indistinctly segmented, with pair of short caudal rami curved dorsally. No abdomen. First antenna 3-segmented; second antenna biramose, exopod (ventral ramus) uncinate; first and second maxillipeds 3-segmented; latter stouter than former, attached inside bases of former. Parasitic on freshwater teleosts. Life cycle (of *S. edwardsi*): copepod hatches into small free-swimming larva which may exist 2 days. Mouth parts bear a peculiar filament for attachment to fish. Larva forces filament into tissue of fish and attaches second maxillae to filament which becomes bulla, thus attaching itself permanently to fish. Entire animal undergoes degeneration, becoming grublike parasite. Male much smaller that female. Copulation occurs 2½ to 3 weeks after attachment; male releases hold on gill and attaches to female. After fertilization, male dies. Each female gives rise to 2 batches of embryonated eggs, after which she dies. Entire life cycle takes about 2½ months (Fasten, 1912). Also see Dedie (1940); Friend (1941); Savage (1935).

Salmincola beani (Wilson, 1908): on *Oncorhynchus tshawytscha, Salmo gairdneri,* Calif., Colo.; Fasten (1921): male; Shaw *et al.* (1934); Wilson (1915): redescription.

S. bicauliculata (Wilson, 1908): on *Salvelinus malma, Salmo mykiss,* Bering Island, Kamchatka, Russia; Wilson (1915): from trout, Ore.; Wilson (1920): *S. malma,* Siberia.

S. californiensis (Dana, 1853). Wilson (1915): redescription, Calif., Idaho.

S. carpenteri (Packard, 1874). Simon (1943): on *Salmo clarki;* Wilson (1915, 1916): redesription, *O. nerka, S. mykiss,* Brit. Col., Colo.

S. edwardsi (Olsson, 1869), (Syn. *Lernaeopoda e.* Olsson; *Basanistes salmonea* Edwards, 1840; *Lernaeopoda fontinalis* Smith, 1874). Primarily on *Salvelinus fontinalis* but also reported from *Oncorhynchus nerka, Salmo clarki, S. gairdneri, S. palja, Salvelinus alpinus,* by Alexander (1960): Ore.; Allison, L. (1950): treatment; Bangham and Adams (1954): Brit. Col.; Barysheva and Bauer (1958): Russia; Bere (1929): N. Bruns.; Davis (1953); DeRoth (1953): Me.; Fasten (1912, 1913, 1916, 1918, 1921): development, biology, treatment, etc., Mich.; Haderlie (1953): Calif.; Hunter and Hunter (1934); MacLulich (1943): Ont.; Meyer, M. (1954): Me.; Mueller (1937); Ricker (1932); Savage (1935): life cycle; Wilson (1915): redescription.

S. extumescens (Gadd, 1901), (Syn. *Lernaeopoda inermis* Wilson, 1911; *S. omuli* Messjatzeff, 1926): N. Europe; Wilson (1908): on *Coregonus nelsonii,* Alaska.

S. falculata (Wilson, 1908): on *Oncorhynchus nerka, Salvelinus, Salmo,* Wash.; Adams (1956): attached to heart of *O. nerka;* Wilson (1920): *S. malma,* N. W. Terr.

S. gibbera (Wilson, 1908): on *Salpina alipes, Salvelinus malma,* Alaska, N. W. Terr.; Wilson (1920): same.

S. gordoni (Gurney, 1933): on *Salmo trutta, Thymallus*; Friend (1938).

S. heintzi (Neresheimer, 1909): on *Salmo salvelinus* (?), Europe.

S. inermis (Wilson, 1911): on *Coregonus harengus*, Can.; Bangham (1955): *C. artedi*, L. Huron; Markewitsch (1957): syn. of *S. extumescens*; Pritchard (1931): *C. hoyi, C. kiyi, C. reighardi*; Tidd (1931): L. Erie; Warren (1952): *C. artedi*, Minn.; Wilson (1911); Wilson (1915): redescription; Wilson (1920): *C. nelsonii*, Alaska; Wright (1894): L. Ontario.

S. oquassa Wilson, 1915: on *Salvelinus oquassa*, Me.

S. salvelini Richardson, 1938: on *Salvelinus fontinalis, S. alpinus*, Labrador.

S. siscowet (Smith, 1874): on *Salvelinus namaycush, Salvelinus* spp., Can.; MacLulich (1943): *S. namaycush*, Ont.; Wilson (1915): *S. alpinus, S. malma, S. namaycush*, redescription, N. W. Terr., Greenland.

S. wisconsinensis Tidd and Bangham, 1945: on *Coregonus artedi*, Wis.; Bangham (1955): *C. artedi, C. hoyi*, L. Huron.

Salmincola sp. Bangham (1951): on *Salmo clarki, Salvelinus namaycush*, Wyo.; Bangham (1955): *Prosopium cylindraceum*, L. Huron; Bangham and Adams (1954): *Cottus asper, Gasterosteus aculeatus, Oncorhynchus nerka, Prosopium williamsii, S. namaycush*, Brit. Col.; Cope (1958): *S. clarki*, Wyo.; Nieland (1952): *S. clarki*, Ore.; Rucker (1957): necessity of discarding heavily infected broodstock; Wales (1964, pers. comm.): Calif. brown trout uninfected, but imported brown trout became infected.

Genus **Cauloxenus** Cope, 1871
(No fig.)

Lernaeopodidea: Cephalothorax stout, elongate, separated from trunk by deep constriction like waist of wasp. Trunk stout, saclike, unsegmented, ovoid; no posterior processes, genital process, abdomen, or caudal rami. Egg sacs ovoid, shorter than trunk. First maxillipeds elongate, attached at about center of cephalothorax, folded back against head, fused through entire length, tapering toward distal end, where they have a broad, disklike bulla. Male unknown. Parasitic on freshwater teleosts.

Cauloxenus stygins Cope, 1872: on upper lip of cave fish, *Amblyopsis spelaea*, Wyandotte Cave, Ind.; original description not illustrated.

Miscellaneous Parasites

The following have rarely been encountered in fish parasite studies.

1. MITES (Arthropoda: Arachnoidea: Acarina)
(Fig. 250)

The larvae of several genera have been reported from the skin, gills, and esophagus of Russian fish (Dubinna, 1962). In North America mites have been noted by Mueller (1936) on the gills of *Amia calva*, Fla. Encysted mites were found in the esophagus of *Lepomis macrochirus* at this laboratory in 1965, but were lost before identification could be made.

2. GLOCHIDIA (Mollusca: Pelecypoda)

The larvae of most, but not all, freshwater clams must go through a parasitic stage on the gills or fins of fish (Coker *et al.*, 1921). The larvae become encysted in the epithelium and grossly resemble metacercarial cysts. The cyst, however, contains a larval clam, which can be recognized because of its thin bivalve shell, which usually has little hooks on its inner edge. There are many species; some are found on only one host, whereas others have little host specificity. If numerous, particularly on the gills, the fish may succumb. Davis (1953) has summarized the subject.

Reported from many species of fish by Arey (1922): immunity; Bangham (1941, 1944, 1951, 1955): Ohio, Wis., Wyo., Mich.; Bangham and Adams (1954): Brit. Col.; Cope (1959): Alaska; Fischthal (1947, 1950, 1956): Wis., N. Y.; Jones (1950), propagation; Lefevre and Curtis (1912); Locke (1963); problem in Maine hatchery; Meyer (1954): Me.; Mueller (1940); Murphy (1942): damaging to trout, Calif.; Pearse (1924): Wis.; Sindermann (1953): Mass.; Wilson (1916): interrelationship with parasitic copepods.

3. SPHAERID CLAMS (Mollusca: Pelecypoda)

Adult sphaerid clams have been found attached to the mouths of small fish (Carbine, 1942), young pike, and (Tanner, 1954) trout fry.

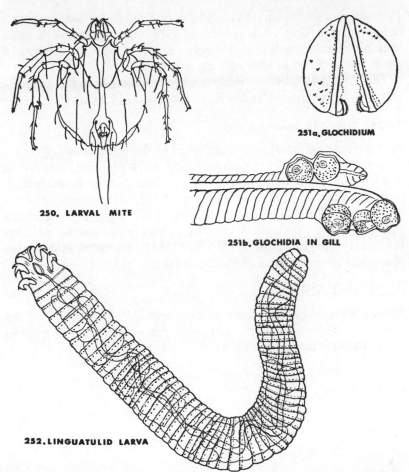

251a. GLOCHIDIUM

250. LARVAL MITE

251b. GLOCHIDIA IN GILL

252. LINGUATULID LARVA

Figs. 250–252. Fig. 250. *Unionicola crassipes,* larval mite (from Mitchell, 1955, in Bykhovskaya-Pavlovskaya *et al.,* 1962). Fig. 251. Glochidia. *a,* glochidium of *Unio* (from Zhadin, 1938, in Bykhovskaya-Pavlovskaya *et al.,* 1962); *b,* gill filaments of rock bass infected with mussel glochidia (from Lefevre and Curtis, 1912, in Davis, 1953). Fig. 252. *Sebekia oxycephala,* linguatulid larva from bullhead and sunfish (from Holl, 1928).

4. LINGUATULID LARVAE (Arthropoda: Pentastomida)

(Fig. 252)

The larva of *Sebekia oxycephala* has been found in the viscera of *Ictalurus natalis* and *Lepomis gibbosus* in North Carolina (Holl, 1928), and in *Amia calva, Chaenobryttus gulosus, Gambusia affinis, Lepomis* spp., *Pomoxis sparoides,* and *Sclerotis punctatus* in Florida

(Venard and Bangham, 1941). The larvae have also been reported from snakes, a lizard, and crocodiles (Heymons, 1935). The adult lives in the lungs, trachea, and pharynx of crocodiles and alligators of South America and southern North America.

The larva of *Leiperia cincinalis* has been reported from the musculature of *Tilapia nilotica* (Southwell and Pillers, 1929).

5. HYDRA (Coelenterata):

Recently hatched larvae of fish are killed by the nematocysts of *Hydra* if present in large numbers. Gudger (1927, 1934) reported that *Hydra* can kill small fish. Cordero (1941) reported damage to the mouth region of a teleost by *Hydra*. Schuberg (1905) demonstrated the nematocysts in the epithelium of recently dead trout, and Beardsley (1904) noted the destruction of trout fry held in the anterior portions of hatchery troughs. Eisler and Simon (1961) demonstrated the death of recently hatched larvae of Pacific salmon.

6. PLANT SEEDS

Small salmon and shiners may become impaled on the barbs of the achenes of *Bidens* sp., an aquatic plant, when attempting to feed on them (Reimers and Bond, 1966).

Fish Parasites Found on or in Other Animals

In addition to some trematodes, cestodes, and nematodes which occur as larvae in fish and adults in other animals, a few parasites parasitize fish and other animals. Some of these probably represent accidental parasitism. The following fish parasites have been reported from the following animals:

Amphipod: *Argulus americanus.*

Salamanders: *Trichodina fultoni, Crepidostomum cooperi, C. cornutum,* *Diplostomulum scheuringi* (?), *Microphallus opacus, Phyllodistomum americanum,* *Gnathostoma spinigerum, Spinitectus carolini.*

Frogs: *Crepidostomum cornutum, Deropegus* sp., *Deropristis inflata,* *Macroderoides spinifera, Microphallus opacus, Pleurogenes,* *Gnathostoma spinigerum, Placobdella montifera, Argulus, Lernaea cyprinacea.*

Snakes: *Crepidostomum, Microphallus opacus,* *Gnathostoma spinigerum, Pomphorhynchus bulbocolli,* *Sebekia oxycephala.*

Turtles: *Crepidostomum cooperi, Microphallus opacus.*

Mammals: *Microphallus opacus* in *Didelphys* and *Procyon.*

* Larval forms.
(?) species identification questionable.

Predators

Natural vertebrate predators of fish include man, mink, otter, bears, herons, kingfishers, osprey, mergansers, loons, ducks on occasion, certain turtles, water snakes, and salamanders. With rare exception, there is no need to attempt to control these animals because of fish predation. In fact, certain predators may be beneficial in removing trash, diseased, and stunted fish.

Invertebrate predators are sometimes very destructive to small fish. The adults and larvae of most Adephaga water beetles (Amphizoidae, Dytiscidae, Gyrinidae, and Noteridae) are predacious; many of these must come to the surface at intervals for air and can be killed with surface toxicants. Of the Polyphage beetles, only larvae of some of the Hydrophilidae are predacious; the adults are vegetarians (Leech and Sanderson, 1959; Pennak, 1953).

Of the hemiptera, the following are predacious and will eat small fish: Notonectidae (back swimmers) and Belostomatidae (giant water bugs) (Pennak, 1953).

Since many of the above-mentioned invertebrates must surface for air, various oils on the surface have been used successfully to control these predators in fish ponds. The oils, as recommended, do not endanger fish. One and a half gallons of kerosene or vegetable oil per acre usually suffice for ponds; apply when the water is relatively calm. Other combinations of diesel fuel, cottonseed oil, and crankcase oil have been used. Insecticides have been used for insect predators which do not surface for air.

Fish and Parasite Check-List*

Contents

Acipenseridae	323	Coregonidae (see Salmonidae)	
Acipenser	323	Cottidae	346
Amiidae	324	*Cottus*	346
Amia	324	Cyprinidae	347
Anabantidae	324	*Acrocheilus*	348
Anguillidae	324	*Apocope*	348
Anguilla	324	*Campostoma*	348
Aphredoderidae	325	*Carassius*	348
Aphredoderus	325	*Chrosomus*	350
Atherinidae	325	*Clinostomus*	350
Labidesthes	325	*Ctenopharyngodon*	350
Menidia	326	*Cyprinus*	350
Catostomidae	326	*Ericymba*	353
Carpiodes	326	*Exoglossum*	353
Catostomus	326	*Gila*	353
Erimyzon	330	*Hesperoleucus*	353
Hypentelium	330	*Hybognathus*	353
Ictiobus	331	*Hybopsis*	354
Minytrema	332	*Lavinia*	354
Moxostoma	332	*Leucosomus* (see *Semotilus*)	
Pantosteus	332	*Mylocheilus*	355
Centrarchidae	333	*Mylopharodon*	355
Ambloplites	333	*Notemigonus*	355
Archoplites	334	*Notropis*	355
Centrarchus	334	*Opsopoedus*	358
Chaenobryttus	334	*Orthodon*	358
Elassoma	335	*Parexoglossum*	358
Enneacanthus	335	*Pimephales*	358
Lepomis	335	*Ptychocheilus*	359
Micropterus	340	*Rhinichthys*	360
Pomoxis	343	*Richardsonius*	361
Sclerotis	345	*Semotilus*	361
Cichlidae	345	*Siphateles*	362
Tilapia	345	*Tinca*	362
Clupeidae	345	Cyprinodontidae	363
Alosa	345	*Belonesox*	363
Clupea	346	*Cyprinodon*	363
Dorosoma	346	*Fundulus*	364

* Wherever possible, the fish names in Amer. Fish. Soc. Spec. Publ. No. 2 were used.

Jordanella	365	*Perca*	382
Lucania	365	*Percina*	384
Zygonectes	365	*Stizostedion*	385
Dallidae (see Umbridae)		Percopsidae	386
Dorosomidae (see Clupeidae)		*Percopsis*	386
Esocidae	366	Petromyzonidae	386
Esox	366	*Lampetra*	386
Etheostomidae (see Percidae)		*Petromyzon*	386
Gadidae	369	Poeciliidae	387
Lota	369	*Gambusia*	387
Microgadus	371	*Heterandria*	387
Gasterosteidae	371	*Mollienesia*	387
Eucalia	371	Polyodontidae	388
Gasterosteus	371	*Polyodon*	388
Pungitius	372	Salmonidae	388
Hiodontidae	373	*Coregonus*	389
Hiodon	373	*Oncorhynchus*	390
Ictaluridae	374	*Prosopium*	393
Ictalurus	374	*Salmo*	394
Noturus	378	*Salvelinus*	399
Pylodictis	379	*Stenodus*	402
Lepisosteidae	379	*Thymallus*	403
Lepisosteus	379	Sciaenidae	403
Mugilidae	380	*Aplodinotus*	403
Osmeridae	380	Serranidae	404
Osmerus	380	*Roccus*	404
Percidae	380	Thymallidae (see Salmonidae)	
Cottogaster	380	Umbridae	405
Etheostoma	380	*Dallia*	405
Hadropterus	382	*Umbra*	405

The fish are listed alphabetically by family and genus for convenience in indexing. The lampreys (Petromyzonidae), although cyclostomes, not fish, are included alphabetically in the fish list. The parasites are listed in their separate categories—Protozoa, Trematoda, etc., following each fish species. Larval forms are denoted by an asterisk and immature adults by a double asterisk, following the system used by R. V. Bangham in his many publications.

Foreign parasites are included for those fishes found on other continents and North America; this includes naturally occurring, as well as transplanted fish species. Foreign records are indicated by (E) = Europe, (Eng) = England, and (R) = Russia; other countries are spelled out. Carp, goldfish, brown trout, grass carp, *Tinca* and *Tilapia* have been introduced to North America and at least some of their parasites have become established. Also, some North American parasites have become established in them. The same holds true for North American fishes which were established in Europe.

Some of the parasites recorded from some fish are probably not primary parasites of those fish; they may have been "accidental" parasites and, in some cases, perhaps erroneous records.

ACIPENSERIDAE (Sturgeons)
Acipenser fulvescens, lake
sturgeon
Trematoda:
  Allocreadium sp.
  Crepidostomum lintoni
  Diclybothrium armatum
  D. hamulatum
  Diplostomulum sp.
  Skrjabinopsolus manteri
Nematoda:
  Cucullanus clitellaris
  Rhabdochona cascadilla

Acanthocephala:
  Echinorhynchus salmonis
  Tanaorhamphus
  ambiguus
Crustacea:
  Argulus canadensis
  A. stizostethi
Acipenser transmontanus,
white sturgeon
Trematoda:
  Nitzschia quadritestes
Nematoda:
  Cystoopsis acipenseri

* Larval forms.

323

Acipenser sp.
Cestoda
Amphilina bipunctata
Scaphirhynchus platorhynchus,
shovelnose sturgeon
Trematoda:
Crepidostomum lintoni
Pristotrema manteri
Nematoda:
Ascaris scaphirhynchi
Leech:
Placobdella montifera
AMIIDAE
Amia calva, bowfin
Trematoda:
Apophallus venustus
Azygia acuminata
A. angusticauda
A. longa
A. tereticolle
Crepidostomum
cornutum
**Crepidostomum* sp.
Diplostomulum sp.
Echinochasmus
donaldsoni
Echinostome (gills)
Leuceruthrus
micropteri
Macroderoides flavus
M. parvus
M. typicus
Microphallus opacus
Posthodiplostomum
minimum
Cestoda:
***Bothriocephalus* sp.
***Haplobothrium*
globuliforme
Proteocephalus
ambloplitis
**P. ambloplitis*
***P. pearsei*
***P. perplexus*
Proteocephalus sp.
***Triaenophorus*
nodulosus

Nematoda:
Camallanus oxycephalus
***Camallanus* sp.
***Contracaecum* sp.
Haplonema immutatum
Spinitectus carolini
Spiroxys sp.
Acanthocephala:
Leptorhynchoides
thecatus
L. thecatus
Neoechinorhynchus
cylindratus
Pomphorhynchus
bulbocolli
Leech:
Illinobdella sp.
Crustacea:
Argulus americanus
A. flavescens
Linguatula
Sebekia oxycephala
ANABANTIDAE (Tropical
Fish)
Helostoma rudolphi, kissing
gourami
Trematoda:
Trianchoratus
acleithrium
ANGUILLIDAE (Eels)
Anguilla rostrata, freshwater
eel
Protozoa:
Cryptobia
markewitchi (R)
Eimeria anguillae (R)
Myxidium giardi (R)
Myxidium illinoisense
Trichodina anguillae (R)
Trichodinella
epizootica (R)
Trematoda:
***Azygia acuminata*
(exper.)
A. longa
A. sebago
Centrovarium lobotes

* Larval forms.
** Immature adults.

Crepidostomum
  brevivitellum
C. cornutum
Deropristis inflata
Gyrodactylus
  anguillae (R)
Microphallus opacus
Plagioporus
  angulatus (R)
Podocotyle atomon (R)
**Posthodiplostomum*
  minimum
Sphaerostoma
  bramae (R)
Cestoda:
  Bothriocephalus
    claviceps
  Bothriocephalus sp.
  *Diphyllobothrium
    latum (R)
  Proteocephalus
    macrocephalus
  *Proteocephalus sp.
  *Triaenophorus
    nodulosus (R)
Nematoda:
  Camallanus lacustris (R)
  C. truncatus (R)
  *Contracaecum
    spiculigerum
  **C. squalii* (R)
  Haplonema aditum
  Ichthyobronema
    gnedini (R)
  Raphidascaris acus (R)
  Spinitectus inermis (R)
Acanthocephala:
  Acanthocephalus
    anguillae (R)
  ***A. jacksoni*
  A. lucii (R)
  Corynosoma
    semerme (R)
  Echinorhynchus
    clavaeceps (?)
  E. coregoni
  E. globulosus (?)

Leptorhynchoides
  thecatus
**L. thecatus*
Metechinorhynchus (R)
Neoechinorhynchus
  agilis (?)
N. cylindratus
N. rutili (R)
Pomphorhynchus
  laevis (R)
Pseudoechinorhynchus
  clavula (R)
Tanaorhamphus
  ambiguus
Crustacea:
  Argulus foliaceus (R)
  A. laticauda
  Ergasilus caeruleus
  E. gibbus (R)
  Lernaea cyprinacea
APHREDODERIDAE
Aphredoderus sayanus, pirate
perch
Protozoa:
  Henneguya monura
Trematoda:
  Crepidostomum
    isostomum
  **Crepidostomum sp.
  Gyrodactylus
    aphredoderi
  Phyllodistomum pearsei
  **Posthodiplostomum*
    minimum
Cestoda:
  *Haplobothrium
    globuliforme
  *Proteocephalus
    ambloplitis
  Proteocephalus sp.
Crustacea:
  Ergasilus caeruleus
ATHERINIDAE
Labidesthes sicculus, brook
silversides
Trematoda:
  Allacanthochasmus
    varius

* Larval forms.
** Immature adults.

Creptotrama funduli
*Diplostomulum sp.
Microphallus opacus
*Posthodiplostomum
minimum
Nematoda:
*Contracaecum
spiculigerum
Menidia beryllina, tidewater
silversides
Trematoda:
*Bucephaloides
strongylurae
*Bucephalus sp.
*Posthodiplostomum
minimum
Cestoda:
*Proteocephalus sp.
Nematoda:
*Contracaecum
spiculigerum
Acanthocephala:
*Neoechinorhynchus
cylindratus
CATOSTOMIDAE (Suckers)
Carpiodes carpio, river
carpsucker
Protozoa:
Amphileptus voracus
Trichodina myakkae
Cestoda:
Biacetabulum
meridianum
Capingens singularis
Hypocaryophyllaeus
paratarius
Monobothrium sp.
*Proteocephalus sp.
Spartoides wardi
Nematoda:
Camallanus oxycephalus
Rhabdochona cascadilla
Acanthocephala:
Leptorhynchoides
thecatus
Neoechinorhynchus
cylindratus
N. prolixus

Pomphorhynchus
bulbocolli
Carpiodes cyprinus, quillback
Protozoa:
Myxosoma rotundum
Cestoda:
Hypocaryophyllaeus
paratarius
Nematoda:
Rhabdochona cascadilla
Carpiodes difformis
Protozoa:
Myxobolus discrepans
(gills)
Trematoda:
Triganodistomum
garricki
Carpiodes thompsoni, lake
carpsucker
Cestoda:
Spartoides wardi
Nematoda:
Camallanus
ancylodirus
Carpiodes velifer, highfin
carpsucker
Protozoa:
Myxobolus obliquus
Cestoda:
Hypocaryophyllaeus
paratarius
Carpiodes sp.
Cestoda:
Glaridacris catostomi
Catostomus ardens, Utah
sucker
Cestoda:
Caryophyllaeus
terbrans
*Ligula intestinalis
Monobothrium terebrans
Acanthocephala:
Echinorhynchus
tuberosus
Catostomus catostomus,
longnose sucker
Trematoda:
Allocreadium commune

* Larval forms.

A. lobatum
*Bolbophorus confusus
*Diplostomum
   flexicaudum
Octomacrum lanceatum
Plagiocirrus sp.
*Posthodiplostomum
   minimum
Triganodistomum
   attenuatum
T. simeri
Triganodistomum sp.
Cestoda:
   Caryophyllaeus terebrans
   Eubothrium sp.
   Glaridacris catostomi
   *Ligula intestinalis
   *Schistocephalus sp.
   *Triaenophorus
      nodulosus
   *Triaenophorus sp.
Nematoda:
   Bulbodacnitis scotti
   Hepaticola bakeri
   Rhabdochona cascadilla
**Rhabdochona sp.
Acanthocephala:
   Echinorhynchus coregoni
   E. leidyi
   E. salmonis
   Leptorhynchoides
      thecatus
   Neoechinorhynchus
      crassus
   N. cristatus
   N. rutili
   N. strigosus
   Octospinifer maculentus
Leech:
   Actinobdella triannulata
Crustacea:
   Ergasilus caeruleus
Catostomus columbianus,
bridgelip sucker
Trematoda:
   *Ornithodiplostomum
      ptychocheilus

Cestoda:
   Monobothrium sp.
Acanthocephala:
   Neoechinorhynchus
      crassus
   N. cristatus
   Neoechinorhynchus sp.
   Pomphorhynchus
      bulbocolli
Crustacea:
   Lernaea cyprinacea
Catostomus commersoni,
white sucker
Protozoa:
   Chloromyxum catostomi
   Chloromyxum sp.
   Cryptobia borreli
   Myxobolus catostomi
   M. subcularis
   Myxosoma bibullatum
   M. catostomi
   M. commersoni
   M. ellipticoides
   Sphaerospora notropis
   Thelohanellus notatus
Trematoda:
   Acolpenteron catostomi
   Allocreadium ictaluri
   A. lobatum
   *Amphimerus
      pseudofelineus
   *Amphimerus sp.
   Anonchohaptor
      anomalum
   *Apophallus venustus
   *Bolbophorus confusus
   *Bucephalus elegans
   *Clinostomum
      marginatum
   Crepidostomum cooperi
   *Diplostomulum corti
   *D. flexicaudum
   *Diplostomulum sp.
   Gyrodactylus spathulatus
   *Metorchis conjunctus
   Murraytrema copulata
   *Neascus sp.

* Larval forms.
** Immature adults.

Octomacrum lanceatum
Phyllodistomum
  etheostomae
P. lysteri
P. superbum
Plagioporus sinitsini
Sanguinicola sp.
*Tetracotyle communis
*Tetracotyle sp.
Triganodistomum
  attenuatum
T. simeri
Cestoda:
  Biacetabulum infrequens
  B. macrocephalum
  Caryophyllaeus terebrans
  Glaridacris catostomi
  G. confusus
  G. intermedius
  G. laruei
  G. oligorchis
  Glaridacris sp.
  Hunterella nodulosa
  *Ligula intestinalis
  Monobothrium hunteri
  M. ingens
  *Proteocephalus sp.
  **Proteocephalus sp.
  *Triaenophorus nodulosus
Nematoda:
  **Camallanus oxycephalus
  Capillaria catenata
  C. catostomi
  **Contracaecum sp.
  Dichylene cotylophora
  *Eustrongylides sp.
  Hepaticola bakeri
  Philometra nodulosa
  *P. nodulosa
  *Philometra sp.
  **Rhabdochona cascadilla
  R. laurentiana (syn. for
    R. cascadilla?)
  Rhabdochona sp.
  *Spiroxys sp.
Acanthocephala:
  Acanthocephalus jacksoni

Echinorhynchus lateralis
E. leidyi
E. salmonis
Fessisentis sp.
Leptorhynchoides
  thecatus
Neoechinorhynchus
  crassus
N. cylindratus
N. prolixoides
N. strigosus
Neoechinorhynchus sp.
Octospinifer macilentus
Pomphorhynchus
  bulbocolli
Leech:
  Actinobdella triannulata
  Illinobdella sp.
  Piscicola punctata
  Placobdella pediculata
Mollusca:
  Glochidia
Crustacea:
  Argulus appendiculosus
  A. biramosus
  A. catostomi
  Ergasilus caeruleus
  E. versicolor
  Ergasilus sp.
  Lernaea cyprinacea

Catostomus fecundus, webug
sucker
Protozoa:
  Ichthyophthirius multifilis
Trematoda:
  Allocreadium lobatum
  *Clinostomum marginatum
  *Diplostomulum
    flexicaudum
  Murraytrema copulata
  *Neascus sp.
  *Posthodiplostomum
    minimum
  *Tetracotyle sp.
  Triganodistomum sp.
Cestoda:
  Caryophyllaeus terebrans

* Larval forms.
** Immature adults.

*Ligula intestinalis
Nematoda:
  *Contracecum
    spiculigerum
  Rhabdochona sp.
Acanthocephala:
  Neoechinorhynchus sp.
  Pomphorhynchus
    bulbocolli
Leech:
  Actinobdella triannulata
Crustacea:
  Ergasilus sp.
Catostomus humboldtianus
Trematoda:
  *Clinostomum marginatum
  *Diplostomulum corti
  *D. flexicaudum
  *Neascus sp.
Cestoda:
  *Ligula intestinalis
Acanthocephala:
  *Neoechinorhynchus
    cristatus
Catostomus macrocheilus,
large-scale sucker
Protozoa:
  Haemogregarina
    catostomi
Trematoda:
  *Nanophyetus salmincola
  Octomacrum lanceatum
  *Ornithodiplostomum
    ptychocheilus
  Plagiocirrus primus
  P. testeus
  Triganodistomum
    attenuatum
Cestoda:
  Caryophyllaeus terebrans
  Caryophyllaeus sp.
  Glaridacris catostomi
  G. confusus
  *Ligula intestinalis
Nematoda:
  *Contracaecum sp.
  *Eustrongylides sp.
  Hepaticola bakeri

Rhabdochona cascadilla
Rhabdochona sp.
*Spiroxys
Acanthocephala:
  Neoechinorhynchus
    crassus
  N. cristatus
  N. rutili
  N. venustus
  Neoechinorhynchus sp.
  Pomphorhynchus
    bulbocolli
Leech:
  Actinobdella triannulata
Crustacea:
  Ergasilus caeruleus
  Lernaea cyprinacea
Catostomus occidentalis,
Sacramento sucker
Trematoda:
  *Diplostomulum
    flexicaudum
  *Diplostomulum sp.
  *Neascus sp.
  Triganodistomum
    polyobatum
Cestoda:
  Glaridacris catostomi
  Glaridacris sp.
Nematoda:
  Cystidicola sp.
Acanthocephala:
  Octospinifer torosus
Crustacea:
  Lernaea cyprinacea
Catostomus rimiculus,
Klamath smallscale sucker
Trematoda:
  *Clinostomum marginatum
  *Diplostomulum
    flexicaudum
  *Neascus sp.
  Triganodistomum
    crassicrurum
Catostomus snyderi, Klamath
largescale sucker
Protozoa:
  Cryptobia borreli

* Larval forms.

Catostomus tahoensis, Tahoe
sucker
  Trematoda:
   *Diplostomulum
    flexicaudum*
   Tetracotyle tahoensis
  Cestoda:
   Glaridacris oligorchis
  Acanthocephala:
   *Neoechinorhynchus
    cristatus*
   Octospinifer torosus
  Mollusca:
   Glochidia
Catostomus teres
  Trematoda:
   Octomacrum lanceatum
Catostomus sp.
  Protozoa:
   Babesiosoma tetragonis
Erimyzon oblongus, creek
chubsucker
  Acanthocephala:
   *Neoechinorhynchus
    prolixoides*
  Crustacea:
   Lernaea cyprinacea
Erimyzon sucetta, lake chub-
sucker
  Protozoa:
   Myxobolus globosus
   M. oblongus
  Trematoda:
   *Clinostomum marginatum
   Neascus sp.
   Octomacrum lanceatum
   Tetracotyle sp.
   *Triganodistomum
    mutabile*
  Cestoda:
   *Biacetabulum
    meridianum*
   Glaridacris catostomi
   **Glaridacris* sp.
   *Proteocephalus
    ambloplitis*
   Proteocephalus sp.

Nematoda:
  ***Camallanus oxycephalus*
  Capillaria sp.
  *Contracaecum
   spiculigerum*
  Contracaecum sp.
  Dorylaimus sp.
  Hedruris tiara
Acanthocephala:
  *Neoechinorhynchus
   crassus*
  N. cylindratus
  *Pomphorhynchus
   bulbocolli*
Crustacea:
  Argulus catostomi
  A. diversus
  Ergasilus caeruleus
  E. centrarchidarum
  E. megaceros
  E. versicolor
  Lernaea tortua
Hypentelium nigricans, north-
ern hogsucker
  Protozoa:
   Trichodina sp.
  Trematoda:
   Acolpenteron catostomi
   *Anonchohaptor
    anomalum*
   Bucephalus elegans
   B. elegans
   *Clinostomum
    marginatum*
   Diplostomulum of eye
   Murraytrema copulata
   Neascus sp.
   Neodactylogyrus apos
   *Posthodiplostomum
    minimum*
   *Triganodistomum
    hypentelii*
   T. simeri
  Cestoda:
   Biacetabulum infrequens
   Glaridacris catostomi
   G. confusus

* Larval forms.
** Immature adults.

Nematoda:
**Contracaecum* sp.
Contracaecum sp.
Philometra sp.
Rhabdochona cascadilla
Acanthocephala:
*Leptorhynchoides
thecatus*
Pomphorhynchus
bulbocolli
Mollusca:
Glochidia
Crustacea:
Argulus catostomi
Ergasilus sp.
Lernaea tortua
Ictiobus bubalus, smallmouth
buffalo
Protozoa:
Amphileptus voracus
Chloromyxum thompsoni
Myxidium sp.
Myxobolus bubalis
M. ovatus
M. symmetricus
M. transovalis
Myxosoma endovasa
M. multiplicatum
M. okobojiensis
M. ovalis
Trichodina myakkae
Trematoda:
Nematobothrium
texomensis
Triganodistomum
translucens
Cestoda:
Biacetabulum giganteum
Capingens singularis
Caryophyllaeus terebrans
Glaridacris confusus
Monobothrium ingens
*Proteocephalidae
*Proteocephalus
ambloplitis
Pseudolytocestus
differtus

Nematoda:
Camallanus ancyloderus
*Contracaecum
spiculigerum
Acanthocephala:
Leptorhynchoides
thecatus
Neoechinorhynchus
longirostris
N. strigosus
Pomphorhynchus
bulbocolli
Crustacea:
Argulus appendiculosus
Ictiobus cyprinellus, bigmouth
buffalo
Trematoda:
Lissorchis gullaris
Nematobothrium
texomensis
Cestoda:
Biacetabulum giganteum
Hypocaryophyllaeus
paratarius
Monobothrium ingens
*Proteocephalus sp.
Spartoides wardi
Nematoda:
Camallanus ancyloderus
Philometra nodulosa
Acanthocephala:
Gracilisentis gracilisentis
Neoechinorhynchus
australis
Pomphorhynchus
bulbocolli
Leech:
Piscicola punctata
Crustacea:
Argulus appendiculosus
A. biramosus
Ictiobus niger, black buffalo
Trematoda:
Lissorchis gullaris
Cestoda:
Capingens singularis
Glaradacris confusus

* Larval forms.
** Immature adults.

Acanthocephala:
  *Pomphorhynchus
    bulbocolli*
Ictiobus sp.
Protozoa:
  Myxobolus discrepans
Cestoda:
  Glaridacris catostomi
Acanthocephala:
  *Neoechinorhynchus
    australis*
  N. distractus
Minytrema melanops, spotted
sucker
Protozoa:
  Myxosoma microthecum
Moxostoma anisurum, silver
redhorse
Protozoa:
  Myxobolus congesticius
  M. gravidus
Trematoda:
  Murraytrema copulata
  Neodactylogyrus ursus
Cestoda:
  Biacetabulum infrequens
Moxostoma breviceps, short-
head redhorse
Protozoa:
  Myxobolus conspicuus
Nematoda:
  Camallanus oxycephalus
Moxostoma duquesnei, black
redhorse
Trematoda:
  Anonchohaptor anomalus
  *Neodactylogyrus
    duquesni*
Moxostoma erythrurum,
golden redhorse
Protozoa:
  Myxobolus conspicuus
  M. moxostomi
  M. vastus
Trematoda:
  Apophallus venustus
  *Clinostomum
    marginatum*

Murraytrema copulata
*Triganodistomum
  hypenteli*
Nematoda:
  **Contracaecum* sp.
  Rhabdochona milleri
  R. ovifilamenta
Acanthocephala:
  *Neoechinorhynchus
    crassus*
  N. strigosus
Leech:
  Piscicola punctata
Moxostoma macrolepidotum,
northern redhorse
Trematoda:
  Plagioporus serotinus
Crustacea:
  Lernaea catostomi
  L. cruciata
Moxostoma rubreques
Trematoda:
  ***Triganodistomum* sp.
Cestoda:
  Biacetabulum infrequens
  **Triaenophorus nodulosus*
Acanthocephala:
  **Leptorhynchoides
    thecatus*
  *Neoechinorhynchus
    crassus*
  *Pomphorhynchus
    bulbocolli*
Crustacea:
  Ergasilus caeruleus
Pantosteus jordani, mountain
sucker
Trematoda:
  **Posthodiplostomum
    minimum*
Pantosteus platyrhynchus
Trematoda:
  **Bolbophorus confusus*
Suckers
Protozoa:
  Cryptobia borreli
  Trichodina myakkae

* Larval forms.
** Immature adults.

Cestoda:
  Glaridacris catostomi
  Proteocephalus
    ambloplitis
Acanthocephala:
  Echinorhynchus
    coregoni
  Neoechinorhynchus
    cylindratus
  Pomphorhynchus
    bulbocolli
Leech:
  Placobdella picta
Crustacea:
  Argulus catostomi
  A. stizostethi
  Lernaea sp.
CENTRARCHIDAE
(Sunfishes)
Ambloplites rupestris,
  rock bass
Protozoa:
  Henneguya rupestris
  Trichodina bursiformis
  T. discoidea
  T. fultoni
  Trichodina sp.
Trematoda:
  Alloglossidium corti
  Azygia angusticauda
  A. longa
  A. loossii
  Bucephalus elegans
  *B. elegans
  Bunodera cornuta
  Caecincola parvulus
  Centrovarium lobotes
  Cleidodiscus alatus
  C. megalonchus
  *Clinostomum
    marginatum
  Crepidostomum cooperi
  C. cornutum
  *Cryptogonimus chyli
  *C. diaphanus
  *Diplostomulum
    scheuringi
  *Diplostomulum sp.

Euparyphium melis
Maritrema medium
M. obstipum
Microphallus obstipus
M. opacus
M. ovatus
*Neascus ambloplitis
*Neascus sp.
*Petasiger nitidus
Phyllodistomum sp.
*Posthodiplostomum
  minimum
Protenteron diaphanum
Proterometra
  macrostoma
*Psilostomum ondatrae
Tetracleidus
  chautauquaensis
*Tylodelphys,
  metacercarie in eyes
Urocleidus
  chautauquaensis
Cestoda:
  Bothriocephalus
    claviceps
  **B. cuspidatus
  **Bothriocephalus sp.
  Proteocephalus
    ambloplitis
  P. fluviatilis
  **P. pearsei
  P. perplexus
  P. pinguis
Nematoda:
  Ascaris lucii
  Ascaris sp.
  Camallanus oxycephalus
  **Camallanus sp.
  Capillaria catenata
  Contracaecum
    brachyurum
  **Contracaecum sp.
  Dichelyne cotylophora
  Philometra cylindracea
  Rhabdochona cascadilla
  Spinitectus carolini
  S. gracilis
  Spinitectus sp.

* Larval forms.
** Immature adults.

Acanthocephala:
　Echinorhynchus
　　lateralis
　E. salmonis
　E. thecatus
　Leptorhynchoides
　　thecatus (in ceca)
　*L. thecatus
　Neoechinorhynchus
　　cylindratus
　N. rutili
　Pomphorhynchus
　　bulbocolli
　*P. bulbocolli
Leech:
　Illinobdella moorei
　Illinobdella sp.
　Piscicolaria sp.
Mollusca:
　Glochidia
Crustacea:
　Achtheres ambloplitis
　A. micropteri
　Argulus biramosus
　A. maculosus
　Ergasilus caeruleus
　E. centrarchidarum
　E. confusus
　E. elegans
　Lernaea cruciata
Archoplites interruptus,
Sacramento perch
Trematoda:
　Plagioporus serotinus
　Urocleidus dispar
Nematoda:
　*Contracaecum
　　spiculigerum
Crustacea:
　Lernaea cyprinacea
Centrarchus macropterus,
flier
Trematoda:
　Phyllodistomum pearsei
　Urocleidus macropterus
　U. wadei
Cestoda:
　*Bothriocephalus sp.

*Proteocephalus
　ambloplitis
P. pearsei
*Proteocephalus sp.
Nematoda:
　*Contracaecum
　　spiculigerum
Acanthocephala:
　*Neoechinorhynchus
　　cylindratus
Crustacea:
　Ergasilus caeruleus
Chaenobryttus gulosus,
warmouth
Trematoda:
　Actinocleidus fergusoni
　A. flagellatus
　A. gulosus
　A. okeechobeensis
　Alloglossidium corti
　Anallocreadium pearsei
　Anallocreadium sp.
　*Ascocotyle tenuicollis
　Cleidodiscus robustus
　Cleidodiscus sp.
　*Clinostomum
　　marginatum
　Crepidostomum cooperi
　**C. cornutum
　Dactylogyrus aureus
　Gyrodactylus
　　macrochiri
　Haplocleidus dispar
　Phyllodistomum sp.
　Pisciamphistoma
　　stunkardi
　*Posthodiplostomum
　　minimum
　Urocleidus chaenobryttus
　U. dolorosae
　U. ferox
　U. grandis
　*Uvulifer ambloplitis
Cestoda:
　Bothriocephalus claviceps
　Bothriocephalus sp.
　Proteocephalus
　　ambloplitis

* Larval forms.
** Immature adults.

Nematoda:
Camallanus oxycephalus
Camallanus sp.
*Camallanus sp.
Capillaria sp.
*Contracaecum
spiculigerum
*Eustrongylides sp.
Spinitectus carolini
Acanthocephala:
Leptorhynchoides
thecatus (in ceca)
**L. thecatus*
Neoechinorhynchus
cylindratus
Pomphorhynchoides
bulbocolli
**P. bulbocolli*
Leech:
Illinobdella moorei
Piscicola punctata
Crustacea:
Actheres micropteri
Argulus flavescens
Ergasilus caeruleus
E. centrarchidarum
E. versicolor
Elassoma zonatum, banded
sunfish
Trematoda:
Gyrodactylus
heterodactylus
Nematoda:
**Camallanus sp.
*Contracaecum
spiculigerum
Enneacanthus gloriosus,
bluespotted sunfish
Trematoda:
Phyllodistomum
pearsei
**Posthodiplostomum*
minimum
Nematoda:
*Contracaecum
spiculigerum

Acanthocephala:
Leptorhynchoides
thecatus
Enneacanthus obesus, banded
sunfish
Trematoda:
*Clinostomum
marginatum
Lepomis auritus, longear
sunfish
Trematoda:
Crepidostomum cooperi
**Posthodiplostomum*
minimum
Nematoda:
Spinitectus carolini
Acanthocephala:
Leptorhynchoides
thecatus
Leech:
Illinobdella moorei
Crustacea:
Ergasilus sp.
Lepomis cyanellus, green
sunfish
Protozoa:
Mitraspora elongata
Myxobolus mesentericus
Myxosoma cartilaginis
Trichodina fultoni
Trematoda:
Actinocleidus fergusoni
A. longus
Anallocreadium sp.
Anchorodiscus
triangularis
Bucephalopsis pusillum
Bucephalus elegans
Cleidodiscus diversus
C. robustus
Crepidostomum cooperi
C. cornutum
Cryptogonimus chyli
*Diplostomulum
scheuringi
Haplocleidus furcatus
**Nanophyetus salmincola*
**Neascus flexicorpa*

* Larval forms.
** Immature adults.

*Neascus wardi
*Neascus of Hobgood
*Neascus sp.
Onchocleidus cyanellus
O. mucronatus
Paramphistomum
stunkardi
Phyllodistomum lohrenzi
*Posthodiplostomum
minimum
Proterometra catenaria
P. hodgesiana
Urocleidus chaenobryttus
U. cyanellus
U. dispar
*Uvulifer ambloplitis
Cestoda:
Bothriocephalus
claviceps
**Bothriocephalus sp.
*Hymenolepis sp.
*Proteocephalus
ambloplitis
P. pearsei
Nematoda:
Camallanus oxycephalus
*Contracaecum
spiculigerum
Contracaecum sp.
Dichelyne cotylophora
Spinitectus carolini
S. gracilis
*Spiroxys sp.
Acanthocephala:
Leptorhynchoides
thecatus
Neoechinorhynchus
cylindratus
Leech:
Piscicola punctata
Mollusca:
Glochidia
Crustacea:
Ergasilus caeruleus
E. centrarchidarum
E. versicolor
Lernaea cyprinacea

Lepomis gibbosus,
pumpkinseed
Protozoa:
Chloromyxum sp.
Henneguya ohioensis
Myxobolus gibbosus
M. osburni
Trichodina sp.
Trematoda:
Actinocleidus aculatus
A. gibbosus
A. incus
A. maculatus
A. oculatus
A. recurvatus
A. scapularis
A. sigmoidea
Allocreadium armatum
Allocreadium sp.
*Apophallus venustus
Azygia angusticauda
A. longa
Bunodera sacculata
Cleidodiscus megalonchus
C. nematocirrus
C. oculatus
C. robustus
*Clinostomum
marginatum
Crepidostomum cooperi
C. cornutum
C. farionis
C. laureatum
*Cryptogonimus chyli
*Diplostomulum cuticola
*D. scheuringi
*Diplostomulum sp.
Haplocleidus affinis
H. dispar
Homalometron armatum
*Neascus of McCoy
*Neascus sp.
Onchocleidus acer
O. dispar
O. ferox
O. mucronatus
O. similis

* Larval forms.
** Immature adults.

O. spiralis
*Opisthorchis tonkae
Phyllodistomum pearsei
P. superbum
Pisciamphistoma
   stunkardi
*Posthodiplostomum
   minimum
Proterometra dickermani,
   exper.
P. macrostoma
P. sagittari
*Psilostomum ondatrae
Rhipidocotyle
   septpapillata
*Tylodelphys (eye)
Urocleidus acer
U. affinis
U. dispar
U. ferox
U. helicis
U. mucronatus
U. procax
U. similis
U. spiralis
U. terox
*Uvulifer ambloplitis
Cestoda:
   Bothriocephalus claviceps
**B. cuspidatus
   Bothriocephalus sp.
   Glaridacris sp. (?)
*Haplobothrium
   globiforme
   Proteocephalus
   ambloplitis
   P. pearsei
*Trianophorus nodulosus
Nematoda:
   Ascaris angulata
   Camallanus oxycephalus
**Camallanus sp.
   Capillarua catenata
   C. eupomotis (liver,
   Italy)
   Contracaecum
   brachyurum

**Contracaecum sp.
   Contracaecum sp.
   Dichelyne cotylophora
*Eustrongylides sp.
   Philometra cylindraceum
   Spinitectus carolini
   S. gracilis
*Spiroxys sp.
Acanthocephala:
   Acanthocephalus
   jacksoni
   Echinorhynchus
   salmonis
   Fessisentis vancleavei
   Leptorhynchoides
   thecatus
*L. thecatus
   Neoechinorhynchus
   cylindratus
   Pomphorhynchus
   bulbocolli
Leech:
   Illinobdella moorei
   Illinobdella sp.
   Myzobdella moorei
Mollusca:
   Glochidia
Crustacea:
   Argulus versicolor
   Ergasilus caeruleus
   E. centrarchidarum
   Ergasilus sp.
   Lernaea cruciata
   L. cyprinacea
Linguatula:
   *Linguatulid
Lepomis humilis, orangespot
sunfish
Protozoa:
   Wardia ovinocua
Trematoda:
   Actinocleidus fergusoni
   Bucephalopsis pusillum
*Clinostomum marginatum
   Crepidostomum cooperi
*Diplostomulum
   huronense

* Larval forms.
** Immature adults.

Homalometron armatum
*Neascus flexicorpa
Onchocleidus mucronatus
*Posthodiplostomum
  minimum
Urocleidus acer
U. dispar
U. ferox
U. mucronatus
Cestoda:
**Bothriocephalus sp.
*Proteocephalus
  ambloplitis
Nematoda:
**Camallanus sp.
*Contracaecum
  spiculigerum
Spinitectus carolini
Acanthocephala:
*Leptorhynchoides
  thecatus
Neoechinorhynchus
  cylindratus
Crustacea:
Ergasilus caeruleus
Lepomis macrochirus, bluegill
Protozoa:
Bodomonas concava
Colponema agitans
Myxobolus osburni
Myxosoma cartilaginis
Scyphidia macropodia
  (See Ambiphrya.)
Trichodina discoidea
T. fultoni
T. nigra (Europe, U. S.)
Trichodina sp.
Trematoda:
Actinocleidus bursatus
A. fergusoni
A. flagellatus
A. gracilis
A. oculatus
Anallocreadium pearsei
Anallocreadium sp.
Anchoradiscus
  anchoradiscus
Asymphylodora sp.

Azygia acuminata
A. angusticauda
**A. sebago
*Bolbophorus confusus
Bucephalus elegans
Cleidodiscus diversus
C. incisor
C. robustus
  (syn. C. incisor)
Cleidodiscus sp.
Clavunculus bursatus
*Clinostomum
  marginatum
Crepidostomum cooperi
C. cornutum
Crepidostomum sp.
Cryptogonimus chyli
Dactylogyrus aureus
*Diplostomulum
  flexicaudum
*D. scheuringi
*Diplostomulum sp.
*Echinochasmus
  donaldsoni
*Euparyphium melis
Gyrodactylus elegans
Haplocleidus dispar
H. furcatus
*Neascus ambloplitis
*Neascus sp.
Onchocleidus
  mucronatus
O. perdix
*Petasiger nitidus
Phyllodistomum lohrenzi
Pisciamphistoma
  reynoldsi
P. stunkardi
*Posthodiplostomum
  minimum
Proterometra dickermani
P. macrostoma
*Psilostomum ondatrae
Pterocleidus biramosus
*Rhipidocotyle papillosum
*Tetracotyle lepomensis
Urocleidus acer
U. attenuatus

* Larval forms.
** Immature adults.

U. biramosus
U. chaenobryttus
U. dispar
U. ferox
U. mucronatus
U. perdix
Cestoda:
Bothriocephalus
claviceps
**B. cuspidatus
**Bothriocephalus sp.
*Hymenolepis sp. (?)
*Proteocephalus
ambloplitis
**P. pearsei
**P. stizostethi
*Triaenophorus nodulosus
Nematoda:
Camallanus oxycephalus
**Camallanus sp.
Capillaria catenata
*Contracaecum
spiculigerum
Contracaecum sp.
Dichelyne cotylophora
*Dichelyne sp.
*Eustrongylides sp.
Spinitectus carolini
S. gracilis
*Spiroxys sp.
Acanthocephala:
Eocollis arcanus
Leptorhynchoides
thecatus
*L. thecatus
Neoechinorhynchus
cylindratus
Pomphorhynchus
bulbocolli
Leech:
Cystobranchus verrelli
Illinobdella alba
I. moorei
Illinobdella sp.
Piscicolaria reducta
Placobdella parasitica
Mollusca:
Glochidia

Crustacea:
Actheres micropteri
Argulus americanus
Ergasilus caeruleus
E. centrarchidarum
E. versicolor
Ergasilus sp.
Lernaea catostomi
L. cyprinicea
L. dolabrodes
L. incisor
L. pomotidis
L. variabilis
Lepomis megalotis, longear
sunfish
Protozoa:
Chloromyxum trijugum
Trematoda:
Actinocleidus aculeatus
A. articularis
A. maculatus
Cleidodiscus articularis
C. bedardi
C. diversus
Crepidostomum cooperi
*Neascus flexicorpa
Onchocleidus
acuminatus
O. distinctus
Pisciamphistoma
stunkardi
Plagioporus lepomis
*Posthodiplostomum
minimum
Urocleidus acuminatus
U. distinctus
U. ferox
*Uvulifer ambloplitis
Cestoda:
**Bothriocephalus
claviceps
*Proteocephalus
ambloplitis
Nematoda:
*Contracaecum
spiculigerum
Spinitectus carolini

* Larval forms.
** Immature adults.

Acanthocephala:
  *Leptorhynchoides
    thecatus*
  *Pomphorhynchus
    bulbocolli*
Crustacea:
  Ergasilus versicolor
Lepomis microlophus, redear
sunfish (shell cracker)
Trematoda:
  Actinocleidus bakeri
  A. bifidus
  A. bifurcatus
  A. crescentis
  A. harquebus
  A. maculatus
  Anallocreadium pearsei
  *Anchoradiscus
    anchoradiscus*
  Clavunculus bifurcatus
  Cleidodiscus nematocirrus
  *Clinostomum marginatum
  Crepidostomum sp.
  Gyrodactylus sp.
  Homalometron armatum
  *Neascus ambloplitis
  Phyllodistomum lohrenzi
  *Pisciamphistoma
    stunkardi*
  **Posthodiplostomum
    minimum*
  Urocleidus attenuatus
  U. parvicirrus
  U. torquatus
  U. variabilis
Cestoda:
  **Proteocephalus
    ambloplitis*
Nematoda:
  **Camallanus sp.
  Capillaria sp.
  *Contracaecum
    spiculigerum*
  Dichelyne sp.
  Spinitectus carolini
Acanthocephala:
  *Leptorhynchoides
    thecatus*

Leech:
  Illinobdella alba
  I. moorei
  I. richardsoni
  Piscicola punctata
Crustacea:
  Argulus flavescens
Lepomis punctatus, spotted
sunfish
Trematoda:
  Gyrodactylus macrochiri
  *Pisciamphistoma
    stunkardi*
  **Posthodiplostomum
    minimum*
  *Uvulifer ambloplitis
Cestoda:
  Bothriocephalus sp.
Nematoda:
  *Camallanus sp.
  **Contracaecum
    spiculigerum*
Acanthocephala:
  **Leptorhynchoides
    thecatus*
Leech:
  Illinobdella richardsoni
Crustacea:
  Lernaea catostomi
Lepomis symmetricus, bantam
sunfish
Trematoda:
  *Actinocleidus
    symmetricus*
  *Anchoradiscus
    triangularis*
  Cleidodiscus diversus
Lepomis sp.
Protozoa:
  Henneguya mictospora
Micropterus dolomieui, small-
mouth bass
Protozoa:
  Chilodonella dentatus
  *Ichthyopthirius
    multifilis*
  Myxidium sp.
  Myxobolus kostiri

* Larval forms.
** Immature adults.

M. osburni
Myxobolus sp.
Scyphidia micropteri
  (See Ambiphrya.)
S. tholiformis
  (See Ambiphrya.)
Trichodina domerguei
T. fultoni
Trichodina sp.
Trichophrya micropteri
Trematoda:
  Acolpenteron
    ureteroecetes
  Actinocleidus bursatus
  A. fusiformis
  Allacanthochasmus
    varius
  Ancyrocephalus sp.
  *Apophallus venustus
  Asymphylodora sp.
  *Asymphylodora sp.
  Azygia angusticauda
  A. longa
  A. loossii
  A. tereticolle
  Bucephalopsis pusilla
  Bucephalus papillosus
  Bunodera cornuta
  Caecincola parvulus
  Centrovarium lobotes
  Clavunculus bursatus
  Cleidodiscus banghami
  C. fusiformis
  C. megalonchus
  *Clinostomum
    marginatum
  Crepidostomum cooperi
  C. cornutum
  Crepidostomum sp.
  Cryptogonimus chyli
  Dactylogyrus extensus
  *Diplostomulum
    scheuringi
  *D. volvens
  Diplostomulum sp.
  Gyrodactylus medius
  Leptocleidus
    megalonchus

Leuceruthrus micropteri
Maritrema medium
M. obstipum
Microphallus opacus
*Neascus sp.
Neochasmus umbellus
Phyllodistomum
  superbum
Pisciamphostoma
  stunkardi
*Psilostomum ondatrae
**Rhipidocotyle
  papillosum
*R. papillosum
R. septpapillata
Sanguinicola huronis
Tetracleidus banghami
*Tetracleidus sp.
Urocleidus principalis
*Uvulifer ambloplitis
Cestoda:
  Bothriocephalus
    claviceps
  **B. cuspidatus
  *Ligula intestinalis
  Proteocephalus
    ambloplitis
  *P. ambloplitis
  P. fluviatilis
  P. microcephalus
  P. osburni
  P. pearsei
  Proteocephalus sp.
  Triaenophorus cooperi
  *T. nodulosus
Nematoda:
  Dacnitoides cotylophora
  Camallanus oxycephalus
  C. truncatus
  **Camallanus sp.
  *Capillaria catenata
  Contracaecum
    brachyurum
  *Philometra cylindracaeum
  Philometra sp.
  Rhabdochona cascadilla
  Rhabdochona sp.
  Spinitectus carolini

* Larval forms.
** Immature adults.

S. gracilis
*S. gracilis
Acanthocephala:
  Echinorhynchus coregoni
  E. lateralis
  E. salmonis
  E. salvelini
  E. thecatus
  Leptorhynchoides
    thecatus (in ceca)
  Neoechinorhynchus
    cylindratus
  N. rutili
  Pomphorhynchus
    bulbocolli
Leech:
  Cystobranchus verrilli
  Illinobdella alba
  I. elongata
  I. moorei
  I. richardsoni
  Illinobdella sp.
  Piscicola punctata
  Piscicolaria sp.
  Placobdella montifera
Mollusca:
  Glochidia
Crustacea:
  Achtheres ambloplitis
  A. micropteri
  Internal copepod!!
  Ergasilus caeruleus
  E. centrarchidarum
  E. confusus
  Ergasilus sp.
  Lernaea cruciata
  L. cyprinacea

Micropterus punctulatus,
spotted bass
Trematoda:
  Actinocleidus fusiformis
  Bucephalus papillosus
  Clavunculus bursatus
  Cleidodiscus banghami
  C. rarus
  Crepidostomum
    cornutum
  *Posthodiplostomum
    minimum

* Larval forms.

Urocleidus furcatus
U. principalis
*Uvulifer ambloplitis
Gordiacea: in visceral cavity
Leech:
  Illinobdella moorei
Crustacea:
  Argulus flavescens
  Ergasilus versicolor
  Lernaea anomala

Micropterus salmoides,
largemouth bass
Protozoa:
  Chilodonella cyprini
  Henneguya mictospora
  Ichthyophthirius
    multifilis
  Myxobolus inornatus
  Myxobolus sp.
  Myxosoma cartilaginis
  Scyphidia micropteri
    (See Ambiphrya.)
  S. tholiformis
    (See Ambiphrya.)
  Trichodina domerguei
  T. fultoni
  T. myakkae
  T. nigra (Europe, U. S.)
  T. pediculus
Trematoda:
  Acolpenteron
    ureteroecetes
  Actinocleidus bursatus
  A. fusiformis
  A. micropteri
  Azygia angusticauda
  A. loossii
  A. micropteri
  A. tereticolle
  Bucephalopsis pusilla
  Bunodera cornuta
  B. lucioperca
  Caecincola parvulus
  Clavunculus bursatus
  C. unguis
  *Clinostomum
    marginatum
  *Crassiphiala ambloplitis
  Crepidostomum cooperi

C. cornutum
C. ictaluri
Crepidostomum sp.
Cryptogonimus chyli
*Diplostomulum
  scheuringi
*Diplostomulum sp.
Gyrodactylus macrochiri
Haplocleidus furcatus
Leuceruthrus micropteri
Microphallus opacus
*Neascus sp.
Neochasmus umbellus
Onchocleidus contortus
O. helicis
O. principalis
Phyllodistomum lohrenzi
Pisciamphostoma
  stunkardi
*Posthodiplostomum
  minimum
Proterometra macrostoma
Rhipidocotyle papillosus
R. septpapillata, expe.
Sanguinicola huronis
*Tetracotyle sp.
Urocleidus dispar
U. furcatus
U. helicis
U. principalis
*Uvulifer ambloplitis
Cestoda:
  Abothrium crassum
  Bothriocephalus claviceps
**B. cuspidatus
  Hymenolepis sp.?
  Proteocephalus
    ambloplitis
*P. ambloplitis
  P. fluviatilis
  P. nodulosa
  P. pearsei
  Proteocephalus sp.
*Triaenophorus nodulosus
Nematoda:
  Camallanus oxycephalus
**Camallanus sp.
  Capillaria catenata

Contracaecum
  brachyurum
*C. spiculigerum
*Contracaecum sp.
Contracaecum sp.
Dacnitoides cotylophora
Dioctophyma sp.
Philometra cylindracea
P. nodulosa
Spinitectus carolini
S. gracilis
Spiroxys sp.
Acanthocephala:
  Echinorhynchus
    salmonis
  Leptorhynchoides
    thecatus (in ceca)
*L. thecatus
  Neoechinorhynchus
    cylindratus
  Pomphorhynchus
    bulbocolli
Leech:
  Illinobdella moorei
  Illinobdella sp.
  Piscicola punctata
  Placobdella montifera
Mollusca:
  Glochidia
Crustacea:
  Achtheres micropteri
  Argulus appendiculosus
  A. flavescens
  Ergasilus caeruleus
  E. centrarchidarum
  E. nigritis
  Ergasilus sp.
  Lernaea anomala
  L. cruciata
  L. cyprinacea
Pomoxis annularis, white
crappie
Protozoa:
  Bodomonas concava
  Colponema agitans
  Euglenosoma branchialis
  Henneguya sp., cartilage
  Lamellasoma bacillaria

* Larval forms.
** Immature adults.

Trematoda:
  Caecincola parvulus
  Cleidodiscus capax
  C. longus
  C. uniformis
  C. vancleavei
  Crepidostomum
    cornutum
  Cryptogonimus chyli
  *Posthodiplostomum
    minimum
  *Uvulifer ambloplitis
Cestoda:
  *Proteocephalus
    ambloplitis
  P. pearsei
Nematoda:
  Camallanus oxycephalus
  **Camallanus sp.
  *Contracaecum
    spiculigerum
  Dacnitoides cotylophora
  Spinitectus gracilis
Acanthocephala:
  Leptorhynchoides
    thecatus
  Pomphorhynchus
    bulbocolli
Leech:
  Illinobdella alba
  I. elongata
  I. moorei
  I. richardsoni
Crustacea:
  Achtheres micropteri
  A. biramosus
  A. longicaudatus
  Ergasilus caeruleus
  E. centrarchidarum
Pomoxis nigromaculatus,
black crappie
Protozoa:
  Bodomonas concava
  Chloromyxum trijugum
  Colponema agitans
  Euglenosoma branchialis
  Lamellosoma bacillaria
  Myxidium melum

Myxobolus discrepans
M. intestinalis
M. iowensis
M. osburni
M. sparoides
Myxosoma sp.
Trichodina discoidea
Trichodina sp.
Trematoda:
  Azygia angusticauda
  Cleidodiscus capax
  C. stentor
  C. vancleavei
  *Clinostomum marginatum
  Crepidostomum cooperi
  C. cornutum
  C. illinoiense
  Cryptogonimus chyli
  *Diplostomulum
    scheuringi
  *Diplostomulum sp.
  *Neascus sp.
  Onchocleidus formosus
  Pisciamphistoma
    stunkardi
  *Posthodiplostomum
    minimum
  Proterometra macrostoma
  Pterocleidus biramosus
  *Tetracotyle sp.
  *Uvulifer ambloplitis
Cestoda:
  *Proteocephalus
    ambloplitis
  **P. pearsi
  **Proteocephalus sp.
  *Proteocephalus sp.
  Triaenophorus
    nodulosus
Nematoda:
  Camallanus oxycephalus
  *C. oxycephalus
  **Camallanus sp.
  Capillaria catenata
  *Contracaecum
    spiculigerum
  *Contracaecum sp.
  Spinitectus carolini

* Larval forms.
** Immature adults.

S. gracilis
Spinitectus sp.
*Spiroxys sp.
Acanthocephala:
Eocollis arcanus
Leptorhynchoides
thecatus
*L. thecatus
Neoechinorhynchus
cylindratus
Pomphorhynchus
bulbocolli
Leech:
Illinobdella richardsoni
Crustacea:
Argulus biramosus
A. versicolor
Ergasilus caeruleus
E. centrarchidarum
Lernaea cyprinacea
Sclerotis miniatus, scarlet
sunfish (red perch)
Trematoda:
Actinocleidus
brevicirrus
A. subtriangularis
Cleidodiscus chelatus
C. venardi
*Clinostomum marginatum
Pisciamphostoma
stunkardi
*Posthodiplostomum
minimum
Urocleidus attenuatus
U. chaenobryttus
U. miniatus
U. parvicirrus
Nematoda:
*Contracaecum
spiculigerum
Cestoda:
*Proteocephalus
ambloplitis
Acanthocephala:
*Leptorhynchoides
thecatus
Crustacea:
Argulus americanus
Ergasilus caeruleus

* Larval forms.

Leech:
Illinobdella richardsoni

CICHLIDAE

This family is included because
of the introduction of foreign
cichlids into North American
fish culture.

Tilapia spp.
Protozoa:
Dactylosoma mariae
(Africa)
Trypanosoma mukasai
(Africa)
Trypanosoma spp.
Trematoda:
*Bolbophorus confusus
(Israel)
Cichlidogyrus spp.
(Israel, Africa)
*Clinostomum sp. (Africa)
Enterogyrus (intestine,
Israel)
Gyrodactylus cf. medius
Plagioporus b. bilaris
(Israel)
*Stictodora
sclerogonocotyla
(Israel)
Acanthocephala: see
Golvan (1957)
Crustacea: see Fryer (1963)
Ergasilus fryeri
Lamproglena monodi
(Africa)
Lernaea laterobrachialis
(Africa)
L. lophiara (Africa)
Leech:
Zeylanicobdella
arugamensis (Ceylon)
Linguatula:
Leiperia cincinalis
CLUPEIDAE (Herrings)
Alosa chrysochloris, skipjack
herring

Crustacea:
  Ergasilus versicola
Alosa pseudoharengus, alewife
Trematoda:
  *Diplostomulum
    flexicaudum
  *Neascus sp.
Nematoda:
  **Contracaecum sp.
Acanthocephala:
  **Leptorhynchoides
    thecatus
Mollusca:
  Glochidia
Crustacea:
  Argulus alosae
Alosa sapidissima, American
shad
Trematoda:
  *Clinostomum
    marginatum
Crustacea:
  Argulus canadensis
Alosa sp.
Acanthocephala:
  *Pomphorhynchus
    bulbocolli*
Cestoda:
  Trypanorhyncha sp.
Clupea vernalis
Crustacea:
  Argulus alosae
Dorosoma cepedianum,
gizzard shad
Protozoa:
  Plistophora cepedianae
Trematoda:
  *Clinostomum sp.
  *Diplostomulum sp.
  *Mazocraeoides
    megalocotyle
  M. olentangiensis
  M. similis
  Mazocraes cepedianum
  Octobothrium sp.
  Pseudanthocotyloides
    banghami*

Cestoda:
  *Glaridacris confusus
  *Proteocephalus sp.
Acanthocephala:
  *Gracilisentis
    gracilisentis
  Neoechinorhynchus
    longirostris
  Tanaorhamphus
    longirostris*
Crustacea:
  *Argulus alosae
  A. appendiculosus
  Ergasilus lanciolatus*
Dorosoma petenense,
threadfin shad
Protozoa:
  Cryptobia sp.
COREGONIDAE (see
SALMONIDAE)
COTTIDAE (Sculpins)
Cottus aleuticas, coastrange
sculpin
Protozoa:
  *Cryptobia lynchi
  C. salmositica*
Cottus asper, prickly sculpin
Fungi:
  Ichthyosporidium, exper.
Trematoda:
  *Crepidostomum
    isostomum
  *Posthodiplostomum
    minimum*
Cestoda:
  Bothriocephalus sp.
  Cyathocephalus truncatus
  Eubothrium sp.
  Proteocephalus sp.
  Schistocephalus sp.
Nematoda:
  *Eustrongylides sp.
  Hepaticola sp.
  Metabronema salvelini
  Rhabdochona cotti*
Acanthocephala:
  *Neoechinorhynchus
    rutili*

* Larval forms.
** Immature adults.

Mollusca:
Glochidia
Crustacea:
Ergasilus sp.
Salmincola sp.

Cottus bairdi, mottled sculpin
Protozoa:
Myxobilatus cotti
Trematoda:
*Bolbophorus confusus
Bucephalus sp.
Crepidostomum cooperi
*Diplostomulum sp.
Gyrodactylus bairdi
*Neascus sp.
Phyllodistomum
brevicecum
P. etheostomae
P. undulans
**Prohemistomum
chandleri, exper.
*Rhipidocotyle papillosum
*Tetracotyle sp.
Cestoda:
**Bothriocephalus sp.
*Proteocephalus pearsei
**P. pearsei
*Schistocephalus solidus
*Schistocephalus sp.
Nematoda:
*Contracaecum sp.
Rhabdochona cascadilla
Acanthocephala:
Leptorhynchoides
thecatus
Neoechinorhynchus
rutili
Pomphorhynchus
bulbocolli
*P. bulbocolli
Mollusca:
Glochidia
Crustacea:
Ergasilus cotti

Cottus cognatus, slimy sculpin
Cestoda:
*Triaenophorus nodulosus

Nematoda:
*Haplonema sp.
Rhabdochona cotti
Cottus perplexus
Trematoda:
*Nanophyetus salmincola
Cottus rhotheus, torrent
sculpin
Protozoa:
Cryptobia lynchi
Cestoda:
Proteocephalus sp.
Acanthocephala:
Neoechinorhynchus
rutili
Cottus semiscaber, Rocky
Mountain bullhead
Trematoda:
Crepidostomum sp.
*Creptotrema sp.
Lebouria cooperi
*Posthodiplostomum
minimum
*Tetracotyle sp.
Nematoda:
Metabronema sp.
Cottus sp.
Protozoa:
Cryptobia borreli
Plistophora sp.
Trematoda:
Plagioporus virens
Cestoda:
Schistocephalus solidus
Triglopsis thompsonii, deep
water sculpin
Acanthocephala:
Echinorhynchus salmonis
CYPRINIDAE (Minnows and
Carps)
Fungi:
Saprolegnia parasitica
Protozoa:
Chilodonella cyprini
Trematoda:
Lebouria cooperi or
Allocreadium
commune (?)

* Larval forms.
** Immature adults.

Acrocheilus alutaceus,
chiselmouth
Trematoda:
Dactylogyrus vancleavei
Nematoda:
Philonema sp.
Rhabdochona cascadilla
Crustacea:
Ergasilus caeruleus
Lernaea cyprinacea
Apocope sp.
Protozoa:
Trichodina guberleti
Campostoma anomalum,
stoneroller
Protozoa:
Henneguya crassicauda
Trichodina sp.
Trematoda:
Allocreadium lobatum
*Clinostomum
marginatum
Dactylogyrus acus
*Neascus sp.
Neodactylogyrus acus
(syn.)
Octobothrium sp.
*Posthodiplostomum
minimum
*Tetracotyle sp.
Cestoda:
Proteocephalus sp.
**P. pearsei
Nematoda:
*Contracaecum sp.
**Contracaecum sp.
Rhabdochona cascadilla
Acanthocephala:
Neoechinorhynchus
crassicus
Mollusca:
Glochidia
Carassius auratus, goldfish
(*C. auratus* and *C. carassius*
here considered synonymous)
Protozoa:
Chloromyxum
carassii (R)

C. fluviatilis (R)
Cryptobia branchialis (R)
C. carassii (external)
C. cyprini
Eimeria aurati
E. carassii (R)
E. nicollae (R)
Foliella subtilis (R)
Glossatella nasalis (R)
(See *Apiosoma*.)
G. piscicola (R)
(See *Apiosoma*.)
Henneguya doneci (R)
Hoferellus schulmani (R)
Myxidium
cuneiforme (R)
M. rhodei (R)
Myxobolus carassii (R)
M. dispar (Europe)
M. dogieli (R)
M. ellipsoides (R)
M. kubanicum (R)
M. mülleri (R)
M. musculi (R)
M. orientalis (R)
M. solidus (R)
M. thelohanellus (R)
Myxosoma acuta (R)
M. sachalinensis (R)
M. sphaerica (R)
Sphaerospora carassii
(Japan, R)
S. cyprini (R)
Thelohanellus
dogieli (R)
Trichodina carassii (E)
T. domerguei (R)
T. reticulata (E, U. S.)
T. (Foliella) subtilis
(E, U. S.)
Trichodinella
epizootica (R)
Tripartiella carassii (R)
Zschokkella nova (R)
Trematoda:
Allocreadium
isoporum (R)
A. transversale (R)

* Larval forms.
** Immature adults.

Asymphylodora
  markewitschi (R)
A. tincae ? (R)
*Bucephaloides clara
  (China)
*Bucephalus
  polymorphus (R)
*Carassotrema koreanum
  (Asia)
*Clinostomum
  complanatum (R)
*Cotylurus pileatus (R)
Dactylogyrus anchoratus
D. arquatus (R)
D. auriculatus (R)
D. baueri (R)
D. crassus (R)
D. dogieli (R)
D. dujardinianus
  (Europe)
D. dulkeiti (R)
D. extensus ? (R)
D. folax
D. formosus (R)
D. inexpectatus (R)
D. intermedius (R)
D. laymani ? (R)
D. vastator (R)
D. wegeneri (R)
*Diplostomulum
  clavatum (R)
D. paradoxum
*D. spathaceum (R)
Diplozoon
  nipponicum (R)
Gyrodactylus anchoratus
G. carassii
G. chinensis (R)
G. elegans (Japan)
G. gurleyi
G. medius (R)
G. mutabilitas (Europe)
G. sprostonae (R)
*Hysteromorpha
  triloba (R)
*Metagonimus
  yokogawai (R)
*Nanophyetus salmincola

Neodactylogyrus
  cryptomeres
*Opisthorchis felineus (R)
*Paracoenogonimus
  ovatus (R)
Phyllodistomum
  elongatum (R)
P. folium (R)
*Posthodiplostomum
  cuticola (R)
Cestoda:
  Caryophyllaeus
    laticeps (R)
  *Cysticercus Paradilepis
    scolecino (R)
  *C. Gryporhynchus
    pusillum (R)
  *Digramma interrupta (R)
  Khawia parva (R)
  K. rossittensis (R)
  *Triaenophorus
    nodulosus (R)
Nematoda:
  *Agamospirura sp. (R)
  Capillaria
    brevispicula (R)
  *Contracaecum
    squalii (R)
  Philometra carassii
  P. sanguinea (R)
  Raphidascaris acus (R)
Acanthocephala:
  Acanthocephalorhyn-
    choides ussuriensis (R)
  Acanthocephalus
    anguillae (R)
  Neoechinorhynchus
    rutili (R)
  Neoechinorhynchus
    sp. (R)
  Paracanthocephalus
    tenuirostris (R)
  Pomphorhynchus
    bulbocolli
  P. laevis (R)
Leech:
  Piscicola geometra (R)
  Trachelobdella
    sinensis (R)

* Larval forms.

Mollusca:
Unionidae gen. sp.
Crustacea:
Argulus foliaceus (R)
A. japonicus
A. lunatus
A. trilineatus
Ergasilus briani (R)
E. sieboldi (R)
Lamproglena carassii
(China)
Lernaea cyprinacea
Neoergasilus
longispinosus
Paraergasilus brevidigitus
P. longidigitus
Sinergasilus
undulatus (R)
Acarina:
*Hydrachnellae gen. sp.

Chrosomus eos, northern
redbelly dace
Trematoda:
Clinostomum
marginatum
Diplostomulum sp.
Neascus sp.
Posthodiplostomum
minimum
Tetracotyle sp.
Nematoda:
**Spiroxys* sp.

Chrosomus erythrogaster,
southern redbelly dace
Trematoda:
Uvulifer ambloplitis
Acanthocephala:
Pomphorhynchus
bulbocolli

Chrosomus neogaeus,
finescale dace
Protozoa:
Myxosoma
parellipticoides
M. pfrille
Thelohanellus notatus

Trematoda:
Echinochasmus
donaldsoni
Uvulifer ambloplitis
Chrosomus sp.
Protozoa:
Myxobolus sp.
Nematoda:
Rhabdochona cascadilla
Clinostomus elongatus,
redside dace
Trematoda:
Dactylogyrus extensus
Neodactylogyrus
confusus
Clinostomus funduloides,
rosyside dace
Protozoa:
Myxobolus transovalis
Ctenopharyngodon idellus,
grass carp
(Introduced into U. S. about
1963. No parasites recorded
from U. S. See Bykhovskaya-
Pavlovskaya *et al.*, 1962 for
Russian parasites.)
Cyprinus carpio, carp
Algae:
Ichthyochtrium
vulgare (E)
Mucophilus cyprini (E)
Fungi:
Branchiomyces sanguinis
in vessels of gills
(Europe)
Dermocystidium koi
(Japan)
Protozoa:
Chilodonella cyprini
Chloromyxum
cyprini (E)
C. koi (E)
C. legeri (E)
Costia necatrix
Cryptobia borreli (E)
C. cyprini (E)
Eimeria carpelli (E)

* Larval forms.
** Immature adults.

E. subepithelialis (E)
Glossatella minuta
  (See Apiosoma.)
Hoferellus cyprini (E)
Ichthyophthirius multifilis
Myxidium
  cuneiforme (R)
M. rhodei (R)
Myxobolus artus (R)
M. bellus
M. cyprini (E)
M. cyprinicola (E)
M. dermatobius (Japan)
M. dispar
M. dogieli (R)
M. ellipsoides (R)
M. koi (Japan, R)
M. kubanicum (R)
M. macrocapsularis (R)
M. oviformis (R)
M. pfeifferi (E)
M. rotundus (R)
M. squamae (R)
M. toyamai (R)
Myxosoma
  branchialis (E)
M. dujardini (E)
M. encephalica (E)
Sphaerospora
  angulata (R)
S. cyprini (E)
Thelohanellus dogieli
T. fuhrmanni
Thelohanellus sp. (R)
Trichodina carassi (E)
T. domerguei
T. megamicronucleata
  (E)
T. myakkae
T. nigra (R)
Trichodinella
  epizootica (R)
Tripartiella carassii (R)
Trypanosoma borreli (E)
T. danilewskyi (E)
T. winchesiense (Eng)
Zschokkella cyprini
  (Eng)

Trematoda:
Allocreadium
  isoporum (R)
Apharyngostrigea
  cornu (R)
Aponurus
  tshugunovi (R)
*Apophallus venustus
Ascocotyle
  coleostoma (R)
Aspidogaster
  amurensis (R)
A. limacoides (R)
Asymphylodora
  kubanicum (R)
*Bolbophorus
  confusus (R)
*Bucephalus
  polymorphis (R)
Bunodera
  luciopercae (R)
Crepidostomum cooperi
Crepidostomum sp.
Dactylogyrus
  achmerovi (R)
D. anchoratus (Israel, E)
D. auriculatus (E)
D. crassus (Poland)
D. cyprini (Java)
D. dujardinianus (E)
D. extensus (E, Israel)
D. falcatus (E)
D. falciformis (R)
D. fallax (R)
D. formosus (E)
D. minutus (E, Israel)
D. mollis
D. solidus (E)
D. vastator (E, Japan,
  Israel)
D. wegeneri (E)
*Diplostomulum
  clavatum (R)
*D. flexicaudum
*D. spathaceum (R)
*Diplostomulum sp.
Diplozoon
  nipponicum (R)

* Larval forms.

*Echinochasmus
    perfoliatus (R)
Gyrodactylus elegans (E)
G. fairporti
G. gracilis (E)
G. medius (E)
G. nagibinae (R)
G. sprostonae (R)
Hysteromorpha
    triloba (R)
Metagonimus
    yokogawai (R)
Neodactylogyrus
    cryptomerus (R)
N. difformis (E)
N. mollis (E)
*Neodiplostomum
    perlatum (E)
*Opisthorchis
    felineus (R)
Phyllodistomum
    dogieli ? (R)
P. elongatum (R)
Plagiocirrus primus
*Posthodiplostomum
    cuticula (E)
Pseudacolpenteron
    pavlovskii (R)
Sanguinicola armata (R)
S. inermis (E)
Sphaerostoma
    bramae (R)
*Tetracotyle
    echinata (R)
*T. sogdiana (R)
Cestoda:
    Archigetes iowensis
    Atractolytocestus
        huronensis
    Biacetabulum
        appendiculatum (R)
    Bothriocephalus
        gowkongensis (R)
    Caryophyllaeus
        fimbriceps (R)
    C. laticeps (E)
    C. terebrans
    Caryophyllaeus sp.
    Corallobothrium sp.

*Cysticercus Paradilepsis
    scolecina (R)
*C. Gryporhynchus
    cheilancristrotus (R)
*Digramma interrupta (R)
*Dilepis sp. (E)
    Khawia iowensis
    K. japonicus (R)
    K. sinensis (R)
    Triaenophorus
        nodulosus (R)
Nematoda:
    Agamospirura sp. (R)
    Anisakis sp. (R)
    Camallanus ancylodirus
    *Contracaecum squali (R)
    Cucullanus cyprini (R)
    C. dogieli (R)
    *Eustrongylides
        excisus (R)
    Philometra sanguinea (E)
    *Porrocaecum
        reticulatum (R)
    Raphidascaris acus
    Rhabdochona cascadilla
    *Spinitectus gracilis
    Spiroxys sp.
Acanthocephala:
    Acanthocephalus
        anguillae (E)
    A. lucii (E)
    *Corynosoma
        strumosum (R)
    Leptorhynchoides
        thecatus
    Neoechinorhynchus
        rutili
    Paracanthocephalus
        curtus (R)
    P. tenuirostris (R)
    Pomphorhynchus
        bulbocolli
    P. laevis (R)
    Pseudoechinorhynchus
        clavula (R)
Leech:
    Hemiclepsis
        marginate (R)
    Piscicola geometra (E)

* Larval forms.

Placobdella montifera
Trachelobdella
  sinensis (R)
Mollusca:
  Unionidae gen. sp.
Crustacea:
  Argulus biramosus (E)
  A. catostomi
  A. coregoni (R)
  A. flavescens
  A. foliaceus (R)
  A. japonicus (E)
  Caligus lacustris
  Ergasilus sieboldi (E)
  Lamproglena
    pulchella (R)
  Lernaea cyprinacea
  L. elegans (Japan)
  L. esocina (E)
  Paraergasilus
    brevidigitus (R)
  P. longidigitus (R)
  Synergasilus
    undulatus (R)
  Tracheliastes
    polycolpus (R)
Ericymba buccata, silverjaw
minnow
Protozoa:
  Myxosoma grandis
Trematoda:
  *Neascus bulboglossa
Exoglossum maxillingua,
cutlips minnow
Trematoda:
  **Allocreadium lobatum
  *Neascus sp.
  Neodactylogyrus scutatus
  *Posthodiplostomum
    minimum
  Rhabdochona cascadilla
Gila atraria, Utah chub
Protozoa:
  Ichthyopthirius
    multifilis
Trematoda:
  Allocreadium lobatum

*Clinostomum
  marginatum
Crepidostomum sp.
*Diplostomulum sp.
Lebouria cooperi
*Neascus sp.
*Posthodiplostomum
  minimum
*Tetracotyle sp.
Triganodistomum
  attenuatum
Cestoda:
  *Diphyllobothrium sp.
  Glaridacris sp.
  *Hymenolepis sp. (?)
  Hypocaryophyllaeus
    gilae
  *Ligula intestinalis
  Proteocephalus
    ptychoceilus
Nematoda:
  Capillaria catenata
  *Contracaecum
    spiculigerum
  *Eustrongylides sp.
  Rhabdochona cascadilla
Acanthocephala:
  Echinorhynchus
    tuberosus
  Neoechinorhynchus
    rutili
Crustacea:
  Ergasilus caeruleus
Hesperoleucus symmetricus,
California roach
Trematoda:
  Dactylogyrus
    occidentalis
  *Neascus sp.
  *Posthodiplostomum
    minimum
Nematoda:
  Cystidicola sp.
Hybognathus hankinsoni,
brassy minnow
Trematoda:
  *Neascus sp.
  Octobothrium sp.

* Larval forms.
** Immature adults.

*Posthodiplostomum
   minimum
*Uvulifer ambloplitis
Hybognathus hayi, cypress
minnow
Trematoda:
   *Posthodiplostomum
      minimum
   *Uvulifer ambloplitis
Hybognathus nuchalis, silvery
minnow
Protozoa:
   Henneguya macrura
Trematoda:
   Dactylogyrus banghami
   D. hybognathus
   D. nuchalis
Cestoda:
   *Ligula intestinalis
Hybopsis amblops, bigeye
chub
Trematoda:
   Neodactylogyrus amblops
Hybopsis gracilis
Trematoda:
   *Bolbophorus confusus
Hybopsis biguttatas, horny-
head chub
Protozoa:
   Chloromyxum sp.
   Myxobolus squamosus
Trematoda:
   Allocreadium lobatum
   *Bucephalus elegans
   *Clinostomum
      marginatum
   *Diplostomulum sp.
   *Neascus sp.
   Phyllodistomum nocomis
   Plagioporus sinitsini
   *Posthodiplostomum
      minimum
   *Tetracotyle sp.
Cestoda:
   Bialovarium nocomis
Nematoda:
   Contracaecum sp.
   *Spiroxys sp.

Acanthocephala:
   Leptorhynchoides
      thecatus
   Pomphorhynchus
      bulbocolli
Hybopsis micropogon,
river chub
Trematoda:
   Gyrodactylus sp.
   *Neascus sp.
   *Posthodiplostomum
      minimum
Hybopsis plumbeus, spotfin
chub
Protozoa:
   Myxobolus couesii
Trematoda:
   Allocreadium lobatum
   Dactylogyrus banghami
   D. mylocheilus
   *Diplostomulum sp.
   Gyrodactylus couesius
   Octomacrum sp.
   *Posthodiplostomum
      minimum
Cestoda:
   Eubothrium sp.
   Hymenolepis sp. (?)
   *Ligula intestinalis
   Proteocephalus sp.
Nematoda:
   Hepaticola bakeri
   Rhabdochona cascadilla
Acanthocephala:
   Echinorhynchus
      salmonis
   Neoechinorhynchus
      rutili
Mollusca:
   Glochidia
Crustacea:
   Ergasilus caeruleus
Hybopsis storeria, silver chub
Trematoda:
   Dactylogyrus texomensis
Lavinia exilicauda, hitch
Trematoda:
   Dactylogyrus
      microlepidotus

* Larval forms.

*Diplostomulum
    flexicaudum
*Posthodiplostomum
    minimum
Crustacea:
    Lernaea cyprinacea
Mylocheilus caurinus, pea-
mouth
Trematoda:
    Allocreadium lobatum
    *Clinostomum
        marginatum
    Dactylogyrus mylocheilus
    Octomacrum lanceatum
    *Posthodiplostomum
        minimum
Cestoda:
    Caryophyllaeus terebrans
    Eubothrium salvelini
    *Ligula intestinalis
Nematoda:
    Bulbodacnitis sp.
    *Eustrongylides sp.
    Hepaticola bakeri
    Rhabdochona cascadilla
    Rhabdochona sp.
Acanthocephala:
    Neoechinorhynchus
        cristatus
    N. rutili
    N. venustus
Crustacea:
    Ergasilus caeruleus
Mollusca:
    Glochidia
Mylopharodon conocephalus,
hardhead
Trematoda:
    *Neascus sp.
Nematoda:
    Cystidicola sp.
Crustacea:
    Lernaea cyprinacea
Notemigonus crysoleucas,
golden shiner
Protozoa:
    Myxobolus notemigoni
    Plistophora ovariae

Trematoda:
    Crepidostomum cooperi
    Dactylogyrus aureus
    D. parvicirrus
    *Diplostomulum corti
    *Diplostomulum sp.
    Gyrodactylus sp.
    *Neascus ambloplitis
    *Neascus sp.
    Octomacrum
        microconfibula
    Plagiocirrus primus
    *Posthodiplostomum
        minimum
    **Proteocephalus
        pearsi
    *Tetracotyle sp.
    Tetraonchus sp.
    *Uvulifer ambloplitis
Cestoda:
    *Hymenolepis sp. (?)
    Pliovitellaria
        wisconsinensis
    Proteocephalus sp.
Nematoda:
    **Contracaecum
        brachyurum
    **Contracaecum sp.
    Hepaticola bakeri
Acanthocephala:
    **Acanthocephalus
        jacksoni
    Neoechinorhynchus
        cylindratus
    Pomphorhynchus
        bulbocolli
Leech:
    Illinobdella alba
    I. moorei
    Piscicolaria reducta
Crustacea:
    Ergasilus megaceros
    Ergasilus sp.
    Lernaea cyprinacea
Notropis anogenus, pugnose
shiner
Protozoa:
    Henneguya brachyura

* Larval forms.
** Immature adults.

Myxobolus aureatus
Notropis atherinoides, emerald shiner
Trematoda:
  Neodactylogyrus archis
  *Posthodiplostomum minimum
Cestoda:
  *Proteocephalus ambloplitis
  P. pinguis, exper.
Notropis bifrenatus, bridle shiner
Trematoda:
  *Centrovarium lobotes
Notropis blennius, river shiner
Protozoa:
  Myxobolus compressus
  Thelohanellus notatus
  Unicauda clavicauda
Notropis cornutus, common shiner
Protozoa:
  Chloromyxum sp.
  Henneguya fontinalis notropis
  Myxobolus grandis
  M. notropis
  M. transversalis
  Myxosoma media
  M. notropis
  M. orbitalis
  M. robustum
  Sphaerospora notropis
  Thelohanellus notatus
Trematoda:
  Allocreadium commune
  A. lobatum
  *Apophallus venustus
  *Bucephalus elegans
  Bunodera sacculata
  *Centrovarium lobotes
  *Clinostomum marginatum
  *Crassiphiala bulboglossa
  Dactylogyrus banghami
  D. bannus
  D. cornutus

D. dubius
D. fulcrum
D. pollex
D. pyriformis
D. vannus
*Diplostomulum perlus
*D. scheuringi
*Diplostomulum sp.
Lebouri cooperi
*Linstowiella szidati
*Neascus sp.
Neodactylogyrus acus
N. bulbus
N. pyriformis
N. vannus
Octomacrum lanceatum
Phyllodistomum notropidus
Plagioporus sinitsini
*Posthodiplostomum minimum
Sanguinicola sp.
*Tetracotyle sp.
*Uvulifer ambloplitis
Cestoda:
  **Bothriocephalus sp.
  *Ligula intestinalis
  Proteocephalus sp.
Nematoda:
  **Contracaecum sp.
  Rhabdochona cascadilla
  *Spiroxys sp.
Acanthocephala:
  Neoechinorhynchus cylindratus
  Pomphorhynchus bulbocolli
Mollusca:
  *Glochidia
Crustacea:
  Ergasilus versicolor
  Lernaea cyprinacea
  L. pectoralis
  L. tortua coquae
Notropis giiberti
Protozoa:
  Myxobolus orbiculatus

* Larval forms.
** Immature adults.

Notropis girardi, Arkansas
river shiner
Trematoda:
Dactylogyrus banghami
Notropis heterodon, blackchin
shiner
Trematoda:
Crepidostomum farionis
**Dilpostomulum* sp.
**Posthodiplostomum*
minimum
Nematoda:
Contracaecum sp.
***Contracaecum* sp.
Philometra cylindraceum
Notropis heterolepis, blacknose
shiner
Protozoa:
Henneguya fontinalis
notropis
Myxobolus notropis
Trematoda:
Cercaria cristafera (prob.
Sanguinicola), exper.
**Diplostomulum* sp.
**Neascus* sp.
**Posthodiplostomum*
minimum
Cestoda:
***Proteocephalus pearsei*
***Proteocephalus* sp.
Nematoda:
**Spiroxys* sp.
Mollusca:
**Glochidia*
Notropis hudsonius, spottail
shiner
Protozoa:
Myxosoma grandis
Trematoda:
Allocreadium lobatum
Bucephalus sp.
**Cryptogonimus chyli*
**Diplostomulum* sp.
**Echinochasmus*
donaldsoni
Lebouria cooperi

**Neascus* sp.
**Petasiger nitidus*
**Posthodiplostomum*
minimum
Sanguinicola lophophora
**Tetracotyle* sp.
**Triganodistomum* sp.
Cestoda:
**Ligula intestinalis*
***Proteocephalus* sp.
Nematoda:
Rhabdochona cascadilla
Spinitectus gracilis
Acanthocephala:
Echinorhynchus
salmonis
Leptorhynchoides
thecatus
Neoechinorhynchus
rutili
**Pomphorhynchus*
bulbocolli
Crustacea:
Lernaea cyprinacea
Notropis lutrensis, red shiner
Trematoda:
Dactylogyrus banghami
D. moorei
**Neascus* sp.
Crustacea:
Lernaea catostomi
Notropis percobromus, plains
shiner
Trematoda:
Dactylogyrus banghami
D. percobromus
Notropis photogens, silver
shiner
Trematoda:
Neodactylogyrus
photegensis
Notropis procne, swallowtail
shiner
Trematoda:
Allocreadium lobatum
**Neascus* sp.

* Larval forms.
** Immature adults.

*Posthodiplostomum
  minimum
Notropis rubellus, rosyface
shiner
Trematoda:
  *Bucephalus elegans
  *Diplostomulum
    scheuringi
  *Neascus sp.
  Neodactylogyrus rubellus
  *Posthodiplostomum
    minimum
  *Tetracotyle sp.
Nematoda:
  *Spiroxys sp.
Notropis spilopterus, spotfin
shiner
Trematoda:
  *Posthodiplostomum
    minimum
Cestoda:
  Protocephalus sp.
Notropis stramineus, sand
shiner
Trematoda:
  Allocreadium lobatum
  *Amphimerus elongatus
  *Centrovarium lobotes
  Dactylogyrus banghami
  D. moorei
  *Diplostomulum sp.
  *Opisthorchis tonkae
  *Ornithodiplostomum
    ptychocheilus
Crustacea:
  Lernaea tortua
Notropis uranoscopus,
stargazing shiner
Trematoda:
  Gyrodactylus protuberus
Notropis venustus, blacktail
shiner
Crustacea:
  Lernaea catostomi
Notropis volucellus, mimic
shiner
Trematoda:
  *Centrovarium lobotes
  Dactylogyrus banghami

*Diplostomulum sp.
*Neascus sp.
  Neodactylogyrus
    distinctus
*Posthodiplostomum
  minimum
Nematoda:
  Rhabdochona cascadilla
Crustacea:
  Lernaea catostomi
Notropis whipplii, steelcolor
shiner
Protozoa:
  Myxobolus teres
Trematoda:
  *Uvulifer ambloplitis
Notropis sp.
Protozoa:
  Myxobolus sp.
Trematoda:
  *Apophallus itascensis
Opsopoedus emiliae, pugnose
minnow
Trematoda:
  *Posthodiplostomum
    minimum
Orthodon microlepidotus,
Sacramento blackfish
Trematoda:
  Dactylogyrus
    microlepidotus
  D. orthodon
  *Diplostomum
    flexicaudum
Crustacea:
  Lernaea cyprinicea
Parexoglossum laurae,
tonguetied minnow
Trematoda:
  Neodactylogyrus scutatus
Pimephales notatus, bluntnose
minnow
Protozoa:
  Chloromyxum sp.
  Myxobolus aureatus
  M. hyborhynchi
  M. mutabilis
  M. nodosus
  M. notatus

* Larval forms.

Myxobolus sp.
Thelohanellus notatus
Trichodina sp.
Trematoda:
*Bucephalus elegans
*Centrovarium lobotes
Dactylogyrus bifurcatus
D. bychowskyi
D. simplex
*Diptostomulum
   scheuringi
*Diptostomulum sp.
*Hysteromurpha triloba,
   exper.
*Neascus sp.
Neodactylogyrus
   bifurcatus
N. simplex
*Posthodiplostomum
   minimum
*Tetracotyle
*Uvulifer ambloplitis
Cestoda:
Bothriocephalus
   cuspidatus
Caryophyllidae
*Ligula intestinalis
Pliovitellaria
   wisconsinensis
Proteocephalus perplexus
Nematoda:
**Contracaecum sp.
Philometra cylindraceum
Rhabdochona cascadilla
Acanthocephala:
Pomphorhynchus
   bulbocolli
Mollusca:
Glochidia
Crustacea:
Lernaea cyprinacea
Pimephales promelas, fathead
minnow
Protozoa:
Myxosoma hoffmanni
Nosema pimephales
Trematoda:
Ancyrocephalus sp.

* Larval forms.
** Immature adults.

*Bolbophorus confusus
Bucephalopsis pusillum
*Clinostomum
   marginatum
*Diplostomulum
   flexicaudum
*Diplostomulum sp.
Gyrodactylus sp.
*Neascus sp.
*Posthodiplostomum
   minimum
*Uvulifer ambloplitis
Cestoda:
**Biacetabulum sp.
*Hymenolepis sp. (?)
*Ligula intestinalis
**Proteocephalus pearsei
*Proteocephalus pearsei
Nematoda:
Contracaecum
   spiculigerum
**Contracaecum sp.
Crustacea:
Lernaea cyprinacea
Pimephales vigilax, bullhead
minnow
Protozoa:
Myxobolus augustus
Thelohanellus notatus
Ptychocheilus grandis,
Sacramento squawfish
Trematoda:
Dactylogyrus
   californiensis
*Neascus sp.
Plagioporus
   macrouterinus
*Posthodiplostomum
   minimum
Cestoda:
Proteocephalus
   cobraeformis
Nematoda:
Cystidicola sp.
Leech:
Austrobdella sp.
Crustacea
Lernaea cyprinacea

Ptychocheilus oregonensis,
northern squawfish
Trematoda:
  Allocreadium lobatum
  Dactylogyrus
    columbiensis
  D. ptychocheilus
  D. tridactylus
  D. vancleavei
Cestoda:
  Eubothrium salvelini
  *Ligula intestinalis
  Proteocephalus
    ptychocheilus
Nematoda:
  Contracaecum
    spiculigerum
  Hepaticola bakeri
  Rhabdochona cascadilla
  R. pellucida
  *Spiroxys sp.
Acanthocephala:
  Neoechinorhynchus
    rutili
Leech:
  Illinobdella sp.
Mollusca:
  *Glochidia
Crustacea:
  Ergasilus caeruleus
  Lernaea cyprinacea
Rhinichthys atratulus,
  (R. atronasus) blacknose
dace
Protozoa:
  Chloromyxum externum
  Myxobolus rhinichthidis
  Myxosoma grandis
  Trichodina fultoni
Trematoda:
  Gyrodactylus atratuli
  *Heterophyid (?)
  Neascus rhinichthysi
  *Neascus sp.
  *Posthodiplostomum
    minimum
  *Tetracotyle sp.

Cestoda:
  *Proteocephalus sp.
Nematoda:
  *Contracaecum sp.
  Rhabdochona cascadilla
  **Rhabdochona sp.
Rhinichthys cataractae,
longnose dace
Protozoa:
  Cryptobia sp.
Trematoda:
  Allocreadium lobatum
  *Bolbophorus confusus
  *Clinostomum
    marginatum
  *Crassiphiala bulboglossa
  Dactylogyrus banghami
  *Neascus sp.
  *N. rhinichthys
  *Posthodiplostomum
    minimum
  Sanguinicola sp.
Cestoda:
  **Proteocephalus sp.
Nematoda:
  Bulbodacnitis sp. and
    cyst
  Contracaecum
    spiculigerum
  **Contracaecum sp.
  Metabronema salvelini
  Rhabdochona cascadilla
Mollusca:
  *Glochidia
Crustacea:
  Lernaea cyprinacea
Rhinichthys osculus, speckled
dace
Protozoa:
  Ichthyopthirius multifilis
  Myxobolus sp.
Trematoda:
  Allocreadium lobatum
  *Clinostomum
    marginatum
  Dactylogyrus maculatus
  D. osculus

* Larval forms.
** Immature adults.

Gyrodactylus
rhinichthius
*Neascus sp.
*Posthodiplostomum
minimum
Nematoda:
Capillaria catenata
*Contracaecum
spiculigerum
*Eustrongylides sp.
Hepaticola bakeri
Neochinorhynchus sp.
Mollusca:
*Glochidia
Crustacea:
Ergasilus caeruleus
Lernaea cyprinacea
Richardsonius balteatus,
redside shiner
Trematoda:
Allocreadium lobatum
*Clinostomum
marginatum
Dactylogyrus banghami
D. richardsonius
*Diplostomulum sp.
Lebouria cooperi
*Nanophyetus salmincola
*Neascus sp.
Neoechinorhynchus
rutili
Octomacrum sp.
*Posthodiplostomum
minimum
Cestoda:
Cestodaria
Eubothrium sp.
*Ligula intestinalis
Proteocephalus
ptychocheilus
*Schistocephalus sp.
Nematoda:
Bulbodacnitis sp.
Capillaria catenata
*Contracaecum
spiculigerum
Hepaticola bakeri
Rhabdochona cascadilla

*Spiroxys sp.
Acanthocephala:
Neoechinorhynchus
rutili
Mollusca:
*Glochidia
Crustacea:
Ergasilus caeruleus
Richardsonius egregius,
Lahontan redside
Trematoda:
Dactylogyrus egregius
Gyrodactylus egregius
G. richardsonius
Pellucidhaptor
pellucidhaptor
Mollusca:
*Glochidia
Richardsonius sp.
Protozoa:
Trichodina guberleti
Semotilus a. atromaculatus,
creek chub
Protozoa:
Trichodina sp.
Trematoda:
Allocreadium isoporum
A. lobatum
Cleidodiscus brachus
*Clinostomum
marginatum
*Crassiphiala bulboglossa
Dactylogyrus
atromaculatus
D. claviformis
D. lineatus
D. microphallus
D. semotilus
D. tenax
*Diplostomulum sp.
Microphallus ovatus
*Neascus sp.
Neodactylogyrus
attenuatus
Plagioporus sinitsini
*Posthodiplostomum
minimum
Triganodistomum sp.

* Larval forms.

*Uvulifer ambloplitis
Cestoda:
  Bothriocephalus formosus
  *Proteocephalus
    ambloplitis
  Proteocephalus sp.
Nematoda:
  **Camallanus oxycephalus
  **Contracaecum sp.
  Hepaticola bakeri
  Philometra sp.
  Rhabdochona cascadilla
Acanthocephala:
  **Acanthocephalus
    jacksoni
  *Leptorhynchoides
    thecatus
  Neoechinorhynchus
    saginatus
  Paulisentis fractus
  Pomphorhynchus
    bulbocolli
Mollusca:
  *Glochidia
Crustacea:
  Ergasilus caeruleus
Semotilus corporalis, fallfish
Trematoda:
  Allocreadium colligatum
  A. lobatum
  Allocreadium sp.
  *Cotylurus communis
  *Crassiphiala bulboglossa
  Dactylogyrus corporalis
  Neascus sp.
  *Neogogatea kentuckiensis
  Plagioporus sinitsini
  *Posthodiplostomum
    minimum
  *Tetracotyle parvulum
Cestoda:
  **Proteocephalus sp.
Nematoda:
  Rhabdochona cascadilla
Acanthocephala:
  Neoechinorhynchus
    saginatus

Crustacea:
  Ergasilus elegans
  E. megaceros
  Ergasilus sp.
Semotilus margarita,
pearl dace
Protozoa:
  Amphileptus voracus
  Chloromyxum externum
  Trichodina bulbosa
  T. platyformis
  T. symmetrica
Trematoda:
  Cleidodiscus brachus
  Gyrodactylus atratuli
  G. margaritae
  *Neascus sp.
  *Posthodiplostomum
    minimum
  *Uvulifer ambloplitis
Nematoda:
  Rhabdochona cascadilla
Siphateles obesus
Trematoda:
  *Diplostomum
    flexicaudum
  *Posthodiplostomum
    minimum
Cestoda:
  **Proteocephalus sp.
Siphateles sp.
Trematoda:
  *Posthodiplostomum
    minimum
Cestoda:
  *Ligula intestinalis
Tinca tinca
Protozoa:
  Chloromyxum
    cristatum (R)
  Cryptobia keisselitzi (R)
  Eimeria minuta (R)
  Haemogregarina
    sp. (R)
  Myxidium pfeifferi (R)
  Myxobolus cyprini (R)
  M. dogieli (R)

* Larval forms.
** Immature adults.

M. ellipsoides (R)
M. muelleri (R)
M. oviformis (R)
Sporozoon tincae (R)
*Thelohanellus
pyriformis* (R)
*Trichodina
domerguei* (R)
T. nigra (R)
*Trichodinella
epizootica* (R)
Trematoda:
*Allocreadium
isoporum* (R)
*Anisakis sp. (R)
*Asymphylodora
kubanicum* (R)
A. tincae (R)
*Cotylurus pileatus (R)
*Crowcrocoecum
skrjabini* (R)
*Dactylogyrus
macracanthus* (R)
D. monocornis
D. similis ?
*Diplostomulum
clavatum* (R)
*D. spathaceum (R)
*Diplozoon
paradoxum* (R)
*Hysteromorpha
triloba* (R)
*Neodiplostomum
pseudattenuatum* (R)
*Opisthorchis felineus (R)
*Paracoenogonimus
ovatus* (R)
*Phyllodistomum
elongatum* (R)
Sanguinicola armata (R)
S. inermis (R)
*Sphaerostoma
bramae* (R)
Nematoda:
*Anisakis sp. (R)
*Contracaecum
squalii* (R)

*Desmidocercella sp. (R)
Raphidascaris acus (R)
Skrjabillanus tincae (R)
Acanthocephala:
*Acanthocephalus
anguillae* (R)
A. lucii (R)
*Corynosoma smerine (R)
*C. strumosum (R)
Echinorhynchus sp.
*Neoechinorhynchus
crassus*
N. rutili (R)
Leech:
Piscicola geometra (R)
Crustacea:
Argulus foliaceus (R)
Ergasilus briani (R)
E. sieboldi (R)
Lernaea cyprinacea (R)
L. esocina (R)

CYPRINODONTIDAE

(Killifishes)

Belonesox belizanus
Trematoda:
*Ascocotyle leighi

Cyprinodon variegatus,
sheepshead minnow
Protozoa:
Myxobolus capsulatus
M. globosus
M. lintoni
Myxosporidea of Rigdon
and Hendricks (1955)
Trematoda:
*Ascocotyle angrense
*A. chandleri
*A. leighi
Gyrodactylus sp.
G. prolongis, exper.
*Parascocotyle diminuta
*Pseudascocotyle diminuta
Copepoda:
Argulus funduli

Cyprinodon sp.
Trematoda:
Gyrodactylus medius

* Larval forms.

Fundulus† catenatus,
northern studfish
Trematoda:
  Urocleidus fundulus
Fundulus chrysotus, golden
topminnow
Trematoda:
  *Ascocotyle angrense
  Creptotrema funduli
  *Diplostomulum sp.
  **Posthodiplostomum
    minimum*
Nematoda:
  **Camallanus sp.
  **Contracaecum
    spiculigerum*
Acanthocephala
  **Leptorhynchoides
    thecatus*
Fundulus diaphanus,
banded killifish
Protozoa:
  Myxosoma funduli
Trematoda:
  Allocreadium commune
  Ancyrocephalus angularis
  *Asymphylodora sp.
  Asymphylodora sp.
  **Clinostomum
    marginatum*
  **Crassiphiala
    bulboglossa*
  Creptotrema funduli
  *Diplostomulum sp.
  Gyrodactylus stegurus
  Homalometron pallidum
  *Neascus sp.
  **Posthodiplostomum
    minimum*
  *Rhipidocotyle
    septpapillata*
  Urocleidus angularis
Cestoda:
  Bothriocephalus rarus
Nematoda:
  *Eustrongylides sp.
Acanthocephala:

  **Leptorhynchoides
    thecatus*
  *Neoechinorhynchus
    cylindratus*
Fundulus grandis, Gulf killifish
Trematoda:
  Ascocotyle angrense
  Dactylogyrus stephanus
  Gyrodactylus prolongis
  *Parascocotyle diminuta
Acanthocephala:
  *Octospiniferoides
    chandleri*
Fundulus heteroclitus,
mummichog
Protozoa:
  Eimeria sp.
  Glugea hertwigi
  Kudoa sp.
  Myxidium folium
  Myxobolus bilineatum
  Myxosoma funduli
  M. hudsonis
  M. subtecalis
  Plistophora sp.
  Sphaerospora renalis
Trematoda:
  *Ascocotyle angrense
  *A. (Phagicola) diminuta
  Crepidostomum cooperi
  *Echinochasmus schwartzi
  Gyrodactylus stephanus
  Gyrodactylus sp.
  Homalometron pallidum
  **Stephanoprora
    denticulata*
Nematoda:
  Dichelyne lintoni
Acanthocephala:
  *Neoechinorhynchus
    cylindratus*
Crustacea:
  Argulus funduli
  Ergasilus lizae
Fundulus jenkinsi, saltmarsh
topminnow

* Larval forms.
** Immature adults.
† See Dillon (1966) for list of parasites occurring on *Fundulus* spp.

Trematoda:
Ascocotyle angrense
Parascocotyle diminuta
Fundulus majalis, striped
killifish
Protozoa:
Chloromyxum renalis
Myxosoma funduli
Trematoda:
Ascocotyle angrense
A. diminuta
Acanthocephala:
Rhadinorhynchus pristis
Crustacea:
Argulus funduli
Fundulus notatus, blackstripe
topminnow
Cestoda:
Proteocephalus
ambloplitis
Nematoda:
Camallanus sp.
Contracaecum
spiculigerum
Acanthocephala:
Leptorhynchoides
thecatus
Fundulus ocellaris
Crustacea:
Argulus funduli
Ergasilus funduli
Fundulus pallidus
Trematoda:
Phagicola legenformis
Fundulus parvipinnis,
California killifish
Trematoda:
Euhaplorchis
californiensis
Mesostephanus
appendiculatus
Parastictodora hancocki
Pygidiopsoides spindalis
Fundulus similis, longnose
killifish
Protozoa:
Myxosporidea of Rigdon
and Hendricks (1955)

Trematoda:
Ascocotyle angrense
Galactosomum spineatum
Gyrodactylus funduli
Parascocotyle diminuta
Crustacea:
Ergasilus lizae
Fundulus sp.
Protozoa:
Myxobolus globosus
Trematoda:
Paramacroderoides
echinus
Cestoda:
Proteocephalus
ambloplitis
Trypanorhynchus sp.
Nematoda:
Philometra
cylindraceum
Acanthocephala:
**Pomphorhynchus*
bulbocolli
Leech:
Placobdella parasitica
Jordanella floridae, flagfish
Trematoda:
Paramacroderoides
echinus
Acanthocephala:
Neoechinorhynchus
doryphorus
Lucania parva, little killifish
Trematoda:
Parascocotyle diminuta
Zygonectes dispar, topminnow
Trematoda:
Diplostomulum sp.
Cestoda:
Proteocephalus sp.
Nematoda:
**Camallanus* sp.
Contracaecum
spiculigerum
Acanthocephala:
Leptorhynchoides
thecatus

* Larval forms.
** Immature adults.

DALLIDAE (see UMBRIDAE)
DOROSOMIDAE (see
  CLUPEIDAE)
ESOCIDAE (Pikes)
  Esox americanus, redfin
    pickerel
    Protozoa:
      Trichodina renicola
    Trematoda:
      Azygia angusticauda
      *Crassiphiala bulboglossa
      Macroderoides flavus
    Cestoda:
      *Proteocephalus
        ambloplitis*
      P. pinguis
    Nematoda:
      *Spiruridae
  Esox lucium × **Esox
    masquinongy** (hybrid)
    Trematoda:
      ***Azygia angusticauda*
      *Neascus sp.
    Cestoda:
      ***Proteocephalus pinguis*
    Acanthocephala:
      ***Leptorhynchoides
        thecatus*
  Esox lucius, northern pike
    (North America, E)
    Fungus:
      *Dermocystidium
        vejdovskyi* (E)
    Protozoa:
      *Brachyspira
        epizootica* (E)
      *Chloromyxum
        esocinum* (R)
      *Cryptobia
        gurneyorum* (E)
      Eimeria esoci (R)
      Glossatella sp. (E)
      *Haemogregarina
        esoci* (R)
      Henneguya lobosa (R)
      H. psorospermica (E)
      H. schizura
      H. zschokkei (R)

Hepatozoon esoci (R)
*Myxidium
  lieberkuhni* (E)
Myxosoma anurus (R)
*Nephrocystidium
  pickii* (E)
Scyphidia sp.
*Trichodina
  domerguei* (R)
T. pediculus (E)
*Trichodinella
  epizootica* (R)
Trypanosoma remaki (E)
Trematoda:
  *Allocreadium
    isoporum* (R)
  *Ancyrocephalus
    monenteron* (E)
  *Apophallus venustus
  **Ascocotyle
    coleostoma* (R)
  Azygia angusticauda
  A. longa
  A. lucii (E)
  A. robusta (R)
  A. tereticolle
  **Bolbophorus
    confusus* (E)
  Bucephalopsis pusilla
  Bucephalus papillosus
  B. polymorphus (R)
  *Bunodera
    luciopercae* (E)
  Centrovarium lobotes
  **Clinostomum
    complanatum* (R)
  **Clinostomum
    marginatum*
  *Cotylurus pileatus
  **Crassiphiala
    bulboglossa*
  Crepidostomum cooperi
  *Crowcrocoecum
    skrjabini* (R)
  ***Cryptogonimus chyli*
  **Diplostomulum
    cuticola* (E)
  *D. scheuringi

* Larval forms.
** Immature adults.

*D. spathaceum (R)
*Diplostomulum sp.
 Diplozoon
   paradoxum (E)
 Gyrodactylus
   elegans (E)
 G. wageneri lucii (E)
*Hysteromorpha
   triloba (R)
 Macroderoides flavus
*Neascus oneidensis
*Neascus sp.
 Neodactylogyrus
   megastoma (E)
*Paracoenogonimus ovatus
 Phyllodistomum
   americanum
 P. folium (E)
 P. superbum
 Phyllodistomum sp.
 Plagiocirrus primus
 Rhipidocotyle illense (R)
 Sanguinicola inermis (R)
 S. volgensis (R)
 Sphaerostoma
   bramae (E)
 Tetracotyle percae-
   fluviatilis (R)
 Tetraonchus
   monenteron (E)
*Tylodelphys clavata
*Uvulifer ambloplitis
Cestoda:
 Bothriocephalus
   cuspidatus
 Cyathocephalus
   truncatus (R)
*Diphyllobothrium latum
 Proteocephalus
   cernuae (R)
 P. esocis (R)
 P. nematosoma
 P. percae (R)
 P. pinguis
 P. stizostethi
 Triaenophorus crassus
 T. nodulosus
 T. robustus

 T. tricuspidatus
Nematoda:
*Anisakis mucron (E)
 Ascaris lucii
 Camallanus lacustris (R)
 C. oxycephalus
 C. truncatus (R)
 Contracaecum
   bidentatum (R)
 C. brachyurum
 C. cayugensis
 C. spiculigerum
*C. spiculigerum
*Eustrongylides
   excisus (R)
*E. mergorum (R)
**Haplonema sp.
 Hedrurus tiara
 Philometra obturans (R)
*Porrocaecum
   capsularia (R)
 Raphidascaris acus (R)
 R. canadensis
 Rhabdochona
   denudata (R)
 Sphaerostoma
   bramae (E)
 Spinitectus carolini
 S. gracilis
*Spiruridae
Acanthocephala:
 Acanthocephalus
   anguillae (R)
 A. lucii (E)
*Corynesomum
   semerme (R)
*C. strumosum (R)
*Corynesoma sp. (E)
 Echinorhynchus
   clavula (E)
 E. gadi (E)
 E. salmonis
 Leptorhynchoides
   thecatus
 Metechinorhynchus
   salmonis (R)
 M. truttae (R)

* Larval forms.
** Immature adults.

Neoechinorhynchus
  cylindratus
N. rutili
N. tenellus
Pomphorhynchus
  bulbocolli
P. laevis (R)
P. proteus (E)
Pseudoechinorhynchus
  clavula (R)
Leech:
  Hemiclepsis
    marginata (R)
  Piscicola geometra (R)
  Illinobdella moorei
  Illinobdella sp.
Mollusca:
  *Glochidia
  *Unionidae gen. sp.
Crustacea:
  Achtheres percarum (E)
  Argulus biramosus
  A. coregoni (R)
  A. foliaceous
  A. versicolor
  Caligus lacustris (R)
  Ergasilus briani (R)
  E. sieboldi (E)
  Lernaea cyprinacea
  L. esocina (E)
Esox masquinongy,
muskellunge
Protozoa:
  Henneguya acuta
  H. nigris
  Myxobolus dentium
  Myxosoma cuneata
  M. muelleri
  Trichodina renicola
  Trichodina sp.
  Vauchomia nephritica
Trematoda:
  Azygia angusticauda
  A. longa
  Cestrahalmins laruei
  Crytogonimus chyli
  *Diplostomulum sp.
  Macroderoides spiniferus

Neascus sp.
Phyllodistomum staffordi
Phyllodistomum sp.
Cestoda:
  *Proteocephalus
    ambloplitis
  P. pinguis
  Triaenophorus nodulosus
  *T. nodulosus
Nematoda:
  ***Camallanus oxycephalus*
  Contracaecum
    brachyurum
  Metabronema salvelini
  *Philometra sp.
  Rhaphidascaris
    canadensis
  *Spiroxys sp.
Acanthocephala:
  Leptorhynchoides
    thecatus
  Neoechinorhynchus
    cylindratus
  N. tenellus
  Pomphorhynchus
    bulbocolli
Mollusca:
  *Glochidia:
Crustacea:
  Argulus americanus
  A. canadensis
  A. maculosus
  A. stizostethi
Esox niger, chain pickerel
Protozoa:
  Henneguya esocis
  H. nigris
  Trichodina renicola
  Trypanosoma remaki
Trematoda:
  Azygia acuminata
  A. angusticauda
  A. longa
  A. sebago
  Bucephalus elegans
  *Clinostomum
    marginatum

* Larval forms.
** Immature adults.

*Crassiphiala
  bulboglossa
Crepidostomum cooperi
*Diplostomulum
  scheuringi
Macroderoides flavus
M. spiniferus
Microphallus opacus
M. ovatus
Phyllodistomum
  superbum
Urocleidus mimus
*Uvulifer ambloplitis
Cestoda:
  Proteocephalus
    ambloplitis
  P. nematosoma
  P. pinguis
  Proteocephalus sp.
  Triaenophorus nodulosus
Nematoda:
  Hedruris tiara
  Philometra sp.
  Raphidascaris sp.
  Spinitectus gracilis
  *Spiruridae
Acanthocephala:
  Leptorhynchoides
    thecatus
  Neoechinorhynchus
    cylindratus
  N. tenellus
Leech:
  Unidentified leech
Crustacea:
  Argulus versicolor
Esox vermiculatus, grass
pickerel
Protozoa:
  Trichodina renicola
Trematoda:
  Azygia angusticauda
  A. longa
  A. loossii
  *Centrovarium lobotes
  Macroderoides flavus
  Microphallus ovatus
  Pisciamphistoma
    stunkardi

Cestoda:
  Glaridacris catostomi
  Proteocephalus pearsei
  P. perplexus
  P. pinguis
Nematoda:
  Ascaris lucii
  Spinitectus gracilis
Acanthocephala:
  Echinorhynchus coregoni
  Leptorhynchoides
    thecatus
Leech:
  Placobdella parasitica
Esox sp.
  Acanthocephala:
    Acanthocephalus jacksoni
ETHEOSTOMIDAE (see
  PERCIDAE)
GADIDAE
Lota lota, burbot
  Protozoa:
    Caudomyxum
      nanum (R)
    Chloromyxum
      dubium (R)
    C. mucronatum (R)
    Glossatella
      megamicronucleata
      (R) (See Apiosoma.)
    Glugea anomala (R)
    Hexamita truttae (R)
    Myxidium
      lieberkühni (R)
    Myxobolus mülleri (R)
    Sphaerospora
      cristata (R)
    S. elegans (R)
    Trichodinella
      epizootica (R)
    T. maior (E)
  Trematoda:
    Allocreadium
      isoporum (R)
    Azygia angusticauda
    A. longa
    A. lucii (R)
    A. robusta (R)

* Larval forms.

Bucephalus
    polymorphus (R)
Bunodera
    luciopercae (R)
Crepidostomum
    brevivitellum
C. farionis (R)
C. metoecus
Crepidostomum sp.
Crowcrocaecum
    skrjabini (R)
Derogenes varicus (R)
Diplostomulum
    clavatum (R)
D. hughesi (R)
D. scheuringi
D. spathaceum (R)
D. volvens
Diplostomulum sp.
Diplozoon
    paradoxum (E)
Gyrodactylus
    elegans (E)
G. lotae (E)
Hemiurus
    appendiculatus (R)
Neodiplostomulum
    sp. (R)
Phyllodistomum
    megalorchis (R)
Posthodiplostomum
    minimum
Sphaerostoma
    bramae (R)
Udonella caligarum (E)
Cestoda:
    Bothriocephalus
        cuspidatus
    Bothriocephalus sp.
    Cyathocephalus
        truncatus (R)
    Diphyllobothrium latum
    Diphyllobothrium sp.
    Eubothrium crassum
    E. rugosum
    **Proteocephalus pearsei*
    Triaenophorus nodulosus

Nematoda:
    Camallanus lacustris (R)
    C. truncatus (R)
    Capillaria
        brevispicula (R)
    Contracaecum
        brachyurum
    Cottocomephronema
        problematica ? (R)
    Cucullanus truttae (R)
    Cystidicola farionis (R)
    Desmidocercella sp. (R)
    Haplonema hammulatum
    Haplonema sp. (R)
    Hepaticola bakeri
    Ichthyobronema
        gnedini (R)
    Raphidascaris acus (R)
    Spinitectus gracilis
Acanthocephala:
    Acanthocephalus
        anguillae
    A. lucii
    Corynosoma
        strumosum (E)
    Echinorhynchus
        cinctulus (E)
    E. clavula
    E. coregoni
    E. gadi (E)
    E. leidyi
    E. salmonis
    Leptorhynchoides
        thecatus
    Neoechinorhynchus
        cylindratus
    N. rutili
    Paracanthocephalus
        tenuirostris (R)
    Pomphorhynchus
        bulbocolli
    P. laevis (R)
    Pseudoechinorhynchus
        clavula (R)
Leech:
    Cystobranchus
        mamillatus (R)

* Larval forms.
** Immature adults.

C. verilli
Piscicola milneri
Mollusca:
*Glochidia
Unionidae gen. sp. (R)
Crustacea:
Achtheres ambloplitis
Ergasilus caeruleus
E. osburni
E. sieboldi (R)
Lernaea esocina (R)
Salmincola latae (R)
Microgadus tomcod, Atlantic
tomcod
Acanthocephala:
**Acanthocephalus jacksoni*
GASTEROSTEIDAE
Eucalia inconstans, brook
stickleback
Protozoa:
Myxosoma eucaliaii
Trichodina sp.
Trematoda:
**Azygia acuminata, exper.*
Bunoderina eucaliae
Crepotrema funduli
Dactylogyrus eucalius
Diplostomulum baeri
eucaliae
*Diplostomulum sp.
*Echinochasmus
donaldsoni
*Neascus sp.
*Posthodiplostomum
minimum
*Tetracotyle sp.
Cestoda:
Proteocephalus filicollis
Proteocephalus sp.
*Proteocephalus sp.
*Schistocephalus solidus
*Triaenophorus nodulosus
Nematoda:
**Contracaecum sp.*
**Rhabdochona cascadilla*
Spinitectus gracilis
*Spiroxys sp.

Acanthocephala:
*Leptorhynchoides
thecatus
Neoechinorhynchus
rutili
Neoechinorhynchus sp.
*Pomphorhynchus
bulbocolli
Leech:
Illinobdella sp.
Mollusca:
*Glochidia
Crustacea:
Argulus stizostethi
Gasterosteus aculeatus,
threespine stickleback
Fungi:
Dermocystidium
gasterostei
Protozoa:
Cryptobia
branchialis (R)
Eimeria gasterostei (R)
Eimeria sp.
Epistylis livoffi
(Czech.)
Glossatella amebae (R)
Glugea anomala (R)
Hemiophrys
branchiarum (R)
Myxidium
gasterostei (R)
Myxobilatus
gasterostei (R)
M. medius (R)
Nosema anomala
Sphaerospora elegans (R)
Trichodina
domerguei (R)
T. gracilis (R)
Trichophrya
intermedia (R)
Trematoda:
Brachyphallus
crenatus (R)
Bunodera eucaliae
(Eng.)

* Larval forms.
** Immature adults.

*Cotylurus pileatus (R)
*Diplostomulum
    spathaceum (R)
*Diplostomulum of
    Cercaria X (Eng.)
Diplozoon
    paradoxum (E)
Gyrodactylus
    aculeati (E)
G. arcuatus (R)
G. bychowski (E)
G. elegans (E)
G. rarus (E)
Gyrodactylus sp.
*Holostephanus lukei
    (Eng)
Lecithaster gibbosus (R)
*Nanophyetus salmincola
*Neascus sp.
Peracreadium
    gasterostei (Den.)
Podocotyle atomon
P. reflexa (R)
*Posthodiplostomum
    cuticola (R)
*P. minimum
Cestoda:
    Bothriocephalus
        claviceps
    *B. scorpi (R)
    *Cyathocephalus
        truncatus (E)
    *Diphyllobothrium
        dendriticum (R)
    *D. norvegicum (R)
    Proteocephalus cernuae
    P. filicollis (R)
    P. pugetensis
    *Schistocephalus solidus
    *Schistocephalus sp.
Nematoda:
    Camallanus lacustris (R)
    C. truncatus (R)
    Cystidicola farionis ?
    *Eustrongylides sp.
    Raphidascaris acus
Acanthocephala:
    Acanthocephalus lucii

*Corynosoma
    semerme (R)
*C. strumosum (R)
Metechinorhynchus
    salmonis (R)
Neoechinorhynchus
    rutili
Pomphorhynchus
    laevis (R)
P. proteus (E)
Pseudechinorhynchus
    clavula (R)
Leech:
    Piscicola geometra (R)
Mollusca:
    *Glochidia
    *Unionidae gen. sp. (R)
Crustacea:
    Argulus alosae
    A. foliaceus (R)
    Caligus lacustris
    Ergasilus auritus
    E. turgidus
    Ergasilus sp.
    Lernaea esocina (R)
    Salmincola sp.
    Thersitina
        gasterostei (R)
Gasterosteus wheatlandi,
twospine stickleback
Cestoda:
    *Schistocephalus solidus
Crustacea:
    Argulus canadensis
Pungitius pungitius,
ninespine stickleback
Fungi:
    Dermocystidium
        gasterostei (Eng)
Protozoa:
    Cryptobia
        branchialis (R)
    Glossatella amoebae (R)
        (See Apiosoma.)
    G. conica (R)
        (See Apiosoma.)
    Glugea anomala (R)
    Hemiophrys
        branchiarum (R)

* Larval forms.

Henneguya pungitii (R)
Myxobilatus
  gasterostei (R)
M. medius (R)
Plistophora typicalis (R)
Sphaerospora elegans (R)
Trichodina
  domerguei (R)
T. gracilis (R)
Tripartiella pungitii (R)
Trematoda:
  Azygia lucii (R)
  Bucephalus
    polymorphus (R)
  Bunodera
    luciopercae (R)
  *Cotylurus pileatus (R)
  *Diplostomulum
    spathaceum (R)
  Gyrodactylus
    armatus (R)
  G. bychowskyi (R)
  G. elegans (E)
  G. rarus (E)
  *Posthodiplostomum
    cuticola (R)
Cestoda:
  *Bothriocephalus
    scorpii (R)
  *Diphyllobothrium
    dendriticum (R)
  *D. norvegicum (R)
  Proteocephalus
    filicollis (R)
  *Schistocephalus
    pungitii (R)
  *S. solidus
  *Triaenophorus
    nodulosus (R)
Nematoda:
  Camallanus lacustris (R)
  Raphidascaris acus (R)
Acanthocephala:
  Acanthocephalus
    lucii (R)
  *Corynosoma
    semerme (R)
  *C. strumosum (R)

Leptorhynchoides
  thecatus
Neoechinorhynchus
  cristatus (R)
N. rutili
Paracanthocephalus
  curtus (R)
Pomphorhynchus laevis
P. minutus
Pseudoechinorhynchus
  clavula (R)
Leech:
  Piscicola geometra (R)
Crustacea:
  Argulus foliaceus (R)
  Thersitina
    gasterostei (R)
Sticklebacks
  Protozoa:
    Glugea anomala
  Trematoda:
    *Echinochasmus milvi
HIODONTIDAE
Hiodon alosoides, goldeye
  Trematoda:
    Crepidostomum
      illinoiense
    Paurorhynchus tergisus
  Cestoda:
    Bothriocephalus
      cuspidatus
    B. texomensis
Hiodon tergisus, mooneye
  Trematoda:
    Crepidostomum
      hiodontos
    C. illinoiense
    Leuceruthrus sp.
    Paurorhynchus tergisus
    Plagioporus serratus
    P. terqisus
    *Tetracotyle sp.
  Cestoda:
    Bothriocephalus
      cuspidatus
  Nematoda:
    Camallanus oxycephalus
    Cystidicola stigmatura
    Rhabdochona cascadilla

* Larval forms.

Acanthocephala:
Leptorhynchoides
thecatus
ICTALURIDAE (Catfishes) †
Silurids, species not reported
Trematoda:
Bunodera cornuta
Generchella tropica
(Argentina)
*Macroderoides typicus
Catfish
Trematoda:
Isoparorchis (in swim
bladder of Assam
catfish)
Crustacea:
Argulus appendiculosus
Ictalurus anguilla, Fulton
catfish
Trematoda:
Phyllodistomum lacustri
Cestoda:
Corallobothrium
fimbriatum
Nematoda:
Dichelyne robusta
Acanthocephala:
Tanaorhamphus
longirostris
Crustacea:
Achtheres pimelodi
Ergasilus caeruleus
E. megaceros
E. versicolor
Ictalurus catus, white catfish
Trematoda:
Alloglossidium corti
Cleidodiscus pricei
*Diplostomulum corti
Cestoda:
Corallobothrium
giganteum
Corallobothrium sp.
Nematoda:
*Contracaecum
spiculigerum
Leech:
Illinobdella sp.

Crustacea:
Lernaea cyprinacea
Ictalurus furcatus, blue catfish
Protozoa:
Henneguya exilis
H. limatula
Myxidium kudoi
Trematoda:
Allocreadium ictaluri
Cleidodiscus floridanus
C. mirabilis
C. pricei
C. vancleavei
Neochasmus ictaluri
Cestoda:
Corallobothrium
procerum
Nematoda:
Cucullanus diplocaecum
Crustacea:
Actheres lacae
Ergasilus versicolor
Ictalurus melas, black bullhead
Protozoa:
Henneguya exilis
H. gurleyi
Myxidium melum
Myxobolus sp.
Scyphidia amiuri
(See *Ambiphrya*.)
Trematoda:
Acetodextra ameuri
Allocreadium ictaluri
Alloglossidium corti
A. geminus
Azygia angusticauda
Bucephalopsis pusillum
Cleidodiscus floridanus
C. longus
C. mirabilis
C. pricei
C. vancleavei
*Clinostomum
marginatum
Crepidostomum cooperi
C. cornutum
C. ictaluri
*Diplostomulum corti

* Larval forms.
† See F. Meyer (1966) for parasitic diseases of catfishes.

*D. spathaceum
*Diplostomulum sp.
Gyrodactylus elegans
G. fairporti
*Hysteromorpha triloba
*Neascus sp.
Phyllodistomum carolini
P. caudatum
**P. lacustri
P. staffordi
Plagioporus sp.
*Posthodiplostomum
  minimum
Cestoda:
Corallobothrium
  fimbriatum
C. giganteum
Corallobothrium sp.
*Proteocephalus
  ambloplitis
*Proteocephalus sp.
Nematoda:
Camallanus sp.
*Contracaecum
  spiculigerum
Dacnitoides cotylophora
Dichelyne robusta
Philometra cylindraceum
Rhabdochona sp.
Spinitectus gracilis
*Spiroxys sp.
Acanthocephala:
Leptorhynchoides
  thecatus
*L. thecatus
Pomphorhynchus
  bulbocolli
Leech:
Illinobdella moorei
Piscicolaria reducta
Piscicolaria sp.
Placobdella montifera
Crustacea:
Achtheres ambloplitis
A. pimelodi
Argulus biramosus
Ergasilus elegans
E. versicolor

* Larval forms.
** Immature adults.

Lernaea tortua
L. variabilis
Ictalurus natalis, yellow
bullhead
Trematoda:
Acetodextra amiuri
Alloglossidium corti
A. geminus
Azygia angusticauda
Centrovarium lobotes
Cleidodiscus floridanus
C. pricei
*Clinostomum
  marginatum
Crepidostomum cooperi
C. cornutum
C. ictaluri
*Diplostomulum
  scheuringi
*Diplostomulum sp.
Macroderoides spiniferus
Phyllodistomum carolini
P. caudatum
P. lacustri
P. staffordi
Pisciamphostoma
  stunkardi
Plagiorchis corti
*Posthodiplostomum
  minimum
Cestoda:
Corallobothrium
  fimbriatum
C. giganteum
Corallobothrium sp.
*Proteocephalus
  ambloplitis
**P. pearsei
*Proteocephalus sp.
Nematoda:
**Camallanus sp.
*Contracaecum
  spiculigerum
**Contracaecum sp.
Dichelyne robusta
Spinitectus carolina
S. gracilis
*Spiroxys sp.

Acanthocephala:
  *Leptorhynchoides
    thecatus*
  **L. thecatus*
  Neoechinorhynchus sp.
  *Pomphorhynchus
    bulbocolli*
  **P. bulbocolli*
Leech:
  Piscicolaria sp.
Mollusca:
  **Glochidia*
Crustacea:
  Achtheres pimelodi
  Argulus diversus
  A. maculosus
  Ergasilus elegans
  E. versicolor
Linguatula:
  Linguatulid
Ictalurus nebulosus, brown
bullhead
Protozoa:
  Henneguya ameiurensis
  H. gurleyi
  Myxobolus sp.
  Scyphidia macropodia
    (See *Ambiphrya*.)
Trematoda:
  Acetodextra amiuri
  Allocreadium ictaluri
  Alloglossidium corti
  A. geminus
  **A. geminus*
  Ancyrocephalus pricei
    (also R)
  **Apophallus venustus*
  Azygia acuminata
  A. angusticauda
  Bucephalus elegans
  Centrovarium lobotes
  Cleidodiscus longus
  C. pricei
  **Clinostomum
    marginatum*
  Crepidostomum cooperi
  C. ictaluri
  Crepidostomum sp.

**Diplostomulum
    scheuringi*
**D. spathaceum*
**Diplostomulum* sp.
**Echinochasmus
    donaldsoni*
**Euparyphium melis*
  Gyrodactylus sp.
**Hysteromorpha triloba*
  Macroderoides spiniferus
  Microphallus opacus
**Neascus* sp.
**Petasiger nitidus*
  *Phyllodistomum
    americanum*
  P. hunteri
  P. staffordi
  Plagiorchis ameiurensis
  P. corti
  Polylekithum halli
**Posthodiplostomum
    minimum*
**Tetracotyle* sp.
  Vietosoma parvum
Cestoda:
  *Bothriocephalus
    cuspidatus*
  *Corallobothrium
    fimbriatum*
  C. intermedium
  C. minutium
  C. parvum
***Haplobothrium
    globuliforme*
  *Proteocephalus
    ambloplitis*
***Proteocephalus* sp.
Nematoda:
***Camallanus oxycephalus*
  **Camallanus* sp.
  *Contracaecum
    brachyurum*
  **C. spiculigerum*
  Contracaecum sp. ·
  **Contracaecum* sp.
  Dacnitoides robusta
  Dichelyne cotylophora
  D. robusta

* Larval forms.
** Immature adults.

*Eustrongylides sp.
Metabronema prevosti
Rhabdochona sp.
Spinitectus carolini
S. gracilis
Spinitectus sp.
*Spiroxys sp.
Acanthocephala:
**Acanthocephalus jacksoni
Echinorhynchus
thecatus
Echinorhynchus sp.
Leptorhynchoides
thecatus
*L. thecatus
Neoechinorhynchus
rutili
Pomphorhynchus
bulbocolli
*P. bulbocolli
Leech:
Illinobdella moorei
I. richardsoni
Illinobdella sp.
Piscicola punctata
Piscicola sp.
Mollusca:
*Glochidia
Crustacea:
Achtheres pimelodi
Argulus americanus
A. appendiculosus
A. maculosus
Ergasilus versicolor
Lernaea cyprinacea
L. pomotides
L. tortua
L. variabilis
**Ictalurus nebulosus
marmoratus,** spotted
bullhead
Trematoda:
Macroderoides spiniferus
*Posthodiplostomum
minimum
Ictalurus punctatus, channel
catfish

* Larval forms.
** Immature adults.

Protozoa:
Amphileptus voracus
Colponema sp.
Costia sp.
Henneguya exilis
H. limatula
H. plasmodia
Ichthyopthirius multifilis
Myxidium bellum
Scyphidia macropodia
(See Ambiphrya.)
Trichodina discoidea
T. symmetrica
T. vallata
Trichophrya ictaluri
T. piscium
Trematoda:
Acetodextra amiuri
Alloglossidium corti
A. kenti
*Apophallus venustus
Azygia angusticauda
*Bolbophorus confusus
Cleidodiscus floridanus
C. mirabilis
C. pricei
*Clinostomum
marginatum
*Crassiphiala ambloplitis
Crepidostomum
cornutum
C. ictaluri
Dactylogyrus sp.
*Diplostomulum sp.
Holostephanus ictaluri
Microphallus opacus
Parastiotrema
ottowenensis
Phyllodistomum lacustri
*Posthodiplostomum
minimum
*Prohemistomum
chandleri, exper.
Vietosoma parvum
Cestoda:
**Bothriocephalus sp.
Corallobothrium
fimbriatum

C. giganteum
C. (M.) thompsoni
Ophiotaenia fragilis
*Proteocephalus
ambloplitis
Nematoda:
Camallanus oxycephalus
**Camallanus sp.
*Contracaecum
spiculigerum
Dacnitoides cotylophora
D. robusta
Dichelene robusta
Rhabdochona sp.
Spinitectus gracilis
Acanthocephala:
*Leptorhynchoides
thecatus
Tanaorphamphus
longirostris
Leech:
Illinobdella moorei
Piscicolaria reducta
Piscicolaria sp.
Crustacea:
Achtheres micropteri
A. pimelodi
Argulus appendiculosus
Ergasilus elegans
E. versicolor
Lernaea cyprinacea
Ictalurus sp.
Trematoda:
*Nanophyetus mustelae
*Psilostomum ondatrae
Cestoda:
*Triaenaphorus nodulosus
Nematoda:
Dacnitoides cotylophora
Philometra cylindraceum
Noturus flavus, stone cat
Trematoda:
Alloglossidium corti
*Clinostomum
marginatum
Crepidostomum ictaluri
*Diplostomulum sp.

Cestoda:
**Corallobothrium
fimbriatum
*Proteocephalus sp.
Nematoda:
*Spiroxys sp.
Acanthocephala:
Leptorhynchoides
thecatus
*L. thecatus
Pomphorhynchus
bulbocolli
*P. bulbocolli
Noturus gyrinus, tadpole
madtom
Protozoa:
Myxobilatus noturi
Trichodina sp.
Trematoda:
*Acetodextra ameuri
Alloglossidium corti
Crepidostomum cooperi,
exper.
C. ictaluri
Plagiorchis corti
*Posthodiplostomum
minimum
Cestoda:
Corallobothrium
fimbriatum
*Proteocephalus
ambloplitis
Nematoda:
*Contracaecum
spiculigerum
*Spiroxys sp.
Acanthocephala:
Leptorhynchoides
thecatus
Pomphorhynchus
bulbocolli
Crustacea:
Ergasilus caeruleus
Noturus insignis, margined
madtom
Trematoda:
Alloglossidium corti

* Larval forms.
** Immature adults.

Crepidostomum ictaluri
Phyllodistomum staffordi
Cestoda:
   Bothriocephalus
     schilbeodis
   Proteocephalus sp.
Acanthocephala:
   Neoechinorhynchus sp.
Noturus miurus, brindled
madtom
Trematoda:
   Alloglossidium corti
   Clinostomum
     marginatum
   Crepidostomum cooperi,
     exper.
   C. ictaluri
Pylodictis olivaris, flathead
catfish
Protozoa:
   Chloromyxum opladeli
Trematoda:
   Allacanthochasmus artus
   Cleidodiscus floridanus
   C. mirabilis
   Phyllodistomum lacustri
   P. staffordi
Cestoda:
   Corallobothrium sp.
   C. fimbriatum
   C. giganteum
   C. (M.) tva
   Marsipometra sp.
   Proteocephalus
     ambloplitis
   P. macrocephalus
Nematoda:
   Contracaecum
     spiculigerum
   **Camallanus* sp.
   C. oxycephalus
Acanthocephala:
   Leptorhynchoides
     thecatus
Leech:
   Cystobranchus verrelli

Mollusca:
   Glochidia:
Crustacea:
   Achtheres pimelodi
   Argulus appendiculosus
   A. flavescens
   Ergasilus versicolor
LEPISOSTEIDAE (Gars)
Lepisosteus osseus, longnose
gar
Trematoda:
   Apophallus venustus
   Macroderoides
     spiniferous
   Plesiocreadium parvum
Cestoda:
   Bothriocephalus sp.
   Proteocephalus
     ambloplitis
   P. ambloplitis
   P. perplexus
   P. singularis
Nematoda:
   Cystidicola lepisostei
Acanthocephala:
   Leptorhynchoides
     thecatus
Crustacea:
   Argulus lepidostei
   A. mississippiensis
   Ergasilus elegans
Lepisosteus platostomus,
shortnose gar
Trematoda:
   Macroderoides spiniferus
   Paramacroderoides
     echinus
Cestoda:
   Proteocephalus
     ambloplitis
   P. perplexus
   P. singularis
Nematoda:
   Contracaecum
     spiculigerum

* Larval forms.
** Immature adults.

Acanthocephala:
  Leptorhynchoides
    thecatus
Crustacea:
  Argulus lepidostei
  A. mississippiensis
  Ergasilus elegans
  Lernaea variabilis
Lepisosteus platyrhincus,
Florida gar
Trematoda:
  *Clinostomum
    marginatum
  Macroderoides spiniferus
  *Odhneriotrema
    incommonum
  Paramacroderoides
    echinus
  *Posthodiplostomum
    minimum
Cestoda:
  Proteocephalus singularis
Nematoda:
  *Camallanus sp.
  *Contracaecum
    spiculigerum
  *Eustrongylides sp.
Acanthocephala:
  *Leptorhynchoides
    thecatus
Crustacea:
  Argulus lepidostii
  *Lernaea sp.
Lepisosteus spatula,
alligator gar
Trematoda:
  Rhipidocotyle lepisostei
Crustacea:
  Ergasilus versicolor
MUGILIDAE (Mullets)
Mugil spp.
Protozoa:
  Kudoa (?)
  Stalked peritrich
Trematoda:
  Carassotrema mugilicola
    (see Shireman, 1964)
  *Mesostephanus
    appendiculatoides

OSMERIDAE (Smelts)
Osmerus dentex, arctic smelt
Cestoda:
  Diphyllobothrium osmeri
Osmerus mordax, American
smelt
Protozoa:
  Glugea hertwigi
Trematoda:
  Azygia longa
  Brachyphallus crenatus
  Derogenes varicus
  Diplostomulum sp.
Cestoda:
  *Diphyllobothrium sebago
  *Ligula intestinalis
  Proteocephalus sp.
Nematoda:
  *Philometra sp.
  *Porrocaecum decipiens
  *Spinitectus gracilis
Acanthocephala:
  *Corynosoma hardweni
  Echinorhynchus salmonis
  *E. salmonis
  Leptorhynchoides
    thecatus
  *Pomphorhynchus
    bulbocolli
Leech:
  Piscicola punctata
Crustacea:
  Argulus alosae
PERCIDAE (Perches)
Cottogaster copelandi,
Copeland's darter
Trematoda:
  Lebouria cooperi
  *Neascus sp.
Cestoda:
  Bothriocephalus
    cuspidatus
Nematoda:
  Camallanus sp.
Etheostoma blennioides,
greenside darter
Trematoda:
  Phyllodistomum
    etheostomae

* Larval forms.

Etheostoma caeruleum,
rainbow darter
Trematoda:
  *Clinostomum
    marginatum
  *Cryptogonimus chyli
  *Diplostomulum sp.
  *Neascus sp.
  *Posthodiplostomum
    minimum
Cestoda:
  *Bothriocephalus
    cuspidatus
Nematoda:
  *Contracaecum sp.
Acanthocephala:
  *Leptorhynchoides
    thecatus
Mollusca:
  *Glochidia
Crustacea:
  Ergasilus cotti

Etheostoma exile, Iowa darter
Protozoa:
  Epistylis sp.
  Myxobolus poecilichthidis
  Trichodina sp.
Trematoda:
  Azygia angusticauda
  Crepidostomum
    laureatum
  *Diplostomulum
    scheuringi
  *Neascus sp.
  *Posthodiplostomum
    minimum
  *Tetracotyle sp.
Cestoda:
  Bothriocephalus formosus
  Hymenolepsis sp. (?)
Nematoda:
  **Camallanus oxycephalus
  Contracaecum
    brachyurum
  Philometra cylindraceum
  Rhabdochona cascadilla
  *Spiroxys sp.

Acanthocephala:
  Leptorhynchoides
    thecatus
  *L. thecatus
  Pomphorhynchus
    bulbocolli
  *P. bulbocolli
Leech:
  Illinobdella sp.
Mollusca:
  *Glochidia

Etheostoma flabellare, fantail
Trematoda:
  *Clinostomum
    marginatum
  *Cryptogonimus chyli
  *Diplostomulum
    scheuringi
  *Diplostomulum sp.
  *Neascus sp.
  Phyllodistomum
    etheostomae
  Plagiocirrus primus
  *Tetracotyle communis
  Urocleidus moorei
Cestoda:
  *Proteocephalus
    ambloplitis
Nematoda:
  *Camallanus oxycephalus
  **Contracaecum sp.
  *Contracaecum sp.
Acanthocephala:
  Leptorhynchoides
    thecatus
  *L. thecatus
  *Neoechinorhynchus
    cylindratus
Mollusca:
  *Glochidia

Etheostoma maculatum,
spotted darter
Trematoda:
  Neascus sp.
  Phyllodistomum
    etheostomae
  Urocleidus malleus

* Larval forms.
** Immature adults.

Nematoda:
  Camallanus oxycephalus
  Contracaecum sp.
Acanthocephala:
  *Leptorhynchoides
  thecatus*
Etheostoma nigrum
Protozoa:
  Epistylis sp.
  Myxosoma neurophila
  Trichodina sp.
Trematoda:
  Allocreadium boleosomi
  Azygia angusticauda
  A. longa
  Bucephalopsis pusillus
  *Clinostomum
  marginatum*
  Crassiphiala bulboglossa
  *Crepidostomum
  canadense*
  C. cooperi
  C. farionis
  C. illinoiense
  C. isostomum
  C. laureatum
  Cryptogonimus chyli
  *Diplostomulum
  scheuringi*
  Diplostomulum sp.
  Lebouria cooperi
  Leuceruthrus sp.
  Neascus sp.
  Neochasmus umbellus
  *Ornithodiplostomum
  ptychocheilus*
  *Phyllodistomum
  etheostomae*
  *Posthodiplostomum
  minimum*
  Tetracotyle sp.
Cestoda:
  *Bothriocephalus
  formosus*
  B. speciosus
  Corallobothrium sp.
  Ligula intestinalis

* Larval forms.
** Immature adults.

*Proteocephalus
ambloplitis*
  Proteocephalus sp.
Nematoda:
  **Contracaecum* sp.
  Contracaecum sp.
  Dichelyne sp.
  Rhabdochona cascadilla
  Spiroxys sp.
Acanthocephala:
  *Leptorhynchoides
  thecatus*
  L. thecatus
  *Neoechinorhynchus
  cylindratus*
  *Pomphorhynchus
  bulbocolli*
  P. bulbocolli
Leech:
  Illinobdella sp.
Mollusca:
  Glochidia
Etheostoma pellucidum,
sand darter
Trematoda:
  Lebouria cooperi
  Neascus sp.
Nematoda:
  Camallanus oxycephalus
Etheostoma phoxocephalum
Leech:
  Piscicolaria reducta
Etheostoma proeliares,
Cypress darter
Trematoda:
  Neochasmus umbellus
Nematoda:
  *Contracaecum
  spiculigerum*
  *Leptorhynchoides
  thecatus*
Hadropterus malleus
Trematoda:
  Cleidodiscus malleus
Perca flavescens, yellow perch
(European *P. fluviatilis* very
similar)

Protozoa:
  Balantidium sp.
  Henneguya doori
  H. percae
  H. wisconsinensis
  *Ichthyophthirius
    multifilis*
  Myxidium percae
  Myxobolus percae
  M. pyriformis
  Myxobolus sp.
  Myxosoma neurophila
  M. scleroperca
  Thelohanellus piriformis
  Trichodina sp.
  Trichophrya piscium
  Trypanosoma percae
  *Trypanosoma percae
    canadensis*
Trematoda:
  Ancyrocephalus sp.
  *Apophallus americanus
  *A. itascensis
  *A. venustas
  *Apophallus sp.
  Asymphylodora sp.
  Azygia acuminata,
    exper.
  A. angusticauda
  A. longa
  A. sebago
  Azygia sp.
  Bucephalopsis pusillum
  Bucephalus elegans
  Bundodera luciopercae
  B. nodulosum
  B. sacculata
  Centrovarium lobotes
  Cleidodiscus sp.
  **Clinostomum
    marginatum*
  *Crassiphiala bulboglossa
  Crepidostomum cooperi
  C. farionis
  C. laureatum (syn. of *C.
    canadense* ?)
  C. solidum

  Cryptogonimus chyli
  **Diplostomulum
    huronense*
  *D. scheuringi
  *Diplostomulum sp.
  Distomum nodulosum
  **Echinochasmus
    donaldsoni*
  *Euparyphium melis
  Gyrodactylus sp.
  Leuceruthrus sp.
  Maritrema medium
  Microphallus medius
  M. opacus
  *Neascus ellipticus
  *N. longicollis
  *N. oneidensis
  *N. pyriformis
  *Neascus sp.
  *Petasiger nitidus
  *Phyllodistomum
    americanum*
  P. superbum
  **Posthodiplostomum
    minimum*
  Ptychogonimus fontanis
  *Sanguinicola
    occidentalis*
  Stephanophiala farionis
  *Tetracotyle diminuta
  *Tetracotyle sp.
  Urocleidus adspectus
  *Uvulifer ambloplitis
Cestoda:
  ***Bothriocephalus
    cuspidatus*
  *Diphyllobothrium latum
  *Ligula intestinalis
  *Proteocephalus
  P. ambloplitis
  P. pearsei
  **P. pearsei
  P. pinguis
  *Proteocephalus sp.
  *Triaenophorus nodulosus
  *T. tricuspidatus

* Larval forms.
** Immature adults.

Nematoda:
**Camallanus oxycephalus*
**Camallanus* sp.
Capillaria catenata
Contracaecum
    brachyurum
C. spiculigerum
Contracaecum sp.
Dacnitoides cotylophora
***D. cotylophora*
**Eustrongylides* sp.
Philometra cylindracea
**P. cylindracea*
Rhabdochona
    ovifilamenta
***Rhabdochona* sp.
Spinitectus carolini
S. gracilis
Spinitectus sp.
**Spiroxys* sp.
Acanthocephala:
Acanthocephalus jacksoni
Echinorhynchus coregoni
E. lateralis
E. salmonis
E. salvelini
Leptorhynchoides
    thecatus
**L. thecatus*
Neoechinorhynchus
    cylindratus
Pomphorhynchus
    bulbocolli
**P. bulbocolli*
Leech:
Actinobdella sp.
Illinobdella alba
I. moorei
Illinobdella sp.
Piscicolaria sp.
Placobdella picta
Mollusca:
**Glochidia*
Crustacea:
Actheres lacae
Argulus appendiculosus
A. biramosus
A. catostomi

A. stizostethi
A. versicolor
Ergasilus caeruleus
A. confusus
Percina caprodes, log perch
Protozoa:
Myxosoma scleroperca
Trichodina sp.
Trematoda:
Allocreadium boleosomi
Azygia longa
Cleidodiscus malleus
Crepidostomum
    isostomum
**Diplostomulum*
    scheuringi
**Euparyphium melis*
**Neascus* sp.
Phyllodistomum
    etheostomae
**Tetracotyle* sp.
Urocleidus malleus
Cestoda:
Bothriocephalus
    cuspidatus
B. formosus
Proteocephalus pearsi
P. punctata
P. stizostethi
Proteocephalus sp.
Nematoda:
Camallanus oxycephalus
Camallanus sp.
***Contracaecum* sp.
Spinitectus gracilis
Acanthocephala:
Leptorhynchoides
    thecatus
**L. thecatus*
Neoechinorhynchus
    cylindratus
**Pomphorhynchus*
    bulbocolli
P. bulbocolli
Leech:
Illinobdella moorei
Illinobdella sp.

* Larval forms.
** Immature adults.

Percina nigrofasciata,
blackbanded darter
Trematoda:
  Gyrodactylus percinae
  Urocleidus nigrofasciata
Stizostedion canadense,
sauger
Trematoda:
  Bucephalus pusillus
  Centrovarium lobotes
  *Cotylurus communis
  *Diplostomulum sp.
  Gasterostomum pusillum
  *Neascus sp.
Cestoda:
  Bothriocephalus claviceps
  B. cuspidatus
  *Diphyllobothrium latum
  Proteocephalus
    luciopercae
  P. stizostethi
  *Triaenophorus sp.
Nematoda:
  Camallanus oxycephalus
Acanthocephala:
  Echinorhynchus salmonis
  Neoechinorhynchus
    cylindratus
Mollusca:
  Glochidia
Crustacea:
  Argulus stizostethi
  Ergasilus caeruleus
  E. centrarchidarum
  Lernaea cruciata
  L. variabilis
  Lernaeocera sp.
Leech:
  Piscicola punctata
Stizostedion vitreum glaucum,
blue pike
Cestoda:
  Bothriocephalus
    cuspidatus
  Proteocephalus
    ambloplitis
  P. stizostethi

Crustacea:
  Argulus stizostethi
Stizostedion v. vitreum, wall-
eye (European *Lucioperca
lucioperca* very similar)
Protozoa:
  Henneguya asymmetricus
  Myxobolus sp.
  Trichodina sp.
Trematoda:
  Ancyrocephalus
    aculeatus
  *Apophallus americanus
  *A. venustus
  Azygia acuminata
  A. angusticauda
  A. bulbosa
  Bucephalopsis pusilla
  Bunodera sacculata
  Centrovarium lobotes
  Cleidodiscus aculeatus
  *Clinostomum
    marginatum
  *Crassiphiala
    bulboglossa
  Crepidostomum sp.
  *Diplostomulum
    scheuringi
  *Diplostomulum sp.
  *Neascus sp.
  Phyllodistomum
    superbum
  *Posthodiplostomum
    minimum
  Sanguinicola occidentalis
  *Tetracotyle communis
  Urocleidus aculeatus
  *Uvulifer ambloplitis
Cestoda:
  Bothriocephalus
    cuspidatus
  *Diphyllobothrium latum
  Proteocephalus
    ambloplitis
  P. luciopercae
  P. macrocephalus
  **P. pearsei
  P. stizostethi

* Larval forms.
** Immature adults.

Proteocephalus sp.
**Triaenophorus
nodulosus*
T. stizostedionis
Nematoda:
Camallanus oxycephalus
Capillaria catenata
*Contracaecum
brachyurum*
Contracaecum sp.
*Dactinitoides
cotylophora*
Oxyurid (accidental?)
Philometra cylindracea
Spinitectus carolini
S. gracilis
Spinitectus sp.
Acanthocephala:
*Leptorhynchoides
thecatus*
*Neoechinorhynchus
cylindratus*
N. tenellus
*Pomphorhynchus
bulbocolli*
Leech:
Cystobranchus verrilli
Iillinobdella moorei
Mollusca:
*Glochidia
Crustacea:
Argulus biramosus
A. appendiculosus
A. stizostethi
Ergasilus caeruleus
E. centrachidarum
E. confusus
E. luciopercarum
PERCOPSIDAE (Trout-
perches)
Percopsis omiscomaycus,
trout-perch
Protozoa:
Myxosoma procerum
Trematoda:
Centrovarium lobotes
*Crepidostomum
isostomum*

**Diplostomulum
huronense*
*D. scheuringi
*Diplostomulum sp.
*Neascus sp.
**Posthodiplostomum
minimum*
*Tetracotyle communis
*T. diminuta
*Tetracotyle sp.
Cestoda:
*Bothriocephalus
claviceps*
B. formosus
Proteocephalus pearsei
*Triaenophorus crassus
*T. stizostedionis
Nematoda:
Camallanus oxycephalus
*Contracaecum
brachyurum*
Spinitectus gracilis
Acanthocephala:
Echinorhynchus salmonis
*Leptorhynchoides
thecatus*
Neoechinorhynchus sp.
**Pomphorhynchus
bulbocolli*
Crustacea:
Argulus sp.
Ergasilus caeruleus
PETROMYZONIDAE
(Lampreys)
Lampetra tridentatas, Pacific
lamprey
Trematoda:
*Nanophyetus salmincola
Cestoda:
Eustrongylides sp.
**Phyllobothrium sp.
Petromyzon marinus, sea
lamprey
Fungi:
Saprolegnia sp.
Protozoa:
*Ichthyophthirius
multifilis*

* Larval forms.
** Immature adults.

Trematoda:
Diplostomulum sp.
Cestoda:
**Abothrium* sp.
Proteocephalus exiguus
P. laruei
Proteocephalus sp.
***Triaenophorus crassus*
Nematoda:
Camallanus sp. (?)
Cystidicola stigmatura
Acanthocephala:
Echinorhynchus coregoni
E. cylindratus
E. leidyi
E. salmonis
Neoechinorhynchus cylindratus
POECILIIDAE (Livebearers)
Gambusia affinis, mosquitofish
Trematoda:
Allacanthochasmus sp.
Ascocotyle mcintoshi
A. tenuicollis
Bolbophorus confusus
Diplostomulum scheuringi
Echinochasmus pelecani
Gyrodactylus gambusiae
Heterophyes aequalis
Nanophyetus salmincola
Paramacroderoides echinus
Posthodiplostomum minimum
Prohemistomulum expeditum (Africa)
Szidatia joyeuxi
Urocleidus seculus
Cestoda:
**Proteocephalus* sp.
Nematoda:
**Camallanus* sp.
Acanthocephala:
Leptorhynchoides thecatus
Neoechinorhynchus cylindratus

* Larval forms.
** Immature adults.

Octospiniforoides chandleri
Crustacea:
Lernaea catostomi
L. cyprinacea
Heterandria formosa, least killifish
Trematoda:
Paramacroderoides echinus
Lebistes reticulatus, guppy
Trematoda:
Azygia sebago
Bolbophorus confusus, exper.
Clinostomum marginatum
Echinochasmus donaldsoni
Gyrodactylus bullatarudis
G. medius (E)
Plagiporus sinitsini, exper.
Posthodiplostomum minimum, exper.
Ribeiroia ondatrae, exper.
Stephanoprora (?), exper.
Urocleidoides reticulatus
Cestoda:
Haplobothrium globuliforme, exper.
Mollienesia latipinna, sailfin molly
Trematoda:
Ascocotyle angrense
A. chandleri
A. leighi
A. mcintoshi
A. tenuicollis
Echinochasmus donaldsoni
Parascocotyle diminuta
Posthodiplostomum minimum

*Pseudoascocotyle
 molliensicola
Saccocoelioides
 sogandaresi (marine)

Mollienesia pelenensis
Crustacea:
 Ergasilus boettgeri

Mollienesia sphenops
Trematoda:
 *Ascocotyle leighi

POLYODONTIDAE
(Paddlefishes)

Polyodon spathula, paddlefish
Trematoda:
 Anallocreadium spathula
 *Clinostomum marginatum
 Cotylaspis cokeri
 Diclybothrium
  hamulatum
 Distomum isoporum
  armatum
 Halipegus perplexus
Cestoda:
 Marsipometra confusa
 M. hastata
 M. parva
Nematoda:
 Camallanus oxycephalus
 *Contracaecum
  spiculigerum
Leech:
 Illinobdella moorei
Crustacea:
 Ergasilus elongatus

SALMONIDAE†

Coregonus alpenae, longjaw
cisco
Trematoda:
 *Diplostomulum sp.
Acanthocephala:
 Echinorhynchus salmonis

Coregonus artedi, cisco
Trematoda:
 *Clinostomum marginatum
 Crepidostomum cooperi
 C. farionis
 *Diplostomulum sp.

Discocotyle salmonis
*Tetracotyle intermedia
Cestoda:
 Abothrium crassum
 *Diphyllobothrium
  oblongatum
 *Diphyllobothrium sp.
 Diplocotyle olrikii
 Eubothrium crassum
 Proteocephalus exiguus
 P. filicollis
 P. laruei
 P. wickliffi
 *Trianeophorus crassus
Nematoda:
 Cystidicola stigmatura
 Hepaticola bakeri
 *Philometra sp.
 Spinitectus gracilis
Acanthocephala:
 Echinorhynchus leidyi
 E. salmonis
 *E. salmonis
 Neoechinorhynchus
  cylindricus
 N. tumidus
Crustacea:
 Achtheres coregoni
 A. corpulentus
 Argulus sp.
 Ergasilus caeruleus
 E. confusus
 Salmincola inermis
 S. wisconsinensis

Coregonus clupeaformis, lake
whitefish
Protozoa:
 Cryptobia gurneyorum
Trematoda:
 Crepidostomum cooperi
 *Diplostomulum sp.
 Discocotyle salmonis
 Phyllodistomum coregoni
Cestoda:
 Abothrium crassum
 Cyathocephalus
  americanus

* Larval forms.
† Dr. Leo Margolis, Biological Station, Nanaimo, B. C., Canada, is preparing a monograph on the parasites of salmon.

C. truncatus
*Diphyllobothrium sp.
Diplocotyle olrikii
Proteocephalus exiguus
P. laruei
P. wickliffii
*Trianeophorus crassus
T. robustus
Nematoda:
Cystidicola serrata
C. stigmatura
*Philometra sp.
Spinitectus gracilis
Acanthocephala:
Echinorhynchus coregoni
E. salmonis
Leptorhynchoides
thecatus
Leech:
Piscicola milneri
P. punctata
Crustacea:
Achtheres ambloplitis
A. coregoni
A. corpulentus
Argulus canadensis
A. stizostethi
Ergasilus caeruleus
E. confusus
Ergasilus sp.
Salmincola inermis
Coregonus hoyi, bloater
Trematoda:
*Diplostomulum sp.
Discocotyle salmonis
Cestoda:
Cyathocephalus
americanus
*Diphyllobothrium sp.
Proteocephalus exiguus
P. laruei
*Triaenophorus crassus
Nematoda:
Cystidicola stigmatura
Acanthocephala:
Echinorhynchus leidyi
E. salmonis

Leech:
Piscicola milneri
Crustacea:
Salmincola inermis
S. wisconsinensis
Coregonus johannae, deep-
water cisco
Nematoda:
*Dacnitoides cotylophora
Crustacea:
Achtheres corpulentus
Coregonus kiyi, kiyi
Cestoda:
*Diphyllobothrium sp.
Crustacea:
Salmincola inermis
Coregonus nelsonii, humpback
whitefish
Crustacea:
Salmincola inermis
Coregonus reighardi,
shortnose cisco
Crustacea:
Salmincola inermis
Coregonus zenithicus,
shortjaw cisco
Cestoda:
Cyathocephalus
truncatus
Protecephalus laruei
Coregonus spp., ciscoes
Protozoa:
Henneguya tegidiensis
(Eng)
Trematoda:
Plagioporus sp. (E)
Cestoda:
Bothriocephalus
cuspidatus
Diphyllobothrium
laruei
Triaenophorus
tricuspidatus
Nematoda:
Philometra cylindraceum
Acanthocephala:
Echinorhynchus
coregoni
E. salvelini

* Larval forms.

Crustacea:
  Argulus coregoni (E)
Leech:
  Piscicola milneri
Oncorhynchus gorbuscha,
pink salmon
Protozoa:
  Cryptobia makeevi (R)
  C. salmositica
  Myxidium oviforme (R)
  Myxosoma cerebralis (R)
  Trichodina truttae (R)
Trematoda:
  Brachyphallus
    crenatus (R)
  Bucephalopsis
    gracilescens (R)
  Crepidostomum
    farionis (R)
  Derogenes varicus (R)
  Genolinea anura (R)
  G. oncorhynchi
  Gyrodactyloides
    strelkowi (R)
  Hemiurus
    appendiculatus (R)
  H. levinseni
    (R, Brit. Col.)
  **Lampritrema nipponicum*
    (Brit. Col.)
  Lecithaster gibbosus (R)
  L. stellatus (R)
  Parahemiurus merus (R)
  Tubulovesicula
    lindbergi (R)
Cestoda:
  Eubothrium crassum (R)
  Nybelinia
    lingualis (?) (R)
  N. surminicola (R)
  Pelichnibothrium
    speciosum (R)
  Scolex pleuronectis (R)
Nematoda:
  Anisakis simplex (R)
  Ascarophis skrjabini (R)
  Contracaecum
    aduncum (R)

  Cucullanus laevis (R)
Acanthocephala:
  Bolbosoma
    caenoforme (R)
  Echinorhynchus gadi (R)
  Paracanthocephalus
    tenuirostris (R)
Leech:
  Piscicola salmostica
Crustacea:
  Argulus coregoni (R)
  Lepeophtheirus salmonis
    (R, N. Amer.)
  Salmincola falculata (R)
Oncorhynchus keeta, chum
salmon
Protozoa:
  Cryptobia makeevi (R)
  Myxidium oviforme (R)
  Myxosoma cerebralis (R)
  M. dermatobia (R)
  M. squamalis
  Henneguya
    zschokkei (R)
  Trichodina truttae (R)
  Tripartiella
    californica (R)
  T. pungitii (R)
Trematoda:
  Brachyphallus
    crenatus (R)
  Bucephalopsis
    gracilescens (R)
  Isoparorchis
    hypselobagri (R)
  Lecithaster gibbosus (R)
  L. stellatus (R)
  Nanophyetus salmincola
  Tubulovesicula
    lindbergi (R)
Cestoda:
  Eubothrium crassum (R)
  Hepatoxylon
    trichiuri (R)
  Nybelinia lingualis (R)
  Pelichnibothrium
    speciosum

* Larval forms.
** Immature adults.

Phyllobothrium ketae
(R, Alaska)
Proteocephalus exiguus
*Scolex pleuronectis (R)
*Triaenophorus crassus
Nematoda:
*Anisakis simplex (R)
Contracaecum
aduncum (R)
Philonema
oncorhynchi (R)
Acanthocephala:
*Bolbosoma
caenoforme (R)
Corynosoma
villosum (R)
Echinorhynchus gadi (R)
Crustacea:
Argulus coregoni (R)
Ergasilus briani (R)
Lepeophtheirus
salmonis (R)
Salmincola lata (R)
S. salmonea (R)
Oncorhynchus kisutch, coho
salmon (silver salmon)
Fungi:
Ichthyosporidium
Protozoa:
Cryptobia borreli
C. lynchi, exper.
C. salmositica
Henneguya salmincola
H. zschokkei (R)
Myxidium minterl
Myxobolus kisutchi
M. squamae
Myxosoma
dermatobia (R)
M. squamalis
Trematoda:
Brachyphallus
crenatus (R)
Crepidostomum farionis
Deropegus aspina
Derogenes sp.
Lecithaster gibbosus (R)
*Nanophyetus salmincola

Podocotyle shawi
Tetraonchus alaskensis
Cestoda:
Cyathocephalus
truncatus (R)
*Diphyllobothrium
cordiceps
*Diphyllobothrium sp.
Eubothrium crassum (R)
*Nybelinia
surminicola (R)
*Pelichnibothrium
speciosum (R)
Proteocephalus sp.
*Scolex pleurnectis (R)
Nematoda:
Contracaecum aduncum
Cystidicola farionis (R)
C. stigmatura
Hepaticola bakeri
Metabronema salvelini
Philonema oncorhynchi
Rhabdochona sp.
Acanthocephala:
Echinorhynchus gadi (R)
E. lateralis
Neoechinorhynchus rutili
Crustacea:
Argulus pugettensis
Oncorhynchus nerka, sockeye
salmon
Fungi:
Ichthyosporidium sp.
Protozoa:
Chloromyxum wardi
Henneguya
zschokkei (R)
Myxidium oviforme (R)
Myxobolus
neurobius (R)
Trichophrya piscium
Trematoda:
Brachyphallus
crenatus (R)
Crepidostomum farionis
Derogenes varicus (R)
**Lampritrema nipponicum
Lecithaster gibbosus (R)

* Larval forms.
** Immature adults.

L. salmonis
Podocotyle shawi
Syncoelium filiferum
Cestoda:
  Cyathocephalus
    tuncatus (R)
  *Diphyllobothrium ursi
  *Diphyllobothrium sp.
  Eubothrium crassum (R)
  E. salvelini
  Eubothrium sp.
  *Pelichnibothrium
    speciosum (R)
  Phyllobothrium sp.
  Proteocephalus arcticus
  Proteocephalus sp.
  *Scolex pleuronectis (R)
  Triaenophorus crassus
Nematoda:
  Contracaecum
    aduncum (R)
  C. spiculigerum
  Dacnitis truttae
  Metabronema salvelini
  Philonema oncorhynchi
  Rhabdochona cascadilla
Acanthocephala:
  Acanthocephalus
    aculeatus
  Bolbosoma caenoforme
  Echinorhynchus gadi
  Neoechinorhynchus rutili
Mollusca:
  *Glochidium (Anodonta,
    Alaska)
Crustacea:
  Ergasilus auritus
  E. nerkae
  E. turgidus
  Lernaeopoda falculata
  Salmincola californiensis
  S. carpenteri
  S. falculata
  Salmincola sp.

**Oncorhynchus nerka
kennerlyi,** landlocked
salmon
Protozoa:
  Plistophora sp.

Trematoda:
  Crepidostomum farionis
  Crepidostomum sp.
  Podocotyle shawi
Cestoda:
  *Dibothriocephalus sp.
  *Diphyllobothrium sp.
  Eubothrium sp.
  Proteocephalus exiguus
  P. laruei
  Schistocephalus sp.
Nematoda:
  Metabronema salvelini
  Philonema oncorhynchi
  Rhabdochona cascadilla
Acanthocephala:
  Neoechinorhynchus
    rutili
  Pomphorhynchus
    bulbocolli
Crustacea:
  Ergasilus caeruleus
  Salmincola edwardsi
  Salmincola sp.

Oncorhynchus tschawytscha,
chinook
Fungi:
  Dermocystidium salmonis
  Ichthyosporidium sp.
Protozoa:
  Chloromyxum majori
  Cryptobia borreli
  Henneguya salmincola
  Myxidium minteri
  M. oviforme (R)
  Myxobolus insidiosus
  Trichodina californica
  Trichodina sp.
  Tripartiella
    californica (R)
Trematoda:
  Brachyphallus crenatus
  Deropegus aspina
  Lecithaster sp.
  Nanophyetus salmincola
Cestoda:
  Eubothrium crassum
  Eubothrium sp.

* Larval forms.

*Nybelina
   surminicola (R)
*Pelichnibothrium
   speciosum (R)
Phyllobothrium salmonis
Nematoda:
   *Anisakis sp.
   Contracaecum
      aduncum (R)
   C. spiculigerum
   *Contracaecum sp.
   Cucullanus truttae
   Rhaphidascaris sp.
Leech:
   Pisicola salmositica
Crustacea:
   Argulus sp.
   Lernaea cyprinacea
   Salmincola beani
   S. falculata
Mollusca:
   Glochidium (Anadonta,
      Alaska)
Oncorhynchus spp.
Protozoa:
   Trichodina domerguei f.
      latispina
   T. gracilis
Trematoda:
   Lecithophyllum spp.
      (marine)
   Plagioporus
      onchorhynchi (Japan)
Cestoda:
   Eubothrium oncorhynchi
   E. salvelini
Acanthocephala:
   Corynosoma sp.
Prosopium cylindraceum
   quadrilaterale
Trematoda:
   Crepidostomum farionis
   *Diplostomulum sp.
   Discotyle salmonis
   Tetraonchus variabilis
Cestoda:
   Proteocephalus exiguus
Nematoda:
   Cystidicola stigmatura

*Philometra sp.
Acanthocephala:
   Echinorhynchus salmonis
   Neoechinorhynchus
      tumidus
Crustacea:
   Achtheres coregoni
   Salmincola sp.
Prosopium quadrilaterale
   (frostfish), river whitefish
Trematoda:
   Crepidostomum cooperi
   C. farionis
   Tetracotyle intermedia
Cestoda:
   Proteocephalus laruei
Crustacea:
   Ergasilus sp.
Prosopium transmontanus
Acanthocephala:
   Neoechinorhynchus
      rutili
Prosopium w. williamsoni [or
   P. w. cismontanus (Jordan)]
Trematoda:
   Allocreadium lobatus
   Bolbophorus confusus
   Crepidostomum farionis
   *Diplostomulum sp.
   Discocotyle salmonis
   Podocotyle shawi
   *Tetracotyle sp.
   Tetraonchus variabilis
Cestoda:
   *Diphyllobothrium sp.
   Eubothrium sp.
   *Ligula intestinalis
   Proteocephalus exiguus
   *P. laruei
   *Schistocephalus sp.
Nematoda:
   Bulbodacnitis globosa
   B. occidentalis
   B. scotti
   Cystidicola stigmatura
   *Eustrongylides sp.
   Hepaticola bakeri
   Metabronema salvelini

* Larval forms.

Philonema
   agubernaculum
*Philonema sp.
Rhabdochona cascadilla
Acanthocephala:
   Neoechinorhynchus rutili
   Pomphorhynchus
      bulbocolli
Crustacea:
   Achtheres coregoni
   Ergasilus sp.
   Salmincola sp.
Salmo aquabonita, golden
trout, California
Trematoda:
   Crepidostomum farionis
Salmo arcturus, Greenland
coast
Crustacea:
   Salmincola arcturi
Salmo carpio, Greenland coast
Crustacea:
   Salmincola carpionis
Salmo clarki, cutthroat trout
Protozoa:
   Costia necatrix
   Ichthyophthirius
      multifilis
   Myxidium sp.
   Octomitus sp.
   Trichodina truttae
Trematoda:
   Allocreadium lobatum
   *Apophallus sp.
   *Clinostomum
      marginatum
   Crepidostomum farionis
   C. transmarinum
   Crepidostomum sp.
   Deropegus aspina
   *"Diplostomulum
      oregonensis"
   Gyrodactylus elegans B.
   *Nanophyetus salmincola
   Plagioporus siliculus
   Podocotyle sp.
   *Posthodiplostomum
      minimum
   Sanguinicola sp.

Cestoda:
   Cyathocephalus truncatus
   Cyathocephalus sp.
   *Diphyllobothrium
      cordiceps
   *Diphyllobothrium sp.
   Eubothrium salvelini
   Proteocephalus arcticus
   P. laruei
   P. primaverus
   P. salmonidicola
   Proteocephalus sp.
Nematoda:
   Ascarophis hardwoodi
   Bulbodacnitis globosa
   B. scotti
   Capillaria catenata
   Capillaria sp.
   Contracaecum sp.
   Cucullanus truttae
   Cystidicola stigmatura
   Cystidicoloides spp .
   *Eustrongylides sp.
   Hepaticola bakeri
   Metabronema salvelini
   Philometra sp.
   Philonema onchorhynchi
   Rhabdochona cascadilla
   Rhabdochona sp.
Acanthocephala:
   Echinorhynchus lateralis
   Neoechinorhynchus rutili
   Neoechinorhynchus sp.
Leech:
   Illinobdella sp.
   Piscicola salmositica
Mollusca:
   *Glochidia
Crustacea:
   Lepeophtheirus salmonis
   Lernaeopoda
      bicauliculata
   Salmincola edwardsii
   Salmincola sp.
Salmo gairdneri, rainbow trout
(steelhead)
Fungi:
   Ichthyophonus sp.

* Larval forms.

Protozoa:
- *Ceratomyxa* sp.
- *Chilodonella cyprini*
- *Chloromyxum majori*
- *C. truttae*
- *Costia necatrix*
- *C. pyriformis*
- *Cryptobia borreli*
- *C. lynchi*
- *Cyclochaeta* sp.
- *Hexamita salmonis*
- *Ichthyopthirius
    multifilis*
- *Myxidium oviforme*
- *M. minteri*
- *Myxosoma cerebralis*
- *M. squamalis*
- *Plistophora salmonae*
- *Schizamoeba salmonis*
- *Trichodina fultoni*
- *Trichophrya piscium*

Trematoda:
- *Allocreadium lobatum*
- *Aponurus* sp.
- *Bolbophorus confusus*
- *Clinostomum
    marginatum*
- *Crepidostomum cooperi*
- *C. farionis*
- *C. laureatum*
- *Deropegus aspina*
- *Diplostomum
    flexicaudum*
- *D. spathaceum*
- *Diplostomulum* sp.
- *Discocotyle salmonis*
- *"Distomulum
    oregonensis"*
- *Echinochasmus milvi*
- *Exocoitocaecum
    wisnienskii* (E)
- *Gyrodactylus "elegans"*
- *Nanophyetus salmincola*
- *Neascus* sp.
- *Phyllodistomum
    lachancei*
- *Phyllodistomum* sp.
- *Plagioporus angusticole*

Podocotyle shawi
Sanguinicola sp.

Cestoda:
- *Abothrium crassum*
- *Cyathocephalus truncatus*
- *Diphyllobothrium* sp.
- *Eubothrium salvelini*
- *Ligula intestinalis*
- *Phyllobothrium* sp.
- *Proteocephalus
    ambloplitis*
- *P. longicollis* (E)
- **P. pinguis*
- *P. salmonidicola*
- *P. tumidocollis*
- *Proteocephalus* sp.
- *Schistocephalus* sp.
- *Triaenophorus
    nodulosus* (R)

Nematoda:
- *Anisakis* sp.
- *Ascaris acus* (Italy)
- *Ascarophis hardwoodi*
- *A. skrjabini* (R)
- *Bulbodacnitis globosa*
- *B. occidentalis*
- *Capillaria eupomotis*
    (liver, Italy)
- *Contracaecum
    spiculigerum*
- *Cucullanus globosus*
- *C. occidentalis*
- *C. truttae*
- *Cystidicola farionis* (R)
- *C. stigmatura*
- *Cystidicoloides
    harwoodi*
- *Dacnitis truttae*
- *Eustrongylides* sp.
- *Goezia ascaroides* (R)
- *Hepaticola bakeri*
- *Metabronema salvelini*
- *Philometra* sp.
- *Philonema
    agubernaculum*
- *P. angusticole*
- *P. oncorhynchi*

* Larval forms.
** Immature adults.

Raphidascaris acus
  (Italy)
Rhabdochona cascadilla
R. denudata (R)
Spinitectus gracilis
Sterliadochona
  tenuissima (Poland)
Acanthocephala:
  Acanthocephalus acerbus
  A. anguillae (E)
  A. jacksoni
  Echinorhynchus leidyi
  E. salmonis
  E. truttae (E)
  Metechinorhynchus
    salmonis (R)
  M. truttae (R)
  Neoechinorhynchus
    rutili
  Pomphorhynchus
    bulbocolli
  Rhadinorhynchus sp.
  Tetrahynchus sp.
Leech:
  Illinobdella sp.
  Piscicola geometra (R)
  P. salmositica
Mollusca:
  Glochidia
  Glochidia of *Margaritana*
    margaritifera
  Pisidium variable
  Pisidium (adults,
    attacking fry)
Gordiacea:
  Chorodes sp. (viscera)
Crustacea:
  Argulus pugettensis
  Ergasilus caeruleus
  E. sieboldi (Poland)
  Lernaea cyprinacea
  L. esocina (E ?)
  Lernaeopoda
    bicauliculata
  Salmincola beani
  S. edwardsii
Salmo mykiss
Trematoda:
  Crepidostomum farionis

C. fausti
C. transmarinum
Tetraonchus alaskensis
Cestoda:
  Diphyllobothrium
    cordiceps
Nematoda:
  Philonema
    oncorhynchi (R)
Acanthocephala:
  Echinorhynchus
    globulosus
  Metechinorhynchus
    salmonis (R)
  Neoechinorhynchus
    crassus (R)
Crustacea:
  Salmincola carpenteri
Salmo salar, Atlantic salmon
Protozoa:
  Chloromyxum
    histolyticum (France)
  Henneguya salmonis
  Hexamita truttae (R)
  Myxidium oviforme
  Myxosoma cerebralis (R)
  Trichophrya intermedia
Trematoda:
  Aphanurus balticus
    (Baltic)
  Azygia lucii (R)
  A. tereticollis
  Brachyphallus crenatus
  Bucephalus
    polymorphus (R)
  Bunocotyle cingulata (R)
  Crepidostomum farionis
    (Eng, R)
  C. metoecus (Eng, R)
  Derogenes varicus
  Diplostomulum
    spathaceum (R)
  Discocotyle sagitta (E)
  D. salmonis
  D. sybillae (E)
  Distomum
    appendiculatum
  D. reflexum
  D. varicum

* Larval forms.

Gyrodactyloides
  bychowskii (R)
Gyrodactylus salaris
  (Sweden)
Gyrodactylus sp. (E)
Hemiurus crenatus
H. levinseni (R)
H. lukei
H. ocreatus (R)
Lampritrema miescheri
  (*Distomum*)
Lecithaster
  bothryophorus
L. confusus (R)
L. gibbus (R)
*Nanophyetus salmincola
Neohemiurus (Poland)
Podocotyle atomon (R)
P. simplex
Cestoda:
  *Bothriocephalus osmeri
  B. proboscideus
  *B. solidus
    (= *Schistocephalus*)
  *Diphyllobothrium
    cordiceps
  D. latum ? (R)
  *D. norvegicum (R)
  *Hepatoxylon trichiuri
    (Ireland)
  Leuckartia sp.
  *Schistocephalus
    dimorphis
  *Scolex pleuronectis (R)
  *S. polymorphus
  *Stenobothrium
    appendiculum
  *Tetrabothrum minimum
  *Tetrahynchobothrium
    bicolor
  *Tetrahynchus grossus
  *T. solidus
  *Triaenophorus
    crassus (R)
  *T. nodulosus
Nematoda:
  *Agomonema capsularia
  *A. commune

Anisakis sp. (R)
Cumullanus lacustris (R)
Contracaecum
  aduncum (R)
Cucullanus elegans
C. serratus (R)
C. truttae (R)
Exocoitocaecum
  wisniewskii (Poland)
Raphidascaris acus (R)
Leech:
  Cystobranchus
    respirans (R)
Acanthocephala:
  Acanthocephalus
    anguillae (E)
  A. lucii (E)
  **Bolbosoma
    heteracanthis (E)
  Echinorhynchus acus
  E. gadi
  E. inflatus
  E. salmonis
  E. truttae (E)
  Metechinorhynchus
    salmonis (R)
  Neoechinorhynchus
    cylindratum
  N. rutili
  Pomphorhynchus
    laeve (E)
  P. proteus
  Pseudoechinorhynchus
    clavula (R)
Crustacea:
  Lepeophtheirus
    pollachii (R)
  L. salmonis
  L. stromii
  Salmincola falculata (R)
  S. salmonae (R)
Leech:
  Cystobranchus respirans
    (*Pisicola*)
  Pisicola geometra
Crustacea:
  Argulus sp.

* Larval forms.
** Immature adults.

Lepeophtheirus pollachii
(Russia)
L. salmonis
L. Strömii (E)
Salmincola falculata (R)
Salmo salar sebago, landlocked
Atlantic salmon
Trematoda:
Azygia longa
A. sebago
Crepidostomum farionis
Cestoda:
*Diphyllobothrium sebago
Eubothrium crassum
E. salvelini
Proteocephalus pusillus
Nematoda:
Philonema
agubernaculum
Acanthocephala:
Leptorhynchoides
thecatus
Leech: ·
Pisicola milneri
Salmo trutta, brown trout
Fungi:
Dermocystidium
branchialis (E)
D. pusula (E)
D. trutta (E)
Dermosporidium
trutta (E)
Saprolegina parasitica
Protozoa:
Chloromyxum truttae
Cryptobia borreli
C. truttae (E)
C. valentini (E)
Eimeria truttae (R, E)
Epistylis sp.
Hexamita intestinalis
H. truttae (R)
Ichthyophthirius multifilis
Myxidium truttae (E)
Myxobolus
neurobius (E)
Myxosoma cerebralis (E)
Trematoda:
*Apophallus brevis, exper.

Azygia longa
A. luci (R)
A. robusta (R)
*Bolbophorus confusus
(Montana)
Brachyphallus
crenatus (R)
Bunocotyle cingulata (R)
Bunodera lucioperca
Coitocaecum spp. (E)
Crepidostomum cooperi
C. farionis
C. metoecus
*Diplostomulum
scheuringi
*D. spathaceum (R)
*D. truttae (Scotland)
Discocotyle sagittata (E)
D. salmonis
Gyrodactylus elegans B.
G. salaris (R)
Hemiurus communis
*Nanophyctus salmincola
*Neascus sp.
Nicolla timoni (Corsica)
Phyllodistomum
megalorchis (R)
P. simile (Eng)
Plagioporus stefanski
(Poland)
Pseudochaetosoma
salmonicola
(Yugoslavia)
Sphaerostoma
globiporum (R)
S. majus (R)
S. salmonis (Poland)
*Tetracotyle
intermedia (R)
Tubulovesicula
lindbergi (R)
Cestoda:
*Cythocephalus
truncatus (E)
*Diphyllobothrium
dendriticum (R)
*D. latum ? (R)
*D. norvegicum (E)

* Larval forms.

*Diphyllobothrium sp.
(Chile)
*Diphyllobothrium sp.
Eubothrium crassum (R)
Proteocephalus
longicollis ? (R)
P. neglectus (R)
P. parallacticus
P. pinguis
Proteocephalus sp.
*Triaenophorus
nodulosus (R)
Nematoda:
Ascarophis skrjabini (R)
*Bulbodacnitis scotti
**Camallanus oxycephalus
Cephalobus ?
Contracaecum
aduncum (R)
Contracaecum sp.
**Contracaecum sp.
Cucullanus globosus
C. truttae (R)
Cystidicola farionis (R)
Cystidicoloides harwoodi
*Eustrongylides gadopsis
Exocoitocaecum
wisnieskii (Poland)
Metabronema salvelini
*Philonema
agubernaculum
Raphidascaris acus (E)
Rhabdochona
denudata (R)
R. filamentosa (R)
Spinitectus gracilis
Sterliadochona ssavini
(Czech.)
Acanthocephala:
Acanthocephalus
anguillae
A. lucii
*Corynosoma
strumosum (E)
Dentitruncus truttae
(Yugoslavia)
Echinorhynchus clavula
E. gadi (R)

E. salmonis
E. truttae (E)
Metechinorhynchus
salmonis (R)
M. truttae (E)
Neoechinorhynchus rutili
Pomphorhynchus
proteus (E)
Pseudoechinorhynchus
clavula (R)
Mollusca:
*Glochidia
Crustacea:
Argulus coregoni (R)
A. foliaceus (R)
Ergasilus sieboldi (R)
Lepeophtheirus
salmonis (R)
Lernaea cruciata
L. esocina (R)
Salmincola salmonea (R)
Leeches:
Acanthobdella
peledina (R)
Piscicola geometra (R)
Salvelinus alpinus, arctic char
Protozoa:
Haemogregarina
irkalupkiki
Henneguya
zschokkei (R)
Myxidium oviforme (R)
Trematoda:
Crepidostomum farionis
C. metoecus
Lecithaster gibbosus (R)
Tetraonchus
alascensis (R)
T. arcticus
Cestoda:
Cyathocephalus
truncatus (R)
*Diphyllobothrium
norvegicum (R)
D. salvelini
Diplocotyle olrikii
Eubothrium crassum
E. salvelini (R)

* Larval forms.
** Immature adults.

Nematoda:
  Contracaecum
    aduncum (R)
  Metabronema salvelini
  Rhabdichona
    denudata (R)
Cestoda:
  *Pelichnibothrium
    speciosum (R)
  *Proteocephalus exiguus
  Raphidascaris acus (R)
  *Scolex pleuronectis (R)
Acanthocephala:
  Metechinorhynchus
    salmonis (R)
  M. truttae (R)
  Neoechinorhynchus rutili
  Pomphorhynchus
    laevis (R)
Leeches:
  Acanthobdella
    peledina (R)
Crustacea:
  Salmincola edwardsi (R)
  S. salmonea (R)
  S. salvelini
  S. thymalli

Salvelinus fontinalis
Fungi:
  Ichthyosporidium sp.
  Lymphosporidium
    truttae (E)
  Saprolegnia invaderis
    (internal)
  S. parasitica
  Intestinal fungisitosis
    (Saprolegnia ferax ?)
Protozoa:
  Chloromyxum leydei
  C. truttae
  Costia pyriformis
  Dactylosoma salvelini
  Eimeria sp.
  Epistylis sp.
  Haemogregarina sp.
  Henneguya fontinalis
  Hexamita salmonis
  Ichthyopthirius multifilis
  Leucocytozoon salvelini

Myxidium minteri
Myxobolus ovoidalis
Myxosoma cerebralis
Schizamoeba salmonis
Trichodina myakkae
Trichodina sp.
Trichophrya sp.
Trypanosoma percae
  canadensis
Zschokkella salvelini
Trematoda:
  Allocreadium lobatum
  *Apophallus brevis
  *A. imperator
  Azygia angusticauda
  *Bolbophorus confusus
  *Clinostomum
    marginatum
  Crepidostomum cooperi
  C. farionis
  C. fausti
  C. transmarinum
  Crepidostomum sp.
  *Diplostomulum sp.
  Discocotyle salmonis
  Gyrodactylus elegans B.
  G. medius
  *Nanophyetus salmincola
  *Neascus sp.
  Phyllodistomum
    lachancei
  P. superbum
  Phyllodistomum sp.
  Pleurogenes sp.
  *Posthodiplostomum
    minimum
  Ptychogonimus fontanus
Cestoda:
  Abothrium crassum
  Cyathocephalus truncatus
  *Diphyllobothrium
    cordiceps
  *D. sebago
  *Diphyllobothrium sp.
  Diplocotyle olrikii
  Eubothrium salvelini
  *Ligula intestinalis
  Proteocephalus arcticus

* Larval forms.

P. parallacticus
P. pinguis
**P. pinguis
P. pusillus
P. tumidicollis
Proteocephalus sp.
*Schistocephalus solidus
Nematoda:
Bulbodacnitis globosa
B. scotti
**Contracaecum
  spiculigerum
*Contracaecum sp.
**Contracaecum sp.
Cystidicola stigmatura
Cystidicoloides harwoodi
Hepaticola bakeri
Metabronema canadense
M. salvelini
Philonema
  agubernaculum
Raphidascaris
  laurentianus
Rhabdochona cascadilla
R. laurentiana
Rhabdochona sp.
Acanthocephala:
Acanthocephalus
  anguillae (E)
A. jacksoni
Echinorhynchus lateralis
Leptorhynchoides
  thecatus
*L. thecatus
Neoechinorhynchus
  cylindratus
N. rutili
Neoechinorhynchus sp.
Leech:
Haemopis grandis
  (accidental)
Macrobdella decora
  (accidental)
Piscicola punctata
Mollusca:
*Glochidia
Crustacea:
Argulus canadensis

A. stizostethi
Lepeophtheirus salmonis
Salmincola edwardsii
S. salvelini
Gordiacea:
Chorodes sp.
**Salvelinus malma, Dolly
Varden
Protozoa:
Trichodina truttae (R)
Tripartiella
  californica (R)
Trematoda:
Aponurus sp.
Brachyphallus
  crenatus (R)
Bucephalopsis
  gracilescens (R)
B. ozarkii
Crepidostomum cooperi
C. farionis
Derogenes varicus (R)
Discocotyle salmonis
Genarches mülleri (R)
Hemiurus levinseni (R)
Lecithaster gibbosus (R)
L. salmonis
Prosorhynchus
  crucibulum (R)
Tetraconchus alaskensis
T. borealis (R)
Cestoda:
Cyathocephalus truncatus
*Diphyllobothrium sp.
Eubothrium crassum (R)
E. salvelini
Nybelinia surmincola (R)
Pelichnibothrium
  speciosum (R)
Phyllobothrium sp.
Proteocephalus
  salmonidicola
Scolex pleuronectis
Nematoda:
Ascarophis malmae (R)
Bulbodacnitis globosa
Contracaecum
  aduncum (R)

* Larval forms.
** Immature adults.

Contracaecum sp.
Cucullanus laevis (R)
Cystidicola farionis (R)
Dacnitis truttae
Eustrongylides sp.
Hepaticola bakeri
Metabronema
  canadense (R)
M. salvelini
Philonema oncorhynchi
Rhabdochona amago (R)
R. salvelini (R)
Acanthocephala:
Echinorhynchus coregoni
E. leidyi
Neoechinorhynchus rutili
Neoechinorhynchus sp.
Rhadinorhynchoides
  mijagawai (Japan)
Crustacea:
Lepeophtheirus salmonis
Salmincola bicauliculata
  (Siberia)
S. edwardsii
S. falculata
S. gibber
S. siscowet
S. smirnovi (R)
Salvelinus namaycush,
lake trout
Protozoa:
Cryptobia gurneyorum
Trichophrya piscium
Trematoda:
Azygia angusticauda
Crepidostomum farionis
Diplostomulum sp.
Gyrodactylus elegans
Nanophyetus salmincola
Cestoda:
Abothrium crassum
Diphyllobothrium
  cordiceps (?)
Diphyllobothrium sp.
Eubothrium salvelini
Proteocephalus
  parallacticus
P. pusillus

Proteocephalus sp.
Nematoda:
Bulbodacnitis scotti
Cystidicola stigmatura
Eustrongylides sp.
Metabronema salvelini
Philonema
  agubernaculum
Philonema sp.
Acanthocephala:
Echinorhynchus coregoni
E. leidyi
E. salmonis
E. salvelini
Leptorhynchoides
  thecatus
Neoechinorhynchus sp.
Leech:
Nephelopsis obscura
  (air bladder, probably
  accidental)
Piscicola milneri
Crustacea:
Achtheres coregoni
Argulus canadense
Ergasilus confusus
Salmincola siscowet
Salmincola sp.
Salvelinus oquassa,
blueback trout
Acanthocephala:
Echinorhynchus coregoni
Crustacea:
Salmincola oquassa
Stenodus leucichthys, Inconnu
Trematoda:
Azygia lucii (R)
A. robusta (R)
Crepidostomum
  farionis (R)
Diplostomulum
  clavatum (R)
D. spathaceum (R)
Lecithaster gibbosus (R)
Phyllodistomum
  conostomum (R)
Tetracotyle
  intermedia (R)

* Larval forms.

Cestoda:
Diphyllobothrium
   latum ? (R)
D. strictum (R)
Discocotyle sagittata (R)
Proteocephalus
   exiguus (R)
P. longicollis (R)
Nematoda:
   Ascarophis skrjabini (R)
   Cucullanus truttae (R)
Leech:
   Acanthobdella
      peledina (R)
Crustacea:
   Argulus foliaceus (R)
   Basanistes enodis (R)
   Coregonicola
      orientalis (R)
   Ergasilus sieboldi (R)
   Salmincola
      nordmanni (R)
Thymallus arcticus,
arctic grayling
Protozoa:
   Chloromyxum
      thymalli (R)
   Myxidium
      ventricosum (R)
Trematoda:
   Ariella baikalensis (R)
   Crepidostomum
      farionis (R)
   Crepidostomum sp.
   *Diplostomum
      spathaceum (R)
   Phyllodistomum
      conostomum (R)
   *Tetracotyle
      intermedia (R)
   Tetraonchus alaskensis
   T. borealis (R)
   T. rauschi (Alaski)
Cestoda:
   Cyathocephalus
      truncatus (R)
   *Diphyllobothrium
      minus (R)

D. strictum (R)
Eubothrium
   salvelini (R)
Proteocephalus
   longicollis (R)
P. thymalli (R)
*Triaenophorus
   nodulosus (R)
Nematoda:
   Ascarophis skrjabini (R)
   Coregonema sibirica (R)
   Cucullanus truttae (R)
   Cystidicola
      farionis (R)
   Rhabdochona
      denudata (R)
Acanthocephala:
   Acanthocephalus
      anguillae (R)
   Neoechinorhynchus
      rutili (R)
   Pseudoechinorhynchus
      clavula (R)
Leech:
   Acanthobdella
      peledina (R)
   Piscicola geometra (R)
Crustacea:
   Salmincola
      baicalensis (R)
   S. carpionis (Iceland)
SCIAENIDAE (Drums)
Aplodinotus grunniens,
freshwater drum
Protozoa:
   Henneguya caudalis
   Myxidium aplodinoti
   M. macrocapsulare
Trematoda:
   Allocreadium armatum
   Anallocreadium pearsei
   *Clinostomum
      marginatum
   Cotylogaster occidentalis
   *Diplostomulum sp.
   Heteraxine cokeri
   Homalometron armatum
   H. pearsei

* Larval forms.

Lintaxine cokeri
Microcotyle eriensis
M. spinicirrus
Microcreadium parvum
Phyllodistomum fausti
Cestoda:
 Proteocephalus
  ambloplitis
 P. pearsei
Nematoda:
 **Camallanus* sp.
 C. oxycephalus
 Contracaecum
  spiculigerum
 Cystidicola serrata
 Rhabdochona
  decaturensis
 Rhabdochona sp.
Acanthocephala:
 Echinorhynchus dirus
 Fessisentis fessus
 Gracilisentis gracilisentis
 Leptorhynchoides
  thecatus
 Pomphorhynchus
  bulbocolli
Leech:
 Illinobdella alba
 I. moorei
 Placobdella pediculata
Crustacea:
 Argulus appendiculosus
 Lernaea tennuis
SERRANIDAE (Sea basses)
Roccus americanus, white
perch
Trematoda:
 Azygia angusticauda
 Bunodera sacculata
 Clinostomum
  marginatum
 Crepidostomum cooperi
 C. cornutum
 Pedocotyle morone
Cestoda:
 Abothrium crassum
 Proteocephalus
  ambloplitis

Nematoda:
 Camallanus truncatus
 Cucullanus sp.
 Dichylene cotylophora
 Metabronema sp.
 Spinitectus carolini
 S. gracilis
 Spinitectus sp.
Acanthocephala:
 Leptorhynchoides
  thecatus
 Neoechinorhynchus
  cylindratus
Leech:
 Piscicolaria sp.
Mollusca:
 **Glochidia*
Crustacea:
 Ergasilus sp.
 Lernaea cruciata
Roccus chrysops, white bass
Protozoa:
 Henneguya magna
Trematoda:
 Allacanthocasmus artus
 A. varius
 Azygia acuminata
 Bucephalus elegans
 Bucephalus sp.
 **Diplostomulum* sp.
 Leucerthrus micropteri
 Onchocleidus mimus
 Urocleidus chrysops
 U. mimus
Cestoda:
 Bothriocephalus
  cuspidatus
 Corallobothrium sp.
 Proteocephalus
  ambloplitis
 **P. ambloplitis*
 **P. pearsei*
 P. perplexus
Nematoda:
 Camallanus oxycephalus
 Philometra cylindraceum
 Spinitectus carolini

* Larval forms.
** Immature adults.

Acanthocephala:
Leptorhynchoides
thecatus
Crustacea:
Argulus appendiculosus
A. stizostethi
Ergasilus caeruleus
E. centrarchidarum
E. versicolor
Roccus mississippiensis,
yellow bass
Trematoda:
Allacanthochasmus artus
A. varius
*A. varius
*Clinostomum
marginatum
*Diplostomulum sp.
Neochasmus umbellus
Onchocleidus interruptus
*Posthodiplostomum
minimum
*Tetracotyle sp.
Urocleidus interruptus
Cestoda:
*Proteocephalus
ambloplitis
*Proteocephalus sp.
Trypanorhyncha (?)
Nematoda:
Camallanus sp.
*Contracaecum
spiculigerum
Acanthocephala:
Leptorhynchoides
thecatus
Neoechinorhynchus
cylindratus
Crustacea:
Ergasilus versicolor
Roccus saxatilis, striped bass
Trematoda:
Aristocleidus hastatus
Diplectanum collinsi
Urocleidus hastatus
Acanthocephala:
Leptorhynchoides
thecatus

Crustacea:
Achtheres lacae
Ergasilus labracis
THYMALLIDAE (See
SALMONIDAE)
UMBRIDAE
Dallia pectoralis, Alaska
blackfish
Protozoa:
Glossatella dallii (R)
(See *Apiosoma.*)
G. robusta (R)
(See *Apiosoma.*)
Plistophora dallii (R)
Trematoda:
Bucephalopsis
gracilescens (R)
Cestoda:
*Diphyllobothrium dalliae
Nematoda:
Cucullanus truttae (R)
Acanthocephala:
Neoechinorhynchus
rutili (R)
Umbra limi, mud minnow
Protozoa:
Henneguya umbri
Myxidium umbri
Myxidium sp.
Trematoda:
Bunoderina eucaliae
*Clinostomum
marginatum
Creptotrema funduli
*Diplostomulum sp.
*Echinochasmus
donaldsoni
Gyrodactylus
cylindriformis
G. limi
*Neascus bulboglossa
*N. grandis
*Neascus sp.
*Petasiger nitidus
Phyllodistomum
brevicecum
P. undulans
*Tetracotyle sp.

* Larval forms.

Cestoda:
 Proteocephalus sp.
Nematoda:
   Contracaecum sp.
   Hepaticola bakeri
   Spiroxys contortus
Acanthocephala:
   *Leptorhynchoides
      thecatus*
   Neoechinorhynchus sp.

 N. rutili
 *Pomphorhynchus
    bulbocolli*
Crustacea:
   Argulus americanus
Umbra pygmaea, eastern
mudminnow
Crustacea:
   Lernaea cyprinacea

* Larval forms.

Bibliography

Abernathy, Clorinne. 1937. Notes on *Crepidostomum cornutum* (Osborn). Tr. Am. Micr. Soc., **56**(2): 206–207.

Adams, J. R. 1956. A parasitic copepod (*Salmincola falculata*) attached to a fish heart. J. Parasitol., **42**(3): 296.

Africa, C. M., W. De Leon, and E. Y. Garcia. 1936. Heterophyidiasis. IV. Lesions found in the myocardium of eleven infested hearts including three cases with valvular involvement. Philippine J. Public Health, **3**(1 & 2): 1–27.

Agarwal, S. M. 1959. Studies on the morphology, systematics and life history of *Clinostomum giganticum* n. sp. (Trematoda: Clinostomatidae). Indian J. Helminth., **11**(2): 75–115.

Alexander, C. G. 1951. A new species of *Proteocephalus* (Cestoda) from Oregon trout. J. Parasitol., **37**(2): 160–164.

———. 1960. A survey of parasites of Oregon trout. Rep. to Oregon State Game Comm., pp. 1–34.

Ali, S. M. 1956. Studies on the nematode parasites of fishes and birds found in Hyderabad state. Indian J. Helminth., **8**(1): 1–83.

———. 1960. On two new species of *Procamallanus* Baylis, 1923, from India, with a key to the species. J. Helminth., **34**(1 & 2): 129–138.

Allison, L. N. 1950. Common diseases of fish in Michigan. Michigan Dept. Conserv., Misc. Publ. no. 5: 1–27.

Allison, R. 1957. A preliminary note on the use of di-n-butyl tin oxide to remove tapeworms from fish. Prog. Fish-Cult., **19**(3): 128.

———. 1962. Pers. comm., Auburn Univ., Auburn, Alabama.

Allum, M. O., and E. J. Hugghins. 1959. Epizootics of fish lice, *Argulus biramosus*, in two lakes of eastern South Dakota. J. Parasitol., **45**(4-2): 33.

Ameel, D. J. 1937. The life history of *Crepidostomum cornutum* (Osborn). J. Parasitol., **23**(2): 218–220.

Anderson, M. G. 1962. *Proterometra dickermani*, sp. nov. (Trematoda: Azygiidae). Tr. Am. Micr. Soc., **81**(3): 279–282.

Anderson, M. G., and F. M. Anderson, 1963. Life history of *Proterometra dickermani* Anderson, 1962. J. Parasitol., **49**(2): 275–280.

Anthony, J. D. 1958. *Atractolytocestus huronensis* n. gen., n. sp. (Cestoda: Lytocestidae) with notes on its morphology. Tr. Am. Micr. Soc., **87**(4): 383.

———. 1963. Parasites of eastern Wisconsin fishes. Wisconsin Acad. Sci., Arts and Lett., **52**: 83–95.

Applegate, V. C. 1950. Natural history of the sea lamprey (*Petromyzon marinus*) in Michigan. Spec. Sci. Rep., Fisheries no. 55: 1–237.

Arey, L. B. 1922. Observations on an acquired immunity to a metazoan parasite. Am. Soc. Zool., **20**: 20.

Arnold, J. G., Jr. 1933. Some trematodes of the common bullhead *Ameiurus nebulosus*. Abstr., J. Parasitol., **20**(2): 136.

———. 1934. Some trematodes of the common bullhead *Ameiurus nebulosus*. Tr. Am. Micr. Soc., **53**(3): 267–276.

Ash, L. R. 1960. Life cycle studies on *Gnathostoma procyonis* Chandler, 1942, a nematode parasite of the raccoon. J. Parasitol., **46**(5-2): 37.

Avdos'ev, B. S. 1962. New methods of using malachite green for ichthyophthiriasis in carp (in Russian). Rybnoe Khoziasist, **38**(7): 27–29.

Babero, B. B. 1953. Studies on the helminth fauna of Alaska. XII. The experimental infection of Alaskan gulls (*Larus glaucescens* Naumann) with *Diphyllobothrium* sp. J. Wash. Acad. Sci., **43**(5): 166–168.

Babero, B. B., and J. R. Shepperson. 1958. Some helminths of raccoons in Georgia. J. Parasitol., **44**(5): 519.

Baer, J. G. 1944. Les trématodes parasites de la musaraigne d'eau *Neomys fodiens* (Schreb.). Bull. Soc. Neuchatel. Sci. Nat., **68**: 33–84.

Bailey, R. M. (chairman). 1960. A list of common and scientific names of fishes from the United States and Canada. Am. Fish. Soc. Spec. Publ. no. 2: 102 pp.

Baldauf, R. J. 1958. Formulation of controls of parasites in state fish hatcheries. Coop. Res., Ann. Rep. Texas Game, Fish Comm., August, 1958.

———. 1961. Another case of parasitic copepods on amphibians. J. Parasitol., **47**(2): 195.

Bangham, R. V. 1925. A study of the cestode parasites of the black bass in Ohio, with special reference to their life history and distribution. Ohio J. Sci., **25**(6): 255–270.

———. 1926a. A new intermediate host of *Proteocephalus pearsei* LaRue. Abst. Am. Soc. Parasitol., 12.

———. 1926b. Parasites other than cestodes in black bass of Ohio. Ohio J. Sci., **26** (3): 117–127.

———. 1927a. Diseases of fish in Ohio hatcheries. Tr. Am. Fish. Soc., **57**: 223–230.

———. 1927b. Life history of the bass cestode, *Proteocephalus ambloplitis*. Tr. Am. Fish. Soc., **57**: 206–209.

———. 1928. Parasites of black bass. Scient. Month., **27**(3): 267–270.

———. 1933. Parasites of the spotted bass, *Micropterus pseudalplites* Hubbs, and summary of parasites of smallmouth and largemouth black bass from Ohio streams. Tr. Am. Fish. Soc., **63**: 220–228.

———. 1937. Parasites of Wayne County (Ohio) stream fish. Tr. Am. Fish. Soc., **66**: 357–358.

———. 1939. Parasites of Centrarchidae from southern Florida. Tr. Am. Fish. Soc., **68**: 263–268.

———. 1940. Parasites of fish of Algonquin Park lakes. Prog. Fish-Cult., Memo. I-131 (52): 37.

———. 1941a. Parasites of fresh-water fish of southern Florida. Proc. Florida Acad. Sci., **5**: 289–307.

———. 1941b. Parasites from fish of Buckeye Lake, Ohio. Ohio J. Sci., **41**(6): 441–448.

———. 1941c. Parasites of fish of Algonquin Park lakes. Tr. Am. Fish. Soc., **70**: 161–171.

———. 1944. Parasites of northern Wisconsin fish. Tr. Wisconsin Acad. Sci., Arts and Lett., **36**: 291–325.

————. 1951. Parasites of fish in the upper Snake River drainage and in Yellowstone Lake, Wyoming. Zoologica, Scient. Contrib. New York Zool. Soc., 36(3): 213–217.

————. 1955. Studies on fish parasites of Lake Huron and Manitoulin Island. Am. Midl. Nat., 53(1): 184–194.

Bangham, R. V., and J. R. Adams. 1954. A survey of the parasites of freshwater fishes from the mainland of British Columbia. J. Fish. Res. Bd. Canada, 11(6): 673–708.

Bangham, R. V., and G. W. Hunter, III. 1936. Studies of fish parasites of Lake Erie. III. *Microcotyle spinicirrus* MacCallum (1918) char. emend. and *M. eriensis* sp. nov. Tr. Am. Micr. Soc., 55(3): 334–339.

————. 1939. Studies on fish parasites of Lake Erie. Distribution studies. Zoologica, New York Zool. Soc., 24(4): 385–448.

Bangham, R. V., and C. E. Venard. 1942. Studies on parasites of Reelfoot Lake fish. IV. Distribution studies and check-list of parasites. J. Tennessee Acad. Sci., 17(1): 22–38.

————. 1946. Parasites of fish of Algonquin Park lakes. Publ. Ontario Fish. Res. Lab. no. 53: 33–46.

Barker, F. D., and S. Parsons. 1914. A new aspidobothrid trematode from Lesseur's terrapin. Tr. Am. Micr. Soc., 33(4): 261–262.

Barreto, A. L. de Barros. 1922. Revision of the family Cucullanidae Barreto, 1916. Mem. Inst. Oswaldo Cruz, 14(1): 68–87.

Barysheva, A. F., and O. N. Bauer. 1958. Parasites of fishes of Lake Ladoga (Russian text, German summary). Izvest. Vsesoiuz. Nauchno-Issled. Inst. Ozern. i Rechn. Rybnoe Khoziasist, 12, Parazity i Bolexni Ryb., 175–226.

Batsch, A. J. G. 1786. Naturgeschichte der Bandwurmgattung überhaupt und ihrer Arten insbesondere, nach den neuern Beobachtungen in einem systematischen Auszuge. Halle. 298 pp.

Bauer, O. N., 1958. Biologie und Bekämpfung von *Ichthyophthirius multifilis* Fouquet. Zeitschr. Fischerei, Neudamm u. Berlin, 7(7 & 8): 575–581.

Bauer, O. N., et al. 1959. Parasites of Freshwater Fish and the Biological Basis for Their Control. Bull. State Sci. Res. Inst. Lake and River Fisheries, 49 (English transl., Office Tech. Services, U. S. Dept. Commerce, 1962, no. 61-31056: 236 pp.).

Bauer, O. N., and N. P. Nikol'skaya. 1952. New data on the intermediate hosts of whitefish parasites. DAN SSSR, 84(5).

Baylis, H. A. 1931a. A species of the nematode genus *Hedruris* occurring in the trout in New Zealand. Ann. Mag. Nat. Hist., (10)7: 105–114.

————. 1931b. On the structure and relationships of the nematode *Capillaria* (*Hepaticola*) *hepatica* Bancroft. Parasitology, 23(4): 533–543.

————. 1935a. Two new parasitic nematodes from Ceylon. Ann. Mag. Nat. Hist., (10) 16: 187–192.

————. 1935b. Four new species of nematodes. Ann. Mag. Nat. Hist., (10) 16: 370–382.

————. 1948. On two nematode parasites of fishes. Ann. Mag. Nat. Hist., (11) 14: 327–335.

Beardsley, A. E. 1904. The destruction of trout fry by hydra. Bull. U. S. Fish. Comm., 22(1902): 157–160.

Beaver, P. 1936. Notes on *Stephanoprora polycestus* (Dietz) from the American crow. Tr. Illinois Acad. Sci., 29(2): 247–250.

————. 1939a. Morphology and life history of *Psilostomum ondatrae* Price, 1931. J. Parasitol., **25**(5): 383–393.

————. 1939b. The morphology and life history of *Petasiger nitidus* Linton (Trematoda: Echinostomidae). J. Parasitol., **25**(3): 269–275.

————. 1941a. The life history of *Echinochasmus donaldsoni* n. sp., a trematode from the pied-billed grebe. J. Parasitol., **27**(4): 347–355.

————. 1941b. Studies on the life history of *Euparyphium melis* (Trematoda: Echinostomidae). J. Parasitol., **27**(1): 35–44.

Beaver, P., and P. H. Simer. 1940. A re-study of the three existing species of the cestode genus *Marsipometra* Cooper (Amphicotylidae) from the spoonbill, *Polydon spathula* (Wal.). Tr. Am. Micr. Soc., **59**: 167–182.

Becker, C. D. 1962. A new haemogregarine from the blood of a freshwater fish, *Catostomus macrocheilus* Girard. J. Parasitol., **48**(4): 596–600.

————. 1964. The parasite-vector-host relationship of the hemoflagellate, *Cryptobia salmositica* Katz, the leech, *Piscicola salmositica* Meyer, and certain freshwater teleosts. Ph.D. thesis, Univ. Washington. 200 pp.

Becker, C. D., and M. Katz. 1965a. Transmission of the hemoflagellate, *Cryptobia salmositica* Katz, 1951, by a rhynchobdellid vector. J. Parasitol., **51**(1): 95–99.

————. 1965b. *Babesiosoma tetragonis* n. sp. (Sporozoa: Dactylosomidae) from a California teleost. J. Protozool., **12**(2): 189–193.

————. 1965c. Infections of the hemoflagellate, *Cryptobia salmositica* Katz, 1951, in freshwater teleosts of the Pacific Coast. Tr. Am. Fish. Soc. **94**(4): 327–333.

————. 1966. Host relationships of *Cryptobia salmositica* (Protozoa: Mastigophora) in a Western Washington stream. Tr. Am. Fish. Soc. **94**(2): 196–202.

Becker, E. 1956. Catalog of Eimeriidae in genera occurring in vertebrates and not requiring intermediate hosts. Iowa State Coll. J. Sci., **31**(1): 85–139.

Beilfuss, E. R. 1954. The life histories of *Phyllodistomum lohrenzi* Loewen, 1935, and *P. caudatum* Steelman, 1938 (Trematoda: Gorgoderinae). J. Parasitol., **40**(5-2): 44.

Beneden, P. J. van. 1849. Notice sur un nouveau genre d'hélminthe cestoide. Bull. Acad. Roy. Sci. Belg., **16**: 182–193.

Benedict, H. M. 1900. On the structure of two fish tapeworms from the genus *Proteocephalus* Weinland, 1858. J. Morphol. **16**(2): 337–368.

Bennington, E., and I. Pratt. 1960. The life history of the salmon-poisoning fluke, *Nanophyetus salmincola* (Chapin). J. Parasitol., **46**(1): 91–100.

Bere, R. 1929. Reports of the Jasper Park lakes investigations. 1925–26. III. The leeches. Contrib. Canad. Biol. and Fish., **4**(14): 175–183.

————. 1931. Copepods parasitic on fish of the Trout Lake region, with descriptions of two new species. Tr. Wisconsin Acad. Sci., Arts and Lett., **26**: 427–436.

————. 1935. Further notes on the occurrence of parasitic copepods on fish of the Trout Lake region, with a description of the male of *Argulus biramosus*. Tr. Wisconsin Acad. Sci., Arts and Lett., **29**: 83–88.

————. 1936. Parasitic copepods from Gulf of Mexico fish. Am. Midl. Nat., **17**(3): 577–625.

Bhalerao, G. D. 1943. On two helminths of *Metacembelus pancalus* (Ham.) including a new record of *Azygia* from India. Proc. Indian Sci. Cong., **29**: 152.

Bishop, E. L., and T. L. Jahn. 1941. Observations on colonial peritrichs (Ciliata; Protozoa) of the Okoboji region. Proc. Iowa Acad. Sci., **48**: 417–421.

Blainville (de), H. M. D. 1818. *In* C. Lamarck. Histoire Naturelle des Animaux sans Vertèbres. Paris. 5: 289.

Bloch, M. E. 1779. Beitrag zur Naturgeschichte der Wurmer, welche in andern Thieren leben. Bcschaft. Berl. Ges. Naturf. Fr., **4**: 534–561.

———. 1782. Abhandlung von der Erzeugung der Eingeweidewurmer und den Mitteln wider dieselben. Berlin. 54 pp.

Bogdanova, E. A., and G. A. Stein. 1963. Infusoria fam. Urceolariidae, infecting the fingerlings of the salmonid fishes. Bull. State Sci. Res. Inst. Lake and River Fisheries, **54**: 48–57.

Bogitsh, B. J. 1958a. *Tetracotyle lepomensis*, n. sp. (Trematoda: Strigeidae), from freshwater in Albemarle County, Virginia. Proc. Helminth. Soc. Washington, **25**(1): 14–16.

———. 1958b. Observations on the seasonal occurrence of a pseudophyllidean tapeworm infecting the alimentary tract of *Lepomis macrochirus* Raf. in Albemarle County, Virginia. Am. Midl. Nat., **60**(1): 97–99.

———. 1961. Histological and histochemical observations on the nature of the cyst of *Neoechinorhynchus cylindratus* in *Lepomis* sp. Proc. Helminth. Soc. Washington, **28**(1): 75–80.

———. 1962. Histochemical observations on larvae of *Posthodiplostomum minimum* (MacCallum, 1921). J. Parasitol., **48**(2): 23.

———. 1963. Histochemical observations on the cercariae of *Posthodiplostomum minimum*. Exper. Parasitol., **14**(2): 193–202.

Bogitsh, B. J., and T. C. Cheng. 1959. *Pisciamphistoma reynoldsi* (Paramphistomatidae), a new trematode parasite of *Lepomis* spp. in Virginia. J. Tennessee Acad. Sci., **34**(3): 159.

Bond, F. F. 1937a. Host specificity of the Myxosporidia of *Fundulus heteroclitus* (Linn.). J. Parasitol., **23**(5): 540–542.

———. 1937b. A microsporidian infection of *Fundulus heteroclitus* (Linn.). J. Parasitol., **23**(2): 229–230.

———. 1937c. *Myxosoma grandis* Kudo in fish from the Hudson River drainage system. J. Parasitol., **23**(2): 231–232.

———. 1937d. A probable constituent of the spore coat of myxosporidian spores. J. Parasitol., **23**(5): 542–543.

———. 1938a. Cnidosporidia from *Fundulus heteroclitus* Linn. Tr. Am. Micr. Soc., **57**(2): 107–122.

———. 1938b. The doubtful relationship of Sporozoa to the ulcers of *Fundulus heteroclitus* (Linn.). J. Parasitol., **24**(3): 207–213.

———. 1938c. Resistance of myxosporidian spores to conditions outside of the host. J. Parasitol., **24**(5): 470–471.

———. 1939a. Experimental studies on myxosporidiosis of *Fundulus heteroclitus* (Linn.) and *F. diaphanus* (Le Sueur). Tr. Am. Micr. Soc., **58**(2): 164–169.

———. 1939b. Myxosporidia from fishes of the genus *Esox*. J. Parasitol., **25**(5): 377–381.

———. 1939c. The seasonal incidence of myxosporidian parasites infecting *Fundulus heteroclitus* (Linn.). Tr. Am. Micr. Soc., **58**(2): 156–163.

Bowen, J. T. 1965. Parasites of freshwater fish. IV. Miscellaneous. 4. Parasitic copepods *Ergasilus, Actheres* and *Salmincola*. Fish Disease Leaflet no. 4 USFWS.

Bowen, J. T., and R. E. Putz. 1965. Parasites of freshwater fish. IV. Miscellaneous. 3. Parasitic copepod *Argulus*. Fish Disease Leaflet no. 3, USFWS.

Bower-Shore, C. 1940. An investigation of the common fish louse, *Argulus foliaceus* (Linn.). Parasitology, 32(4): 361–371.

Branson, B. A., and B. G. Amos. 1961. The leech *Placobdella pediculata* Hemingway parasitizing *Aplodinotus grunniens* in Oklahoma. Southwestern Nat., 6(1): 53.

Braun, M. 1893. Vermes. Abhandlung I. (a) Trematodes. Bronn's Klassen und Ordnungen des Tierreichs.

Brinkmann, A. (Jr.). 1952. Fish trematodes from Norwegian waters. Univ. Bergen Aarbok, Natur. Rek., (1): 134.

Britt, H. G. 1947. Chromosomes of digenetic trematodes. Am. Naturalist, 81: 276–296.

Brown, E. M. 1934. On *Oodinium ocellatum* Brown, a parasitic dinoflagellate causing epidemic disease in marine fish. Proc. Zool. Soc. London, (3): 583–607.

Brown, E. M., and R. Hovasse. 1946. *Amyloodinium ocellatum* (Brown), a peridinian parasitic on marine fishes. Proc. Zool. Soc., 116(1): 33–46.

Bruhl, C. B. 1860. *Lernaeocera gasterostei*, ein Schmarotzerkrebs aus der Familie der Pennellina, mit zwölf Ruderfüssen, zwei Stummelfüssen, und Schwanzfurca. Mitth. K.K. Zool. Inst. Univ. Pest, (1), 18 pp.

Bullock, W. L. 1957. *Octospiniferoides chandleri* n. gen., n. sp., a neoechinorhynchid acanthocephalan from *Fundulus grandis* Baird and Girard on the Texas coast. J. Parasitol., 43(1): 97–100.

———. 1958. The blood protozoa of the marine fish of southern New England. J. Parasitol., 44(4): 24–25.

———. 1962. A new species of *Acanthocephalus* from New England fishes, with observations on variability. J. Parasitol., 48(3): 442–451.

———. 1963a. Intestinal histology of some salmonid fishes with particular reference to the histopathology of acanthocephalan infections. J. Morphol., 112(1): 23–44.

———. 1963b. *Neoechinorhynchus prolixoides* n. sp. (Acanthocephala) from North American fishes. Proc. Helminth. Soc. Washington, 30(1): 92–96.

———. 1964. *Octospiniferoides chandleri* Bullock, 1957 (Neoechinorhynchidae), a common parasite of *Gambusia affinis* in central Florida. Abst., J. Parasitol., 50(3–2): 32.

———. 1965. Histochemical observations on the neoechinorhynchid apical organ. Abst., J. Parasitol., 51(2–2): 20.

Burmeister, H. 1835. Beschreibung einiger neuen oder wenig bekannten Schmarotzerkrebse, nebst allgemeinen Betrachtungen über die Gruppe, welcher sie angehoren. Nova Acta Acad. Nat. Curios., 17(1): 269–336.

Burton, P. R. 1956. Morphology of *Ascocotyle leighi*, n. sp. (Heterophyidae), an avian trematode with metacercaria restricted to the conus arteriosus of the fish *Mollienesis latipinna* Le Sueur. J. Parasitol., 42(6): 660.

———. 1958. A review of the taxonomy of the trematode genera *Ascocotyle* (Looss) and *Phagicola* (Faust) of the family Heterophyidae. Proc. Helminth. Soc. Washington, 25(2): 117–122.

Bütschli, O. 1882. Myxosporidia. Bronn's Klass. u. Ordnung., Protozoa, 1: 590–603.

Bykhovskaya-Pavlovskaya, I. E., *et al.* 1962. Key to Parasites of Freshwater Fish of the USSR. Zool. Inst., Acad. Sci. U.S.S.R. (English transl. TT 64-11040, OTS, Dept. Commerce, Washington, D. C., 919 pp.).

Bychowsky, B. E. 1957. Monogenetic Trematodes, Their Systematics and Phylogeny (transl. from Russian into English by P. C. Oustinoff, ed. W. H. Hargis, 1961. AIBS. 627 pp.).

Byrd, E. E., and C. E. Venard. 1940. The excretory system in Trematoda. I. Studies on the excretory system in the trematode subfamily Gorgoderinae Looss, 1899. J. Parasitol., **26**(5): 407-420.

Caballero, E. 1940. Sanguijuelas del Lago de Patzcuaro y descripcion de una nueva especie, *Illinobdella patzcuarensis*. Ann. Inst. Biol., **11**(2): 449-464.

Caballero, Y. E., and H. A. Winter. 1954. Metacercariae of *Diplostomum spathaceum* (Rudolph, 1819) Braun, 1893. *In* Freshwater fishes of Mexico, Ciencia, **14**(4-6): 77-80.

Cable, R. M. 1935. *Cercaria kentuckiensis* n. sp., first representative of the *Vivax* group known to occur in the United States. J. Parasitol., **21**: 441.

―――. 1952. On the systematic position of the genus *Deropristis*, of *Dihemistephanus sturionis* Little, 1930, and of a new digenetic trematode from a sturgeon. Parasitology, **42**(1 & 2): 85-91.

―――. 1955. Taxonomy of some digenetic trematodes from sturgeons. J. Parasitol., **41**(4): 441.

―――. 1956. *Opistholebes diodontis* n. sp.: its development in the final host, the affinities of some amphistomatous trematodes from marine fishes and the allocreadioid problem. J. Parasitol., **46**(1): 1-13.

―――. 1960, 1966. Pers. comm., Biol. Dept., Purdue Univ., Lafayette, Indiana.

Cable, R. M., and A. V. Hunninen. 1942. Studies on *Deropristis inflata* (Molin): its life history and affinities to trematodes of the family Acanthocolpidae. Biol. Bull., **82**(2): 292-312.

Calentine, R. L. 1962. *Archigetes iowensis* sp. n. (Cestoda: Caryophyllaeidae) from *Cyprinus carpio* L. and *Limnodrilus hoffmeisteri* Claparede. J. Parasitol., **48**(4): 513-524.

―――. 1964. The life cycle of *Archigetes iowensis* (Cestoda: Caryophyllaeidae). J. Parasitol., **50**(6): 454-458.

Calentine, R. L., and M. J. Ulmer. 1961. *Khawia iowensis* n. sp. (Cestoda: Caryophyllaeidae) from *Cyprinus carpio* L. in Iowa. J. Parasitol., **47**(5): 795-805.

Cameron, T. W. M. 1944. The morphology, taxonomy, and life history of *Metorchis conjunctus* (Cobbold, 1860). Canad. J. Research, **22**: 6-16.

―――. 1945. Fish-carried parasites in Canada. Canad. J. Comp. Med., **9**: 245-254, 283-286, 302-311.

Campana-Rouget, Y. 1957. Sur quelques espèces de Cucullanidae. Révision de la sousfamille. Bull. l'Inst. Franc. d'Afr. Noire, **19**(2): 417-473.

Camper, J. 1962. Pers. comm., Pisgah Forest Nat. Fish Hatchery, North Carolina.

Canavan, W. P. 1928. A new species of *Phyllobothrium* van Ben. from an Alaska dog salmon, with a note on the occurrence of *Crossobothrium angustum* Linton in the thresher shark. J. Helminth., **6**: 51-55.

Carbine, W. F. 1942. Sphaeriid clams attached to the mouth of young pike. Copeia, no. 3: 187.

Carl, G. C. 1937. Flora and fauna of brackish water. Ecology, **18**: 446–453.

Causey, D. 1957. Parasitic Copepoda from Louisiana freshwater fish. Am. Midl. Nat., **58**(2): 378–382.

Chandler, A. C. 1931. New genera and species of nematode worms. Proc. U. S. Nat. Mus., **78**: 1–11.

———. 1935. Parasites of fishes in Galveston Bay. Proc. U. S. Nat. Mus., **83**(2977): 123–157.

———. 1942. The helminths of raccoons in east Texas. J. Parasitol., **28**(2): 135–140.

———. 1951. Studies on metacercariae of *Perca flavescens* in Lake Itasca, Minnesota. Am. Midl. Nat., **45**(3): 711–721.

———. 1952. Key to the furcocercous cercariae. Rice Univ., Houston, Texas. Mimeographed.

———. 1955. Introduction to Parasitology. 9th ed., Wiley & Sons, N. Y. 799 pp.

Chapin, E. A. 1926. A new genus and species of trematode, the probable cause of salmon-poisoning in dogs. North Am. Vet., **7**(4): 36–37.

Chen, Chih-leu. 1955. The protozoan parasites of *Ctenopharyngodon idellus*. Acta Hydrobiol. Sinica, (2): 123–164.

Cheng, T. C. 1957. A study of the metaceraria of *Crepidostomum cornutum* (Osborn, 1903), (Trematoda: Allocreadiidae). Proc. Helminth. Soc. Washington, **24**(2): 107–108.

Chiriac, E. 1959. A case of *Lernaea cyprinacea* on the triton. Comunicarile AR, **9**(3): 259–263.

Chitwood, M. B., and A. McIntosh, 1950. An American host record for the Russian sturgeon nematode, *Cystoopsis acipenseri* Wagner, 1868. J. Parasitol., **36**(6–2): 29.

Choquette, L. P. E. 1947. *Phyllodistomum lachancei* sp. nov., a trematode from the ureters of *Salvelinus fontinalis* (Mitchill), with a note on its pathogenicity. Canad. J. Research, D, **25**: 131–134.

———. 1948*a*. On the species of the genus *Metabronema* Yorke and Maplestone, 1926, parasitic in trout and char. Canad. J. Research, D, **26**: 329–333.

———. 1948*b*. Parasites of freshwater fish. IV. Internal helminths parasitic in speckled trout (*Salvelinus fontinalis* Mitchill) in rivers and lakes of the Laurentide Park, Quebec, Canada, Canad. J. Research, D. **26**: 204–211.

———. 1951*a*. Parasites of freshwater fish. V. Parasitic helminths of the muskellunge, *Esox m. masquinongy* Mitchill, in the St. Lawrence watershed. Canad J. Zool., **29**: 290–295.

———. 1951*b*. On the nematode genus *Rhabdochona* Railliet, 1916 (Nematoda: Spiruroidea). Canad. J. Zool., **29**: 1–16.

———. 1954. A note on the intermediate hosts of the trematode, *Crepidostomum cooperi* Hopkins, 1931, parasitic in speckled trout (*Salvelinus fontinalis* (Mitchill) in some lakes and rivers of the Quebec Laurentide Park. Canad. J. Zool., **32**: 375–377.

———. 1955. The life history of the nematode *Metabronema salvelini* (Fujita, 1920) parasitic in the speckled trout, *Salvelinus fontinalis* (Mitchill), in Quebec. Canad. J. Zool., **33**: 1–4.

Chubb, J. C. 1963. Seasonal occurrence and maturation of *Triaenophorus nodulosus* (Pallas, 1781) (Cestoda: Pseudophyllidea) in the pike, *Esox lucius* L., of Llyn Tegid. Parasitology, **53**: 419–433.

————. 1965. Report on the parasites of freshwater fishes of Lancashire and Cheshire. Lancashire and Cheshire Fauna Comm., 35th Rep., 1965: 1–5.

Clark, A. S. 1952. Maturation of the plerocercoid of the pseudophyllidean cestode, *Schistocephalus solidus*, in alien hosts. Exper. Parasitol., **2**(3): 223–229.

Clemmens, H. P., and K. E. Sneed. 1958. The chemical control of some diseases and parasites of channel catfish. Prog. Fish-Cult., **20**(1): 8–15.

Cobbold, T. S. 1858. Observations on Entozoa with notices of several new species including an account of two experiments in regard to the breeding of *Taenia serrata* and *Taenia cucumerina*. Tr. Linn. Soc. London, **22**: 155–172.

Cockerell, T. D. A. 1926. A parasite of the goldfish. Science, **64**: 623.

Coil, W. H. 1954. Contributions to the life cycles of gorgoderid trematodes. Am. Midl. Nat., **52**(2): 481–500.

Coker, R. E., A. F. Shira, H. W. Clark, and A. D. Howard. 1921. Natural history and propagation of freshwater mussels. Bull. U. S. Bur. Fish., **37**: 75–182.

Collet-Meygret, G. F. H. 1802. Mémoire sur un ver trouvé dans le rein d'un chien. J. Phys., **55**: 458–464.

Colley, F. C., and A. C. Olson. 1963. *Posthodiplostomum minimum* (Trematoda: Diplostomidae) in fishes of Lower Otay reservoir, San Diego County, California. J. Parasitol., **49**(1): 148.

Connor, R. S. 1953. A study of the seasonal cycle of a proteocephalan cestode, *Proteocephalus stizostethi* Hunter and Bangham, found in the yellow pike-perch, *Stizostedion vitreum vitreum* (Mitchill). J. Parasitol., **39**(6): 621–624.

Cooper, A. R. 1914. On the systematic position of *Haplobothrium globuliforme* Cooper. Tr. Roy. Soc. Canada, Sect. 4, 3, vol. **8**: 1–5.

————. 1915. Trematodes from marine and fresh-water fishes, including one species of ectoparasitic turbellarian. Tr. Roy. Soc. Canada, Sect. 4, 3, vol. **9**: 181–205.

————. 1917. A morphological study of bothriocephalid cestodes from fishes. J. Parasitol., **4**(1): 33–39.

————. 1919. North American pseudophyllidean cestodes from fishes. Illinois Biol. Monogr., **4**(4): 288–541.

————. 1920. *Glaridacris catostomi* gen. nov., sp. nov., a cestodarian parasite. Tr. Am. Micr. Soc., **39**(1): 5–24.

— ——. 1921. Trematodes and cestodes of the Canadian Arctic Expedition 1913–18. Rep. Canad. Arctic Exped. 1913–18, **9**: 3–27.

Cope, O. B. 1958. Incidence of external parasites on cutthroat trout in Yellowstone Lake. Utah Acad. Proc., **35**: 95–100.

————. 1959. New parasite records from stickleback and salmon in an Alaska stream. Tr. Am. Micr. Soc., **75**(2): 157–162.

Cope. 1871. Proc. Acad. Nat. Sci. Phila., **23**: 297.

Cordero, E. H. 1941. Observaciones sobre algunas especies Sud Americanas del genero Hydra. III. (1) Hydra en Venezuela; (2) La accion de *Hydra iheringi* sobre las larvas de ciertos peces del nordeste del Brasil. Ann. Acad. Brasil Sci., **13**(3): 195–201.

Ćorić, B. 1963. *Cyathocephalus truncatus* kod peša (*Cottus gobio*). Veterinaria, Sarajevo, **12**(1): 117–118.

Corkum, K. C., R. Ringhouse, and S. Roesch. 1958. A parasite of the swim bladder of black catfish. Tr. Illinois State Acad. Sci. **50**: 301–302.

Corliss, J. O. 1961. The Ciliated Protozoa, Characterization, Classification, and Guide to the Literature. Pergamon Press, New York. 310 pp.

Cort, W. W. 1919. A new Cercariaeum from North America. J. Parasitol., 5(2): 86–91.

Cort, W. W., D. J. Ameel, and Anne Van der Woude. 1950. The germinal mass in the rediae of *Triganodistomum mutabile* (Cort) (Trematoda: Lissorchiidae). J. Parasitol., 36(2): 145–151.

————. 1951. Early developmental stages of strigeid mother sporocysts. Proc. Helminth. Soc. Washington, 18(1): 5–9.

Cort, W. W., and S. Brackett. 1937. Precocious development of the metacercaria stage of *Diplostomum flexicaudum* in the snail intermediate host. J. Parasitol., 23(5): 545–546.

Cort, W. W., K. L. Hussey, and D. J. Ameel. 1957. Variations in infections of *Diplostomum flexicaudum* (Cort and Brooks, 1928) in snail intermediate hosts of different sizes. J. Parasitol., 43(2): 221–232.

Cort, W. W., and L. J. Olivier. 1941. Early development stages of strigeid trematodes in the first intermediate host. J. Parasitol., 27(6): 493–504.

Crawford, W. W. 1937. A further contribution to the life history of *Alloglossidium corti* (Lamont), with especial reference to dragonfly naiads as second intermediate hosts. J. Parasitol., 23(4): 389–399.

————. 1943. Colorado trematode studies. I. A further contribution to the life history of *Crepidostomum farionis* (Müller). J. Parasitol., 29(6): 379–384.

Creplin, F. C. H. 1825. Observationes de Entozois. 86 pp.

————. 1829. Novae Observationes de Entozois. 134 pp.

————. 1839. Eingeweidewurmer, Binnenwurmer, Thierwurmer Allg. Encycl. Wissensch. u. Kunste (Ersch u. Gruber), 32(1): 277–302.

Cross, S. X. 1934. A probable case of non-specific immunity between two parasites of ciscoes of the Trout Lake region of northern Wisconsin. J. Parasitol., 20(4): 244–245.

————. 1938. A study of the fish parasite relationships in the Trout Lake region of Wisconsin. Tr. Wisconsin Acad. Sci., Arts and Lett., 31: 439–456.

Culbertson, J. R. and R. W. Hull. 1962. Species identification in *Trichophrya* (Suctorida) and the occurrence of melanin in some members of the genus. J. Protozool., 9(4): 455–459.

Curtin, C. B. 1956. The numbers and distribution of gill trematodes on the carp, *Cyprinus carpio* L., collected from three different localities. Diss. Abstr., 16(12): 2561–2562.

Dana, J. D. 1852–1853. United States Exploring Expedition During the Years 1838–42 under the Command of Charles Wilkes, U.S.N. 13. Crustacea. Part I (1852), 685 pp.; Part II (1853), pp. 686–1618. Synopsis of characters of parasitic copepods. *In* Proc. Am. Acad. Arts and Sci., 2(1852): 9–61.

Dana, J. D., and E. C. Herrick. 1837. Description of *Argulus catostomi*, a new parasitic crustaceous animal. Am. J. Sci. Arts, 39(2): 297–308.

Davis, D. J. 1936. Report on the preparation of an histolytic ferment present in the bodies of cercariae. J. Parasitol., 22(1): 108–110.

Davis, H. S. 1917. The Myxosporidia of the Beaufort region, a systematic and biologic study. Bur. Fish. Doc., (855): 201–243.

————. 1922. A new myxosporidian parasite of the channel catfish, *Ictalurus punctatus*. J. Parasitol., 8(3): 118–122.

————. 1923. Studies on sporulation and development of the cysts in a new species of Myxosporidia, *Lentospora ovalis*. J. Morphol., **37**(3): 425–456.

————. 1924. A new myxosporidian parasite, the cause of "wormy" halibut (Appendix VIII, Rep. of U. S. Commissioner of Fisheries for 1923). Bur. Fish. Doc., (957): 1–5.

————. 1926. *Schizamoeba salmonis*, a new ameba parasitic in salmonid fishes. Bur. Fish. Doc., (987): 1–8.

————. 1942. A suctorian parasite of the smallmouth black bass, with remarks on other suctorian parasites of fishes. Tr. Am. Micr. Soc., **61**(4): 309–327.

————. 1943. A new polymastigine flagellate, *Costia pyriformis*, parasitic on trout. J. Parasitol., **29**(6): 385–386.

————. 1944. A revision of the genus *Henneguya* (Myxosporidia) with descriptions of two new species. Tr. Am. Micr. Soc., **63**(4): 311–320.

————. 1946. Care and diseases of trout. Fish and Wildlife Research Rep. no. 12: 1–98.

————. 1947. Studies of the protozoan parasites of freshwater fishes. Fish and Wildlife Service Fishery Bull. **51**(41): 1–29.

————. 1953. Culture and Diseases of Game Fishes. Univ. Calif. Press., Berkeley. 332 pp.

Davis, W. 1956. American shad, *Alosa sapidissima*, parasitized by *Argulus canadensis* in the Connecticut River. J. Parasitol., **42**(3): 315.

Davison, R. C., W. Breese, and M. Katz. 1954. The hemoflagellate, *Cryptobia salmositica*, in Oregon salmon. J. Parasitol., **40**(6): 703–704.

Dawes, B. 1946. The Trematoda with Special Reference to British and Other European Forms. Cambridge Univ. Press. 644 pp.

Dechtiar, A. O. 1965a. A new distribution record for *Myxosoma scleroperca* Guilford, 1963 (Sporozoa; Myxosomatidae) in yellow perch of Lake Erie. Canadian Fish Cult. **34**: 31–34.

————. 1965b. Preliminary observations on *Glugea hertwigi* Weissenberg, 1911 (Microsporidia; Glugeidae) in American smelt, *Osmerus mordax* (Mitchill) from Lake Erie. Canadian Fish Cult. **34**: 35–38.

————. 1966. A new species of *Phyllodistomum* (Trematoda: Gorgoderidae) from *Coregonus clupeaformis* (Mitchill) from Lake of The Woods (Ontario). Canadian J. Zool. **44**: 135–140.

Dedie, O. 1940. Étude de *Salmincola mattheyi* n. sp., Copepode parasite de l'omble-chevalier (*Salmo salvelinus*). Rev. Suisse Zool., **47**(1):1–63.

DeGiusti, D. L. 1939. Preliminary note on the life cycle of *Leptorhynchoides thecatus*, an acanthocephalan parasite of fish. J. Parasitol., **25**(2): 180.

————. 1949. The life cycle of *Leptorhynchoides thecatus* (Linton), an acanthocephalan of fish. J. Parasitol., **35**(5): 437–460.

————. 1962. Ecological and life history notes on the trematode *Allocreadium lobatum* (Wallin, 1909), and its occurrence as a progenetic form in amphipods. J. Parasitol., **48**(2-2): 22.

DeGiusti, D. L., and J. Budd. 1959. A three-year survey of the infection rate of *Echinorhynchus coregoni* and *Cyathocephalus truncatus* in their intermediate host, *Pontoporeia affinis*, from South Bay Mouth, Ontario. J. Parasitol., **45**(4-2): 25.

Delco, E. A. 1962. Observations on the freshwater fish parasite *Lernaea catostomi*. Texas J. Sci., **14**(3): 365–367.

Dence, W. A. 1958. Studies on Ligula-infected common shiners (*Notropis cornutus frontalis* Agassiz) in the Adirondacks. J. Parasitol., **44**(3): 334–338.

DeRoth, G. C. 1953. Some parasites from Maine freshwater fishes. Tr. Am. Micr. Soc., **72**(1): 49–50.

Dickerman, E. E. 1934. Studies on the trematode family Azygiidae. I. The morphology and life cycle of *Proterometra macrostoma* Horsfall. Tr. Am. Micr. Soc., **53**(1): 8–21.

———. 1945. Studies on the trematode family Azygiidae. II. Parthenitae and cercariae of *Proterometra macrostoma* (Faust). Tr. Am. Micr. Soc., **64**(2): 138–144.

———. 1946. Studies on the trematoda family Azygiidae. III. The morphology and life cycle of *Proterometra sagittaria* n. sp. Tr. Am. Micr. Soc., **65**(1): 37–44.

———. 1948. On the life cycle and systematic position of the aspidogastrid trematode, *Cotylogaster occidentalis* Nickerson, 1902. J. Parasitol., **34**(2): 164.

———. 1954. *Paurorhynchus hiodontis*, a new genus and species of Trematoda (Bucephalidae: Paurorhynchinae n. subfam.) from the mooneye fish, *Hiodon tergisus*. J. Parasitol., **40**(3): 311–314.

Diesing, K. M. 1850. Systema Helminthum. Vindobonae. 679 pp.

———. 1858. Revision der Myzhelminthen. Abtheilung: Trematoden. Sitzungsb. K. Akad. Wissensch. Wien, Math.-Naturw., **32**(23): 307–390.

———. 1859. Revision der Rhyngodeen. Sitzungsb. K. Akad. Wissensch. Wien,-Naturw. Cl., **37**(21): 719–782.

———. 1863. Revision der Cephalocotyleen. Abtheilung: Paramecocotyleen. Abstr., Sitzungsb. K. Akad. Wissensch. Wien, Math.-Naturw., **48**(1): 200–345.

Dietz, E. 1909. Die Echinostomiden der Vogel. Zool. Anz., Leipzig, **34**(6): 180–192.

Dillon, W. A. 1966. Provisional list of parasites occurring on *Fundulus* spp. Virginia Inst. Marine Sc., Gloucester Point, Contrib. no. 204: 21–31.

Dobrovolny, C. G. 1939a. Life history of *Plagioporus sinitsini* Mueller and embryology of new cotylocercous cercariae (Trematoda). Tr. Am. Micr. Soc., **58**(2): 121–155.

———. 1939b. The life history of *Plagioporus lepomis*, a new trematode from fishes. J. Parasitol., **25**(6): 461–470.

Doflein, F. 1898. Studien zur Naturgeschichte der Protozoen. III. Ueber Myxosporidien. Zool. Jahrb., Anat., **11**: 281–350.

Dogiel, V. A., G. K. Petrushevski, and Yu. I. Polyanski. 1958. Parasitology of Fishes. Leningrad Univ. Press (English transl. Z. Kabata. Oliver and Boyd, Edinburgh. 384 pp).

Dolley, J. S. 1933. Preliminary notes on the biology of the St. Joseph River. Am. Midl. Nat., **14**(3): 193–227.

———. 1940. A new lernaean (parasitic copepod) from minnows in Lafayette County, Mississippi. Tr. Am. Micr. Soc., **59**(1): 70–77.

Dollfuss, R. Ph. 1949. Sur une cercaire ophthalmoxiphidiocerque, *Cercaria isopori* A. Looss, 1894, et sur la délimitation des Allocreadioidea. Ann. Parasitol., **24**(5 & 6): 424–435.

Dombroski, E. 1955. Cestode and nematode infection of sockeye smelts from Babine Lake, British Columbia. J. Fish. Res. Bd. Canada, **12**(1): 93–96.

Dorier, A. 1926. Dermite mortella à infusoires observée chez des alevin de truite arc-en-ciel. Travaux du laboratoire de pisciculture de l'université de Grenoble, Bd. 17, S. 155–158.

Doss, Mildred A., Katharine F. Roach, and Virginia L. Breen. 1963–64. Trematoda and trematode diseases. *In* Index-Catalogue of Medical and Veterinary Zoology. U. S. Dept. Agric.

Douglass, J. R. 1951. New parasite records from California dogs. Cornell Vet., **41**(4): 342–346.

Dubinina, M. N. 1957. Experimental study of the life cycle of *Schistocephalus solidus* (Cestoda: Pseudophyllidea), (Russian text, English summary). Zool. Zhurnal, **36**(11): 1647–1658.

————. 1962. Acarina. *In* Bykhovskaya-Pavlovskaya *et al.* Key to Parasites of Freshwater Fish. Zool. Inst., Acad. Sci. U.S.S.R. Keys to the Fauna of the U.S.S.R. no. 80 (English transl. TT 64–11040, OTS, Dept. Commerce, Washington, D.C. 919 pp.).

Dubois, G. 1935. Contribution à l'étude de quelques parasites de l'ordre des Strigeatoidea. Rev. Suisse Zool., **42**(1): 1–19.

————. 1936. Nouveaux principes de classification des trématodes du groupe des Strigeida. Rev. Suisse Zool., **43**(3): 507–515.

————. 1938. Monographie des Strigeida (Trematoda). Mem. Soc. Neuchatel Sci. Nat., **6**: 535 pp.

————. 1953. Systématique des Strigeida. Complément de la monographie. Mem. Soc. Neuchatel Sci. Nat., **8**(2): 141 pp.

————. 1964. Pers. comm. to Dr. Elizabeth Boyd, Mt. Holyoke College, South Hadley, Massachusetts.

Dubois, G., and J. Mahon. 1959. Étude de quelques trématodes Nord-Américains (avec note sur la position systématique de Parorchis Nicoll 1907) suivie d'une révision des genres *Galactosomum* Looss 1899 et *Ochetosoma* Braun 1901. Bull. Soc. Neuchatel. Sci. Nat., **82**: 191–229.

Dujardin, F. 1845. Histoire naturelle des hélminthes ou vers intestinaux. 654 pp.

Dunagan, T. T. 1960. Cercariae belonging to the Opisthorchioidea. Proc. Helminth. Soc. Washington, **27**(1): 44–52.

Dunkerly, J. S. 1915. *Agarella gracilis*, a new genus and species of myxosporidian parasitic in *Lepidosiren paradoxa*. Proc. Roy. Phys. Soc. Edinburgh, **19**(8): 213–219.

Du Plessis, S. S. 1948. A gyrodactyloid parasite from the ureters of largemouth bass at the Jonkershoek Inland Fish Hatchery, South Africa. Tr. Am. Fish. Soc., **75**: 105–109.

Duvernoy, G. L. 1842. Note sur un nouveau genre de ver intestinal, de la famille des tenioides, le bothrimone de l'esturgeon (*Bothrimonus sturionis*, Nob.). Ann. Sci. Nat., Zool., **18**: 123–126.

Dyk, V. 1958. Katedra parasitologie veterinarni fakulty Vysoke skoly zemedelske, Brno. Dynamika motolice *Crepidostomum farionis* (O. F. Müller 1784) a jeji vztahy k hostitelum a prostredi. Československa Parasitologie, **5**(2): 51–57.

Earp, B. J., and R. L. Schwab. 1954. An infestation of leeches on salmon fry and eggs. Prog. Fish-Cult., **16**(3): 122–124.

Eckmann, F. 1932. Beiträge zur Kenntnis der Trematodenfamilie Bucephalidae. Zeitschr. Parasitenk., Berlin, **5**(1): 94–111.

Eisler, R., and R. C. Simon. 1961. Destruction of salmon larvae by *Hydra oligactis*. Tr. Am. Fish. Soc., **90**(3): 329–332.

Ekbaum, Ella K. 1933. *Philonema oncorhynchi* nov. gen. et spec. Contrib. Canad. Biol. and Fish., **8**(4): 89–98.

Elkan, E. 1962. *Dermocystidium gasterostei* n. sp., a parasite of *Gasterosteus aculeatus* L. and *Gasterosteus pungitius* L. Nature, **196**(4858): 958–960.

Elliot, A. M., and Louise R. Russert. 1949. Some condition characteristics of a yellow perch population heavily parasitized by *Clinostomum marginatum*. J. Parasitol., **35**(2): 183–190.

Enders, H. E., and S. A. Rifenburgh. 1928. A preliminary report on *Lernaea*, a parasite of goldfish. Proc. Indiana Acad. Sci., **38**: 333–334.

Engelbrecht, H. 1961. Wpływ środowiska na rozwój robaków pasożytniczych. Wiadomości Parazytologiczne, **7**(2): 109–113.

Erasmus, D. A., and Bennet. 1965. A study of some of the factors affecting excystation *in vitro* of the metacercarial stages of *Holostephanus bushiensis* Khan, 1962 (Strigeida: Trematoda). J. Helminthol. **39**(2 & 3): 185–196.

Erickson, D. G., and F. G. Wallace. 1959. Studies on blood flukes of the genus *Sanguinicola*. J. Parasitol., **45**(3): 310–322.

Erickson, J. D. 1965. Report on the problem of *Ichthyosporidium* in rainbow trout. Prog. Fish-Cult. **27**(4): 179–184.

Essex, H. E. 1927. The structure and development of *Corallobothrium*. J. Parasitol., **14**(2): 130–131.

———. 1928*a*. *Crepidobothrium fragile*, a new species of tapeworm of the channel-cat (*Ictalurus punctatus*). J. Parasitol., **15**(2): 137.

———. 1928*b*. On the life-history of *Bothriocephalus cuspidatus* Cooper, 1917, a tapeworm of the wall-eyed pike. Tr. Am. Micr. Soc., **47**(3): 348–355.

———. 1928*c*. The structure and development of *Corallobothrium* with descriptions of two new fish tapeworms. Illinois Biol. Monogr., **11**(3): 261–328.

———. 1929*a*. *Crepidobothrium fragile* n. sp., a tapeworm of the channel catfish. Parasitology, **21**(1 & 2): 164–167.

———. 1929 *b*. The life-cycle of *Haplobothrium globuliforme* Cooper, 1914. Science, **69**: 677–678.

Essex, H. E., and G. W. Hunter, III. 1926. A biological study of fish parasites from the central states. Tr. Illinois State Acad. Sci., **19**: 151–181.

Etges, F. T. 1961. Contributions to the life history of the brain fluke of newts and fish, *Diplostomulum scheuringi* Hughes, 1929 (Trematoda: Diplostomatidae). J. Parasitol., **47**(3): 453–458.

Evans, W. S. 1963. *Amphimerus pseudofelineus* (Ward, 1901), (Digenea: Opisthorchidae) and its second intermediate host in Manitoba. Canad. J. Zool., **41**: 650–651.

Fantham, H. B. 1930. Some parasitic protozoa found in South Africa. 13. South African J. Sci., **27**: 376–390.

Fantham, H. B., and Annie Porter. 1943. *Plasmodium struthionis*, sp. n., from Sudanese ostriches and *Sarcocystis salvelini*, sp. n., from Canadian speckled trout (*Salvelinus fontinalis*), together with a record of a *Sarcocystis* in the eel pout (*Zoarces angularis*). Proc. Zool. Soc., **113**(1 & 2): 25–30.

———. 1947. The parasitic fauna of vertebrates in certain Canadian fresh waters, with some remarks on their ecology, structure and importance. Proc. Zool. Soc., **117**(4): 609–649.

Fantham, H. B., Annie Porter, and L. R. Richardson. 1939. Some Myxosporidia found in certain fresh-water fishes in Quebec Province, Canada. Parasitology, 31(1): 1–77.

―――. 1940. Some more Myxosporidia observed in Canadian fishes. Parasitology, 32(3): 333–353.

―――. 1941. Some microsporidia found in certain fishes and insects in eastern Canada. Parasitology, 33(2): 186–208.

―――. 1942. Some haematozoa observed in vertebrates in eastern Canada. Parasitology, 34(2): 199–226.

Farrell, R. K., Marilyn A. Lloyd, and B. Earp. 1964. Persistence of *Neorickettsiae helminthoeca* in an endoparasite of the Pacific salmon. Science, 145(3628): 162–163.

Fasten, N. 1912. The brook trout disease at Wild Rose and other hatcheries. Bien. Rep. Comr. Fish. Wisconsin, 12–22.

―――. 1913. The behavior of a parasitic copepod, *Lernaeopoda edwardsii* Olsson. J. Animal Behavior, 3(1): 36–60.

―――. 1916. The eye of the parasitic copepod, *Salmincola edwardsii* Olsson (*Lernaeopoda edwardsii* Olsson). Biol. Bull., 31(6): 407–418.

―――. 1918. Trout and fish lice. Publ. Puget Sound Biol. Sta., 2(35–38): 73–77.

―――. 1921a. Another male copepod of the genus *Salmincola* from the gills of the chinook salmon. Biol. Bull., 41(3): 121–124.

―――. 1921b. Studies on parasitic copepods of the genus *Salmincola*. Am. Naturalist, 55: 449–456.

―――. 1922. The tapeworm infection in Washington trout and its related biological problems. Am. Naturalist, 56(64): 439–447.

Faust, E. C. 1918. Life history studies on Montana trematodes. Illinois Biol. Monogr., 4(1): 120 pp.

―――. 1920. Two new Proteocephalidae. J. Parasitol., 6(2): 79–83.

Fellows, C. S. 1888. A description of *Ergasilus chautauquaensis*: A new species of copepoda, and a list of other entramostraca found at Lake Chautauqua in August, 1886. Proc. Am. Soc. Micr., 10: 246–249.

Ferguson, M. S. 1943. Development of eye flukes of fishes in the lenses of frogs, turtles, birds, and mammals. J. Parasitol., 29(2): 136–142.

Ferguson, M. S., and R. A. Hayford. 1941. The life history and control of an eye fluke. An account of a serious hatchery disease caused by a parasitic worm. Prog. Fish-Cult., Memo. 1–131, (54): 1–13.

Fischer von Waldheim, G. 1798. Sur un nouveau genre des vers intestins, *Cystidicola farionis*, suivi de quelques remarques sur les milieux dans lesquels les vers intestins vivent. J. Phys., Chim. et Hist. Nat., Paris, 4: 304–309; Arch. Phys., 3(1): 95–100.

Fischthal, J. H. 1942a. Three new species of *Phyllodistomum* (Trematoda: Gorgoderidae) from Michigan fishes. J. Parasitol., 28(4): 269–275.

―――. 1942b. *Triganodistomum hypentelii* n. sp. (Trematoda: Lissorchiidae) from the hog sucker, *Hypentelium nigricans* (Le Sueur). J. Parasitol., 28(5): 389–393.

―――. 1943. A description of *Phyllodistomum etheostomae* Fischthal, 1942 (Trematoda: Gorgoderidae) from percid fishes. J. Parasitol., 29(1): 7–9.

―――. 1947a. Parasites of northwest Wisconsin fishes. I. The 1944 survey. Tr. Wisconsin Acad. Sci., Arts and Lett., 37: 157–220.

―――. 1947b. Parasites of Brule River fishes. Brule River survey: Rep. no. 6. Tr. Wisconsin Acad. Sci., Arts and Lett., **37**: 275–278.

―――. 1949a. *Sanguinicola huronis* n. sp. (Trematoda: Sanguinicolidae) from the blood system of the largemouth and smallmouth basses. J. Parasitol., **35**(6): 566–568.

―――. 1949b. *Epistylis*, a peritrichous protozoan on hatchery brook trout. Prog. Fish-Cult., **11**(2): 122–124.

―――. 1950a. Additional hosts and geographical distribution records for the common fish acanthocephalan, *Leptorhynchoides thecatus*. J. Parasitol., **36**(1): 88.

―――. 1950b. A new genus and species of Caryophyllaeidae (Cestoda) from fishes. J. Parasitol., **36**(6): 28.

―――. 1950c. Parasites of northwest Wisconsin fishes. II. The 1945 survey. Tr. Wisconsin Acad. Sci., Arts and Lett., **49**(1): 87–113.

―――. 1951. *Pliovitellaria wisconsinensis* n. g., n. sp., (Cestoda: Caryophyllaeidae) from Wisconsin cyprinid fishes. J. Parasitol., **37**(2): 190–194.

―――. 1952. A redescription of *Phyllodistomum lysteri* Miller, 1940 (Trematoda: Gorgoderidae), from the common white sucker. J. Parasitol., **38**(3): 242–244.

―――. 1953. *Hypocaryophyllaeus gilae* n. sp. (Cestoda: Caryophyllaeidae) from the Utah chub, *Gila straria* in Wyoming. Proc. Helminth. Soc. Washington, **20**(2): 113–117.

―――. 1954. *Bialovarium nocomis* Fischthal, 1953 (Cestoda: Caryophyllaeidae) from the hornyhead chub, *Nocomis biguttatus* (Kirtland). Proc. Helminth. Soc. Washington, **21**(2): 117–120.

―――. 1956. Observations on the occurrence of parasites in the fishes of certain south-central New York streams. New York Fish and Game J., **3**(2): 225–233.

―――. 1957. *Cestrahelmins laruei* n. g., n. sp., a digenetic trematode from the muskellunge, *Esox m. masquinongy* Mitchill. J. Parasitol., **43**(4): 484–487.

Fischthal, J. H., and L. N. Allison. 1940. *Acolpenteron ureteroecetes* n. g., n. sp., a monogenetic trematode from the ureters of black basses. J. Parasitol., **26**(6): 34–35.

―――. 1941. *Acolpenteron ureteroecetes* Fischthal and Allison, 1940, a monogenetic trematode from the ureters of the black basses, with a revision of the family Calceostomatidae (Gyrodactyloidea). J. Parasitol., **27**(6): 517–524.

―――. 1942. *Acolpenteron catostomi* n. sp. (Gyrodactyloidea: Calceostomatidae), a monogenetic trematode from the ureters of suckers, with observations on its life history and that of *A. ureteroecetes*. Tr. Am. Micr. Soc., **61**(1): 53–56.

Fish, F. F. 1939. Observations on *Henneguya salminicola* Ward, a myxosporidian parasitic in Pacific salmon. J. Parasitol., **25**(2): 169–172.

Foote, L. E., and B. P. Blake. 1945. Life history of the eastern pickerel in Babcock pond, Connecticut. J. Wildlife Management, **9**(2): 89–96.

Forel, F. A. 1868. Préparations microscopiques d'une nouvelle espèce de *Triaenophorus*. Bull. Soc. Vaudoise Sci. Nat., Lausanne, **9**: 696.

Fouquet, D. 1876. Note sur une espèce d'infusoires parasites des poissons d'eau douce. Arch. Zool. Exper. et Gen., **5**(2): 159–165.

Fourment, L. 1883. Sur les filaments ovulaires chez les nématodes. Compt. Rend. Soc. Biol., Paris, **35**(5): 575–578.

Fox, A. 1961. Pers. comm., Dept. Zool. and Entomol., Montana State Coll., Bozeman, Montana.

———. 1962. Parasite incidence in relation to size and condition of trout from two Montana lakes. Tr. Am. Micr. Soc., **81**(2): 179–184.

———. 1965a. Some effects of strigeid metacercariae on rainbow trout (*Salmo gairdneri*). Tr. Am. Micr. Soc., **84**(1): 153.

———. 1965b. The life cycle of *Bolbophorus confusus* (Krause, 1914) Dubois, 1935 (Trematoda: Strigeoidea), and the effects of the metacercaria on fish hosts. Ph.D. diss., Montana State Univ.

Fox, E., and R. E. Olson. 1965. The life cycle of the digenetic trematode *Bolbophorus confusus*, and effects of water temperatures on the developmental stages. Tr. Am. Micr. Soc., **84**(1): 153–154.

Frank, W. 1962. Histologische untersuchungen bei *Carassius carassius auratus* L. (Pisces, Teleostei) nach starkem Befall durch *Trichodina domerguei* Wallengreen, 1897 (Protozoa, Euciliata). Z. f. Parasitenkunde **21**: 446–456.

Freeman, R. S. 1964. Flatworm problems in fish. Canad. Fish. Cult., **32**: 11–18.

Frese, V. Y. 1965. Proteocephalata—(Cestoda-Helminthes) of fish, amphibians and reptiles. Acad. Nauk. U.S.S.R. Osnovy Cestodologii. Tom V, Izdatel'stvo (Nauka), Moscow, 538 pp. (Eng. transl. in progress, OTS, U. S. Dept. Commerce, Washington).

Friend, G. F. 1939. Gill parasites of brown trout in Scotland. Scottish Naturalist, (239): 123–126.

———. 1941. Tr. Roy. Soc. Edinburgh, 60(2), no. 15: 503–541.

Fritsch. 1886. *Corallobothrium solidum*. Tagebl. 59. Versamml. Deutsch. Naturf. u. Aerzte, (9): 371–372.

Fritts, D. H. 1959. Helminth parasites of the fishes of northern Idaho. Tr. Am. Micr. Soc., 78(2): 194.

Fryer, G. 1961. Variation and systematic problems in a group of lernaeid copepods. Crustaceana, **2**(4): 275–285.

———. 1963. Crustacean parasites from cichlid fishes of the genus *Tilapia* in the Musée Royal de l'Afrique centrale. Rev. Zool. Bot. Afr., **68**(3 & 4): 386–392.

Fujita, T. 1921. On parasites of Japanese fishes. IV. 3. Nematoda (Japanese text, English summary). Dobuts. Zasshi, Tokyo, (395), **33**: 292–300.

———. 1922. On the parasites of Japanese fishes. Dobuts. Zasshi, Tokyo, (403), **34**: 577–584.

———. 1939. On the nematoda-parasites of the Pacific salmon. J. Fac. Agric., Hokkaido Imp. Univ., **42**(3): 239–266.

Gadd, P. 1901. Nagra forut obeskrifna, parasitiskt lefvande copepoder (Ueber einige neue parasitische Copepoden). Medd. Soc. pro Fauna et Flora Fennica, **27**: 98–100, 181–182.

Garoian, G. S. 1960. *Schistocephalus thomasi*, n. sp. (Cestoda: Diphyllobothriidae), from fish-eating birds. Proc. Helminth. Soc. Washington, **27**(2): 199–202.

Gauthier, M. 1926. Endoparasites de la truite indigène (*T. fario* L.) en Dauphine. Assoc. Franç. Avance. Sci., 442–444.

Gebhart, G. A., R. E. Millemann, S. E. Knapp, and P. Nyberg. 1966. "Salmon poisoning" disease. II. Second intermediate host susceptibility studies. J. Parasitol. **52**(1): 54–59.

Ghittino, P. 1961. Su una capillariosi epatica in trote di allevamento e in altri teleostei delle acque libere del bacino del po in Piemonte, con descrizione di una nuova specie (*Capillaria eupomotis*), (English transl., U. S. Dept. Interior). Estratto dalla Rivista di Parassitologia, **22**(3): 193–204.

Gnadeberg, W. 1949. Beiträge zur Biologie und Entwicklung des *Ergasilus sieboldii* v. Nordmann (Copepoda Parasitica). Zeitschr. Parasitenk., **14** (1 & 2): 103–180.

Gnanamuthu, C. P. 1951. Notes on the life history of a parasitic copepod, *Lernaea chackoensis*. Parasitology, **41**(3 & 4): 148–155.

Goeze, J. A. E. 1782. Versuch einer Naturgeschichte der Eingeweidewurmer Thierischer Korper. Blankenburg. 471 pp.

——. 1792. Verzeichniss der Naturalian meines Kabinets, besonders aus dem Thierreich, Mehrentheils in Weingeist, mit naturhistorischen Anmerkungen. Nachweisung des Systems und Anzeige der besten Abbildungen. Leipzig. 80 pp.

Goin, C. J., and L. H. Ogren. 1956. Parasitic copepods (Argulidae) on amphibians. J. Parasitol., **42**(2): 172.

Goldberger, J. 1911. Some known and three new endoparasitic trematodes from American freshwater fish. Hyg. Lab., U. S. Pub. Health and Mar.-Hosp. Serv., Bull. 71: 7–35.

Golvan, Y. J. 1957. Acanthocephala des poissons. Résultats scientifiques. Exploration hydrobiologique des lacs Kivu, Edouard et Albert (1952–1954). Brussels, **3**(2): 55–64.

——. 1960. Le phylum des Acanthocephala. Troisième note. La classe de Palaeacanthocephala (Meyer, 1931), (à suivre). Ann. Parasitol., **35**(1 & 2): 138–165, 350–386, 574–593.

Gower, W. C. 1938. Seasonal abundance of some parasites of wild ducks. J. Wildlife Management, **2**(4): 223–232.

Grabda, J. 1956. Badania nad rozwojem *Lernaea esocina* (Burm., 1833) i *Lernaea cyprinacea* L., 1758. Wiadomości Parazytologiczne, **2** Suppl.

——. 1958. Developmental cycle of *Lernaea cyprinacea* L. Wiad. Parazytol., **4**(5/6).

——. 1963. Life cycle and morphogenesis of *Lernaea cyprinacea* L. Acta Parasitologica Polonica, **11**(14): 169–198.

Grabiec, S., A. Guttowa, and W. Michajlow. 1963. Effect of light stimulus on hatching of coracidia of *Diphyllobothrium latum* (L.). Acta Parasitologica Polonica, **11**(14/18): 229–238.

Gracia, R. A. 1960. Contribucion al conocimiento de los acantocefalos parasitos de los peces en Venezuela. Rev. Vet. Venezolana, (42) **8**: 3–32.

Griffith, Ruth. 1953. Preliminary survey of the parasites of fish of the Palouse area. Tr. Am. Micr. Soc., **72**(1): 51–57.

Guberlet, J. E. 1928. Notes on a species of *Argulus* from goldfish. Univ. Washington Publ. Fish., **2**(3): 31–42.

——. 1929. Notes on parasitic life history studies. J. Parasitol., **16**(2): 98–99.

Gudger, E. W. 1927. Hydras as enemies of young fishes. Nat. His., **27**(3): 270–274.

——. 1934. Coelenterates as enemies of young fishes. I. Hydras and sessile colonial hydroids as fish-eaters. Ann. Mag. Nat. Hist., **13**(74): 192–212.

Guidice, J. J. 1950. Control of *Lernaea carassii* Tidd, a parasitic copepod infesting goldfish in hatchery ponds with related observations on crayfish and the "fish louse" *Argulus* sp. Thesis, Univ. Missouri. 54 pp.

Guilford, H. G. 1954. Parasites found in the sea lamprey, *Petromyzon marinus*, from Lake Michigan. J. Parasitol., **40**(3): 364.

———. 1963. New species of myxosporidia found in percid fishes from Green Bay (Lake Michigan). J. Parasitol., **49**(3): 474–478.

———. 1965. New species of myxosporidia from Green Bay (Lake Michigan). Tr. Am. Micr. Soc. **84**(4): 566–573.

Gurley, R. R. 1893. On the classification of Myxosporidia, a group of protozoan parasites infesting fishes. Bull. U. S. Fish. Comm., **11**: 407–420.

———. 1894. The Myxosporidia, or psorosperms of fishes, and the epidemic produced by them. Rep. U. S. Fish. Comm., **26**: 65–304.

Gurney, R. 1933. Notes on some copepoda from Plymouth. J. Marine Biol. Assoc., United Kingdom, **19**: 299–304.

Gustafson, P. V. 1939. Life cycle studies on *Spinitectus gracilis* and *Rhabdochona* sp. (Nematoda: Thelaziidae) J. Parasitol., **25**(6): 12–13.

———. 1942a. A peculiar larval development of *Rhabdochona* spp. (Nematoda: Spiruroidea). J. Parasitol., **28**(6-2): 30.

———. 1942b. Some parasites of fresh water fish of the Spokane area. Northwest Sci., **16**(1): 28.

———. 1949. Description of some species of *Rhabdochona* (Nematoda: Thelaziidae). J. Parasitol., **35**(5): 543–540.

Gustafson, P. V., and R. R. Rucker. 1956. Studies on an Ichthyosporidium infection in fish: transmission and host specificity. USFWS, Spec. Sci. Rep., Fisheries no. 166: 1–7.

Guttowa, A. 1961. Potential intermediate hosts (Copepoda) of the broad tapeworm of man *Diphyllobothrium latum* (L.) in Norway. Nytt Magasin for Zoologi, Oslo, **10**, 57–62.

Haas, G. 1933. Beitrage zur Kenntnis der Cytologie von *Ichthyophthirius multifilis* Fouq. Arch. Protistenk., **81**(1): 88–137.

Haderlie, E. C. 1953. Parasites of the fresh-water fishes of northern California. Univ. Calif. Publ. Zool., **57**(5): 303–440.

Hahn, G. W. 1913. Sporozoon parasites of certain fishes in the vicinity of Woods Hole, Massachusetts. Bull. U. S. Bur. Fish., **33**: 191–214.

Haider, G. 1964. Monographie der Familie Urceolariidae (Ciliata, Peritricha, Mobilia) mit besonderer Berücksichtigung der im Süddentaschen Raum verkommen den Arten. Parasitolog. Schr., Reihe **17**. G. Fischer, Jena. 251 pp.

Haley, A. J. 1954a. Microsporidian parasite, *Glugea hertwigi*, in American smelt from the Great Bay region, New Hampshire. Tr. Am. Fish. Soc., **83**: 84–90.

———. 1954b. Further observations on *Glugea hertwigi* Weissenberg, 1911, 1913 (Microsporidia), in fresh water smelt in New Hampshire. J. Parasitol., **40**(4): 482–483.

Haley, A. J., and W. L. Bullock. 1953. A new species of Acanthocephala from the sunfish, *Lepomis gibbosus* (Linnaeus), with a redescription of the family Fessisentidae Van Cleave 1931. Am. Midl. Nat., **59**(1): 202–205.

Haley, A. J., and H. E. Winn. 1959. Observations on a lernaean parasite of freshwater fishes. Tr. Am. Fish. Soc., **88**(2): 128–129.

Hall, M. C. 1916. Nematode parasites of mammals of the orders Rodentia, Lagomorpha, and Hyracoidea. Proc. U. S. Nat. Mus., **50**: 1–258.

Hallberg, C. W. 1952. *Dioctophyma renale* (Goeze, 1782). A study of the migration routes to the kidneys of mammals and resultant pathology. Diss. Abst., **12**(2): 232.

Harding, J. P. 1950. On some species of *Lernaea* (Crustaeca, Copepoda: parasites of freshwater fish). Bull. British Mus. (Nat. Hist.) Zool., **1**(1): 1–27.

Hardy, A. D. 1907. Nature (London), **75**: 599.

Hare, R. C. 1943. An ecological study of the worm parasites of Portage Lakes fishes. Ohio J. Sci., **43**(5): 201–208.

Hargis, W. J. (Jr.). 1952a. Monogenetic trematodes of Westhampton Lake fishes. I. Two new forms. Am. Midl. Nat., **47**(2): 471–477.

———. 1952b. Monogenetic trematodes of Westhampton Lake fishes. II. A list of species and key to the genera encountered. Virginia J. Sci., n.s., **3**(2): 112–115.

———. 1952c. A revision of the genera of the subfamily Tetraonchinae. Proc. Helminth. Soc. Washington, **19**(1): 40–44.

———. 1953a. Monogenetic trematodes of Westhampton Lake fishes. III. Part 1. Comparative morphology of the species encountered. J. Parasitol., **39**(1): 88–105.

———. 1953b. Chloretone as a trematode relaxer, and its use in mass-collecting techniques. J. Parasitol., **39**(2): 224–225.

———. 1955a. Monogenetic trematodes of Gulf of Mexico fishes. Part I. The superfamily Gyrodactyloidea. Biol. Bull., **108**(2): 125–137.

———. 1955b. Monogenetic trematodes of Gulf of Mexico fishes. Part II. The superfamily Gyrodactyloidea. J. Parasitol., **41**(2): 185–193.

———. 1958a. The parasite *Argulus laticauda* as a fortuitous human epizoon. J. Parasitol., **44**(1): 45.

———. 1958b. Parasites and fishery problems. Proc. Gulf and Caribbean Fish. Inst., **11**: 70–75.

Harkema, R., and G. C. Miller. 1961. Observations on parasitism of the raccoon of Cape Island, S. C. J. Parasitol., **47**(4-2): 41.

———. 1962. Parasitic helminths from the raccoon, mink, and muskrat. J. Parasitol., **48**(2-2): 22

Harms, C. E. 1959. Checklist of parasites from catfishes of northeastern Kansas. Tr. Kansas Acad. Sci., **62**(4): 262.

———. 1960. Some parasites of catfishes from Kansas. J. Parasitol., **46**(6): 695–701.

———. 1963. The development and cultivation of the acanthocephalan, *Octospinifer macilentis* (*macilentus*) Van Cleave, 1919. Diss. Abst., **23**(7): 2632–2633.

Harrises, A. E. 1962. New Dactylogyridae (Trematoda: Monogenea) from Mississippi. Am. Midl. Nat., **67**(1): 199–203.

Harwood, P. D. 1935. *Maculifer chandleri*, n. sp. (Allocreadiidae), a trematode from Texas catfish. Proc. Helminth. Soc. Washington **2**(2): 75–76.

Hausmann, L. 1896. Ueber Trematoden der Süsswasserfische. Vorläufige Mitteilung. Centralbl. Bakt., 1. Abt., **19**(11): 389–392.

Hedrick, L. R. 1935. The life history and morphology of *Spiroxys contortus* (Rudolphi); Nematoda: Spiruridae. Tr. Am. Micr. Soc., **54**(4): 307–335.

Heitz, F. A. 1917. *Salmo salar* Lin., seine Parasitenfauna und seine Ernahrung im Meer und im Süsswasser. Eine parasitologischbiologische Studie. Inaug.-Diss., 1–137.

Hemingway, E. E. 1912. Anatomy of *Placobdella pediculata* (erroneously titled *P. parasitica*). *In* The leeches of Minnesota. Geol. and Nat. Hist. Surv. Minnesota, Zool. Ser. no. 5.

Henderson, H. E. 1938. The cercaria of *Crepidostomum cornutum* (Osborn). Tr. Am. Micr. Soc., **57**(2): 165–172.

Henderson, J. T. 1927. Description of a copepod gill parasite of pike-perches in lakes of northern Quebec, including an account of the free-swimming male and some developmental stages. Contrib. Canad. Biol. and Fish., **3**: 235–245.

Hermann, J. 1783. Helmintologische Bemerkungen. Der Naturforscher, no. 19: 31–59.

Herrick, J. A. 1936. Two new species of *Myxobolus* from fishes of Lake Erie. Tr. Am. Micr. Soc., **55**(2): 194–198.

———. 1941. Some myxosporidian parasites of Lake Erie fishes. Tr. Am. Micr. Soc., **60**(2): 164–170.

Heymons, R. 1935. Pentastomida. *In* Bronn's Klass. u. Ordnung. Tierreichs, **5**, Abt. 4, Buch 1, Lief. 1:1–160.

Hilliard, D. K. 1960. Studies on the helminth fauna of Alaska. 38. The taxonomic significance of eggs and coracidia of some diphyllobothriid cestodes. J. Parasitol., **46**(6): 703–716.

Hirschmann, H., and K. Partsch. 1953. Die Karpfenlaus (Ueberarbeitung von *Argulus pellucidus* Wagler). Mikrokosmos, **43**: 217–223.

Hobgood, J. O. 1938. The metacercaria of *Cercaria flexicorpa* Collins. Tr. Am. Micr. Soc., **57**(2): 158–164.

Hoff, E. C., and H. E. Hoff. 1929. *Proteocephalus pugetensis*, a new tapeworm from a stickleback. Tr. Am. Micr. Soc., **48**(1): 54–61.

Hoffman, G. L. 1949. Isolation of *Saprolegnia* and *Achlya* with penicillin-streptomycin and attempts to infect fish. Prog. Fish-Cult., **11**(3): 171–174.

———. 1953*a*. Parasites of fish of Turtle River, North Dakota. Proc. North Dakota Acad. Sci., **7**: 12–19.

———. 1953*b*. *Scaphanocephalus expansus* (Crepl.), a trematode of the osprey in North America. J. Parasitol., **39**(5): 568.

———. 1954*a*. The occurrence of *Ornithodiplostomum ptychocheilus* (Faust), (Trematoda: Strigeida) in fish and birds. J. Parasitol., **40**(2): 232–233.

———. 1954*b*. Polyvinyl alcohol fixative-adhesive for small helminths and protozoa. Tr. Am. Micr. Soc., **73**(3): 328–329.

———. 1955*a*. Notes on the life cycle of *Bunodera eucaliae* Miller (Trematoda: Allocreadiidae) of the stickleback, *Eucalia inconstans*. Proc. Iowa Acad. Sci., **62**: 638–639.

———. 1955*b*. *Neascus nolfi* n. sp. (Trematoda: Strigeida) from cyprinid minnows with notes on the artificial digest recovery of helminths. Am. Midl. Nat., **53**(1): 198–204.

———. 1956*a*. The life cycle of *Crassiphiala bulboglossa* (Trematoda: Strigeida), development of the metacercaria and cyst, and effect on the fish hosts. J. Parasitol., **42**(4): 435–444.

———. 1956*b*. Unpublished research.

———. 1957*a*. Studies on the life cycle of *Cryptocotyle concavum* from the common sucker and experimentally in the chick. Proc. North Dakota Acad. Sci., **11**: 55–56.

———. 1957*b*. Unpublished research.

———. 1958*a*. Studies on the life-cycle of *Ornithodiplostomum ptychocheilus*

(Faust), (Trematoda: Strigeoidea) and the "self cure" of infected fish. J. Parasitol., **44**(4): 416–421.

————. 1958*b*. Experimental studies on the cercaria and metacercaria of a strigeoid trematode, *Posthodiplostomum minimum*. Exper. Parasitol., **7**(1): 23–50.

————. 1958*c*. Experimental infection with strigeoid cercariae. J. Parasitol., **44**(2): 229.

————. 1959*a*. Studies on the life cycle of *Apatemon gracilis pellucidus* (Yamag.). Tr. Am. Fish. Soc., **88**: 96–99.

————. 1959*b*. Recommended treatment for fish parasitic diseases. BSFW, Fishery Leaflet 486.

————. 1959*c*. Unpublished research.

————. 1960. Synopsis of Strigeoidea (Trematoda) of fishes and their life cycles. BSFW, Fishery Bull. 175, **60**: 439–469.

————. 1962. Unpublished research.

————. 1963. Parasites of freshwater fish. I. Fungi. 1. Fungi (*Saprolegnia* and relatives) of fish and fish eggs. BSFW, Fishery Leaflet 564: 1–6.

————. 1964. Unpublished research.

————. 1965*a*. *Eimeria aurati* n. sp. (Protozoa: Eimeriidae) from goldfish (*Carassius auratus*) in North America. J. Protozool., **12**(2): 273–275.

————. 1965*b*. Unpublished research.

————. 1965*c*. The control of fish parasites. *In* Biology Problem in Water Pollution, 3d Seminar, 1962: 283–285.

Hoffman, G. L., H. Bishop, and C. E. Dunbar. 1960. Algal parasites in fish. Prog. Fish-Cult., **22**(4): 180.

Hoffman, G. L., and C. E. Dunbar. 1961. Mortality of eastern brook trout caused by plerocercoids (Cestoda: Pseudophyllidea: Diphyllobothriidae) in the heart and viscera. J. Parasitol., **47**(3): 399–400.

————. 1963. Studies on *Neogogatea kentuckiensis* (Cable, 1935) n. comb. (Trematoda: Strigeoidea: Cyathocotylidae). J. Parasitol., **49**(5): 737–744.

Hoffman, G. L., C. E. Dunbar, and A. Bradford. 1962. Whirling disease of trouts caused by *Myxosoma cerebralis* in the United States. BSFW, Spec. Sci. Rep., Fisheries no. 427: 1–15.

Hoffman, G. L., and J. B. Hoyme. 1958. The experimental histopathology of the "tumor" on the brain of the stickleback caused by *Diplostomum baeri eucaliae* Hoffman and Hundley, 1957 (Trematoda: Strigeoidea). J. Parasitol., **44**(4): 374–378.

Hoffman, G. L., and J. B. Hundley. 1957. The life-cycle of *Diplostomum baeri eucaliae* n. subsp. (Trematoda: Strigeida). J. Parasitol., **43**(6): 613–637.

Hoffman, G. L., G. W. Prescott, and C. R. Thompson. 1965. *Chlorella* (Alga: Chlorophyta) parasitic in bluegills. Prog. Fish-Cult., **27**(3): 175.

Hoffman, G. L., and R. E. Putz. 1963. Unpublished research.

————. 1964. Studies on *Gyrodactylus macrochiri* n. sp. (Trematoda: Monogenea) from *Lepomis macrochirus*. Proc. Helminth. Soc. Washington, **31**(1): 76–82.

————. 1965. The black-spot (*Uvulifer ambloplitis*: Trematoda: Strigeoidea) of centrarchid fishes. Tr. Am. Fish. Soc., **94**(2): 143–151.

Hoffman, G. L., R. E. Putz, and C. E. Dunbar. 1965. Studies on *Myxosoma cartilaginis* n. sp. (Protozoa: Myxosporidea) of centrarchid fish and a synopsis of the *Myxosoma* of North American freshwater fishes. J. Protozool., **12**(3): 319–332.

Hoffman, R. L. 1964. A new species of *Cystobranchus* from southwestern Virginia (Hirudinea: Piscicolidae). Am. Midl. Nat., **72**(2): 390–395.

Hoffman, J. 1959. A propos d'une maladie de captivité des Cystobranches. Arch. Inst. Grand-Ducal Luxembourg, Sect. Sci. Nat., Phys. et Math., **26**: 237–243.

Holl, R. J. 1928*a*. A linguatulid parasite from North American fishes. J. Parasitol., **15**(1): 63–66.

———. 1928*b*. Two new nematode parasites. J. Elisha Mitchell Sci. Soc., **43**(3 & 4): 184–186.

———. 1929. The phyllodistomes of North America. Tr. Am. Micr. Soc., **48**(1): 48–53.

Holliman, R. B., and W. H. Leigh. 1953. Life cycle of *Paramacroderoides echinus* Venard, 1941, a parasite of the Florida gar, *Lepisosteus platyrhincus*. J. Parasitol., **39**(4-2): 21.

Hollis, E. H, and C. M. Coker. 1949. A trematode parasite of the genus *Clinostomum* new to the shad, *Alosa sapidissima*. J. Parasitol., **34**(6): 493–495.

Holloway, H. L. 1953. Notes of the occurrence of *Neoechinorhynchus cylindratus* in fishes of Westhampton Lake. Virginia J. Sci., **4**(4): 231.

———. 1957. The distribution of *Neoechinorhynchus cylindratus* Van Cleave in North America. Virginia J. Sci., **8**(4): 296–297.

Holloway, H. L., and B. J. Bogitsh. 1964. Helminths of Westhampton lake fish. Virginia J. Sci., **15**(1): 41–44.

Honigberg, B. M., *et al.* 1964. A revised classification of the phylum Protozoa. J. Protozool., **11**(1): 7–20.

Hopkins, S. H. 1931*a*. Studies on *Crepidostomum*. I. *Crepidostomum isostomum* n. sp. J. Parasitol., **17**(3): 145–150.

———. 1931*b*. Studies on *Crepidostomum*. II. The *Crepidostomum laureatum* of A. R. Cooper. J. Parasitol., **18**(2): 79–91.

———. 1933. Note on the life history of *Clinostomum marginatum* (Trematoda). Tr. Am. Micr. Soc., **52**(2): 147–149.

———. 1934*a*. The papillose Allocreadiidae. Illinois Biol. Monogr., **13**(2): 45–124.

———. 1934*b*. The parasite inducing pearl formation in American freshwater Unionidae. Science, n.s. (2052), **79**: 385–386.

———. 1934*c*. Studies on *Crepidostomum*. III. *Crepidostomum brevivitellum* n. sp. J. Parasitol., **20**(5): 295–298.

———. 1937. A new type of allocreadiid cercaria: the cercariae of *Anallocreadium* and *Microcreadium*. J. Parasitol., **23**(1): 94–97.

———. 1953. Pers. comm., Biol. Dept., A & M College of Texas, College Station, Texas.

———. 1954. The American species of trematode confused with *Bucephalus* (*Bucephalopsis*) *haimeanus*. Parasitology, **44**(3 & 4): 353–370.

Horsfall, Margery W. 1934. Studies on the life history and morphology of the cystocercous cercariae. Tr. Am. Micr. Soc., **53**(4): 311–347.

Hoshina, T., and Y. Sahara. 1950. A new species of the genus *Dermocystidium*, *D. koi* sp. nov., parasitic in *Cyprinus carpio* L. Bull. Japan. Soc. Sci. Fish., **15**(12): 825–829.

Hsü, Hsi-Fan. 1935. Contributions à l'étude des céstodes de Chine. Rev. Suisse Zool., **42**(4): 477–570.

Hugghins, E. J. 1954a. Life history of a strigeid trematode, *Hysteromorpha triloba* (Rudolphi, 1819) Lutz, 1931. I. Egg and miracidium. Tr. Am. Micr. Soc., **73**(1): 2–15.

———. 1954b. Life history of a strigeid trematode, *Hysteromorpha triloba* (Rudolphi, 1819) Lutz, 1931. II. Sporocyst through adult. Tr. Am. Micr. Soc., **73**(3): 221–236.

———. 1956. Ecological studies on a trematode of bullheads and cormorants at Spring Lake, Illinois. Tr. Am. Micr. Soc., **75**(3): 281–289.

———. 1958. Studies on parasites of fishes in South Dakota. J. Parasitol., **44**(4, Sect. 2): 33.

———. 1959. Parasites of fishes in South Dakota. South Dakota Exper. Sta. Bull. 484: 1–73.

Hughes, R. C. 1927. Studies on the trematode family Strigeidae (Holostomidae). No. 8. A new metacercaria, *Neascus ambloplitis*, sp. nov., representing a new larval group. Tr. Am. Micr. Soc., **46**(4): 248–267.

———. 1928a. Studies on the trematode family Strigeidae (Holostomidae). No. 9. *Neascus vancleavei* (Agersborg). Tr. Am. Micr. Soc., **47**(3): 320–341.

———. 1928b. Studies on the trematode family Strigeidae (Holostomidae). No. 10. *Neascus bulboglossa* (van Haitsma). J. Parasitol., **15**(1): 52–57.

———. 1928c. Studies on the trematode family Strigeidae (Holostomidae). No. 13. Three new species of *Tetracotyle*. Tr. Am. Micr. Soc., **47**(4): 414–433.

———. 1929a. Studies on the trematode family Strigeidae (Holostomidae). No. 14. Two new species of *Diplostomula*. Occas. Papers Mus. Zool. Univ. Michigan, no. 202: 1–29.

———. 1929b. Studies on the trematode family Strigeidae (Holostomidae). No. 19. *Diplostomulum scheuringi* sp. nov. and *D. vegrandis* (LaRue). J. Parasitol., **15**(4): 267–271.

Hughes, R. C., and P. G. Berkhout. 1929. Studies on the trematode family Strigeidae (Holostomidae). No. 15. *Diplostomulum gigas*, sp. nov. Papers Michigan Acad. Sci., Arts and Lett., **10**: 483–488.

Hughes, R. C., and Lucilla J. Hall. 1929. Studies on the trematode family Strigeidae (Holostomidae). No. 16. *Diplostomulum huronense*. Papers Michigan Acad. Sci., Arts and Lett., **10**: 489–494.

Hundley, J. B. 1957. Body and anchor development in *Lernaea cyprinacea* (Copepoda: Lernaeidae). M.S. thesis, Univ. North Dakota. 15 pp.

Hunninen, A. V. 1936. Studies of fish parasites in the Delaware and Susquehanna watersheds. *In* A biological survey of the Delaware and Susquehanna watersheds. 25 Ann. Rep. New York State Conserv. Dept., Suppl., 237–245.

Hunninen, A. V., and G. W. Hunter, III. 1933. On the species of *Crepidostomum* in trout. Tr. Am. Micr. Soc., **52**(2): 150–157.

Hunter, G. W. III. 1927. Notes on the Caryophyllidae of North America. J. Parasitol. **14**(1): 16–26.

———. 1929a. Life-history studies on *Proteocephalus pinguis* LaRue. Parasitology, **21**(4): 487–496.

———. 1929b. New Caryophyllidae from North America. J. Parasitol. **15**(3): 185–192.

———. 1930a. Studies on the Caryophyllaeidae of North America. Illinois Biol. Monogr., **11**(4): 370–556.

————. 1930b. Contribution to the life cycle of *Centrovarium lobotes* (Mac-Callum, 1895). J. Parasitol., **17**(2): 108.

————. 1932. A new trematode (*Plesiocreadium parvum*, sp. nov.) from fresh water fish. Tr. Am. Micr. Soc., **51**(1): 16–21.

————. 1933. The strigeid trematode, *Crassiphiala ambloplitis* (Hughes, 1927). Parasitology, **25**(4): 510–517.

————. 1942. Studies on the parasites of freshwater fishes of Connecticut. Connecticut Geol. and Nat. Hist. Survey, Bull. (63): 228–288.

Hunter, G. W. III and R. V. Bangham. 1932. Studies on fish parasites of Lake Erie. I. New trematodes (Allocreadiidae). Tr. Am. Micr. Soc., **51**(2): 137–152.

————. 1933. Studies on the fish parasites of Lake Erie. II. New cestoda and nematoda. J. Parasitol., **19**(4): 304–311.

Hunter, G. W. III and H. C. Dalton. 1939. Studies on *Clinostomum*. 5. The cyst of the yellow grub of fish (*Clinostomum marginatum*). Proc. Helminth. Soc. Washington, **6**(2): 73–76.

Hunter, G. W. III and Wanda S. Hunter. 1929. Further experimental studies on the bass tapeworm, *Proteocephalus ambloplitis* (Leidy). *In* A biological survey of the Erie-Niagara system. 18 Ann. Rep. New York State Conserv. Dept., Suppl., 198–207.

————. 1930. Studies on the parasites of fishes of the Lake Champlain watershed. *In* A biological survey of the Champlain watershed. 19 Ann. Rep. New York State Conserv. Dept., Suppl., 241–260.

————. 1931. Studies on fish parasites in the St. Lawrence watershed. *In* A biological survey of the St. Lawrence watershed. 20 Ann. Rep. New York State Conserv. Dept., Suppl., 197–216.

————. 1932. Studies on parasites of fish and of fish-eating birds. *In* A biological survey of the Oswegatchie and Black River systems. 21 Ann. Rep. New York State Conserv. Dept., Suppl., 252–271.

————. 1934. Studies on fish and bird parasites. *In* A biological survey of the Racquette watershed. 23 Ann. Rep. New York State Conserv. Dept., Suppl., 245–254.

Hunter, G. W., III, and J. S. Rankin, Jr. 1940. Parasites of northern pike and pickerel. Tr. Am. Fish. Soc. **69**: 269–272.

Hunter, Wanda. 1933. A new strigeid metacercaria, *Neascus rhinichthysi*, n. sp. Tr. Am. Micr. Soc., **52**(3): 255–258.

Hussey, Kathleen L. 1941. Comparative embryological development of the excretory system in digenetic trematodes. Tr. Am. Micr. Soc., **60**(2): 171–210.

————. 1943. Further studies on the comparative embryological development of the excretory system in digenetic trematodes. Tr. Am. Micr. Soc., **62**(3): 271–279.

Hutton, R. F. 1957. Preliminary notes on Trematoda (Heterophyidae and Strigeoidea) encysted in the heart and flesh of Florida mullet, *Mugil cephalus* L. and *M. curema* Cuvier and Valenciennes. Florida State Bd. Conserv. Marine Lab., Contrib. no. 4.

————. 1964. A second list of parasites from marine and coastal animals of Florida. Tr. Am. Micr. Soc., **83**(4): 439–447.

Hutton, R. F., and F. Sogandares-Bernal. 1958. Variation in the number of oral spines of *Phagicola longicollis* Kuntz and Chandler, 1956, and the description of *P. inglei* n. sp. (Trematoda: Heterophyidae). J. Parasitol., **44**(6): 627–632.

————. 1959a. Further notes on trematoda encysted in Florida mullets. Quart. J. Florida Acad. Sci., 21(4): 329–334.

————. 1959b. Studies on the trematode parasites encysted in Florida mullets. Florida State Bd. Conserv., Spec. Sci. Rep. no. 1: 88 pp. Mimeographed.

————. 1960. Preliminary notes on the life-history of *Mesostephanus appendiculatoides* (Price, 1934) Lutz, 1935. Bull. Marine Sci. Gulf and Caribbean, 10(2): 234–236.

Ikezaki, F. M., and G. L. Hoffman. 1957. *Gyrodactylus eucaliae* n. sp. (Trematoda: Monogenea) from the brook stickleback, *Eucalia inconstans*. J. Parasitol., 43(4): 451–455.

Ingles, L. G., 1936. Worm parasites of California amphibia. Tr. Am. Micr. Soc., 55(1): 73–92.

Irving, R. B. 1954. Ecology of the cutthroat trout in Henrys Lake, Idaho. Tr. Am. Fish. Soc., 84: 275–296.

Ishii, S. 1934. On a Filaria parasitic in the caudal fin of *Carassius auratus* L. from Japan. Proc. 5, Pacific Sci. Cong., 5: 4141–4143.

Iversen, E. S. 1954. A new myxosporidian, *Myxosoma squamalis*, parasite of some salmonoid fishes. J. Parasitol., 40(4): 397–404.

Jacobs, D. L. 1946. A new dinoflagellate from freshwater fish. Tr. Am. Micr. Soc. 65(1): 1–17.

Jägerskiöld, L. A. K. E. 1902. *Dichelyne fossor* n. g., n. sp., in *Lates niloticus* angetroffen. Zool. Anz., Leipzig, (677) 25: 564–565.

————. 1904. *Scaphanocephalus expansus* (Crepl.) Eine Genitalnapftragende Distomide. Results Swedish Zool. Exped. Egypt and White Nile 1901 (Jägerskiöld), pt. 1: 16 pp.

————. 1909. Zur Kenntnis der Nematoden-Gattungen *Eustrongylides* und *Hystrichis*. Upsala. 48 pp.

Jaiswal, G. P. 1957. Studies on the trematode parasites of fishes and birds found in Hyderabad state. Zool. Jahrb. Syst., 85(1/2): 1–72.

Jakowska, Sophie, and R. F. Nigrelli. 1956. *Babesiosoma* gen. nov. and other babesioids in erythrocytes of cold-blooded vertebrates. Ann. New York Acad. Sci., 64(2): 112–127.

Jameson, A. P. 1931. Notes on California Myxosporidia. J. Parasitol., 18(2): 59–68.

Jarecka, L. 1959. On the life-cycle of *Bothriocephalus claviceps* (Goeze, 1782). Acta Parasitologica Polonica, 7(7): 527–533.

————. 1960. Life-cycles of tapeworms from lakes Goldapiwo and Mamry Polnocne. Acta Parasitologica Polonica, 8(4): 48–66.

Jensen, T. 1953. The life cycle of the fish acanthocephalan, *Pomphorhynchus bulbocolli* (Linkins) Van Cleave, 1919, with some observations on larval development *in vitro*. Diss. Abstr., 12(4): 607.

Jirovec, O. 1939. *Dermocystidium vejdovskyi* n. sp., ein neuer Parasit des Hechtes, nebst einer Bemerkung über *Dermocystidium daphniae* (Ruhberg). Arch. Protistenk., 92(1): 137–146.

Jones, A. W., C. Kerley, and K. E. Sneed. 1956. New species and a new subgenus of *Corallobothrium* (Cestoda, Proteocephala) from catfishes of the Mississippi Basin. J. Tennessee Acad. Sci., 31(3): 179–185.

Jones, K. L., and D. M. Hammond. 1960. A study of the parasites from rainbow trout of a commercial fish farm in Cache Valley, Utah. Proc. Utah Acad. Sci., 37: 157–158.

Jones, R. O. 1950. Propagation of fresh-water mussels. Prog. Fish-Cult., **12**(1): 13.

Jordan, Helen E., and W. T. Ashby. 1957. Liver fluke (*Metorchis conjunctus*) in a dog from South Carolina. J. Am. Vet. Med. Assoc., **131**(5): 239–240.

Kahl, A. 1935. Urtiere oder Protozoa. I. Wimpertiere oder Ciliata (Infusoria). Eine Bearbeitung der freilebenden und ectocommensalen Infusorien der *Erde*, unter Ausschluss der marinen Tintinnidae. 4. Peritricha und Chonotricha. Tierwelt Deutschlands (Dahl), Teil 30: 651–886.

Kanouse, Bessie B. 1932. A physiological and morphological study of *Saprolegnia parasitica*. Mycologia, **24**(5): 431–452.

Karmanova, E. M. 1961. The first report of *Dioctophyma renale* in fish in the USSR. Akad. Nauk SSSR., **11**: 118–121.

———. 1963. The life-cycle of *Dioctophyma renale*. Meditsinskaya Parasitologiya i Parazitarnie Bolezni, Moscow, **32**(3): 331–334.

Kasahara, S. 1962. Studies on the biology of the parasitic copepod *Lernaea cyprinacea* Linnaeus and the methods for controlling this parasite in fish-culture ponds. Contrib. Fish. Lab. Fac. Agr., Univ. Tokyo. No. 3: 103–196 (Eng. synopsis).

Kathariner, L. 1895. Die Gattung *Gyrodactylus* v. Nrdm. arb. Zool.-Zootom. Inst. Wurzburg, **10**: 125–164.

Katz, M. 1951. Two new hemoflagellates (Genus *Cryptobia*) from some western Washington teleosts. J. Parasitol., **37**(3): 245–250.

Kearn, G. C. 1963. Feeding in some monogenean skin parasites: *Entobdella soleae* on *Solea solea* and *Acanthocotyle* sp. on *Raia clavata*. J. Marine Biol. Assoc. United Kingdom, **43**(3): 749–766.

Keleher, J. J. 1952. Growth and *Triaenophorus* parasitism in relation to taxonomy of Lake Winnipeg Ciscoes (*Leucichthys*). J. Fish. Res. Bd. Canada, **8**(7): 469–478.

Kellicott, D. S. 1878. Description of a new species of *Argulus*. Am. J. Micr., **3**(1): 1–3.

———. 1879. On certain Crustacea, parasitic on fish from the Great Lakes. Am. J. Micr., **4**(10/12): 208–210.

———. 1880a. *Argulus stizostethii*, n.s. Am. J. Micr., **5**(3): 53–58.

———. 1880b. Observations on *Lerneocera cruciata*. Proc. Am. Soc. Micr., August, 1879, 64–68.

———. 1882. On certain crustaceous parasites of freshwater fishes. Proc. Am. Soc. Micr., **5**: 75–78.

Kellogg, Stephen J., and A. C. Olson (Jr.). 1963. Some factors influencing the infectivity of the metacercariae of *Posthodiplostomum minimum* (Trematoda: Diplostomidae). J. Parasitol., **49**(5): 744.

Kessler, K. T. 1868. Beiträge zur zoologischen Kenntniss des Onegassee's und dessen Umgebung. Beilage z. d. Arb. d. 1. russischen naturf. Versamml. St. Petersburg. 183 pp.

Keysselitz, G. R. E. 1908. Ueber durch Sporozoen (Myxosporidien) hervorgerufene pathologische Veranderungen. Verhandl. Gesellsch. Deutsch. Naturf. u. Arzte., **2**(2): 542–543.

Khajuria, H., and W. R. Pillay. 1950. On a new species of *Zoothamnium* Stein (Protozoa: Vorticellidae) from the grey mullet, *Mugil tade* Forsk. Rec. Indian Mus., **48**(3 & 4): 55–58.

Klass, E. E. 1963. Ecology of the trematode, *Clinostomum marginatum*, and its hosts in eastern Kansas. Tr. Kansas Acad. Sci., **66**(3): 519–538.

Kniskern, V. B. 1950. *Rhipidocotyle septpapillata* Krull, 1934 (Trematoda): the cercaria and notes on the life history. J. Parasitol., **36**(2): 155–156.

———. 1952*a*. Studies on the trematode family Bucephalidae Poche, 1907. Part I. A systematic review of the family Bucephalidae. Tr. Am. Micr. Soc., **71**(3): 253–266.

———. 1952*b*. Studies on the trematode family Bucephalidae Poche, 1907. Part II. The life history of *Rhipidocotyle septpapillata* Krull, 1934. Tr. Am. Micr. Soc., **71**(4): 317–340.

Kostylew, N. N. 1924. Le genre Leptorhynchoides, nouveau genre d'acanthocephale parasite des poissons. Ann. Parasitol., **2**(3): 214–223.

Krabbe, H. 1874. *Diplocotyle olrikii*, en uledde Baendelorm af bothriocephalernes Gruppe. Vidensk. Medd. Naturh. Forening Kjobenhavn, **26**(3): 22–25.

Krause, R. K. L. 1914. Beitrag zur Kenntnis der Hemistominen. Zeitschr. Wissensch. Zool., **112**(1): 93–238.

Krøyer, R. 1863. Bidrag til Kundskab om Snyltekrebsene. Naturh. Tidsskr., **2**(1 & 2): 75–320.

Krueger, R. F. 1954. A survey of the helminth parasites of fishes from Van Buren Lake and Rocky Ford Creek. Ohio J. Sci., **54**(4): 277.

Krull, W. H. 1934*a*. *Cercaria bessiae* Cort and Brooks, 1928, an injurious parasite of fish. Copeia, no. 2: 69–73.

———. 1934*b*. Egg albumen as a mounting medium in the study of living helminths. Proc. Helminth. Soc., Washington **1**(1): 5–6.

Kudo, R. R. 1918. Contributions to the study of parasitic Protozoa. 4. Note on some Myxosporidia from certain fish in the vicinity of Woods Hole. J. Parasitol., **5**(1): 11–16.

———. 1920. Studies on Myxosporidia. A synopsis of genera and species of Myxosporidia. Illinois Biol. Monogr., **5**(3–4): 1–265.

———. 1921. On some protozoa parasitic in fresh-water fishes of New York. J. Parasitol. **7**(4): 166–174.

———. 1924. A biologic and taxonomic study of the Microsporidia. Illinois Biol. Monogr., **9**(2 & 3): 1–268.

———. 1926. On *Myxosoma catostomi* Kudo 1923, a myxosporidian parasite of the sucker, *Catostomus commersonii*. Arch. Protistenk., **56**(1): 90–115.

———. 1929. Histozoic Myxosporidia found in freshwater fishes of Illinois, U. S. A. Arch. Protistenk., **65**(3): 364–378.

———. 1930*a*. Myxosporidia. *In* Hegner, R. W., and J. M. Andrews. Problems and Methods of Research in Protozoology. New York. pp. 303–324.

———. 1930*b*. Microsporidia. *In* Hegner, R. W. and J. M. Andrews. Problems and Methods of Research in Protozoology. New York. pp. 325–347.

———. 1933. A taxonomic consideration of Myxosporidia. Tr. Am. Micr. Soc., **52**(3): 195–216.

———. 1934. Studies on some Protozoan parasites of fishes of Illinois. Illinois Biol. Monogr., **13**(1): 44 pp.

———. 1954. Protozoology. Thomas, Springfield. 966 pp.

Kudo, R. R., and E. W. Daniels. 1963. An electron microscope study of the spore of a microsporidian, *Thelohania californica*. J. Protozool., **10**(1): 112–120.

Kuhlow, F. 1953. Bau und Differentialdiagnose heimischer *Diphyllobothrium* Plerocercoidc. Zeitschr. Trop.-Med. u. Parasit., **4**(2): 186–202.

Kuitunen-Ekbaum, Ella. *See* Ekbaum, Ella K.

Kulakovskaya, O. P. 1964. Life cycle of Caryophyllaeidae (Cestoda) in the conditions of Western Ukraine. Czech. Parasit., **11**: 177–186.

Kulda, J., and J. Lom. 1964. Remarks on the diplomastigine flagellates from the intestine of fishes. J. Parasitol., **54**: 753–762.

Kulwiec, Z. 1927. Badania nad gatunkami rodzaju *Dactylogyrus* Diesing. (Untersuchungen an Arten des Genus *Dactylogyrus* Diesing). Bull. Internat. Acad. Polon. Sci. et Lettres, Cracovie, Cl. Sci. Math. et Nat., s B: Sci. Nat., 1 & 2: 113–144.

Kuntz, R. E. 1951. Embryonic development of the excretory system in a Psilostome cercaria, a Gymnocephalous (Fasciolid) cercaria and in three Monostome cercariae. Tr. Am. Micr. Soc., **70**(2): 95–118.

Kuntz, R. E., and A. C. Chandler. 1956. Studies on Egyptian trematodes with special reference to the heterophyids of mammals. I. Adult flukes, with descriptions of *Phagicola longicollis* n. sp., *Cynodiplostomum namrui* n. sp., and a *Stephanoprora* from cats. J. Parasitol., **42**(4): 445–459.

Labbe, A. 1899. Sporozoa. Berlin (Das Tierreich. Berl., 5. Lief., 180 pp.).

Lacey, R. J. 1965. The histological structure of *Crepidostomum isostomum* Hopkins, 1931. J. Parasitol., **51**(2–2): 24.

Lahav, M., and S. Sarig. 1964. Observations on the biology of *Lernaea cyprinacaea* L. in fish ponds in Israel. Bamidgeh, **16**(3): 77–86.

Lahav, M., S. Sarig, and M. Shilo. 1964. The eradication of *Lernaea* in storage ponds of carps through destruction of the copepodidal stage by Dipterex. Bamidgeh, **16**(3): 87–94.

Laird, J. A., and G. C. Embody. 1931. Controlling the trout gill worm (*Discocotyle salmonis* Schaffer). Tr. Am. Fish. Soc., **61**: 189–192.

Laird, M. 1961. Parasites from northern Canada. II. Haematozoa of fishes. Canad. J. Zool., **39**(4): 541–548.

Laird, M., and E. Meerovitch. 1961. Parasites from northern Canada. I. Entozoa of Fort Chimo Eskimos. Canad. J. Zool., **39**(1): 63–67.

Lamont, M. E. 1920. Two new parasitic flatworms. Occas. Papers Mus. Zool. Univ. Michigan, (93): 1–6.

Lane, C. 1916. The genus *Dacnitis* Dujardin, 1845. Indian J. Med. Research, **4**(1): 93–104.

Larsh, J. E. 1941. *Corallobothrium parvum* n. sp., a cestode from the common bullhead, *Ameiurus nebulosus* Le Sueur. J. Parasitol., **27**(3): 221–227.

Larson, O. 1961a. The distribution of the progenetic trematode, *Asymphylodora amnicolae* Stunkard, 1959. J. Parasitol., **47**(3): 371.

———. 1961b. Larval trematodes of freshwater snails of Lake Itasca, Minnesota. Proc. Minnesota Acad. Sci., **29**: 252–254.

———. 1964. Pers. comm, Biology Dept., Univ. North Dakota, Grand Forks, North Dakota.

———. 1965. *Diplostomulum* (Trematoda: Strigeoidea) associated with herniations of bullhead lenses. J. Parasitol., **51**(2): 224–229.

LaRue, G. R. 1911. A revision of the cestode family Proteocephalidae. Zool. Anz., Leipzig, **38**(22 & 23): 473–482.

———. 1914. A revision of the cestode family Proteocephalidae. Illinois Biol. Monogr., **1**(1 & 2): 1–350.

————. 1919. A new species of tapeworm of the genus *Proteocephalus* from the perch and rock bass. Occas. Papers Mus. Zool. Univ. Michigan, (67): 10 pp.

————. 1927. Studies on the trematode family Strigeidae (Holostomidae). 5. *Proalaria huronensis*, sp. nov. Tr. Am. Micr. Soc., **46**(1): 26–35.

————. 1932. Morphology of *Cotylurus communis* Hughes (Trematoda: Strigeidae). Tr. Am. Micr. Soc., **51**(1): 28–74.

————. 1957. The classification of digenetic Trematoda: A review and a new system. Exper. Parasitol., **6**: 306–349.

Lawler, G. H. 1961. Heming Lake experiment. Fish. Res. Bd. Canada, Prog. Rep. Biol. Sta., Tech. Unit, London, Ontario.

Lawler, G. H., and W. B. Scott. 1954. Notes on the geographical distribution and the hosts of the cestode genus *Triaenophorus* in North America. J. Fish. Res. Bd. Canada, **11**(6): 884–893.

Lawler, G. H., and N. H. F. Watson. Measurements of immature stages of *Triaenophorus*. J. Fish. Res. Bd. Canada, **20**(4): 1089–1093.

Layman, E. M. 1949. Manual of Fish Diseases. Food Industry Publ., Moscow. 306 pp.

Lee, H., and B. S. Seo. 1959. A new large-tailed echinostome cercaria from *Amnicola limosa* (Say) in the Douglas Lake region of Michigan. Tr. Am. Micr. Soc., **78**(2): 215–219.

Leech, H. B., and M. A. Anderson. 1959. Coleoptera. *In* Edmondson, W. T. Fresh-water Biology. Wiley & Sons, New York. 1248 pp.

Lefevre, G., and W. C. Curtis. 1912. Studies on the reproduction and artificial propagation of freshwater mussels. Bull. U. S. Bur. Fish., **30**: 105–201.

Léger, L. 1906. Myxosporidies nouvelles, parasites des poissons. I. Sur une nouvelle maladie myxosporidienne de la truite indigène. II. Sur une nouvelle myxosporidie de la tanche commune. Ann. Univ. Grenoble, **18**: 267–272.

————. 1914. Sur un nouveau protiste du genre *Dermocystidium* parasite de la truite. Compt. Rend. Acad. Sci., Paris, **158**(11): 807–809.

Leidy, J. 1851. Descriptions of new species of Entozoa. Proc. Acad. Nat. Sci., **5**(7): 155–156.

————. 1856. A synopsis of Entozoa and some of their ectocongeners observed by the author. Proc. Acad. Nat. Sci., **8**(1): 42–58.

————. 1857. Observations on Entozoa found in the Naiades. Proc. Acad. Nat. Sci., **9**(2): 18.

————. 1858. Contributions to helminthology. Proc. Acad. Nat. Sci., **10**(2): 110–112.

————. 1871. Remarks on *Toenia mediocanellata*. Proc. Acad. Nat. Sci., **23**(3): 53–55.

————. 1886. Notices of nematoid worms. Proc. Acad. Nat. Sci., **38**(3): 308–313.

————. 1887. Notice of some parasitic worms. Proc. Acad. Nat. Sci., **39**(3): 20–24.

————. 1888. Parasites of the pickerel. Proc. Acad. Nat. Sci., **40**(3): 169.

Leigh, W. H. 1956*a*. The life-history of *Macroderoides spiniferus* Pearse, 1924, a trematode of the Florida spotted gar, *Lepisosteus platyrhincus*. J. Parasitol., **42**(4-2): 38.

————. 1956*b*. Observations on life-histories of members of the genus *Ascocotyle* Looss (Heterophyidae). J. Parasitol., **42**(4-2): 39.

————. 1958. The life-history of *Macroderoides spiniferus* Pearse, 1924, a trematode of the Florida spotted gar, *Lepisosteus platyrhinchus*. J. Parasitol., **44**(4): 379–387.

Leigh, W. H., and R. B. Holliman. 1956. The life-history of *Paramacroderoides echinus* Venard, 1941, a trematode of the Florida gar, *Lepisosteus platyrhincus*. J. Parasitol., **42**(4): 400–407.

Leigh-Sharp, W. H. 1925. *Lernaea* (*Lernaeocera*) *elegans* n. sp., a parasite copepod of *Anguilla japonica*. Parasitology, **17**(3): 245–251.

Leitritz, E. 1960. Trout and salmon culture. Fish. Bull. 107, State of Calif., Dept. of Fish and Game, 169 pp.

LeSueur, C. A. 1824. On three new species of parasitic vermes, belonging to the Linnaean genus *Lernaea*. J. Acad. Nat. Sci. Phila., **3**(2): 286–293.

Leuckart, F. S. 1835. *Diclybothrium armatum*. Abstr. of rep. before Versamml. Deutsch. Naturf. u. Aerzte Bonn. Notiz. Geb. Nat.-u. Heilk., **46**(6): 88.

Leuckart, K. G. F. R. 1878. *Archigetes sieboldi,* eine geschlechtsreife Cestodenamme. mit Bemerkungen über die Entwicklungsgeschichte der Bandwurmer. Zeitschr. Wissensch. Zool., **30**(3): 593–606.

Lewis, W. M., and Sue D. Lewis. 1963. Control of epizootics of *Gyrodactylus elegans* in golden shiner populations. Tr. Am. Fish. Soc., **92**(1): 60–62.

Lewis, W. M., and J. Nickum. 1964. The effect of *Posthodiplostomum minimum* upon the body weight of the bluegill. Prog. Fish-Cult., **26** (3): 121–123.

Lillis, W. G., and R. Nigrelli. 1965. *Fundulus heteroclitus* (killifish), second intermediate host for *Echinochasmus schwartzi* (Price, 1931), (Trematoda: Echinostomidae). J. Parasitol., **51**(2-2): 23.

Lincicome, D. R., and H. J. Van Cleave. 1949. Distribution of *Leptorhynchoides thecatus*, a common acanthocephalan parasitic in fishes. Am. Midl. Nat., **21**(2): 421–431.

Linnaeus, C. 1746. Fauna Suecica Sistens Animalia Sueciae Regni: Quadrupedia, Aves, Amphibia, Pisces, Insecta, Vermes, Distributa per Classes et Ordines, Genera et Species. Stockholmiae. 411 pp.

————. 1758. Systema Naturae per Regna Tria Naturae, Secundum Classes, Ordines, Genera, Species, cum Characteribus, Differentiis, Synonymis, Locis. Editio Decima, Reformata. Holmiae. 823 pp.

————. 1761. Fauna Suecica Sistens Animalia Sueciae Regni: Mammalla, Aves, Amphibia, Pisces, Insecta, Vermes, Distributa per Classes et Ordines, Genera et Species. Editio Altera, Auctior. Stockholmiae. 578 pp.

von Linstow. 1878a. Compendium der Helminthologie. Ein Verzeichniss der bekannten Helminthen, die frei oder in thierischen Korpern leben, geordnet nach ihren Wohnthieren, unter Angabe der Organe, in denen sie gefunden sind, und mit Beifügung der Litteraturquellen. Hannover. 382 pp.

————. 1878b. Neue Beobachtungen an Helminthen. Arch. Naturg., Berlin, **44**(2): 218–245.

Linton, E. 1893. On fish Entozoa from Yellowstone National Park. Rep. U. S. Comr. Fish., **17**: 545–564.

————. 1897. Notes on larval cestode parasites of fishes. Washington pp. 787–824.

————. 1898. Notes on larval cestode parasites of fishes. Zool. Centralbl., **5**(2): 46–47.

————. 1901. Parasites of fishes of the Woods Hole region, Bull. U. S. Fish. Comm. (1899), **19**: 405–492.

————.1928. Notes on trematode parasites of birds. Proc. U. S. Nat. Mus., **73**(1): 1–36.

————. 1940. Trematodes from fishes mainly from the Woods Hole region, Massachusetts. Proc. U. S. Nat. Mus., **88**: 172.

————. 1941. Cestode parasites of teleost fishes of the Woods Hole Region, Massachusetts. Proc. U. S. Nat. Mus., **90**: 417–442.

Little, P. A. 1930. A new trematode parasite of *Acipenser sturio* L. (royal sturgeon), with a description of the genus *Dihemistephanus* Lss. Parasitology, **22**(4): 399–413.

Lloyd, L. C., and J. E. Guberlet. 1936. *Syncoelium filiferum* (Sars) from the Pacific salmon. Tr. Am. Micr. Soc., **55**(1): 44–48.

Locke, D. 1963. Pers. comm., Dept. Inland Fisheries, Augusta, Maine.

Loewen, S. L. 1929. A description of the trematode *Catoptroides lacustri* n. sp., with a review of the known species of the genus. Parasitology, **21**(1 & 2): 55–62.

————. 1935. A new trematode of the family Gorgoderidae. J. Parasitol., **21**(3): 194–196.

Lom, J. 1956. Nález *Trichodinella epizootica* (Raabe, 1950) V ČSR. Zoologické Listy, Folia Zoologica, Ročnik V (XIX), Číslo **2**: 120–124.

————. 1958. A contribution to the systematics and morphology of endoparasitic trichodinids from amphibians, with a proposal of uniform specific characteristics. J. Protozool., **5**: 251–263.

————. 1959. On the systematics of the genus *Trichodinella* Sramek-Husek (= *Brachyspira* Raabe). Acta Parasitologica Polonica, **7**(32): 573–590.

————. 1963. The ciliates of the family Urceolariidae inhabiting gills of fishes (the Trichodinella-group). Acta Soc. Zool. Bohemoslov, **27**(1): 7–19.

————. 1964. Pers. comm., Protozool. Dept., Inst. Parasitol., Czechoslovak Acad. Sci., Praha, Czechoslovakia.

————. 1966. Sessiline peritrichs from the surface of some freshwater fishes. Folia Parasitol. (Praha), **13**(1): 36–56.

Lom, J., and G. L. Hoffman. 1964. Geographic distribution of some species of trichodinids (Ciliata: Peritricha) parasitic on fishes. J. Parasitol., **50**(1): 30–35.

Lom, J., and G. A. Stein. 1966. Trichodinids from sticklebacks and a remark on the taxonomic position of *Trichodina domerguei* (Wall). Acta Soc. Zool. Bohemoslov. **30**(1): 39–48.

Lom, J., and J. Vavra. 1961. *Epistylis lwoffi* (?) from the skin of perches. Acta Zool. Bohemoslov, **25**(4): 273–276.

Lom, J., and J. Vavra. 1963. Mucous envelopes of spores of the subphylum Cnidospora (Doflein, 1901). Acta Soc. Zool. Bohemoslov, **27**(1): 4–6.

Long, S., and W. C. Lee. 1958. Parasitic worms from Tai Hu fishes: digenetic trematodes. II. Opisthorchiidae and other families, with a description of a new species of *Opisthorchis* (in Chinese, English summary). Acta Zool. Sinica, Peking, **10**(4): 369–376.

Looss, A. 1899. Weitere Beiträge zur Kenntnis der Trematoden-Fauna Aegyptens, zugleich Versuch einer natürlichen Gliederung des Genus *Distomum* Retzius. Zool. Jahrb., Jena, Abt. Syst., **12**(5 & 6): 521–784.

Lopukhina, A. M. 1961. The influence produced by *Triaenophorum nodulosus* Pallas on young of *Salmo gairdneri* hatched during the same year. Dokl. Akad. Nauk SSSR, **137**(1): 244–247.

Luehe, M. F. L. 1901. Ueber Hemiuriden. (Ein Beitrag zur Systematik der digenetischen Trematoden.) Zool. Anz., Leipzig, 24(8): 394–403.

———. 1911. Acanthocephalen. Register der Acanthocephalen und parasitischen Plattwurmer, geordnet nach ihren Wirten. Süsswasserfauna Deutschlands Brauer, 16: 1–116.

Lumsden, R. D. 1963. A new heterophyid trematode of the *Ascocotyle* complex of species encysted in Poeciliid and Cyprinodont fishes of southwest Texas. Proc. Helminth. Soc. Washington, 30(2): 293–296.

Lumsden, R. D., and Carol Ann Winkler. 1962. The opossum, *Didelphis virginiana* (Kerr), a host for the cyathocotylid trematode *Linstowiella szidati* (Anderson, 1944) in Louisiana. J. Parasitol., 48(3): 503.

Lumsden, R. D. and J. A. Zischke. 1963. Studies on the trematodes of Louisiana birds. Zeitschr. f. Parasitenk., 22: 316–366.

Lundahl, W. S. 1939. Life history of *Caecincola parvulus* Marshall and Gilbert (Trematoda: Heterophyidae) J. Parasitol., 25(6): 27–28.

———. 1941. Life history of *Caecincola parvulus* Marshall and Gilbert (Cryptogonimidae, Trematoda) and the development of its excretory system. Tr. Am. Micr. Soc., 60(4): 461–484.

Lynch, J. E. 1936. New species of Neoechinorhynchus from the western sucker *Catostomus macrocheilus* Girard. Tr. Am. Micr. Soc., 55(1): 21–43.

Lyster, L. L. 1939. Parasites of freshwater fish. I. Internal trematodes of commercial fish in the central St. Lawrence watershed. Canad. J. Research, 17(7): 154–168.

———. 1940a. *Apophallus imperator* sp. nov., a heterophyid encysted in trout, with a contribution to its life history. Canad. J. Research, 18(3): 106–121.

———. 1940b. Parasites of freshwater fish. 2. Parasitism of speckled and lake trout and the fish found associated with them in Lake Cammandant, Que. Canad. J. Research, 18(2): 66–78.

MacCallum, G. A. 1920. Notes on the genus *Microcotyle*. 3. Studies Dept. Path. Coll. Phys. and Surg. Columbia Coll., 17: 71–78.

———. 1921. Studies in Helminthology. Part 1. Trematodes. Part 2. Cestodes. Part 3. Nematodes. Zoopathologica, 1(6): 135–284.

MacCallum, G. A., and W. G. MacCallum. 1913. Four species of *Microcotyle*, *M. pyragraphorus, macroura, eueides* and *acanthophallus*. Zool. Jahrb., Jena, Abt. Syst., 34(3): 223–244.

MacCallum, W. G. 1895. On the anatomy of two distome parasites of freshwater fish. Vet. Mag., 2(7): 401–412.

McCauley, J. E., and I. Pratt. 1961. A new genus *Deropegus* with a redescription of *D. aspina* (Ingles, 1936) nov. comb. Tr. Am. Micr. Soc., 80(3): 373–377.

McCoy, O. R. 1928. Life history studies on trematodes from Missouri. J. Parasitol., 14(4): 207–228.

McCrae, R. C. 1962. *Biacetabulum macrocephalum* sp. n. (Cestoda: Caryophyllaeidae) from the white sucker *Catostomus commersoni* (Lacepede) in northern Colorado. J. Parasitol., 48(6): 807–811.

McDaniel, J. S. 1963. Parasites from the genus *Lepomis* (Centrarchidae) in Lake Texoma, Oklahoma. Tr. Am. Micr. Soc., 82(4): 423–425.

McGregor, E. A. 1963. Publications on fish parasites and diseases, 330 B.C.— A.D. 1923. BSFW, Spec. Sci. Rep., Fisheries no. 474: 1–84.

Machado, D. A. 1959. *Neoechinorhynchus spectabilis* sp. n. (Neoechinorhynchidae, Acanthocephala). Rev. Brasil. Biol., 19(2): 190–194.

McIntosh, A. 1939. A new allocreadiid trematode, *Podocotyle shawi*, n. sp., from the silver salmon. Proc. Wash. Acad. Sci., **29**(9): 379–381.

McIntosh, A., and J. T. Self. 1955. *Nematobothrium texomensis* n. sp. from a fresh-water fish, *Ictiobus bubalus* (Rafinesque, 1819). J. Parasitol., **41**(6-2): 36–37.

Mackiewicz, J. 1961. Studies on the Caryophyllaeidae (Cestoidea) of *Catostomus commersoni*. Diss. Abst., **21**: 3566–3567.

———. 1962. Distribution and vertebrate hosts (pisces) of *Glaridacris catostomi* Cooper, 1920 (Cestoidea: Caryophyllaeidae). J. Parasitol., **48**(2-2): 45.

———. 1963. *Monobothrium hunteri* sp. n. (Cestoidea: Caryophyllaeidae) from *Catostomus commersoni* (Lacepede) (Pisces: Catostomidae) in North America. J. Parasitol., **49**(5): 723–730.

———. 1964. Variations and host-parasite relationships of Caryophyllaeids (Cestoidea) from fish of Lake Texoma, Marshall County, Oklahoma. J. Parasitol., **50**(3-2): 31.

———. 1965. Pers. comm., State Univ. New York, Coll. Educa., Albany, New York.

Mackiewicz, J., and R. McCrae. 1962. *Hunterella nodulosa* gen. n., sp. n. (Cestoidea: Caryophyllaeidae), from *Catostomus commersoni* (Lacepede) Pisces: Catostomidae) in North America. J. Parasitol., **48** (6): 798–806.

Mackin, J. G. 1961. Oyster diseases caused by *Dermocystidium marinum* and other microorganisms in Louisiana. Publ. Inst. Marine Sci., Univ. Texas, **7**: 132–229.

McLain, A. L. 1951. Diseases and parasites of the sea lamprey, *Petromyzon marinus*, in the Lake Huron Basin. Tr. Am. Fish. Soc., **81**: 94–100.

MacLennan, R. F. 1935a. Dedifferentiation and redifferentiation in *Ichthyophthirius*. 1. Neuromotor system. Arch. Protistenk., **86**(2): 191–210.

———. 1935b. Observations on the life cycle of *Ichthyophthirius*, a ciliate parasitic on fish. Northwest Sci., **9**(3): 12–14.

———. 1939. The morphology and locomotor activities of *Cyclochaeta domerguei* Wallengren (Protozoa). J. Morphol., **65**(2): 241–255.

———. 1943. Centrifugal stratification of granules in the ciliate *Ichthyophthirius*. J. Morphol., **72**(1): 1–25.

MacLulich, D. A. 1943a. *Proteocephalus parallacticus*, a new species of tapeworm from lake trout, *Cristivomer namaycush*. Canad. J. Research, **21**(5): 145–149.

———. 1943b. Parasites of trout in Algonquin Provincial Park, Ontario. Canad. J. Research, **21**(D-12): 405–412.

McMullen, D. B. 1935. The life histories and classification of two allocreadiid-like plagiorchids from fish, *Macroderoides typicus* (Winfield) and *Alloglossidium corti* (Lamont). J. Parasitol., **21**(5): 369–380.

———. 1937. A discussion of the taxonomy of the family Plagiorchiidae Luhe, 1901, and related trematodes. J. Parasitol., **23**(3): 244–258.

———. 1938. Observation on precocious metacercarial development in the trematode superfamily Plagiorchioidea. J. Parasitol., **24**(3): 273–280.

McNeil, P. L., Jr. 1961. The use of benzene hexachloride as a copepodicide and some observations on lernaean parasites in trout rearing units. Prog. Fish-Cult., **23**(3): 127–133.

McPhee, C. 1961. An experimental study of competition for food in fish. Ecology, **42**(4): 668–681.

Magath, T. B. 1918. The morphology and life history of a new trematode parasite, *Lissorchis fairporti* nov. gen., et nov. spec. from the buffalo fish, *Ictiobus*. J. Parasitol., **4**(2): 58–69.

Malmberg, G. 1956*a*. On the presence of *Gyrodactylus* in Swedish fishes (with description of species and a summary in English). Sartryck ur Skrifter av Sodra Sveriges Fiskeriforening Arsskrift, 19–76 (English transl. of pp. 19–50, U. S. Dept. Interior).

———. 1956*b*. On a new genus of viviparous monogenetic trematodes. Utgivet av K. Svenska Vetensk.-Akad., **2**(10): 317–329.

Mann, K. H. 1962. Leeches (Hirudinea): Their Structure, Physiology, Ecology and Embryology. Pergamon Press, New York. 201 pp.

Manter, H. W. 1926. Some North American fish trematodes. Illinois Biol. Monogr., **10**(2): 127–264.

———. 1947. The digenetic trematodes of marine fishes of Tortugas, Florida. Am. Midl. Nat. **38**: 247–416.

———. 1949. The trematode *Cathaemasia pulchrosoma* (Travassos, 1916) n. comb. from the body cavity of a kingfisher (*Megaceryle alcyon*) in Nebraska. J. Parasitol., **35**(2): 221.

———. 1954. Trematoda of the Gulf of Mexico. *In* Gulf of Mexico, its origin, waters and marine life. USFWS, Fish. Bull. (89): 335–350.

———. 1962. Notes on the taxonomy of certain digenetic trematodes of South American freshwater fishes. Proc. Helminth. Soc. Washington, **29**(2): 97–102.

Manter, H. W., and D. F. Prince. 1953. Some monogenetic trematodes of marine fishes from Fiji. Proc. Helminth. Soc. Washington, **20**(2): 105–112.

Margolis, L. 1961. Pers. Comm., Pacific Biol. Sta., Nanaimo, British Columbia, Canada.

———. 1963. Parasites as indicators of the geographical origin of sockeye salmon, *Oncorhynchus nerka* (Walbaum), occurring in the North Pacific Ocean and adjacent seas. Contrib. Fish, Res. Bd. Canada. Int. North Pacific Fish. Comm., Dec. 466, Bull. 11: 101–156.

———. 1964. *Paurorhynchus hiodontis* Dickerman, 1954 (Trematoda: Bucephalidae): a second record involving a new host and locality in Canada. Canad. J. Zool., **42**: 716.

Margolis, L., and J. R. Adams. 1956. Description of *Genolinea oncorhynchi* n. sp. (Trematoda: Hemiuridae) from *Oncorhynchus gorbuscha* in British Columbia with notes on the genus. Canad. J. Zool., **34**: 573–577.

Markevich, A. P. 1940. Diseases of Fresh-Water Fishes (Ukrainian text). Kiev. 167 pp.

———. 1957. Parasitic Copepoda of Fish of SSSR (transl. in progress, OTS, Dept. Commerce). Kiev. 259 pp.

Marshall, W. S., and N. C. Gilbert. 1905. Notes on the food and parasites of some freshwater fishes from the lakes at Madison, Wis. Rep. U. S. Bur. Fish. (1904): 513–522.

Martin, W. E. 1950*a*. *Euhaplorchis californiensis* n.g., n. sp., *Heterophyidae*, Trematoda, with notes on its life-cycle. Tr. Am. Micr. Soc., **64**(2): 194–209.

———. 1950*b*. *Parastictodora hancocki* n. gen., n. sp. (Trematoda: Heterophyidae), with observations of its life cycle. J. Parasitol., **36**(4): 360–370.

———. 1951. *Pygidiopsoides spindalis* n. gen., n. sp. (Heterophyidae: Trematoda), and its second intermediate host. J. Parasitol., **37**(3): 297–300.

————. 1961. Life cycle of *Mesostephanus appendiculatus* (Ciurea, 1916) Lutz, 1935 (Trematoda: Cyathocotylidae). Pacific Sci., **15**(2): 278–281.

————. 1964. Life cycle of *Pygidiopsoides spindalis* Martin, 1951 (Heterophyidae: Trematoda). Tr. Am. Micr. Soc., **83**(2): 270–272.

Mathers, C. K. 1948. The leeches of the Okoboji region. Thesis, Univ. Iowa, Iowa City. 56 pp.

Mathias, P. 1936. Sur le cycle évolutif d'un trématode digénetique, *Allocreadium angusticolle* (Hausmann). Compt. Rend. Soc. Biol., Paris, **122**(25): 1175–1176.

Matthey, R. 1963. Rapport sur les maladies des poissons en Suisse. Bull. de l'Office Internat. des Epiz., **59**(1 & 2): 121–126.

Mavor, J. W. 1915a. On the occurrence of a trypanoplasm, probably *Trypanoplasma borreli* Laveran et Mesnil, in the blood of the common sucker, *Catostomus commersonii*. J. Parasitol., **2**(1): 1–6.

————. 1915b. Studies on the Sporozoa of the fishes of the St. Andrew's region. Contrib. Canad. Biol., **1**: 25–38.

————. 1916. On the occurrence of a parasite of the pike in Europe, *Myxidium lieberkuhni* Butschli, in the pike on the American continent and its significance. Biol. Bull., **31**(5): 373–378.

Mavor, J. W., and W. Strasser. 1916. On a new myxosporidian, *Henneguya wisconsinensis* n. sp., from the urinary bladder of the yellow perch, *Perca flavescens*. Tr. Wisconsin Acad. Sci., Arts and Lett., **18**(2): 676–692.

Meehean, O. L. 1940. A review of the parasitic crustacea of the genus *Argulus* in the collections of the United States National Museum. Proc. U. S. Nat. Mus., **88**(3087): 459–522.

Meglitsch, P. A. 1937. On some new and known Myxosporidia of the fishes of Illinois. J. Parasitol., **23**(5): 467–477.

————. 1942a. *Myxosoma microthecum* n. sp., a myxosporidian inhabiting the mesenteries of *Minytrema melanops*. Tr. Am. Micr. Soc., **61**(1): 33–35.

————. 1942b. On two new species of Myxosporidia from Illinois fishes. J. Parasitol., **28**(1): 83–89.

————. 1947. Studies on Myxosporidia of the Beaufort region. 1. Observations on *Chloromyxum renalis*, n. sp., and *Chloromyxum granulosum* Davis. J. Parasitol., **33**(3): 265–270.

————. 1949. On *Kudoa funduli* (Hahn). Tr. Am. Micr. Soc., **67**(3): 272–274.

————. 1963. On *Myxosoma hoffmanni*, sp. nov., inhabiting the eye of *Pimephales notatus* (Raf.). Tr. Am. Micr. Soc., **82**(4): 416–417.

Mehra, H. R. 1962. Revision of Allocreadioidea Nicoll, 1934, 1. Families: Lepocreadiidae Nicoll, 1934, Deropristiidae n. fam., Homalometridae n. fam. and Maseniidae Gupta, 1953. Proc. Nat. Acad. Sci. India, **32**(1): 1–22.

Meinkoth, N. A. 1947. Notes on the life-cycle and taxonomic position of *Haplobothrium globuliforme* Cooper, a tapeworm of *Amia calva* L. Tr. Am. Micr. Soc., **66**: 256–261.

Merritt, S. V., and I. Pratt. 1964. The life history of *Neoechinorhynchus rutili* and its development in the intermediate host (Acanthocephala: Neoechinorhynchidae). J. Parasitol., **50**(3): 394–400.

Meserve, F. G. 1938. Some monogenetic trematodes from the Galapagos Islands and the neighboring Pacific. Rep. Hancock Pacific Exped., **2**(5): 29–88.

Messjatzeff, I. I. 1928. Parasitische Copepoden aus dem Baikal-See. Arch. Naturg., Berlin, **92**(4): 120–134.

Meyer, A. 1932. Acanthocephala. *In* Bronn's Klass. u. Ordnung. Tierreichs, 4 Abt. 2, Buch 1, Lief 1: 1–332.

———. 1933. Acanthocephala (concluded). *In* Bronn's Klass. u. Ordnung. Tierreichs, 4 Abt. 2, Buch 2, Lief 2: 333–582.

Meyer, F. 1958. Helminths of fishes from Trumbull Lake, Clay County, Iowa. Proc. Iowa Acad. Sci., 65: 477–516.

———. 1960, 1964. Pers. comm., USFWS, Fish Farming Exper. Sta., Stuttgart, Arkansas.

———. 1966. *Ichthyophthirius multifilis.* Fish Disease Leaflet 2, USFWS, 4 pp.

———. 1966. Parasites of catfish. Fish Disease Leaflet 5, 7 pp. USFWS.

Meyer, M. C. 1940. A revision of the leeches (Piscicolidae) living on fresh-water fishes of North America. Tr. Am. Micr. Soc., 59(3): 354–376.

———. 1946a. Further notes on the leeches (Piscicolidae) living on fresh-water fishes of North America. Tr. Am. Micr. Soc., 65(3): 237–249.

———. 1946b. A new leech, *Piscicola salmositica* n. sp. (Piscicolidae), from steelhead trout (*Salmo gairdneri gairdneri* Richardson, 1838). J. Parasitol., 32(5): 467–476.

———. 1949. On the parasitism of the leech, *Piscicola salmositica* Meyer, 1946. J. Parasitol., 35(2): 215.

———. (1954). 1962. The larger animal parasites of the fresh-water fishes of Maine. State of Maine, Dept. Inland Fish and Game, Fish. Res. and Management Div. Bull. no. 1: 88 pp.

———. 1958a. Studies on *Philonema agubernaculum,* a dracunculoid nematode infecting salmonids. J. Parasitol., 44(4-2): 42.

———. 1958b. Pers. comm., Zool. Dept., Univ. Maine, Orono, Maine.

———. 1960. Notes on *Philonema agubernaculum* and other related dracunculoids infecting salmonids. Sobretiro del libro Homenaje al Doctor Eduardo Caballero y Caballero. Caballero Jubilee Vol., pp. 487–492.

———. 1965a. Fish leeches (Hirudinea) from tropical West Africa. Atlantide Rep. no. 8: 237–245.

———. 1965b. Pers. comm., Dept. Zool., Univ. Maine, Orono, Maine.

Meyer, M. C., and R. V. Bangham. 1950. Erratic hirudiniasis in a lake trout (*Cristovomer namaycush*). J. Parasitol, 36(6-2): 20.

Meyer, M. C., and J. P. Moore. 1954. Notes on Canadian leeches (Hirudinea), with the description of a new species. Wasmann J. Biol., 12(1): 63–96.

Meyer, M. C., and L. R. Penner. 1962. Laboratory Essentials of Parasitology. Brown, Dubuque. 134 pp.

Meyer, M. C., and E. S. Robinson. 1963. Description and occurrence of *Diphyllobothrium sebago* (Ward, 1910). J. Parasitol., 49(6): 969–973.

Miller, G. C. 1959. Studies on the genus *Homalometron* Stafford, 1904 (Trematoda: Lepocreadiidae), with a redescription of *H. armatum* (MacCallum, 1895). J. Parasitol., 45(5): 539–542.

Miller, M. J. 1936. *Bunoderina eucaliae* gen. et sp. nov., a new papillose Allocreadiidae from the stickleback. Canad. J. Research, 14(2): 11–14.

———. 1940. Parasites of freshwater fish. 3. Further studies on the internal trematodes of fish in the central St. Lawrence watershed. Canad. J. Research, 18(12): 423–434.

———. 1941. A critical study of Stafford's report on "Trematodes of Canadian fishes" based on his trematode collection. Canad. J. Research, 19(1): 28–52.

————. 1942. Black spot disease of speckled trout (French summary). Rev. Canad. Biol., **1**(4): 464–471.

Miller, R. B. 1945. Studies on cestodes of the genus *Triaenophorus* from fish of Lesser Slave Lake, Alberta. 4. The life of *Triaenophorus crassus* Forel in the second intermediate host. Canad. J. Research, D, **23**: 105–115.

————. 1952. A review of the Triaenophorus problem in Canadian Lakes. Fish. Res. Bd. Canada, Bull. 95.

————. 1954. Tapeworm infection in Lesser Slave Lake. Prog. Fish-Cult., **16**(4): 184.

Miller, R. B., and H. B. Watkins. 1946. An experiment in the control of the cestode, *Triaenophorus crassus* Forel. Canad. J. Research, D, **24**: 175–179.

Minakata, K. 1908. An alga growing on fish. Nature (London), **79**: 99.

Minckley, W. L., and J. E. Deacon. 1959. Biology of the flathead catfish in Kansas. Tr. Am. Fish. Soc., **88**(4): 344–355.

Mizelle, J. D. 1936. New species of trematodes from the gills of Illinois fishes. Am. Midl. Nat., **17**(5): 785–806.

————. 1937a. Ectoparasites of the blunt-nosed minnow (*Hyborhynchus notatus*). Am. Midl. Nat., **18**(4): 612–621.

————. 1937b. Notes on ectoparasitic trematodes of fishes. Tr. Illinois State Acad. Sci., **30**(2): 311–312.

————. 1938a. Comparative studies on trematodes (Gyrodactyloidea) from the gills of North American fresh-water fishes. Illinois Biol. Monogr., **17**(1): 81 pp.

————. 1938b. New species of monogenetic flukes from Illinois fishes. Am. Midl. Nat., **19**(2): 465–470.

————. 1940a. Studies on monogenetic trematodes. II. New species from Tennessee fishes. Tr. Am. Micr. Soc., **59**(3): 285–289.

————. 1940b. Studies on monogenetic trematodes. III. Redescriptions and variations in known species. J. Parasitol., **26**(3): 165–178.

————. 1941a. Studies on monogenetic trematodes. IV. *Anchoradiscus*, a new dactylogyrid genus from the bluegill and the stump-knocker sunfish. J. Parasitol., **27**(2): 159–163.

————. 1941b. Studies on monogenetic trematodes. V. Tetraonchinae of the stump-knocker sunfish, *Eupomotis microlophus* (Guenther). Am. Midl. Nat., **26**(1): 98–104.

————. 1955. Studies on monogenetic trematodes. XIX. The status of North American Dactylogyrinae and Tetraonchinae. Proc. Indiana Acad. Sci., **64**: 260–264.

————. 1962. Studies on monogenetic trematodes. XXII. *Dactylogyrus californiensis* sp. n. from the Sacramento squawfish. J. Parasitol., **48**(4): 555–557.

————. 1963. Studies on monogenetic trematodes. XXIV. A new dactylogyrid genus from *Acanthurus olivacens* Bloch. J. Parasitol., **49**(5): 752–753.

Mizelle, J. D., and J. A. Arcadi. 1945. Studies on monogenetic trematodes. XIII. *Urocleidus seculus*, a new species of Tetraonchinae from the viviparous top minnow, *Gambusia affinis affinis* (Baird and Girard). Tr. Am. Micr. Soc., **65**(4): 293–296

Mizelle, J. D., and W. J. Brennan. 1942. Studies on monogenetic trematodes. VII. Species infesting the bluegill sunfish. Am. Midl. Nat., **27**(1): 135–144.

Mizelle, J. D., and J. P. Cronin. 1943. Studies on monogenetic trematodes. X. Gill parasites from Reelfoot Lake fishes. Am. Midl. Nat., 30(1): 196–222.

Mizelle, J. D., and Sr. M. Angela Donahue. 1944. Studies on monogenetic trematodes. XI. Dactylogyridae from Algonquin Park fishes. Am. Midl. Nat., 31(3): 600–624.

Mizelle, J. D. and R. C. Hughes. 1938. The North American freshwater Tetraonchinae. Am. Midl. Nat., 20(2): 341–353.

Mizelle, J. D., and B. J. Jaskoski. 1942. Studies on monogenetic trematodes. VIII. Tetraonchinae infesting *Lepomis miniatus* Jordan. Am. Midl. Nat., 27(1): 145–153.

Mizelle, J. D., and A. R. Klucka. 1953. Studies on monogenetic trematodes. XIV. Dactylogyridae from Wisconsin fishes. Am. Midl. Nat., 49(3): 720–733.

Mizelle, J. D., D. R. LaGrave, and R. P. O'Shaughnessy. 1943. Studies on monogenetic trematodes. IX. Host specificity of *Pomoxis* Tetraonchinae. Am. Midl. Nat., 29(3): 730–731.

Mizelle, J. D., and C. E. Price. 1963. Additional haptoral hooks in the genus *Dactylogyrus*. J. Parasitol., 49(6): 1028–1029.

————. 1964a. Studies on monogenetic trematodes. XXV. Six new species of Ancyrocephalinae from the gills of *Zanclus canescens* (*Linnaeus*) Ancyrocephalinae. J. Parasitol., 50(1): 81–89.

————. 1964b. Studies on monogenetic trematodes. XXVII. Dactylogyrid species with the proposal of *Urocleidoides* gen. n. J. Parasitol., 50(4): 579–584.

Mizelle, J. D., and B. R. Regensberger. 1945. Studies on monogenetic trematodes. XII. Dactylogyridae from Wisconsin fishes. Am. Midl. Nat., 34(3): 673–700.

Mizelle, J. D., and A. Seamster. 1939. Studies on monogenetic trematodes. I. New species from the Warmouth bass. J. Parasitol., 25(6): 501–507.

Mizelle, J. D., P. S. Stokely, B. J. Jaskoski, A. P. Seamster, and L. H. Monaco. 1956. North American freshwater Tetraonchinae. Am. Midl. Nat., 55(1): 162–179.

Mizelle, J. D., R. J. Toth, and H. Wolf. 1961. On the life cycle of *Cleidodiscus pricei* Mueller, 1936. J. Parasitol., 47(4): 634.

Mizelle, J. D., and F. O. Webb. 1953. Studies on monogenetic trematodes. XV. Dactylogyridae from Alaska, Wisconsin, and Wyoming. Am. Midl. Nat., 50(1): 206–217.

Molin, R. 1859. Prospectus helminthum, quae in parte secunda prodromi faunae helminthologicae venetae continentur. Sitzungsb. K. Akad. Wissensch. Wien, Math.-Naturw., 33(26): 287–302.

Monaco, L. H., and J. D. Mizelle. 1955. Studies on monogenetic trematodes. XVII. The genus *Dactylogyrus*. Am. Midl. Nat., 53(2): 455–477.

Monod, T. 1932. Contribution à l'étude de quelques copépodes parasites de poissons. Ann. Parasitol., 10(4): 345–380.

Monticelli, F. S. 1892. *Cotylogaster michoelis* n. g., n. sp., e revisione degli Aspidobothridae. Festschr. 70. Geburtst. R. Leuckart's 7: 168–214.

Moore, Emmeline. 1923. Description of *Octomitus salmonis*. 12 Ann. Rep. New York State Conserv. Dept., 69–76.

————. 1925. Diseases of fish. 14 Ann. Rep. New York State Conserv. Dept., 83–97.

————. 1926. Fish diseases. *In* Problems in the freshwater fisheries. 15 Ann. Rep. New York State Conserv. Dept., 139–146.

————. 1938. Bureau of biological survey. Parasitology unit. 27 Ann. Rep. New York State Conserv. Dept., 251–252.

Moore, J. P. 1912. Classification of the leeches of Minnesota. *In* The leeches of Minnesota. Geol. and Nat. Hist. Survey Minnesota. Zool. Ser. no. 5.

————. 1924. The leeches (Hirudinea) of Lake Nipigon. Univ. Toronto Studies, Ontario Fish. Res. Lab. Publ. no. 23.

————. 1959. Hirudinea. *In* Edmondson, W. T. Fresh-water Biology, pp. 542–557. Wiley, New York. 1248 pp.

Moulton, J. M. 1931. A new species of *Haplonema* Ward and Magath 1916 from the stomach of *Lota Maculosa*. J. Parasitol., **18**(2): 105–107.

Mueller, J. F. 1930. The trematode genus *Plagiorchis* in fishes. Tr. Am. Micr. Soc., **49**(2): 174–177.

————. 1932. *Trichodina renicola* (Mueller, 1931) a ciliate parasite of the urinary tract of *Esox niger*. Roosevelt Wild Life Ann., **3**(2): 139–154. Bull. New York State Coll. Forest., **5**(2c).

————. 1934a. Parasites of Oneida Lake fishes. Part IV. Additional notes on parasites of Oneida Lake fishes, including descriptions of new species. Roosevelt Wild Life Ann., **3**(3 & 4): 335–373. Bull. New York State Coll. Forest., **7**(1).

————. 1934b. Two new trematodes from Oneida Lake fishes. Tr. Am. Micr. Soc., **53**(3): 231–236.

————. 1936a. New gyrodactyloid trematodes from North American fishes. Tr. Am. Micr. Soc., **55**(4): 457–464.

————. 1936b. Notes on some parasitic copepods and a mite, chiefly from Florida freshwater fishes. Am. Midl. Nat., **17**(5): 807–815.

————. 1936c. Studies on North American Gyrodactyloidea. Tr. Am. Micr. Soc., **55**(1): 55–72.

————. 1937a. The Gyrodactylidae of North American freshwater fishes. Fish Culture, New York, **3**(1): 1–14.

————. 1937b. Parasitic copepods of the Syracuse region (Russian summary). Rabot. Gel'mint., 412–417.

————. 1937c. Some species of *Trichodina* (Ciliata) from freshwater fishes. Tr. Am. Micr. Soc., **56**(2): 177–184.

————. 1937d. Further studies on North American Gyrodactyloidea. Am. Midl. Nat. **18**(2): 207–219.

————. 1938a. Additional species of North American Gyrodactyloidea (Trematoda). Am. Midl. Nat., **19**(1): 220–235.

————. 1938b. A new species of *Trichodina* (Ciliata) from the urinary tract of the muskalonge, with a repartition of the genus. J. Parasitol., **24**(3): 251–258.

————. 1940. Parasitism and disease in fishes of the Lake Ontario watershed. 29 Ann. Rep. New York State Conserv. Dept., 211–225.

Mueller, J. F., and H. J. Van Cleave. 1931. What is *Diplostomulum scheuringi*? J. Parasitol., **18**(2): 126–127.

————. 1932. Parasites of Oneida Lake fishes. II. Descriptions of new species and some general taxonomic considerations, especially concerning the trematode family Heterophyidae. Bull. New York State Coll. Forest., **5**(2c). Roosevelt Wild Life Ann., **3**(2): 79–137.

Mugard, H. 1949. Contribution à l'étude des infusoires hyménostomes histiophages. Ann. Sci. Nat., Zool. (ser. 11), **10**: 171–268.

Mullin, Catherine A. 1926. Study of the leeches of the Okoboji Lake region. Ph.D. thesis, Univ. Iowa.

Murphy, G. 1942. Relationship of the freshwater mussel to trout in the Truckee River. Calif. Fish and Game, **28**: 89–102.

Myer, D. 1960*a*. Pers. comm., Southern Illinois Univ., Alton, Illinois.

————. 1960*b*. On the life history of *Mesostephanus kentuckiensis* (Cable, 1935) n. comb. (Trematoda: Cyathocotylidae). J. Parasitol., **46**(6): 819–832.

Nagaty, H. F. 1937. Trematodes of Fishes from the Red Sea. Part 1. Studies on the Family Bucephalidae Poche, 1907. Fac. Med., Egypt. Univ., Publ. (12): 172 pp.

Nakai, N. 1927. On the development of a parasitic copepod, *Lernaea elegans* Leigh-Sharpe, infesting on *Cyprinus carpio* L. J. Imp. Fish. Inst., Tokyo, **23**(3): 35–39.

Nakai, N., and E. Kokai. 1931. On the biological study of a parasitic copepod, *Lernaea elegans* Leigh-Sharpe, infesting on Japanese fresh water fishes (Japanese text, English abstract). Suisan Shikenjo Hoboku, **2**: 93–128.

Neiland, K. A. 1952. A new species of *Proteocephalus* Weinland, 1858 (Cestoda), with notes on its life history. J. Parasitol., **38**(6): 540–545.

Neresheimer, E. R. 1909. Die parasitischen Copepoden. Süsswasserfauna Deutschlands (Brauer), **11**: 70–84.

Newsom, I. E., and E. N. Stout. 1933. Proventriculitis in chickens due to flukes. Vet. Med., **28**(11): 462–463.

Nickerson, W. S. 1902. *Cotylogaster occidentalis* n. sp. and a revision of the family Aspidobothridae. Zool. Jahrb., Jena, Abt. Syst., **15**(6): 597–624.

Nicoll, W. 1907. A contribution towards a knowledge of the Entozoa of British marine fishes. Part 1. Ann. and Mag. Nat. Hist., **7**(109), **19**: 66–94.

————. 1909. Studies on the structure and classification of the digenetic trematodes. Quart. J. Micr. Sci., n.s. (211), **53**(3): 391–487.

————. 1934. Vermes. Zool. Rec. (1933): 70, Div. VI, 137 pp.

Nigrelli, R. F. 1935. Some tropical fishes as hosts for the metacercaria of *Clinostomum marginatum*. J. Parasitol., **21**(6): 438–439.

————. 1936*a*. The morphology, cytology and life-history of *Oodinium ocellatum Brown*, a dinoflagellate parasite on marine fishes. Zoologica: Scient. Contrib. New York Zool. Soc., **21**(3): 129–164.

————. 1936*b*. Some tropical fishes as host for the metacercaria of *Clinostomum complanatum* (Rud. 1814). (= *C. marginatum* Rud. 1819). Zoologica: Scient. Contrib. New York Zool. Soc., **21**(4): 251–256.

————. 1940. Studies on the trematodes of the subfamily Neoschasminae Van Cleave and Mueller, 1932. Anat. Rec., **78**(4): Suppl., 178.

————. 1943. Causes of diseases and death of fishes in captivity. Zoologica: Scient. Contrib. New York Zool. Soc., **28**(4): 203–216.

————. 1948. Prickle cell hyperplasia in the snout of the redhorse sucker (*Moxostoma aureolum*) associated with an infection by the myxosporidian *Myxobolus moxostomi* sp. nov. Zoologica, **33**: 133–136.

Nigrelli, R. F., J. J. A. McLaughlin, and Sophie Jakowska. 1958. Histozoic algal growth in fish. Copeia, no. 4: 331–333.

Nigrelli, R. F., and G. M. Smith. A papillary cystic disease affecting the barbels of *Ameiurus nebulosus* (Le Sueur), caused by the myxosporidian *Henneguya ameiurensis* sp. nov. Zoologica: Scient. Contrib. New York Zool. Soc., **25**(1): 89–96.

Nitzsch, C. L. 1821. *Ascaris.* Allg. Encycl. Wissensch. u. Kunste (Ersch u. Gruber), **6**: 44–49.

Noble, E. R. 1943. Nuclear cycles in the protozoan parasite *Myxidium gasterostei* n. sp. J. Morphol., **73**(2): 281–289.

———. 1950. On a myxosporidian (protozoan) parasite of California trout. J. Parasitol., **36**(5): 457–460.

Nordlie, Frank. 1960. Some ecological aspects of the helminth parasitism of the American smelt, *Osmerus mordax* (Mitchell). Proc. Minnesota Acad. Sci., **25–26**: 239–241.

von Nordmann, A. 1832. Mikographische Beiträge zur Naturgeschichte der wirbellosen Thiere, **1**: 118 pp.

Nybelin, O. 1922. Anatomisch-systematische Studien über Pseudophyllideen. Goteborgs K. Vetensk.-O. Vitterhets. Samh. Handl., **26**(1): 1–228.

———. 1924. *Dactylogyrus vastator* n. sp. Ark. Zool., Stockholm, **16**(3): 1–2.

Odening, K. 1962. Furcocercarien (Trematoda: Strigeata und Schistosomatata, larvae) aus Brandenburg und Sachsen. Sond. Monatsb. Deutsch. Akad. Wissensch. Berlin, **4**(6): 384–392.

Odhner, T. 1902. Mitteilungen zur Kenntnis der Distomen. 1. Centralbl. Bakt., Abt. 1, Orig., **31**(2): 58–69.

———. 1905. Die Trematoden des arktischen Gebietes. *In* Römer and Schaudinn. Fauna Arctica, **4**(2): 291–372.

Odlaug, T. O. 1954. Parasites of some birds of Minnesota. An annotated bibliography. The Flicker, **26**(2): 59–65.

———. 1956. Helminth parasites reported from vertebrates in Minnesota. The Flicker, **28**(4): 138–148.

Odlaug, T. O., E. G. Arseneau, and G. H. Brownell. 1962. Intestinal helminths of fish from Basswood Lake, Superior National Forest, Minnesota. J. Parasitol., **48**(1): 31.

Okada, Y. K. 1927. Copépode parasite des amphibiens; nouveau parasitisme de *Lernaea cyprinacea* L. Annot. Zool. Jap., **11**(2): 185–187.

Olivier, L. J. 1942. Four new species of strigeid cercariae from northern Michigan and the metacercaria of one of them. Tr. Am. Micr. Soc., **61**(2): 168–179.

Olivier, L. J., and W. W. Cort. 1941. An experimental test of the life cycle described for *Cotylurus communis* (Hughes). J. Parasitol., **28**(1): 75–81.

Olsen, O. W. 1937. A systematic study of the trematode subfamily Plagiorchiinae Pratt, 1902. Tr. Am. Micr. Soc., **56**(3): 311–339.

Olson, R. E. 1966. Some experimental fish hosts of the strigeid trematode *Bolbophorus confusus* and effects of temperature on the cercaria and metacercaria. J. Parasitol. **52**(2): 327–334.

Olsson, O. W. 1869. Nova genera parasitantia copepodorum et platyhelminthium. Lunds Univ. Arsskr., Afd. Math. o. Naturv., **6**: 6 pp.

———.1876. Bidrag till skandinaviens helminthfauna. 1. K. Svenska Vetensk.-Akad. Handl., Stockholm (1875), **14**(1): 35 pp.

Osborn, H. L. 1903*a*. On *Cryptogonimus* (n.g.) *chyli* (n. sp.), a fluke with two ventral suckers. Zool. Anz., Leipzig, **26**(2): 315–318.

———. 1903*b*. On *Phyllodistomum americanum* (n. sp.): A new bladder distome from *Amblystoma punctatum*. Biol. Bull., **4**(5): 252–258.

———. 1903*c*. *Bunodera cornuta* sp. nov.: A new parasite from the crayfish and certain fishes of Lake Chautauqua, N. Y. Biol. Bull., **5**(2): 63–73.

————. 1910. On the structure of *Cryptogonimus* (nov. gen.) *Chyli* (n. sp.), an aberrant distome from fishes of Michigan and New York. J. Exper. Zool., 9(3): 517–536.

————. 1911. On the distribution and mode of occurrence in the United States and Canada of *Clinostomum marginatum*, a trematode parasitic in fish, frogs and birds. Biol. Bull., 20(6): 350–366.

————. 1912. On the structure of *Clinostomum marginatum*, a trematode parasite of the frog, bass and heron. J. Morphol., 23(2): 189–228.

————. 1919. Observations on *Microphallus ovatus* sp. nov. from the crayfish and black bass of Lake Chautauqua, N. Y. J. Parasitol., 5(3): 123–127.

Osborn, Paul. 1962. Pers. comm., Ozark Fisheries, Stoutland, Missouri.

Otto, G. R., and T. L. Jahn. 1943. Internal myxosporidian infections of some fishes of the Okoboji region. Proc. Iowa Acad. Sci., 50: 323–335.

Ozaki, Y. 1925. Preliminary notes on a trematode with anus. J. Parasitol., 12(1): 51–53.

Packard, A. S. 1875a. Life-histories of the crustacea and insects. Am. Nat., 9: 583–622.

————. 1875b. *Achtheres carpenteri* sp. nov., East River, Colorado, on trout. Ann. Rep. U. S. Territory, (612).

Palmer, E. D. 1939. Diplostomiasis, a hatchery disease of fresh-water fishes new to North America. Prog. Fish-Cult., Memo. 1–131 (45): 41–47.

Paperna, I. 1964. The metazoan parasite fauna of Israel inland water fishes. Bamidgeh, 14(1 & 2): 1–66.

Parisi, B. 1912. Primo contributo alla distribuzione geografica die missosporidi in Italia. Atti Soc. Ital. Sci. Nat. Milano, 59(4): 283–299.

Park, J. T. 1937. A revision of the genus *Podocotyle* (Allocreadiinae), with a description of eight new species from tide pool fishes from Dillon's Beach, California. J. Parasitol., 23(4): 405–422.

————. 1938. A new fish trematode with single testis from Korea. Keijo J. Med., 9(4): 290–298.

Parker, M. V. 1941. The trematode parasites from a collection of amphibians and reptiles. J. Tennessee Acad. Sci., 16(1); 27–45.

Pearse, A. S. 1924a. Observations on parasitic worms from Wisconsin fishes. Tr. Wisconsin Acad. Sci., Arts and Lett., 21: 147–160.

————. 1924b. The parasites of lake fishes. Tr. Wisconsin Acad. Sci., Arts and Lett., 21: 161–194.

Peckham, R. S., and C. F. Dineen. 1957. Ecology of the central mudminnow *Umbra limi* (Kirtland). Am. Midl. Nat., 58(1): 222–231.

Pennak, R. W. 1953. Fresh-water Invertebrates of the United States. Ronald Press, N.Y. 769 pp.

Perez, C. 1907. *Dermocystis pusula*, organisme nouveau parasite de la peau des tritons. Compt. Rend. Soc. Biol., Paris, 63(32): 445–446.

Perkins, K. W. 1956. Studies on the morphology and biology of *Acetodextra amiuri* (Stafford), (Trematoda: Heterophyidae). Am. Midl. Nat., 55(1): 139–161.

Peters, L. E. 1957. An analysis of the trematode genus *Allocreadium looss* with the description of *Allocreadium neotenicum* sp. nov. from water beetles. J. Parasitol., 43(2): 136–142.

————. 1959. Pers. comm., Northern Michigan Coll., Marquette, Mich.

————. 1961a. The allocreadioid problem with reference to the excretory system in four types of cercariae. Proc. Helminth. Soc. Washington, 28(2): 102–108.

————. 1961b. The genus *Skrjabinopsolus* (Trematoda: Digenea), with reference to the allocreadioid problem. Am. Midl. Nat., 65(2): 436–445.

————. 1963. *Crepidostomum illinoiense* (Trematoda): its excretory system and miracidium. J. Parasitol., 49(1): 147.

Petrochenko, V. I. 1956. Acanthocephala of Domestic and Wild Animals (in Russian). Moscow. 435 pp.

Petrushevskaya, M. G. 1962. On the systematics of flukes of the genus *Azygia* found in fishes in SSSR (Russian text; English summary). Vestnik Leningrad. Univ., 17(3): 72–92.

Pitt, C. E., and A. W. Grundman. 1957. A study into the effects of parasitism on the growth of the yellow perch produced by the larvae of *Ligula intestinalis* (Linnaeus, 1758) Gmelin, 1790. Proc. Helminth. Soc. Washington, 24(2): 73–80.

Plehn, Marianne. 1924. Praktikum der Fischkrankheiten. E. Schwerzerbart'sche (Erwin Nagele). Stuttgart. 479 pp.

Plehn, Marianne, and K. Mulsow. 1911. Der Erreger der "Taumelkrankheit" der Salmoniden. Centralbl. Bakt. Parasitol., 59: 63–68.

Post, G. 1965. A review of advances in the study of diseases of fish: 1954–1964. Prog. Fish-Cult., 27(1): 3–12.

Prakash, A., and J. R. Adams. 1960. A histopathological study of the intestinal lesions induced by *Echinorhynchus lageniformis* (Acanthocephala-Echinorhynchidae) in the starry flounder. Canad. J. Zool., 38(5): 895–897.

Pratt, H. S. 1919. A new cystocercous Cercaria. J. Parasitol., 5(3) 128–131.

————. 1923. Preliminary report on the parasitic worms of Oneida Lake, New York. Roosevelt Wild Life Bull., Syracuse, N. Y., 2(1): 55–71.

Pratt, I., and Dianna V. Matthias. 1962. *Stephanoprora polycestus* (Dietz, 1909), (Trematoda: Echinostomatidae) recovered from a western grebe, *Aechmophorus occidentalis* (Aves). J. Parasitol., 49(2): 275.

Pratt, I., S. E. Knapp, and R. E. Milleman. 1964. Life cycle of the "salmon poisoning" fluke. J. Parasitol., 50(3-2): 46–47.

Price, E. W. 1929. Two new species of trematodes of the genus *Parametorchis* from fur-bearing animals. Proc. U. S. Nat. Mus. (2809), 76(12): 1–5.

————. 1931a. A new species of trematode of the family Heterophyidae, with a note on the genus *Apophallus* and related genera. Proc. U. S. Nat. Mus. (2882), 79(17): 1–6.

————. 1931b. Occurrence of *Apophallus donicus* (syn. *Rossicotrema donicum*) in wild rats. J. Parasitol., 18(1): 55.

————. 1935. Descriptions of some heterophyid trematodes of the subfamily Centrocestinae. Proc. Helminth. Soc. Washington, 2(2): 70–73.

————. 1936a. A new heterophyid trematode of the genus *Ascocotyle* (Centrocestinae). Proc. Helminth. Soc. Washington, 3(1): 31–32.

————. 1936b. North American monogenetic trematodes. George Washington Univ. Bull., Summaries Doct. Theses (1934–1936): 10–13.

————. 1937a. A new monogenetic trematode from Alaskan salmonoid fishes. Proc. Helminth. Soc. Washington, 4(1): 27–29.

————. 1937b. North American monogenetic trematodes. I. The superfamily Gyrodactyloidea. J. Wash. Acad. Sci., 27(3): 114–130.

————. 1938. A new species of *Dactylogyrus* (Monogenea: Dactylogyridae), with the proposal of a new genus. Proc. Helminth. Soc. Washington, 5(2): 48–49.

————. 1942. North American monogenetic trematodes. V. The family Hexabothriidae, n.n. (Polystomatoidea). Proc. Helminth. Soc. Washington, 9(2): 39–56.

————. 1943. North American monogenetic trematodes. VII. The family Discocotylidae (Diclidophoroidea). Proc. Helminth. Soc. Washington, 10(1): 10–15.

————. 1958. Some new monogenetic trematodes from the gizzard shad, *Dorosoma cepedianum* (Le Sueur). J. Alabama Acad. Sci., 30(2): 9–10.

————. 1959. A proposed reclassification of the gastrocotyloid Monogenea. J. Parasitol., 45(4): 22–23.

————. 1962. North American monogenetic trematodes. X. The family Axinidae. Proc. Helminth. Soc. Washington, 29(1): 1–18.

Price, C. E., and W. S. Berry. 1966. *Trianchoratus*, a new genus of Monogenea. Proc. Helminth. Soc. Washington 33(2): 201–203.

Price, C. E., and J. D. Mizelle. 1964. Studies on monogenetic trematodes. XXVI. Dactylogyrinae from California with the proposal of a new genus, *Pellucidhaptor*. J. Parasitol., 50(4): 572–578.

Pritchard, A. L. 1931. Taxonomic and life history studies of the ciscoes of Lake Ontario. Univ. Toronto Studies, Biol. Ser. (35): 78 pp.

Putz, R. E., and J. T. Bowen. 1964. Parasites of freshwater fishes. IV. Miscellaneous. The anchor worm (*Lernaea cyprinacea*) and related species. BFSW. Fishery Leaflet 575.

Putz, R. E., and G. L. Hoffman. 1963. Two new *Gyrodactylus* (Trematoda: Monogenea) from cyprinid fishes with synopsis of those found on North American fishes. J. Parasitol., 49(4): 559–566.

————. 1964. Studies on *Dactylogyrus corporalis* n. sp. (Trematoda: Monogenea) from the fallfish *Semotilus corporalis*. Proc. Helminth. Soc. Washington, 31(2): 139–143.

————. 1965. Unpublished research, Eastern Fish Disease Lab., BSFW, Kearneysville, West Virginia.

Putz, R. E., G. L. Hoffman, and C. E. Dunbar. 1965. Two new species of *Plistophora* (Microsporidea) from North American fish with a synopsis of Microsporidea of freshwater and euryhaline fishes. J. Protozool., 12(2): 228–236.

Qadri, S. S. 1962. An experimental study of the life cycle of *Trypanosoma danilewskyi* in the leech, *Hemiclepsis marginata*. J. Protozool., 9(3): 254–258.

Raabe, Z. 1950. Uwagi o Urceolariidae (Ciliata-Peritricha) skrzel ryb. Ann. Univ. Mariae Curie-Sklodowska, 5: 292.

————. 1952. *Ambiphrya miri* g. n., sp. n.: Eine Uebergangsform zwischen Peritricha-Mobilia und Peritricha-Sessilia. Ann. Univ. Mariae Curie-Sklodowska, Lublin, Polonia 10 (Sect. C): 339–358.

————. 1959. *Trichodina pediculus* (O. F. Muller, 1786, Ehrenberg, 1838) et *Trichodina domerguei* (Wallengren, 1897). Acta Parasitologica Polonica, 6: 189–202.

Railliet, A., and A. C. L. Henry. 1915. Sur les nématodes du genre *Camallanus* Raill. et Henry, 1915 (*Cucullanus* Auct., non Mueller, 1777). Bull. Soc. Path. Exot., 9(7): 446–452.

Raj, P. J. S. 1962. Concerning *Ozobranchus branchiatus* (Menzies, 1791), (Piscicolidae: Hirudinea) from Florida and Sarawak. Tr. Am. Micr. Soc., 81(4): 364–371.

Ramsey, J. S. 1965. *Barbulostomum cupuloris* gen. et sp. n. (Trematoda: Lepocreadiidae) from sunfishes (*Lepomis* spp.) in Lake Pontchartrain, Louisiana. J. Parasitol., 51(5): 777–780.

Ransom, B. H. 1920. Synopsis of the trematode family Heterophyidae with descriptions of a new genus and five new species. Proc. U. S. Nat. Mus. Washington, 527–573.

Rasheed, S. 1963. A revision of the genus Philometra Costa, 1845. J. Helminth., 37(1/2): 89–130.

Rausch, R. L. 1947. Some observations on the host relationships of *Microphallus opacus* Ward, 1894 (Trematoda: Microphallidae). Tr. Am. Micr. Soc., 66(1): 59–63.

––––––. 1954. Studies on the helminth fauna of Alaska. XXI. Taxonomy, morphological variation, and ecology of *Diphyllobothrium ursi* n. sp. provis. on Kodiak Island. J. Parasitol., 40(5): 540–563.

––––––. 1956. Studies on the helminth fauna of Alaska. XXVIII. The description and occurrence of *Diphyllobothrium dalliae* n. sp. (Cestoda). Tr. Am. Micr. Soc., 75(2): 180–187.

Rawson, D. 1960. Fish. Rep. no. 5, Dept. Nat. Resources, Prov. Saskatchewan.

Reed, R. J. 1955. Occurrence of the white liver grub, *Posthodiplostomum minimum*, in fishes of Slippery Rock Creek, Pennsylvania. Copeia, no. 3: 240–241.

Rehder, D. D. 1959. Some aspects of the life of the carp, *Cyprinus carpio*, in the Des Moines River, Boone County, Iowa. Iowa State Coll. J. Sci. 34: 11–26.

Reichenbach-Klinke, H. 1950. Der Entwicklungskreis der Dermocystidien sowie Beschreibung einer neuen Haplosporidienart *Dermocystidium percae* n. sp. Verhandl. Deutsch. Zool.: Ang., Suppl. 14: 126–132.

––––––. 1954. Untersuchungen über die bei Fischen durch Parasiten hervorgerufenen System und deren Wirkung auf den Wirtskorper. Zeitschr. Fischerei, Neudamm u. Berlin, 3(6/8): 565–636.

––––––. 1955. Untersuchungen über die bei Fischen durch Parasiten hervorgerufenen System und deren Wirkung auf den Wirtskorper. Zeitschr. Fischerei, Hilfswissenschaft., 4(1/2): 27–29.

––––––. 1956. Die Dinoflagellatenart *Oodinium pillularis* Schäeperclaus als Bindegewebsparasit von Süsswasserfischen (Dinoflagellata, Gymnodinidae), (English summary). Gior. Microbiol. 1(4): 263–265.

––––––. 1957. Krankheiten der Aquarienfische, Stuttgart. 215 pp.

––––––. 1960. Die Discus-Krankheit und ihre Ursachen. Die Aquarien- Terrarien- Zeitschr. (Datz), 13(10): 303–305.

Reimers, P. E. and C. E. Bond. 1966. Occurrence of the Bidens (sp.) achene in the snout of chinook salmon and redside shiners. Prog. Fish-Cult. 28(1): 62.

Remley, L. W. 1942. Morphology and life history studies of *Microcotyle spinicirrus* MacCallum, 1918, a monogenetic trematode parasitic on the gills of *Aplodinotus grunniens*. Tr. Am. Micr. Soc., 61(2): 141–155.

Rice, V. J., and T. L. Jahn. 1943. Myxosporidian parasites from the gills of some fishes of the Okoboji region. Proc. Iowa Acad. Sci., 50: 313–321.

Richardson, L. R. 1937a. *Raphidascaris laurentianus* sp. n. (Ascaroidea) from *Salvelinus fontinalis* (Mitchill) in Quebec. Canad. J. Rescarch, 15(5): 112–115.

———. 1937b. Observations on the parasites of the speckled trout in Lake Edward, Quebec. Tr. Am. Fish Soc., 66: 343–356.

———. 1938. An account of a parasitic copepod *Salmincola salvelini* sp. nov., infecting the speckled trout. Canad. J. Research, 16(8): 225–229.

Rigdon, R. H., and J. W. Hendricks. 1955. Myxosporidia in fish in waters emptying into Gulf of Mexico. J. Parasitol. 41(5): 511–518.

Roberts, L. S. 1957. Parasites of the carp, *Cyprinus carpio* L., in Lake Texoma, Oklahoma. J. Parasitol., 43(1): 54.

———. 1963. *Ergasilus nerkae* n. sp. (Copepoda: Cyclopoida) from British Columbia with a discussion of the copepods of the *E. caeruleus* group. Canad. J. Zool., 41: 115–124.

Rodrigo, Gracia. *See* Gracia, Rodrigo Angel.

Rogers, W. A., and T. L. Wellborn. 1965. Studies on *Gyrodactylus* (Trematoda: Monogenea) with descriptions of five new species from the southeastern U.S. J. Parasitol., 51(6): 977–982.

Ross, A. J., and T. J. Parisot. 1958. Record of the fungus *Ichthyosporidium* Caullery and Mesnil, 1905, in Idaho. J. Parasitol., 44(4): 453–454.

Rucker, R. R. 1957. Some problems of private trout hatchery operators. Tr. Am. Fish. Soc., 87: 375–379.

Rucker, R. R., and P. V. Gustafson. 1953. An epizootic among rainbow trout. Prog. Fish-Cult., 15(4): 179–181.

Rudolphi, C. A. 1802. Fortsetzung der Beobachtungen über die Eingeweidewurmer. Arch. Zool. u. Zoot., 2(2): 1–67.

———. 1808. Entozoorum Sive Vermium Intestinalium Historia Naturalis. Amstelaedami. 1: 527 pp.

———. 1809. Entozoorum Sive Vermium Intestinalium Historia Naturalis. Amstelaedami. 2(1): 457 pp.

———. 1810. Entozoorum Sive Vermium Intestinalium Historia Naturalis. Amstelaedami. 2(2): 386 pp.

———. 1819. Entozoorum Synopsis cui Accedunt Mantissa Duplex et Indices Locupletissimi. Berolini. 811 pp.

Rupp, R. S., and M. C. Meyer. 1954. Mortality among brook trout, *Salvelinus fontinalis*, resulting from attacks of freshwater leeches. Copeia, no. 4: 294–295.

Sars, G. O. 1885. Report on the Schizopoda collected by H.M.S. Challenger during the Years 1873–76. Rep. Scient. Results Voyage H.M.S. Challenger 1873–76. Zool., 37(37): 228 pp.

Savage, J. 1935. Copepod infection of speckled trout. Tr. Am. Fish. Soc., 65: 334–339.

Schäperclaus, W. 1954. Fischkrankheiten. Akademie-Verlag, Berlin. 708 pp.

Schneider, A. F. 1866. Monographie der Nematoden. Berlin. 357 pp.

Schneider, G. S. 1902. *Caryophyllaeus fennicus* n. sp. Arch. Naturg., Berlin, 68. j., 1(1): 65–71.

Schrank, F. von P. 1790. Fortekning, pa nagra hittils obeskrifne intestinal-krak. K. Vetensk. Acad. N. Handl., Stockholm, 11: 118–126.

Schuberg, A. 1905. Süsswasserpolypen als Forellenfeinde. Allg. Fisch.-Beitung, 30(11): 201–203.

Schuberg, A., and O. Schroeder. 1905. Myxosporidien aus dem Nervensystem und der Haut der Bachforelle (*Myxobolus neurobius* n. sp. u. *Henneguya nusslini* n. sp.). Arch. Protistenk., **6**(1): 47–60.

Schumacher, R. F. 1952. *Argulus* outbreaks in Minnesota lakes. Prog. Fish-Cult., **14**(2): 70.

Schwartz, Frank J. 1956. First record of infestation and death in the ictalurid catfish, *Schilbeodes miwus*, by the parasite *Clinostomum marginatum*. Copeia, no. 4: 250.

Scott, W. W. 1964. Fungi associated with fish diseases. Developments in Indust. Microbiol., **5**: 109–123.

Scott, W. W., and A. H. O'Bier. 1962. Aquatic fungi associated with diseased fish and fish eggs. Prog. Fish-Cult., **24**(1): 3–15.

Scott, W. W., J. R. Powell, and R. L. Seymour. 1963. Pure culture techniques applied to the growth of *Saprolegnia* spp. on a chemically defined medium. Virginia J. Sci., **14**(2): 42–46.

Scott, W. W., and C. O. Warren, Jr. 1964. Studies of the host range and chemical control of fungi associated with diseased tropical fish. Virginia Agric. Exper. Sta., Tech. Bull. 171: 1–23.

Seamster, A. 1938a. Gill trematodes from Oklahoma fishes. Proc. Oklahoma Acad. Sci., **18**: 13–15.

———. 1938b. Studies on gill trematodes from Oklahoma fishes. Am. Midl. Nat., **20**(3): 603–612.

———. 1948a. Gill parasites from Louisiana fishes with a description of *Urocleidus wadei* n. sp. Am. Midl. Nat., **39**(1): 165–168.

———. 1948b. Two new Dactylogyridae (Trematoda: Monogenea) from the golden shiner. J. Parasitol., **34**(2): 111–113.

———. 1960. A new species of *Dactylogyrus* from the silver chub. Sobretiro del libro Homenaje al Doctor Eduardo Caballero y Caballero. Caballero Jubilee Vol., pp. 269–270.

Seitner, P. G. 1951. The life history of *Allocreadium ictaluri* Pearse, 1924 (Trematoda: Digenea). J. Parasitol., **37**(3): 223–244.

Self, J. T. 1954. Parasites of the goldeye, *Hiodon alosoides* (Raf.), in Lake Texoma. J. Parasitol., **40**(4): 386–389.

Self, J. T., and J. W. Campbell. 1956. A study of the helminth parasites of the buffalo fishes of Lake Texoma with a description of *Lissorchis gullaris*, n. sp. (Trematoda: Lissorchiidae). Tr. Am. Micr. Soc., **75**(4): 397–401.

Self, J. T., L. E. Peters, and C. E. Davis. The egg, miracidium, and adult of Nematobothrium texomensis. J. Parasitol., **49**(5): 731–736.

Self, J. T., and H. F. Timmons. 1955. The parasites of the river carpsucker (*Carpiodes carpio* Raf.) in Lake Texoma. Tr. Am. Micr. Soc., **74**(4): 350–352.

Senk, O. 1956. *Cyathocephalus truncatus*, Pallas: Uticaj na resplodne elemente potocnih pastrmki (*Salmo trutta fario*). Veterinaria, Sarajevo, **5**(4): 607–615.

Shaw, J. N. 1947. Some parasites of Oregon wild life. Oregon Agric. Exper. Sta., Sta. Tech. Bull. **11**: 3–15.

Shaw, J. N., B. T. Simms, and O. H. Muth. 1934. Some diseases of Oregon fish and game and identification of parts of game animals. Oregon Agric. Exper. Sta., Sta. Bull. (322): 23 pp.

Shields, R. J., and W. M. Tidd. 1963. The tadpole as a host for larval *Lernaea Cyprinacea*. J. Parasitol., **49**(5–2): 43.

Shilo, M., S. Sarig, and R. Rosenberger. 1960. Ton scale treatment of *Lernaea* infected carps. Bamidgeh, **12**(2): 37–42.

Shireman, J. V. 1964. *Carassotrema mugilicola*, a new haploporid trematode from the striped mullet, *Mugil cephalus*, in Louisiana. J. Parasitol., **50**(4): 555–556.

Shpolyanskaya, A. Iu. 1953. A study of the leucocyte count of the blood of fish as influenced by the tapeworm *Ligula*. Dokl. Akad. Nauk SSSR, **90**(2): 319–320.

Shulman, S. S. 1948. Novj vid kruglch cervej, parazitirujuscij v peceni ryb. Isvestija vsesojuznovo naucno-isledovatelskovo instituta ozenorvo i recnovo rybnovo chozjajstva, **27**: 235–238.

Shulman, S. S. 1964. Evolution and phylogeny of Myxosporidia. Zool. Inst., Acad. Sc. USSR, Leningrad, 9 p.

Sillman, E. I. 1953. Notes on the life history of *Opisthorchis tonkae* Wallace and Penner, 1939 (Trematoda: Opisthorchiidae). J. Parasitol., **39**(4): 21.

———. 1962. The life history of *Azygia longa* (Leidy 1851), (Trematoda: Digenea), and notes on *A. acuminata* Goldberger 1911. Tr. Am. Micr. Soc., **81**(1): 43–65.

Simer, P. H. 1929. Fish trematodes from the lower Tallahatchie River. Am. Midl. Nat., **11**(12): 563–588.

———. 1930. A preliminary study of the cestodes of the spoonbill, *Polyodon spathula* (Wal.). Tr. Illinois State Acad. Sci., **22**: 139–145.

Simon, J. R. 1934. A study of four parasites of Yellowstone Lake trout. J. Colorado-Wyoming Acad. Sci., **1**(6): 75.

———. 1935. A new species of nematode, *Bulbodacnitis scotti*, from the trout, *Salmo lewisi* (Giard). Univ. Wyoming Publ., **2**(2): 11–15.

Simon, J. R., and F. Simon. 1936. *Philonema agubernaculum* sp. nov. (Dracunculidae), a nematode from the body cavity of fishes. Parasitology, **28**(3): 440–442.

Sindermann, C. J. 1953. Parasites of fishes of north central Massachusetts. Fisheries Rep. for Lakes of North Central Massachusetts, 5–28.

Sindermann, C. J., and Alva E. Farrin. Ecological studies of *Cryptocotyle lingua* (Trematoda: Heterophyidae) whose larvae cause "pigment spots" of marine fish. Ecology, **43**(1): 69–75.

Sindermann, C. J., and L. W. Scattergood. 1954. Diseases of fishes of the western North Atlantic. II. Ichthyosporidium disease of the sea herring (*Clupea harengus*). Dept. Sea and Shore Fish., Res. Bull. **19**: 1–39.

Sinitsin, D. F. 1931. Studien über die Phylogenie der Trematoden. IV. The life histories of *Plagioporus siliculus* and *Plagioporus virens*, with special reference to the origin of Digenea. Zeitschr. Wissensch. Zool., **138**(3): 409–456.

Skinker, Mary S. 1930. *Cystidicola canadensis*, a new species of nematode from fishes. J. Parasitol., **16**(3): 167.

———. 1931. A redescription of *Cystidicola stigmatura* (Leidy), a nematode parasitic in the swim bladder of salmonoid fishes, and a description of a new nematode genus. Tr. Am. Micr. Soc., **50**(4): 372–379.

Smedley, Enid M. 1933. Nematode parasites from Canadian marine and freshwater fishes. Contrib. Canad. Biol. and Fish., **8**(14): 169–179.

------. 1934. Some parasitic nematodes of Canadian fishes. J. Helminth., **12**(4): 205–220.

Smith, R. F. 1949. Notes on *Ergasilus* parasites from the New Brunswick, New Jersey, area with a check list of all species and hosts east of the Mississippi River. Zoologica, **34**(3): 127–132.

Smith, Septima C. 1932. Two new cystocercous cercariae from Alabama. J. Parasitol., **19**(2): 173–174.

------. 1935. Life-cycle studies and distribution of cystocercous cercariae. J. Alabama Acad. Sci., **6**: 18.

------. 1936. Life-cycle studies of *Cercaria hodgesiana* and *Cercaria melanophora*. J. Alabama Acad. Sci., **8**: 30–32.

Smith, S. I. 1874. The crustacea of fresh waters of the United States. B. The crustacean parasites of fresh-water fishes of U. S. Rep. Comm. Fish. for 1872, **2**: 661–665.

Sneed, K. W. 1950. The genus *Corallobothrium* from catfishes in Lake Texoma, Oklahoma, with a description of two new species. J. Parasitol., **6**(2): 43.

------. 1961. A description of anomalous and atypically developed tapeworms (Proteocephalidae: *Corallobothrium*) from catfishes (*Ictalurus*). J. Parasitol., **47**(5): 809–812.

Snieszko, S. F., and G. L. Hoffman. 1963. Control of fish diseases. Lab. Animal Care, **13**(3): 197–206.

Sogandares-Bernal, F. 1955. Some helminth parasites of fresh and brackish water fishes from Louisiana and Panama. J. Parasitol., **41**(6): 587–594.

------. 1965. Parasites from Louisiana crayfishes. Tulane Studies in Zool., **12**(3): 79–85.

Sogandares-Bernal, F., and J. F. Bridgman. 1960. Three *Ascocotyle* complex trematodes (Heterophyidae) encysted in fishes from Louisiana, including the description of a new genus. Tulane Studies in Zool., **8**(2): 31–39.

Sogandares-Bernal, F., and R. F. Hutton. 1958. Pers. comm., Tulane Univ., New Orleans, Louisiana.

------. 1960. Notes on the probable partial life-history of *Galactosomum spinetum* (Braun, 1901) (Trematoda) from the west coast of Florida. Proc. Helminth. Soc. Washington, **27**(1): 75–77.

Sogandares-Bernal, F., and R. D. Lumsden. 1963. The generic status of the heterophyid trematodes of the *Ascocotyle* complex, including notes on the systematics and biology of *Ascocotyle angrense* Travassos, 1916. J. Parasitol., **49**(2): 264–274.

------. 1964. The heterophyid trematode *Ascocotyle* (*A.*) *leighi* Burton, 1956, from the hearts of certain poeciliid and cyprinodont fishes. Zeitschr. Parasitenk., **24**: 3–12.

Southwell, T., and A. W. N. Pillers. 1929. A note on a nymphal linguatulid, *Leiperia cincinalis*, from the musculature of the fish *Tilapia nilotica*. Ann. Trop. Med. and Parasitol., **23**(1): 130.

Spaeth, F. W. 1951. The influence of acanthocephalan parasites and radium emanations on the sexual characters of *Hyalella* (Crustacea: Amphipoda). J. Morphol., **88**(2): 361–383.

Sparks, A. K. 1951. Some helminth parasites of the largemouth bass in Texas. Tr. Am. Micr. Soc., **70**(4): 351–358.

Spassky, A. A., and V. A. Roytman. 1958. [*Salmonchus skrjabini* nov. gen., nov.

sp. (Monogenoidea): new parasite of salmonid fishes] (in Russian). Rabot. Gel'mint., **80**, Let. Skrj.: 354–459.

Sprague, V. 1965. Pers. comm., Chesapeake Biol. Lab., Solomons, Maryland.

Sproston, Nora G. 1944. *Ichthyosporidium hoferi* (Plehn and Mulsow, 1911), an internal fungoid parasite of the mackerel. J. Marine Biol. Assoc. United Kingdom, **26**: 72–98.

————. 1946. A synopsis of the monogenetic trematodes. Tr. Zool. Soc. London, **25**(4): 185–600.

Šrámek-Hušek, R. 1953. Dva novi nalevnici z kuze kapra Rozpr. Cesk. Akad. Ved., **63**(5): 37–43.

Srivastava, C. B. 1963. On three new species of the genus *Bucephalus* Baer, 1827 (Trematoda: Bucephalidae Poche, 1907), with remarks on the systematic position of *B. indicus* Srivastava, 1938. Indian J. Helminth., **15**(1): 36–44.

Sroufe, S. A., Jr. 1958. *Mazocraeoides olentangiensis*, n. sp., a monogenetic trematode parasitic on the gills of the gizzard shad, *Dorosoma cepedianum* (Le Sueur). J. Parasitol., **44**(6): 643–646.

Stafford, J. 1904. Trematodes from Canadian fishes. Zool. Anz., Leipzig, **27** (16 & 17): 481–495.

————. 1905. Trematodes from Canadian vertebrates. Zool. Anz., Leipzig, **28**(21 & 22): 681–694.

Steelman, G. M. 1938. A description of *Phyllodistomum caudatum* n. sp. Am. Midl. Nat., **19**(3): 613–616.

Steen, E. B. 1938. Two new species of *Phyllodistomum* (Trematoda: Gorgoderidae) from Indiana fishes. Am. Midl. Nat., **20**(1): 201–210.

Stolyarov, V. P. 1936. Zur Kenntnis des Entwicklungszyklus von *Lernaea cyprinacea*. Trudy Leningrad. Obsh. Estestvois., Otdel, Zool., **65**(2): 239–253.

Strandine, J. 1943. Variations in *Microphallus*, a genus of trematodes, from fishes of Lake Lelanau, Michigan. Tr. Am. Micr. Soc., **62**(3): 293–300.

Strout, R. C. 1961. Pers. comm., University of New Hampshire, Durham, New Hampshire.

Stunkard, H. W. 1956. The morphology and life-history of the digenetic trematode, *Azygia sebago* Ward, 1910. Biol. Bull., **111**(2): 248–268.

————. 1959. The morphology and life-history of the digenetic trematode, *Asymphylodora amnicola* n. sp.; the possible significance of progenesis for the phylogeny of the Digenea. Biol. Bull., **117**(3): 562–581.

Stunkard, H. W., and R. M. Cable. 1931. Notes on a species of *Lernaea* parasitic in the Larvae of *Rana clamitans*. J. Parasitol., **18**(2): 92–97.

Stunkard, H. W., and C. B. Haviland. 1924. Trematodes from the rat. Am. Mus. Novitates (126): 1–10.

Stunkard, H. W., and J. R. Uzmann. 1962. The life-cycle of the digenetic trematode, *Stephanoprora denticulata* (Rudolphi, 1802) Odhner, 1910. J. Parasitol., **48**(2-2): 23.

Summerfelt, R. C. 1964. A new microsporidian parasite from the golden shiner, *Notemigonus crysoleucas*. Tr. Am. Fish. Soc. **93**(1): 6–10.

Summers, W. A. 1937. A new species of Tetraonchinae from *Lepomis symmetricus*. J. Parasitol., **23**(4): 432–434.

Summers, W. A., and H. J. Bennett. 1938. A preliminary survey of the trematodes from the gills of Louisiana fishes. Proc. Louisiana Acad. Sci., **4**(1): 247–248.

Surber, E. W. 1928. *Megalogonia ictaluri*, a new species of trematode from the channel catfish, *Ictalurus punctatus*. J. Parasitol., **14**(4): 269–271.

———. 1940. *Scyphidia micropteri*, a new protozoan parasite of largemouth and smallmouth black bass. Prog. Fish-Cult., Memo. 1–131 (50): 42.

———. 1942. *Scyphidia tholiformis*, a peritrichous protozoan found on the gills and external surfaces of *Micropterus dolomieu* and *Micropterus salmoides*. Tr. Am. Fish. Soc., **72**: 197–203.

Swezy, O. 1919. The occurrence of *Trypanoplasma* as an ectoparasite. Tr. Am. Micr. Soc., **38**(1): 20–24.

Szidat, L. 1932. Ueber cysticerke Riesencercarien, insbesondere *Cercaria mirabilis* M. Braun und *Cercaria splendens* n. sp., und ihre Entwicklung im Magen von Raubfischen zu Trematoden der Gattung *Azygia* Looss. Zeitschr. Parasitenk., Berlin, **4**(3): 477–505.

Tanner, V. M. 1954. Small clam attacks young trout. Great Basin Nat., **14**(1 & 2): 23–25.

Thélohan, P. 1895. Recherches sur les myxosporidies. Bull. Scient. France et Belgique (1894), **26**(4): s., 5, 29 Aout, 100–394.

Thiele, J. 1900. Diagnosen neuer Arguliden-Arten. Zool. Anz., Leipzig (606), 23: 46–48.

Thomas, J. D. 1956. Life-history of *Phyllodistomum simile* Nybelin. Nature, 178: 1004.

Thomas, L. J. 1929a. Notes on the life history of *Haplobothrium globuliforme* Cooper, a tapeworm of *Amia calva*. Anat. Rec., **44**(3): 262.

———. 1929b. *Philometra nodulosa* nov. spec. with notes on the life history. J. Parasitol., **15**(3): 193–198.

———. 1937a. Environmental relations and life history of the tapeworm, *Bothriocephalus rarus* Thomas. J. Parasitol., **23**(2): 133–152.

———. 1937b. Life cycle of *Raphidascaris canadensis* Smedley, 1933, a nematode from the pike, *Esox lucius*. J. Parasitol., **23**(6): 572.

———. 1946. New pseudophyllidean cestodes from the Great Lakes region. I. *Diphyllobothrium oblongatum* n. sp. from gulls. J. Parasitol., **32**(1): 1–6.

———. 1947a. The life cycle of *Diphyllobothrium oblongatum* Thomas, a tapeworm of gulls. J. Parasitol., **33**(2): 107–117.

———. 1947b. Notes on the life cycle of *Schistocephalus* sp., a tapeworm from gulls. J. Parasitol., **33**(6): 10.

Thompson, Sally, D. Kirkegaard, and T. L. Jahn. 1947. *Schyphidia ameiuri* n. sp., a peritrichous ciliate from the gills of the bullhead, *Ameiurus melas melas*. Tr. Am. Micr. Soc., **66**(4): 315–317.

Thomsen, R. 1944. Copépodos parásitos de los peces marinos del Uruguay. Comun. Zool. Mus. Hist. Nat. Montevideo, 3(54): 1–41.

Tidd, W. M. 1931. A list of parasitic copepods and their fish hosts from Lake Erie. Ohio J. Sci., **31**(16): 453–454.

———. 1933. A new species of *Lernaea* (parasitic Copepoda) from the goldfish. Ohio J. Sci., **33**(6): 465–469.

———. 1934. Recent infestations of goldfish and carp by the "anchor parasite" *Lernaea carassii*. Tr. Am. Fish. Soc., **64**: 176–180.

———. 1962. Experimental infestations of frog tadpoles by *Lernaea cyprinacea*. J. Parasitol., **49**(6): 870.

———. 1965. Transfer of larvae of *Lernaea cyprinacea* from goldfish to the leopard frog, *Rana pipiens*. J. Parasitol., **51**(2-2): 61.

Tidd, W. M., and R. V. Bangham. 1945. A new species of parasitic copepod, *Ergasilus osburni*, from the burbot. Tr. Am. Micr. Soc., **64**(3): 225–227.

Tonn, R. J. 1955. The white sucker, *Catostomus commersonii* (Lacepede), a new host of *Caryophyllaeus terebans* (Linton, 1893). J. Parasitol., **41**(2): 219.

Tornquist, N. 1931. Die Nematodenfamilien Cucullanidae und Camallanidae. Goteborgs K. Vetensk.-O. Vitterhets. Samh. Handl., Ser. B., **2**(3): 441 pp.

Travassos, L. P. 1916. Trematodeos novos. Brazil-Med., **30**(40): 313–314.

————. 1930. Revisao do genero *Ascocotyle* Looss, 1899 (Trematoda: Heter-ophyidae) (Neubearbeitung der Gatung *Ascocotyle* Looss, 1899), (Trema-toda: Heterophyidae). Mem. Inst. Oswaldo Cruz, **23**(2): 61–79.

Tripathi, Y. R. 1954. Studies on parasites of Indian fishes. III. Protozoa. 2. (Mastigophora and Ciliophora). Rec. Indian Mus., **52**(2/4): 221–230.

Turnbull, Eleanor R. 1956. *Gyrodactylus bullatarudis* n. sp. from *Lebistes reticulatus* Peters with a study of its life cycle. Canad. J. Zool., **34**: 583–594.

Ulmer, M. J. 1960. Passeriform birds as experimental hosts for *Posthodiplostomum minimum*. J. Parasitol., **46**(5-2): 18.

Uspenskaya, A. V. 1957. The ecology and spreading of the pathogen of trout whirling disease—*Myxosoma cerebralis* (Hofer, 1903: Plehn, 1905)—in the fish ponds of the Soviet Union. *In* Petrushevskii, G. K. Parasites and Diseases of Fish. Bull. All-Union Sci. Res. Inst., Fresh-water Fish., Leningrad, **42**: 338 pp.

Uzmann, J. R. 1958–1963. Pers. comm., Bur. Commer. Fish., Biol. Lab., Booth-bay Harbor, Maine.

Uzmann, J. R., and S. H. Hayduk. 1963a. Larval *Echinochasmus* (Trematoda: Echinostomatidae) in rainbow trout, *Salmo gairdneri*. J. Parasitol., **50**(4): 586.

————. 1963b. In vitro culture of the flagellate protozoan *Hexamita salmonis*. Science, **140**(3564): 290–292.

Uzmann, J. R., and M. N. Hesselholt. 1957. New host and locality record for *Triaenophorus crassus* Forel (Cestoda: Pseudophyllidea). J. Parasitol., **43**(2): 205.

Uzmann, J. R., and J. W. Jesse. 1963. The *Hexamita* (= *Octomitus*) problem: a preliminary report. Prog. Fish-Cult., **25**(3): 141–143.

Uzmann, J. R., G. J. Paulik, and S. H. Hayduk. 1965. Experimental hexamitia-sis in juvenile coho salmon (*Oncorhynchus kisutch*) and steelhead trout (*Salmo gairdneri*). Tr. Am. Fish. Soc., **94**(1): 53–61.

Uzmann, J. R., and A. P. Stickney. 1954. *Trichodina myicola* n. sp., a peri-trichous ciliate from the marine bivalve, *Mya arenaria* L. J. Protozool., **1**: 149–155.

Van Beneden, P. J., and C. E. Hesse. 1863. Recherches sur les bdellodes ou hirudinées et les trématodes marines. Mem. Acad. Roy. Belg., **34**, 1–142.

Van Cleave, H. J. 1913. The genus *Neorhynchus* in North America. Zool. Anz., Leipzig, **43**(4): 177–190.

————. 1919a. Acanthocephala from fishes of Douglas Lake, Michigan. Occas. Papers Mus. Zool. Univ. Michigan, **72**: 12 p.

————. 1919b. Acanthocephala from the Illinois River, with descriptions of species and a synopsis of the family Neoechinorhynchidae. Bull. Illinois Nat. Hist. Survey, **13**(8): 225–257.

————. 1919c. Preliminary survey of the Acanthocephala from fishes of the Illinois river. Tr. Illinois State Acad. Sci., **12**: 151–156.

——. 1921a. Acanthocephala from the eel. Tr. Am. Micr. Soc., **40**(1): 1–13.

——. 1921b. Notes on two genera of ectoparasitic trematodes from freshwater fishes. J. Parasitol., **8**(1): 33–39.

——. 1922. A new genus of trematodes from the white bass. Proc. U. S. Nat. Mus. (2430), **61**(9): 1–8.

——. 1923. Acanthocephala from the fishes of Oneida Lake, New York. Roosevelt Wild Life Bull., Syracuse, New York, **2**(1): 73–84.

——. 1924. A critical study of the Acanthocephala described and identified by Joseph Leidy. Proc. Acad. Nat. Sci. Phila., **76**: 279–334.

——. 1928. Acanthocephala from China. 1. New species and new genera from Chinese fishes. Parasitology, **20**(1): 1–9.

——. 1931. New Acanthocephala from fishes of Mississippi and a taxonomic reconsideration of forms with unusual numbers of cement glands. Tr. Am. Micr. Soc., **50**(4): 348–363.

——. 1941a. Hook patterns on the acanthocephalan proboscis. Quart. Rev. Biol., **16**(2): 157–172.

——. 1941b. Relationships of the Acanthocephala. Am. Naturalist (756), **75**: 31–47.

——. 1947. The Eoacanthocephala of North America, including the description of *Eocollis arcanus*, new genus and new species, superficially resembling the genus *Pomphorhynchus*. J. Parasitol., **33**(4): 285–296.

——. 1948. Expanding horizons in the recognition of a phylum. J. Parasitol., **34**(1): 1–20.

——. 1949. The acanthocephalan genus *Neoechinorhynchus* in the catostomid fishes of North America, with descriptions of two new species. J. Parasitol. **35**(5): 500–512.

——. 1952a. Some host-parasite relationships of the Acanthocephala, with special reference to the organs of attachment. Exper. Parasitol., **1**(3): 305–330.

——. 1952b. Speciation and formation of genera in Acanthocephala. System. Zool., **1**(2): 72–83.

Van Cleave, H. J., and R. V. Bangham. 1949. Four new species of the acanthocephalan family Neoechinorhynchidae from freshwater fishes of North America, one representing a new genus. J. Wash. Acad. Sci., **39**(12): 398–409.

Van Cleave, H. J. and E. C. Haderlie. 1950. A new species of the acanthocephalan genus *Octospinifer* from California. J. Parasitol., **36**(2): 169–173.

Van Cleave, H. J. and D. R. Lincicome. 1940. A reconsideration of the acanthocephalan family Rhadinorhynchidae. J. Parasitol., **26**(1): 75–81.

Van Cleave, H. J. and J. E. Lynch. 1949. Preliminary report on the circumpolar distribution of *Neoechinorhynchus rutili* (Acanthocephala) in fresh water fishes. Science (2835), **109**: 446.

——. 1950. The circumpolar distribution of *Neoechinorhynchus rutili*, an acanthocephalan parasite of fresh-water fishes. Tr. Am. Micr. Soc., **69**(2): 156–171.

Van Cleave, H. J. and J. F. Mueller. 1932. Parasites of the Oneida Lake fishes. Part I. Descriptions of new genera and new species. Roosevelt Wild Life Ann., **3**(1): 5–71.

——. 1934. Parasites of Oneida Lake fishes. Part III. A biological and ecological survey of the worm parasites. Roosevelt Wild Life Ann., **3**(3 & 4): 161–334.

Van Cleave, H. J., and H. F. Timmons. 1952. An additional new species of the acanthocephalan genus *Neoechinorhynchus*. J. Parasitol., **38**(1): 53–56.

Van Cleave, H. J., and L. H. Townsend. 1936. On the assignment of *Echinorhynchus dirus* to the genus *Acanthocephalus*. Proc. Helminth. Soc. Washington, **3**(2): 63.

Van Duijn, C. (Jr.). 1956. Diseases of Fishes: Water Life. Dorset House, London. 174 pp.

Van Haitsma, J. P. 1925. *Crassiphiala bulboglossa* nov. gen., nov. spec., a holostomatid trematode from the belted kingfisher, *Ceryle alcyon* Linn. Tr. Am. Micr. Soc., **44**(3): 121–131.

———. 1930. Studies on the trematode family Strigeidae (Holostomidae). XX. *Paradiplostomum ptychocheilus* (Faust). Tr. Am. Micr. Soc. **49**(2): 140–153.

———. 1931. Studies on the trematode family Strigeidae (Holostomidae). XXIII. *Diplostomum flexicaudum* (Cort and Brooks) and stages in its life-history. Papers Michigan Acad. Sci., Arts and Lett., **13**: 483–516.

van Thiel, P. H., F. C. Kuipers, and R. Th. Roskam. 1960. A nematode parasitic to herring causing acute abdominal syndromes in man. Trop. Geograph. Med., **12**: 97–113.

Velasquez, Carmen C. 1958. Notes on *Azygia pristipomai* Tubangui, the genus *Azygia* and related genera (Digenea: Azygiidae). Proc. Helminth. Soc. Washington, **25**(2): 91–94.

Venard, C. E. 1940. Studies on parasites of Reelfoot Lake fish. I. Parasites of the large-mouth black bass, *Huro salmoides* (Lacépède). J. Tennessee Acad. Sci., **15**(1): 43–63.

———. 1941*a*. Studies on parasites of Reelfoot Lake fish. II. Parasites of the warmouth bass, *Chaenobryttus gulosus* (Cuvier and Valenciennes). J. Tennessee Acad. Sci., **16**(1): 14–16.

———. 1941*b*. Studies on parasites of Reelfoot Lake fish. III. A new genus and new species of trematode (Plagiorchioidea: Macroderoididae) from *Lepisosteus platostomus*. J. Tennessee Acad. Sci., **16**(4): 379–383.

Venard, C. E., and R. V. Bangham. 1941. *Sebekia oxycephala* (Pentastomida) from Florida fishes and some notes on the morphology of the larvae. Ohio J. Sci., **41**(1): 23 28.

Venard, C. E., and J. H. Warfel. 1953. Some effects of two species of Acanthocephala on the alimentary canal of the largemouth bass. J. Parasitol., **39**(2): 187–190.

Vergeer, T. 1942. Two new pseudophyllidean tapeworms of general distribution in the Great Lakes region. Tr. Am. Micr. Soc., **61**(4): 373–382.

Vernberg, Winona B. 1952. Studies on the trematode family Cyathocotylidae Poche, 1926, with the description of a new species of *Holostephanus* from fish and the life history of *Prohemistomum chandleri* sp. nov. J. Parasitol. **38**(4): 327–340.

Verrill, A. E. 1872. Description of North American fresh-water leeches. Am. J. Sci., **3**: 126–159.

Vik, R. 1954. Investigations on the pseudophyllidean cestodes of fish, birds and mammals in the Anoya water system in Trondelag. 1. *Cyathocephalus truncatus* and *Schistocephalus solidus*. Nytt Magasin for Zoologi, **2**: 5–51.

———. 1958. Studies of the helminth fauna of Norway. 2. Distribution and life cycle of *Cyathocephalus truncatus* (Pallas, 1781), (Cestoda). Nytt Magasin for Zoologi, **6**: 97–110.

————. 1959. Studies of the helminth fauna of Norway. 3. Occurrence and distribution of *Triaenophorus robustus* Olsson, 1892, and T. nodulosus (Pallas, 1760) (Cestoda) in Norway. Nytt Magasin for Zoologi, **8**: 64–73.

————. 1963. Studies of the helminth fauna of Norway. 4. Occurrence and distribution of *Eubothrium crassum* (Bloch, 1779) and *E. salvelini* (Schrank, 1790), (Cestoda) in Norway, with notes on their life cycles. Nytt Magasin for Zoologi, **11**: 47–73.

————. 1964a. Notes on the life history of *Philonema agubernaculum* Simon et Simon, 1936 (Nematoda). Canad. J. Zool., **42**: 1–9.

————. 1964b. Studies of the helminth fauna of Norway. 5. Plerocercoids of *Diphyllobothrium* spp. from the Rossaga water system, Nordland County. Nytt Magasin for Zoologi, **12**: 1–9.

————. 1964c. Penetration of stomach wall by Anisakis-type larvae in porpoises. Canad. J. Zool., **42**: 513.

Vinyard, W. C. 1955. Epizoophytic algae from mollusks, turtles and fish in Oklahoma. Proc. Oklahoma Acad. Sci., **34**: 63–65.

Vishniac, Helen S., and R. F. Nigrelli. 1957. The ability of the Saprolegniaceae to parasitize platyfish. Zoologica, **42**(4): 131–134.

Vogel, H. 1934. Der Entwicklungszyklus von *Opisthorchis felineus* (Riv.) nebst Bemerkungen über die Systematik und Epidemiologie. Zoologica, Stuttgart, Heft 86: v. 33, Lief. 2–3, pp. 1–103.

Vojtek, J. 1958. *Urocleidus* Mueller 1934, novy rod zabrohlistu (Trematoda, Monogenea) pro CSR. (*Urocleidus* Mueller, 1934, eine neue Gattung niederer Saugwurmer (Trematoda, Monogenea) für die CSR.) Biologia, Bratislava, **13**(8): 612–615.

Von Linstow. *See* Linstow.

Wagener, G. R. 1957. Beiträge zur Entwicklungs-Geschichte der Eingeweidewurmer. Eine von der Hollandischen Societat de Wissenschaften zu Haarlem i. J. 1855 gekronte Preisschrift. Natuurk. Verhandel. Holland, Maatsch. Vetensch. Haarlem, Verzamel. 2, Deel 13: 112 pp.

Wagh, P. V. 1961. Transplantation of a myxosporidian, *Myxosoma ovalis* from *Ictiobus bubalus* (small-mouth buffalo) to *Notemigonus crysoleucas* (golden shiner). J. Biol. Sci., **4**(2): 47–51.

Wagler, E. 1935. Die deutschen Karpfenlause. Zool. Anz., Leipzig, **110** (1 & 2): 1–10.

Wagner, E. D. 1953. A new species of *Proteocephalus* Weinland, 1858 (Cestoda), with notes on its life history. Tr. Am. Micr. Soc., **72**(4): 364–369.

————. 1954. The life history of *Proteocephalus tumidocollus* Wagner, 1953 (Cestoda), in rainbow trout. J. Parasitol., **40**(5): 489–498.

Wales, J. 1958. Two new blood fluke parasites of trout. Calif. Fish and Game, **44**(2): 125–136.

————. 1964. Pers. comm., Dept. Fish and Game Management, Oregon State Coll., Corvallis, Oregon.

Wales, J., and H. Wolf. 1955. Three protozoan diseases of trout in California. Calif. Fish and Game, **41**(2): 183–187.

Walkey, M. 1962. Observations on the life history of *Neoechinorhynchus rutili* (O. F. Müller, 1776). Parasitology, **52**(3 & 4): 18–19.

Wallace, F. G. 1935. A morphological and biological study of the trematode, *Sellacotyle mustelae* n. g., n. sp. J. Parasitol., **21**(3): 143–164.

———. 1939. The metacercaria of *Amphimerus elongatus* Gower (Trematoda: Opisthorchiidae). J. Parasitol., **25**(6): 491–494.

———. 1940. Some fish-borne parasites of birds and mammals. Proc. Minnesota Acad. Sci., **8**: 50.

Wallace, F. G., and L. R. Penner. 1939. A new liver fluke of the genus *Opisthorchis*. J. Parasitol., **25**(5): 437–440.

Wallace, H. E. 1941. Life history and embryology of *Triganodistomum mutabile* (Cort), (Lissorchiidae, Trematoda). Tr. Am. Micr. Soc., **60**(3): 309–326.

Wallin, I. E. 1909. A new species of the trematode genus *Allocreadium*. With a revision of the genus and a key to the sub-family Allocreadiinae. Tr. Am. Micr. Soc., **29**(1): 50-66.

Walton, A. C. 1928. A revision of the nematodes of the Leidy collections. Proc. Acad. Nat. Sci. Phila. (1927), **79**: 49–163.

Ward, Helen. 1937. Acanthocephala from the burbot "eel" (*Lota vulgaris*), with special reference to variations in the female of *Echinorhynchus coregoni*. Tr. Am. Micr. Soc., **56**(3): 355–363.

———. 1940. Studies on the life history of *Neoechinorhynchus cylindratus* (Van Cleave, 1913), (Acanthocephala). Tr. Am. Micr. Soc., **59**(3): 327–347.

———. 1951. The species of acanthocephala described since 1933. I. J. Tennessee Acad Sci., **26**(4): 282–311.

Ward, H. B. 1894. On the parasites of the lake fish. 1. Notes on the structure and life history of *Distoma opacum* n. sp. Proc. Am. Micr. Soc., **15**(30): 173–182.

———. 1901. Notes on the parasites of the lake fish. 3. On the structure of the copulatory organs in *Microphallus* nov. gen. Tr. Am. Micr. Soc., **22**: 175–187.

———. 1910. Internal parasites of the Sebago salmon. Bull. Bur. Fish., **28**(2): 1151–1194.

———. 1918. Parasitic roundworms. *In* Ward, H. B., and G. C. Whipple. Fresh-water Biology, 506–552.

———. 1920. Notes on North American Myxosporidia. J. Parasitol., **6**(2): 49–64.

Ward, H. B., and T. B. Magath. 1916. Notes on some nematodes from fresh-water fishes. J. Parasitol., **3**(2): 57–64.

Ward, H. B., and J. F. Mueller. 1926. A new pop-eye disease of trout-fry. Arch. Schiffs- u. Tropen-Hyg., **30**(9): 602–609.

Ward, H. B., and G. C. Whipple. 1918. Fresh-water Biology. Wiley, New York. 1111 pp.

Wardle, R. A. 1932a. The Cestoda of Canadian fishes. I. The Pacific Coast region. Contrib. Canad. Biol. and Fish., **7**(18): 221–243.

———. 1932b. The Cestoda of Canadian fishes. II. The Hudson Bay drainage system. Contrib. Canad. Biol. and Fish., **7**(30): 377–403.

———. 1932c. The limitations of metromorphic characters in the differentiation of Cestoda. Tr. Roy. Soc. Canada, **26**(5): 193–204.

———. 1933. Significant factors in the plerocercoid environment of *Diphyllobothrium latum* (Linn.). J. Helminth., **11**(1): 25–44.

Wardle, R. A., and J. A. McCleod. 1952. The Zoology of Tapeworms. Univ. Minnesota Press, Minneapolis. 780 pp.

Warren, B. 1952. Report of parasites from the Lake Superior cisco, *Leucichthys artedi arcturus.* J. Parasitol., **38**(5): 495.

———. 1953. A new type of metacercarial cyst of the genus *Apophallus*, from the perch, *Perca flavescens*, in Minnesota. Am. Midl. Nat., **50**(2): 397–401.

Watson, N. H. F. 1963. A note on the upper lethal temperature of eggs of two species of *Triaenophorus.* J. Fish. Res. Bd. Canada, **20**(3): 841–844.

Watson, N. H. F., and G. H. Lawler. 1963. Temperature and rate of hatching of *Triaenophorus* eggs. J. Fish. Res. Bd. Canada, **20**(1): 249–251.

Watson, N. H. F., and J. L. Price. 1960. Experimental infections of cyclopid copepods with *Triaenophorus crassus* Forel and *T. nodulosus* (Pallas). Canad. J. Zool., **38**: 345–356.

Weinland, D. F. 1858. Human Cestodes: An Essay on the Tapeworms of Man. . . . Cambridge. 93 pp.

Weissenberg, R. 1911. Ueber einige Mikrosporidien aus Fischen (*Nosema lophii* Doflein, *Glugea anomala* Moniez, *Glugea hertwigii* nov. spec.). Sitzungsb. Gesellsch. Naturf. Fr. Berlin, **8**: 344–351.

Welberry, A. E., and W. Pacetti. 1954. Intestinal fluke infestation in a native Negro child, Dade County (Florida). Med. Bull., Jan., 34–35.

Welker, G. W. 1962. Helminth parasites of the common grackle *Quiscalus quiscula versicolor* Bieillot in Indiana. Diss. Abst., **23**(2): 761.

Wellborn, Tom. 1964. Pers. comm., USFWS, Auburn Univ., Auburn, Alabama.

Wellborn, T. L., and W. A. Rogers. 1966. A key to the common parasitic protozoans of North American fishes. Auburn Univ. Zool.-Ent. Series, Fish. no. 4: 17 pp.

Weller, Thomas H. 1938. Description of *Rhabdochona ovifilamenta* n. sp. (Nematoda: Thelaziidae) with a note on the life history. J. Parasitol., **24**(5): 403–408.

Wenrich, D. H. 1931. A trypanoplasm on the gills of carp from the Schuylkill River. J. Parasitol., **18**(2): 133

West, A. J. 1964. The acanthor membranes of two species of Acanthocephala. J. Parasitol., **50**(6): 731–734.

White, F. M., and R. M. Cable. 1942. Studies on the morphology of *Cystidicola cristivomeri* sp. nov. (Nematoda: Thelaziidae) from the swim bladder of the lake trout, *Cristivomer namaycush* (Walbaum). Am. Midl. Nat., **28**(2): 416–423.

Whitman, C. O. 1889. Some new facts about the Hirudinea. J. Morphol., **2**: 586–599.

Wigdor, M. 1918. Two new nematodes common in some fishes of Cayuga Lake. J. Parasitol., **5**(1): 29–34.

Williams, H. H. 1965. Roundworms in fishes and so-called "herring-worm disease." British Med. J., **1**: 964–967.

Wilson, C. B. 1902. North American parasitic copepods of the family Argulidae, with a bibliography of the group and a systematic review of all known species. Proc. U. S. Nat. Mus., 635–742.

———. 1904. The fish parasites of the genus *Argulus* found in the Woods Hole region. Proc. U. S. Nat. Mus., 115–131.

———. 1907. Additional notes on the development of the Argulidae, with description of a new species. Proc. U. S. Nat. Mus., **32**(1531): 411–424.

———. 1908. North American parasitic copepods: A list of those found upon the fishes of the Pacific Coast, with descriptions of new genera and species. Proc. U. S. Nat. Mus., 35(1652): 431–481.

————. 1911a. North American parasitic copepods belonging to the family Ergasilidae. Proc. U. S. Nat. Mus., 39(1788): 42–219.

————. 1911b. North American parasitic copepods: Description of new genera and species. Proc. U. S. Nat. Mus., 39(1805): 625–634.

————. 1911c. North American parasitic copepods. Part 9. The Lernaepodidae. Proc. U. S. Nat. Mus., 39(1783): 189–226.

————. 1912. Descriptions of new species of parasitic copepods in the collections of the United States National Museum. Proc. U. S. Nat. Mus., 42(1900): 233–243.

————. 1915. North American parasitic Copepoda belonging to the Lernaeopodidae, with a revision of the entire family. Proc. U. S. Nat. Mus., 47(2063): 565–729.

————. 1916. Copepod parasites of fresh-water fishes and their economic relations to mussel glochidia. Bull. U. S. Bur. Fish., 34(1914): 331–374.

————. 1917a. The economic relations, anatomy, and life history of the genus Lernaea. Bur. Fish. Doc., 854: 165–198.

————. 1917b. North American parasitic Copepoda belonging to the Lernaeidae, with a revision of the entire family. Proc. U. S. Nat. Mus., 53(2194): 1–150.

————. 1920a. Food and parasites of the fishes. In Evermann and Clark. Lake Maxinkuckee, a physical and biological survey. 1: 291–305, Indiana.

————. 1920b. Report on the parasitic Copepoda collected during the Canadian Arctic Expedition, 1913–18. Rep. Canad. Arctic Exped. 1913–18., 7(L): 3–16.

————. 1922. Parasitic copepods in the collection of the Zoological Museum, Kristiania. Medd. Zool. Mus., Kristiania, 4: 7 pp.

————. 1924. New North American parasitic copepods, new hosts and note on copepod nomenclature. Proc. U. S. Nat. Mus., 64(2507): 1–22.

————. 1932. The copepods of the Woods Hole region, Massachusetts. U. S. Nat. Mus., Bull. (158): 635 pp.

————. 1936. Argulus canadensis from Cape Breton Island. J. Biol. Bd. Canada, 2(4): 355–358.

————. 1944. Parasitic copepods in the United States National Museum. Proc. U. S. Nat. Mus., 94(3177): 529–582.

Wilson, W. D. 1957. Parasites of fishes from Leavenworth County State Lake, Kansas. Tr. Kansas Acad. Sci., 60(4): 393–399.

Winfield, G. F. 1929. Plesiocreadium typicum, a new trematode from Amia calva. J. Parasitol., 16(2): 81–87.

Wisniewski, L. W. 1932. Cyathocephalus truncatus, its development, morphology and biology. Nature, 130(3284): 555.

————. 1958. The development cycle of Bunodera luciopercae (O. F. Müller). Acta Parasitologica Polonica, 6(11): 289–307.

Witenberg, G. G. 1929. Studies on the trematode-family Heterophyidae. Ann. Trop. Med. and Parasitol., 23(2): 131–239.

Wolf, H. 1960. Pers. comm., State Dept. Fish and Game, Sacramento, California.

Wood, E. M., and W. T. Yasutake. 1956. Histopathology of fish. IV. A granuloma of brook trout. Prog. Fish-Cult., 18(3): 108–112.

Wood, R. A., and J. D. Mizelle. 1957. Studies on monogenetic trematodes. XXI. North American Gyrodactylinae, Dactylogyrinae and a new host record for *Urocleidus dispar* (Mueller, 1936). Am. Midl. Nat., **57**(1): 183–202.

Woodbury, L.A. 1934. Notes on some parasites of three Utah reptiles. Copeia, no. 1: 51–52.

Woodhead, A. E. 1929. Life history studies on the trematode family Bucephalidae. Tr. Am. Micr. Soc., **48**(3): 256–275.

———. 1930. Life history studies on the trematode family Bucephalidae. II. Tr. Am. Micr. Soc., **49**(1): 1–17.

———. 1950. Life history cycle of the giant kidney worm, *Dioctophyma renale* (Nematoda), of man and many other animals. Tr. Am. Micr. Soc., **69**(1): 21–46.

Wootton, D. M. 1957. The life history of *Cryptocotyle concavum* (Creplin, 1825) Fischoeder, 1903 (Trematoda: Heterophyidae). J. Parasitol., **43**(3): 271–279.

Worley, D. E., and R. V. Bangham. 1952. Some parasites of fishes of the upper Gatineau River valley. Ohio J. Sci., **52**(4): 210–212.

Wright, R. R. 1879. Contributions to American helminthology. No. 1. Proc. Canad. Inst., **1**(1): 54–75.

———. 1882. Notes on American parasitic Copepoda. No. 1. Proc. Canad. Inst., **1**(3): 243–254.

Wu, K. 1938. Progenesis of *Phyllodistomum lesteri* sp. nov. (Trematoda: Gorgoderidae) in freshwater shrimps. Parasitology, **30**(1): 4–19.

Wyatt, E. J., and I. Pratt. 1963. *Myxobolus insidiosus*, sp. n. a myxosporidian from the musculature of *Oncorhynchus tshawytscha* (Walbaum). J. Parasitol., **49**(6): 951.

Yakimoff, V. L., and V. F. Gousseff. 1935. Kokzidien bei Fischen (*Carassius carassius*). Zeitschr. Infektionskr. Haustiere, **48**(3): 149–150.

Yamaguti, S. 1933. Studies on the helminth fauna of Japan. Part 1. Trematodes of birds, reptiles and mammals. Japan J. Zool., **5**(1): 1–134.

———. 1934. Studies on the helminth fauna of Japan. Part 2. Trematodes of fishes. Japan J. Zool., **5**(3): 249–541.

———. 1942. Studies on the helminth fauna of Japan. Part 38. Larval trematodes of fishes. Japan J. Med. Sci., VI, Bacteriol. and Parasitol., **2**(3): 131–160.

———. 1953. Systema Helminthum. Part I. Digenetic Trematodes of Fishes. Tokyo. 405 pp.

———. 1958. Systema Helminthum. Vol. I, Part II. The Digenetic Trematodes of Vertebrates. Interscience, New York. 1575 pp.

———. 1959. Systema Helminthum. Vol. II. The Cestodes of Vertebrates. Interscience, New York. 860 pp.

———. 1961. Systema Helminthum. Vol. III. The Nematodes of Vertebrates. Interscience, New York. pp. 681–1261.

———. 1963a. Systema Helminthum. Vol. IV. Monogenea and Aspidocotylea. Interscience, New York. 699 pp.

———. 1963b. Systema Helminthum. Vol. V. Acanthocephala. Interscience, New York. 421 pp.

———. 1963c. Parasitic Copepoda and Branchiura of Fishes. Interscience, New York. 1104 pp.

Yashouv, A. 1959. On the biology of *Lernaea* in fish ponds. Bamidgeh, **11**(4): 80–89.

Yasutake, W. T., D. R. Buhler, and U. E. Shanks. 1961. Chemotherapy of hexamitiasis in fish. J. Parasitol., **47**(1): 81–86.

Yasutake, W. T., and E. M. Wood. Some Myxosporidia found in Pacific Northwest salmonids. J. Parasitol., **43**(6): 633–642.

Yeatman, H. C. 1965. Redescription of the freshwater branchiuran crustacean, *Argulus diversus* Wilson, with a comparison of related species. J. Parasitol., **51**(1): 100–107.

Yeatman, H. C. 1966. *Argulus* parasitic on killifish. American Killifish Assoc. Spring, 1966, 4 pp.

Yeh, L. S. 1955. A new tapeworm, *Diphyllobothrium salvelini* sp. nov., from a salmon, *Salvelinus alpinus*, in Greenland. J. Helminth., **29**(1 & 2): 37–43.

Yin, W. Y. 1949. Three new species and a new genus of parasitic copepods (Ergasilidae) from Chinese pond fishes. Sinensia, **20**(1/6): 32–42.

————. 1956. Studies on the Ergasilidae (parasitic Copepoda) from the freshwater fishes of China. Acta Hydrobiol. Sinica, **2**: 209–270.

Yin, W., and Nora G. Sproston. 1948. Studies on the monogenetic trematodes of China. Sinensia, **19**: 58–85.

Yorke, W., and P. A. Maplestone. 1926. The Nematode Parasites of Vertebrates. Hafner, New York. 536 pp.

Zandt, F. 1935. *Achtheres pseudobasanistes* n.n., syn. *Basanistes coregoni* (Neresheimer). Die postembryonale Entwicklung und geographische Verbreitung eines Lernaeopodiden. Zool. Jahrb., Jena, Abt. Anat., **60**(3 & 4): 289–344.

Zhukov, E. V. 1963. On the fauna of parasites of fishes of the Chukotsk Peninsula and adjoining seas. II. Endoparasitic worms of marine and freshwater fishes (Russian text, English summary). Acad. Sci. S.S.S.R., **21**: 96–139.

Zischke, J. A., and C. M. Vaughn. 1962. Helminth parasites of young-of-the-year fishes from the Fort Randall reservoir. Proc. South Dakota Acad. Sci., **41**: 97–100.

ADDENDUM

P. 407   Arey, L. B. (1921). An experimental study on the glochidia and the factors underlying encystment.
         J. Exp. Zool. 33: 463–499.

P. 410   Becker, C. D., and W. D. Brunson. 1966. Transmission of *Diplostomum flexicaudum* to trout by ingestion of precocious metacercariae in mollusks. J. Parasitol. 52(4): 829–830.

P. 411   Bowen, J. T. 1965. Should be 1966.

P. 415   Clarke, A. H., Jr., and C. O. Berg. 1959. The freshwater mussels of central New York. Cornell Univ. Agric. Exp. Sta., N.Y. State Coll. Agric., Mem. 367, 79 p.

P. 417   Davis, H. S., and Estelle C. Lazar. 1940. A new fungus disease of trout. Tr. Amer. Fish. Soc. 70: 264–271.

P. 422   Fischthal, J. H. 1952. Parasites of northwest Wisconsin fishes. III. The 1946 survey. 41: 17–58.

P. 423   Frederickson, L. H., and M. J. Ulmer. 1965. Caryophyllaeid cestodes from two species of redhorse (*Moxostoma*). Iowa Acad. Sci. 72: 444–461.

P. 429 Howard, A. D. 1914. Experiments in the propagation of freshwater mussels of the Quadrula group. Rept. U.S. Comm. Fish. 193, Doc. 801, 51 p.

———— ———— 1922. Experiments in the cultivation of freshwater mussels. Bull. U.S. Bur. Fish. 38: 63–89.

P. 437 Lewis, F. J. 1935. The trematode genus *Phyllodistomum* Braun. Tr. Amer. Microscop. Soc. 54(2): 103–117.

P. 439 Lumsden, R. D. 1963. *Saccocoelioides sogandaresi* sp. n., a new haploporid trematode from the sailfin molly *Mollienisia latipinna* Le Sueur in Texas. J. Parasitol. 49(2): 281–284.

———— ———— and J. A. Zischke. 1961. Seven trematodes from small mammals in Louisiana. Tulane Studies Zool. 9(2): 77–85.

P. 440 Mackiewicz, J. S. 1965. *Isoglaridacris bulbocirrus* gen et sp. n. (Cestoidea Caryophyllaeidae) from *Catostomus commersoni* in North America. J. Parasitol. 51(3): 377–381.

———— ———— 1965. Redescription and distribution of *Glaridacris catostomi* Cooper, 1920 (Cestoidea: Caryophyllaeidae). J. Parasitol. 51(4): 554–560.

P. 441 Malmberg, G. 1962 (1964). Taxonomical and ecological problems in *Gyrodactylus* (Trematoda: Monogenea), p. 203–230. *In* R. Ergens and B. Rysavy (ed.), Parasitic worms and aquatic conditions, Proc. Symp. held in Prague. Publ. Czech. Acad. Sci.

P. 442 Meggitt, F. J. 1914. The structure and life-history of a tapeworm (*Ichthyotaenia filicollis* Rud.) parasitic in the stickleback. Proc. Zool. Soc. (London) 8: 113–138.

P. 448 Ortmann, A. E. 1911. A monograph of the Najades of Pennsylvania. Mem. Carnegie Mus. 4: 279–347.

P. 452 Reichenbach-Klinke, H., and E. Elkan. 1965. The Principal Diseases of Lower Vertebrates. Academic Press, London and New York, 600 p.

P. 452 Reuling, F. H. 1919. Acquired immunity to an animal parasite. J. Infect. Dis. 24: 337–346 (Glochidia).

P. 453 Sahay, U., and D. Prasad. 1965. On a new species of a nematode (Thelaziidae, Spinitectinae, *Spinitectus,* Fourment, 1883) with a key to the species of the genus *Spinitectus*. Jap. J. Med. Sci. & Biol. 18(3): 143–150.

P. 457 Stammer, Y. 1959. Beiträge fur Morphologie, Biologie und Bekämpfung der Karpfenläuse. Zeitsch. Parasiten. 19: 135–208.

P. 459 Tucker, M. E. 1928. Studies on the life-cycle of two species of freshwater mussels belonging to the genus *Anodonta*. Biol. Bull. 54: 117–127.

P. 465. Witenberg, G. 1944. What is the cause of the parasitic laryngopharyngitis in the Near East ("Halzoun")? Act. Med. Orient (Jerusalem). 3: 191–192.

P. 467 Yeatman ref. J. Amer. Killifish Assoc. 3(1): 8–11.

Index*

abdominal syndrome, 4
aboral membranella, 39
acanthella, 271
acanthocephala, 8, 9, 13, 14, *270*
Acanthocephalus, 273, *281*
  acerbus, 283
  aculeatus, 283
  anguillae, 283
  dirus, 283
  jacksoni, 283
  lateralis, 283
Acanthocolpidae, 165
acanthor, 271
Acarina, 316
accidental parasite, 134, 159
Acetodextra, 119, 199
  ameiuri, 12, 116, *119*, 120, 160, 180, 199
achene, 318
Achlya, 17
Achtheres, 11, 301, *311*, 313
  ambloplitis, 311
  coregoni, 313
  corpulentus, 313
  lacae, 313
  micropteri, 313
  pimelodi, 313
Acnidosporidia, 46, *47*
Acolopenteron, 12, 76, *77*
  catostomi, 77
  ureteroecetes, 77
Actinobdella triannulata, 289, 291, 294, *295*
Actinocleidus, 86, *96*
  articularis, 96
  bakeri, 96
  bifidus, 96
  bifurcatus, 96
  brevicirrus, 96
  bursatus, 96
  crescentis, 96
  fergusoni, 96
  flagellatus, 97
  fusiformis, 97
  gibbosus, 97
  gracilis, 97

  harquebus, 97
  incus, 97
  longus, 97
  maculatus, 97
  oculatus, 97
  okeechobeensis, 97
  recurvatus, 97
  scapularis, 97
  sigmoideus, 97
  subtriangularis, 97
  triangularis, 97
  unguis, 97
Actinomycetes, 14
Adephaga, 320
adhesions, 205
adhesive units, 99
AFA fixative, 8
Agarella, 52, *60*
  gracilis, 60
Agrion, 254
air bladder, 12, 160, 245, 259
alae: caudal, 244, 245, 250, 253, 255, 259, 261, 262, 263
alcohol-phenol, 10
algae, 14, 16; blue-green, 16
Allacanthochasmus, *116*, *193*
  artus, 117
  varius, 117, 193
alligators, 265, 318
Allocreadiidae, 109, *142*, 143, 160, 165
Allocreadioidea, 165
Allocreadium, 12, 143, *149*
  armatum, 150
  boleosomi, 150
  colligatum, 150
  commune, 144, *150*, 158
  halli, 150
  ictaluri, 150
  lobatum, 150
  shawi, 151
Alloglossidium, 141, *142*
  corti, 142
  kenti, 142
Alloplagiorchis garricki, 124
Alopex, 231
alveoli, 104

Ambiphrya, 39
Amblystoma, 134
Amnicola, 121, 133, 155, 185, 190–191, 197
Amphibia, 28, 37, 56, 57, 220, 231, 258, 265, 283, 301
Amphicotylidae, 224, *225*
Amphileptus, 11, 34, *39*
  voracus, 39
Amphilina, 206
Amphimerus, 181, *190*
  elongatus, 191
  pseudofelineus, 191
amphipods, 39, 236, 271, 281, 284, 285, 319
amphistomate, 111
Amphizoidae, 320
anadromous, 2, 108, 109, 301
Anallocreadium, 155
Anaplocamus, 178
Anchoradiscus, 87, *96*
  anchoradiscus, 96
  triangularis, 96
Ancyracanthus, 259
Ancyrocephalinae, 85
Ancyrocephalus, 86, *97*
anesthetic, fish, 6; leech, 289; monogenea, 72
Anisakis, 4, 245, *251*
Annelida, 209, 289
Anonchohaptor, 76, *77*
  anomalum, 77
anus, fish, 244, 253, 300
Apatemon, cercariae, 167
Apatemon gracilis pellucidus, 174
Aphanomyces, 17
Apiosoma, 34, *41*
Aponurus, 12, 135, *137*
Apophallus, 4, 191
  brevis, 191, 298
  donicus, 191
  imperator, 193
  itascensis, 193
  venustus, 193
appendix, esophageal, 243, 245, 250, 251
  aquarium fish, 20
Arachnoidea, 316
Archigetes, 207, *209*
  iowensis, 209
areoli, 102
Argulidae, 301
Argulus, 11, 15, 300, *301*
  americanus, 301
  appendiculosus, 301
  biramosus, 301
  canadensis, 302
  catostomi, 302
  coregoni, 302
  diversus, 302
  flavescens, 302
  floridensis, 302
  foliaceus, 302

funduli, 302
ingens, 302
japonicus, 302, 303
laticauda, 302
lepidostei, 302
longicaudatus, 302
lunatus, 303
maculosus, 302, 303
mississippiensis, 303
nobilis, 303
pellucidus, 302, 303
pugettensis, 303
stizostethi, 303
trilineatus, 303
versicolor, 303
Aristocleidus, 86, *91*
Arthropoda, 128, 144, 148, 150, *299*, 316
Ascaridea, 249
"*Ascaris*," 243, *249*
Ascocotyle, 179, *193*, 194
  angrense, 195
  chandleri, 194
  diminuta, 194
  leighia, 194
  longa, 195
  mcintoshi, 194
  molienesicola, 195
  tenuicollis, 194
Asellus, 158, 283
Aspidocotylea, 102
Asymphylodora, 121
  amnicolae, 121
Atractolytocestus, 208, *218*
  huronensis, 218
autospores, 17
Azygia, 12, 130, *131*, 164
  acuminata, 131
  angusticauda, 133
  bulbosa, 133
  longa, 133
  loossii, 133
  sebago, 133
  tereticollis, 133
Azygiidae, 108, 130

Babesiosoma, 11, *48*
back swimmers, 320
bacteria, 289; filamentous, 14
bacteriology, 2
Balantidium, 33, *36*
Barbulostomum, 154, *156*
Basanistes, 313
bear, 189, 231, 320
Belostomatidae, 320
Biacetabulum, 207, *208*
  giganteum, 209
  infrequens, 209
  macrocephalum, 209
  meridianum, 209
Bialovarium, 207, 209
  nocomis, 211
Bidens sp., 318

bile ducts, 12
birds, 191, 194, 197, 198, 231, 250; fish-eating, 231, 232
Bittium, 126
blackbird, 176
"black-spot," 175–177, 181, 191, 197
bladder-inhabiting trematode, 77, 128
blood, 6, 11, 23, 38, *55*
blood flagellates, 9
blood picture, 310
blood protozoa, 9
blood smears, 6
blood vessel, 110, 159
blue slime, 23
blue-green algae, 17
Bodomonas, 11, *30*
  concava, 27, *30*
body cavity, 11, 159
Bolbophorus confusus, 169
borax carmine, 9
bothridia, leaf-like, 239
Bothriocephalus, 225, *229*
  claviceps, 230
  cuspidatus, 230
  formosus, 230
  rarus, 230
  schilbeodis, 230
  speciosus, 230
  texomensis, 230
Bothriomonas, 236
Bothrioscolex, 209
Bouin's fixative, 8, 9
Brachyphallus, 135, *136*
  crenatus, 137
Brachyspira, 44
Brachyurus, 213
brackish-water fishes, 2
brain, fish, 7, 170, 180
branchial cavity, 11, 108, 127
Branchiura, 230, 301
brine, 5
buccal capsule, 242, 244, 253, 263; pseudo, 244
buccal cavity, 140
buccal denticles, 247
Bucephalidae, 108, *113*, 159, 165, 178, *182*
Bucephaloidea, 182
Bucephaloides, *113*, 178, *283*
  clara, 283
  ozakii, 113
  pusillus, 113
  strongylurae, 283
Bucephalus, 113, *114*, 178, *183*
  elegans, 114, 183
  papillosum, 114
Bulbodacnitis, 244, *254*
  globosa, 254
  occidentalis, 254
  scotti, 255
bulla, 311, 313, 314
Bunodera, 143, *147*
  armatum, 148

  cornuta, 148
  luciopercae, 148
  nodulosa, 148
  sacculata, 148
Bunoderina, 143, *148*
  eucaliae, 148
  sacculata, 148
bursa, 269

caddis fly, 128
Caecincola, 12, 116, 120, 181, *190*, 201
  parvulus, 121, 201
Cainocreadium, 157
calcareous: concretions, 14, 169, 170, 173, 177; granules, 174
Calceostomatidae, 74, *76*
Calligidea, 307
Camallanidae, 253
Camallanus, 244, *253*, 254
  ancylodirus, 253
  lacustris, 254
  oxycephalus, 254
  trispinosus, 254
  truncatus, 254
Cambarus, 134
Campeloma, 133, 189
canary, 185, 187
Capillaria, 243, 247
  catenata, 247
  catostomi, 247
  eupomotis, 247
  petruschewskii, 247
Capingens, 207, *211*
  singularis, 213
carapace, 310
Carassotrema, 123
Carchesium, 11, 34, *43*
Cardicola, 107
Carnivora, 189
"carrier," 229
cartilage, 13
Caryophyllaeidae, 207
Caryophyllaeus, 208, *217*
  laticeps, 217
  terebrans, 217
Caryophyllidea, 207
cat, 178, 189–191, 193, 197, 231
Cathaemasia, 179, *186*
  pulchrosoma, 186
Cathaemasiidae, 186
Catoptroides, 127
caudal alae, 244, 245, 247, 250, 253, 259, 261, 262, 263
caudal appendage, 63
caudal filament, 234
caudal fin, 268
Cauloxenus, 315
  stygins, 315
ceca: intestinal, 250; nematode, 243, 245
ceca, pyloric, 12
centripetal thorn, 43
Centrovarium, 12, 116, *120*, 181, *201*

lobotes, 120, 201
cephalic claspers, 300
cephalic glands, 76, 99, 257
cephalic horns, 307
cephalic papillae, 245
cephalothorax, 313
Ceratomyxa, *53*
  shasta, 53
Ceratomyxidae, 51
Cercaria, Apatemon, 167
C., *Clinostomum*, 168
C. *Cotylurus communis*, 171
C. *Cotylurus cornutus*, 167
C. *Cotylurus flabelliformis*, 167
C., *Cyathocotyle*, 168
C. *Diplostomum spathaceum*, 167
C. *dubia*, 166
C. *flexicorpa*, 167, 176
C. *isomi*, 168
C. *louisianae*, 168
C. *marginatae*, 167
C. *multicellulata*, 168
C. *Neogogatea kentuckiensis*, 168
C. *Paracaenogonimus szidati*, 167
C. *paramulticellulata*, 168
C. *posthodiplostomum minimum*, 168
C. *Prohemistomum vivax*, 167
C. *Prosostephanus industrius*, 167
C. *ranae*, 166
C. *Szidatia joyeuxi*, 167
C. *tatei*, 167
C. *tauiana*, 168
C. *Uvulifer ambloplitis*, 167
cercariae, amphistomate, 165
cercariae, cotylocercous xiphidiocercaria, 158
cercariae of cyathocotylids, 167
cercariae, cysticercous, 130, 133
cercariae, cystophorous, 165
cercariae, echinostomate, 165
cercariae, furcocercous, 164, *166*; gasterostomate, 182
cercariae, furcocysticercous, 164
cercariae, gasterostomate, 165
cercariae, longifurcate, 169
cercariae, lophocercous, 168, 190, 191,
  197; xiphidiocercaria, 151, 152
cercariae, macrocercous, 128
cercariae, microcercous, 188, 189
cercariae, monostome, 195, 197
cercariae, Multicellulata group, 168
cercariae, Novena group, 168
cercariae, opthalmo, 154
cercariae, opthalmogymnocephalus,
  195
cercariae, parapleurolophocercous,
  165, 197
cercariae, pleurolophocercous, 121,
  165, 191, 195, 197
cercariae, Prohemistomatinae, 167
cercariae, Sanguinicolidae, 168
cercariae, Schistosomatoidea, 168
cercariae, Spirorchidae, 168

cercariae, *Tetis* group, 167
cercariae, *Vivax* group, 167
cercariae, xiphidiocercaria, 145, 147,
  148, 151, 188
Cerithidea, 178, 195, 197, 198
Ceryle (see kingfisher)
Cestoda, 9, 14, 206
Cestodaria, 205
cestodes, 3, 4, 8, 9, 14, 205
Cestrahelmins, 125, *126*
  laruei, 127
cheek "galleries," 243, 268
Chelydra, 145
chicks, 170, 171, 174, 176, 177, 187,
  191, 194–197
Chilodon, 39
Chilodonella, 11, *39*
  cyprini, 39
  dentatus, 39
Chironomus, 124
chitinoid: hooks, 225; ring, 45
chlorazol black E, 9
Chlorella, 16, 17
Chlorococcales, 16
Chloromyxidae, 52
Chloromyxum, 52, *54*, 55
  catostomi, 54
  externum, 54
  funduli, 54
  gibbosum, 55
  majori, 55
  opladeli, 55
  renalis, 55
  thompsoni, 55
  trijugum, 55
  truttae, 55
  wardi, 55
Chlorophyta, 17
chloroplast, 17
chromatophores, 25, 26, 28
Ciliates, 31; nonsuctorian, 36
Ciliophora, 24, *31*
circumoral spines, 115–117, 125, 178,
  179, 186, 188, 193–195
cirrus, spined, 75, 108, 123
Cladophora, 16
clamps, cuticular, 73
clams, 115, 151, 155, 316; clams,
  larval, 15; sphaerid, 145, 147,
  151, 316
Claspers, cephalic, 300
Clavunculus, 86, *95*
  bifurcatus, 95
  bursatus, 95
  unguis, 95
Cleidodiscus, 86, *88*, 97, 98
  aculeatus, 88
  alatus, 88
  articularis, 88
  banghami, 88
  bedardi, 88
  brachus, 88
  capax, 88

chelatus, 88
diversus, 88
floridanus, 88
fusiformis, 89
incisor, 89
longus, 89
malleus, 89
megalonchus, 86, *89*
mirabilis, 89
nematocirrus, 89
oculatus, 89
pricei, 89
rarus, 89
robustus, 89
stentor, 89
uniformis, 89
vancleavei, 89
venardi, 89
Clinostomatoidea, 164, *181*
Clinostomum, 13, 164, 168, *182*
marginatum, 4, *182*
Cloaca, 186
Cnidospora, 46, *49*
Coccidia, 47
Coccomyxa, 52, *57*
morovi, 57
Coccomyxidae, 57
Coelenterata, 318
coelestin blue B, 9
coelozoic, 51, 56
Coleoptera, 15
Colponema, 11, 27, *30*
agitans, 27, *30*
concentrating helminths, 6, 7
concretions: calcareous, 169–171, 177;
cecal, 179, 195
Conidia, 17
connective tissue, 13
connective tissue proliferation, 3
Contracaecum, 243, 246, *250*
brachyurum, 250
collieri, 250
spiculigerum, 250
contracted helminths, 8
contractile vacuole, 25
control, 3
cooking, 5
Copepoda, 300, *303*
copepodid, 305, 307, 311
copepods, 9, 11, 15, 148, 165, 219,
223, 224, 226, 229–234, 253, 265,
267, 268, 271; parasitic, 15, 300,
316
copulatory sheath, 243
Corallobothrium, 205, 218, *219*, 220
fimbriatum, 219
giganteum, 219
intermedium, 219
minutium, 219
parvum, 219
procerum, 219
thompsoni, 219
tva, 220

cormorant, 171, 195, 250
Corynosoma hardweni, 287
Cosmerodius, 194
Costia, 11, 23, 25, 27, *30*
necatrix, 30
pyriformis, 30
cotylae, 173
Cotylaspis, 12, 104
cokeri, 104
Cotylogaster, 104
Cotylophallus, 191, 193
Cotylurus, cercariae, 255
Cotylurus communis, 171, *174*
cranial cavity, 175
crassiphiala bulboglossa, 176
crayfish, 134, 144, 145, 148, 158, 159
Crepidostomum, 143, *144*, 160
ambloplitis, 144
brevivitellatum, 144
canadense, 145
cooperi, 12, 145
cornutum, 145
farionis, 12, 145
fausti, 145
hiodontos, 145
ictaluri, 145
illinoiense, 147
isostomum, 147
laureatum, 147
lintoni, 147
petalosum, 147
solidum, 147
transmarinum, 147
Creptotrema, *143*, 144
funduli, 144
crippling fish, 2
crocodile, 318
Crustacea, 145, 220, 231, 274, 298
Cryptobia, 6, 11, 23, *29*, 289, 297
borreli, 29
branchialis, 27, 30
carassii, 29
gurneyorum, 29
lynchi, 29
salmositica, 29
Cryptocotyle, 197
concavum, 197
lingua, 197
Cryptogonimidae, 108, *115*, 116, 165,
199
Cryptogonimus, 116, *119*, 180, *199*
chyli, 119, 199
diaphanus, 119
Crysemys, 134
Cucullanidae, 244, 254
Cucullanus, 244, 246, *255*
clitellarius, 255
globosus, 255
occidentalis, 255
scotti, 255
truttae, 255
cuticular clamps, 73
cuticular head bulb, 246, 265

Cyathocephalidae, 235
Cyathocephalus, 235
  americanus, 235
  truncatus, 235
Cyathocotyle, cercariae, 167
Cyathocotyloidea, 164
Cyclochaeta, 11, 35, *45*
  domerguei, 35
Cyclophyllidea, 240
Cyclopidea, 303
Cyclops, 223, 227, 253, 263; -like, 303
Cypria, 274
cyst: host, 20; hyaline, 169
Cystidicola, 12, 245, *259*
  canadensis, 259
  cristivomeri, 259
  lepisostei, 259
  serratus, 259
  stigmatura, 259
Cystidicoloides, 245, *263*
Cystobranchus, 295
  verrilli, 292, 295
Cystoopsis, 243, *247*
  acipenseri, 249
cytostome, 25
cysts, 6

Dacnitis, 255
Dacnitoides, 244, 246, *255*, 257
  cotylophora, 255
  robusta, 257
Dactylogyridae, 74, *77*
Dactylogyrinae, 77, *79*
Dactylogyrus, 79
  acus, 80
  amblops, 80
  anchoratus, 80
  apos, 80
  atromaculatus, 80
  attenuatus, 80
  aureus, 80
  banghami, 80
  bifurcatus, 80
  bulbosus, 80
  bulbus, 80
  bychowskyi, 80
  californiensis, 80
  claviformis, 80
  columbiensis, 80
  confusus, 80
  cornutus, 80
  corporalis, 80
  distinctus, 80
  dubius, 80
  duquesni, 80
  egregius, 80
  eucalius, 80
  extensus, 80
  fulcrum, 81
  hybognathus, 81
  leptobarbus, 81
  lineatus, 81

  maculatus, 81
  microlepidotus, 81
  microphallus, 81
  moorei, 81
  mylocheilus, 81
  nuchalis, 81
  occidentalis, 81
  orchis, 81
  orthodon, 81
  osculus, 81
  parvicirrus, 81
  percobromus, 81
  perlus, 81
  photogenis, 81
  pollex, 81
  ptychocheilus, 81
  pyriformis, 81
  rhinichthius, 81
  richardsonius, 81
  rubellus, 81
  scutatus, 81
  semotilus, 81
  simplex, 81
  tenax, 81
  texomonensis, 81
  tridactylus, 81
  ursus, 81
  vancleavei, 81
  vannus, 82
  vastator, 82
  wegeneri, 82
Dactylosoma, 11, *48*, 289
  salvelini, 48
damage to fish, 3
denticle, 61; buccal, 247
dead fish, 6
Dermocystidium, 11, 47, *68*
  branchialis, 68
  gasterostei, 69
  koi, 69
  percae, 69
  salmonis, 69
  vejdovskyi, 69
Derogenes, 12, 135, *139*
  aspina, 139
  varicus, 139
Deropristiidae, 108, *125*
Deropristis, 125
  inflata, 126
Dichadena, 135
Dichelyne, 244, 246, *257*
  cotylophora, 257
  diplocaecum, 257
  lintoni, 257
Didymozoidae, 107, *110*, 159
digenetic trematodes, 15, 105
digestion technique, 7
Dinoflagellida, 16, *27*
Dioctophyma, 269
  renale, 4, 246, *269*
Dioctophymidea, 268
Diphyllobothriidae, 230
Diphyllobothrium, 13, 205, 225, *231*

cordiceps, 231
dalliae, 231
ditremum, 231
laruei, 231
latum, 4, 231
oblongatum, 231
osmeri, 231
salvelini, 231
sebago, 231
ursi, 231
Diplectaninae, 79, *83*
Diplectanum, 79, *83*
  collinsi, 83
Diplocotyle, 237
  olrikii, 239
Diplocotylidae, 236
Diplostomatidae, 164, *169*, 174
Diplostomula, 166
Diplostomulum, 13, 164, *169*
    Bolbophorus confusus, 170
    corti, 170
    Diplostomum baeri eucaliae, 170
    Diplostomum huronense, 170
    Diplostomum spathaceum, 167, 171
    emarginatae, 171
    flexicaudum, 171
    gigas, 171
    Hysteromorpha triloba, 171
    ictaluri, 171
    indistinctum, 171
Dipterex, 309
Discocotyle, 99, *102*
  sagittata, 102
  salmonis, 102
disease, "whirling," 59
Distomulum oregonensis, 189
diverticula, intestinal, 243
dog, 177, 188–190, 193, 231, 269
dorso-cephalic tubercle, 244, 254
dragonfly, 265
ducks, 170, 171, 176, 187, 191, 197, 320
Dytiscidae, 320

ears, frog, 140
Echinochasmus, 179, *185*
  donaldsoni, 185
  schwartzi, 185
Echinorhynchidea, 280
Echinorhynchus, 273, *280*
  carpionis, 283
  clavaeceps, 281
  coregoni, 281
  dirus, 281, 283
  globolusus, 280, *281*
  lateralis, 281
  leidyi, 281
  linstowi, 281
  maraenae, 281
  murenae, 281
  oricola, *281*, 284
  pachysomus, 281

paucihamatus, 274, *281*
phoenix, 281
propinquus, 281
proteus, 281
salmonis, 12, *281*
salvelini, 281
thecatus, *281*, 284
tuberosus, 274, *281*
Echinostomata, 183
Echinostomatidae, 178, *185*
Echinostomatoidea, 165
Echinostomida, 165, 183
ecology, 2, 3
egg shell, striated, 249
eggs: bipolar filaments, 245; fish, 11, 14, 17, 297; opercula, 258; polar filaments, 259, 262, 263; polar plugs, 247, 249, 262
egret, 195
Eimeria, 6, 12, 23, *47*
  aurata, 47
  carassii, 47
  nicollei, 47
Elongidae, N. N., 64
Emyda, 145
Emys, 134
Enchelys parasitica, 37
enzyme, 7
Eocollis, 272, 277
  arcanus, 277
Epistylis, 11, *41*
Episymbionts, 43
Epitheliocystidia, 187
epithelium, 73
epizoon, human, 301
epizootic, 1
eradication, 3
Ergasilus, 11, 15, 300, *303*
  auritus, 305
  caeruleus, 305
  celestis, 305
  centrarchidarum, 305
  chautauquaensis, 305
  confusus, 305, *306*
  cotti, 306
  elegans, 306
  elongatus, 306
  fragilis, *306*
  funduli, 306
  labracis, 306
  lanceolatus, 306
  lizae, 2, *306*
  luciopercarum, 306
  magnicornis, 306
  manicatus, 306
  megaceros, 306
  nerkae, 306
  nigratus, 306
  osburni, 306
  skrjabini, 305, *306*
  turgidus, 306
  versicolor, 305, *306*
Erpobdellidae, 290

esophageal appendix, 244, 245, 250, 251
esophageal bulb, 246
esophagus, 140, 259, 316; fish, 7, 12, 130
esthetic, 5
Eubothrium, 226
  crassum, 226
  oncorhynchi, 226
  rugosum, 226
  salvelini, 12, *226*
Euglenida, 28
euglenoid, 28
Euglenosoma, 25, *28*
  branchialis, 28
Euhaplorchis, 13, 180, *197*
  californiensis, 197
Euparhyphium melis, 186
euryhaline, 2
Eurysporea, 53
Eustoma, 4
Eustrongylides, 245, *269*
examination of fish, 6
external parasites, 3, 6
eye, 7, 12, 16, 171; fish, 170, 174, 176
eye spots, 119, 194, 197, 199

fast green, 9
ferret, 189
Fessisentis, 273, *284*
  fessus, 284
  vancleavei, 284
filamentous bacteria, 14
Filaria, 267
Filariidea, 243, 245, *267*
Filipodia, 25
fin, 6, 11; caudal, 268
fish farming, 1
fish hatchery, 2
fish kills, 3
fish management, 234
fixation, 8; leech, 289
Flagellata, *25*
flagellum transverse, 26
Fluminicola, 110, 159, 189
food, 73
foreign fish parasites, 2
formalin, 8
Fossaria, 171
fox, 189, 193, 231
freezing, 5
frogs, 134, 135, 140, 145, 167, 171, 182, 188, 262, 265, 274, 294
fungi, 14, 17; fungi, culture, 18; Imperfecti, 20

Galactosomum, 180, *198*
  spineatum, 199
gall bladder, 6, 12, 38, 49, 51, 52, 139, 261
gametocytes, 48
Gammarus, 145, 158, 236, 259, 283
gastrointestinal, 6

gasterostomate, 182
Gemma, 155
Genarchella, 140
genital cone, 135
Genolinea, 12, 135, *137*
  oncorhynchi, 138
geographic range, 2, 19
Giemsa's, 9
gillrakers, 305
gills, 6, 11, 24
gill vessel, 110, 159
Glaridacris, 208, *215*
  catostomi, 215
  confusus, 216
  hexacotyle, 216
  intermedius, 216
  laruei, 216
  oligorchis, 216
Glaucoma, 38
Glochidia, 14, 316
Glossatella, 41
  micropteri, 41
Glossidium, 141
  geminum, 141
Glossiphoniidae, 290, 291, *294*
Glugea, 67
  hertwigi, 2, *67*
"*Glugea*" cysts, 67
glycerine, 8
glycerine jelly, 9
Gnathostoma, 246, *265*
  procyonis, 265
  spinigerum, 265
Gnathostomidae, 265
Goniobasis, 131, 158, 178
gonotyl, 108, 179, 193, 197–199
Gordiacea, 14
Gorgoderidae, 108, *127*, 160
Gracilisentis, 272, *279*
  gracilisentis, 279
granuloma, 20
grebe, 186
Guara, 195
gubernaculum, 245, 255, 257, 263
guinea pig, 171, 189
gulls, 171, 176, 186, 191, 193, 195, 197, 231, 250
gut, rhabdocoele, 118
Gymnostomatida, 38
Gyraulus, 171
Gyrinidae, 320
Gyrodactylidae, 74, *75*
Gyrodactylinae, 75
Gyrodactylus, 2, 11, 73, *75*
  atratuli, 75
  bairdi, 75
  bullatarudis, 73, *75*
  couesius, 75
  cylindriformis, 75
  egregius, 75
  elegans, 75
  elegans muelleri, 75
  elegans salmonis, 75

eucaliae, 75
fairporti, 76
funduli, 76
gurleyi, 76
limi, 76
macrochiri, 73, *76*
margaritae, 76
medius, 76
micropogonus, 76
prolongis, 76
rhinichthius, 76
richardsonius, 76
spathulatus, 76
stegurus, 76
stephanus, 76

Haemogregarina, 11, 23, *48*, 289
catostomi, 48
irkalukpiki, 48
myoxocephali, 48
Haemopsis grandis, 289
Haemosporidia, 47, *48*
halibut, "wormy," 52, 56
Halipegus, 135, 140
perplexus, 140
hamster, 189, 195
Haplobothriidae, 225, *233*
Haplobothrium, 12, 225, *233*
globuliforme, 233
"*Haplocleidus*," 79, 91, *93*, 97
Haplonema, 12, *257*
aditum, 258
hamulatum, 258
immutatum, 258
Haplonematidae, 257
Haploporidae, 121
Haplosporidia, 47
haptor, 73
hatcheries, 3
hatching temperature, 234
hawk, 193; Cooper's, 187
head: bulb, cuticular, 246; collar, 185;
      lappets, 76; organ, 74, 76; sinuses,
      13; spines, 265
heart, 245
Hedruridae, 258
Hedruris, 243, 244, 246, *258*
spinigera, 258
tiara, 258
Helisoma, 152, 154, 171, 176, 177, 182,
      185, 188
hematoxylin, 9
Hemiptera, 15, 320
Hemiuridae, 109, *134*, 165
Hemiuroidea, 165
Hemiurus, 12, 135, *136*
appendiculatus, 137
levinseni, 137
Henneguya, 13, 53, *63*
acuta, 63
ameiurensis, 63
amiae, 63
brachyura, 13

doori, 63
esocis, 63
exilis, 63
fontinalis, 63
gurleyi, 63
limatula, 63
macrura, 63
magna, 63
monura, 63
nigris, 63
percae, 63
salminicola, 63
salmonis, 64
schizura, 13, *64*
umbri, 405
hepatic bile duct, 12
Hepaticola, 243, *249*
bakeri, 249
herons, 4, 171, 176, 177, 182, 193–195,
      320
Heterocheilidae, 250
Heterophyes heterophyes, 4
Heterophyidae, 160, 165, 179
heterophyids, 4
Hexamita, 6, 12, 23, 25, 27, *31*
intestinalis, 31
salmonis, 31
Hirudinea, 288
hirudiniasis, erratic, 289
histochemistry, acanthocephala, 277
histopathology, 2, 235; acanthoceph-
      ala, 271, 274, 283
histozoic, 56, 57
Hoferellus, 53, *64*
cyprini, 64
hog, 189
Holostephanus ictaluri, 177
Holotrichia, 31, *37*
Homalometron, 154, *155*
armatum, 155
boleosomi, 154, *155*
pallidum, 152
pearsei, 152
hooklets, 73
hooks (anchors), 73; scolex, 234
host: cyst, 20; list, 321; specificity, 2, 4
human epizoon, 301
Hunterella, 208, 217
nodulosa, 217
Hyalella, 158, 261
Hydra, 318
Hydranassa, 194
Hydrobia, 155
Hydrophilidae, 320
Hymenolepidae, 240
Hymenolepis, 240
Hymenostomatida, 33, *37*
Hypha, 18
Hypocaryophyllaeus, 207, *213*
gilae, 213
paratarius, 213
Hysteromorpha triloba, 170, *171*
Hysterothylacium, 250

Ichthyobronema, 250
Ichthyonema, 268
Ichthyophonus hoferi, 2, 3, 12, 14, 18, 20
Ichthyophthirius multifilis, 3, 11, 23, 33, *37*
Ichthyosporidium, 20
    hoferi, 20
Identification aid, 14
Illinobdella, 292, 293, *298*
    alba, 293, *298*
    elongata, 292, *298*
    moorei, 292, *298*
    patzcuarensis, 292, *298*
    richardsoni, 292, *298*
immunity, 2, 38, 315
Imperfect stage, 17
importance, 2
Infusoria, 31
inner ray, 45
insect predators, 15, 320
internal infections, 3
interstate commerce, 5
intestinal, 6, 12
intestinal: ceca, 250; mucus, 7
intestine: red, 246; X-shaped, 110
invertebrate predators, 320
invertebrates, 37
iodinophilous vacuole, 51, 53, 56
iron hematoxylin, 9
isotonic, 7
isopods, 158, 271

Khawia, 207, 211
    iowensis, 211
kidney, 4, 6, 12, 49; tubules, 49
killing fish, 6
kingfisher, 176, 177, 186, 320
Kudoa, 11, *55*

lacto-phenol, 10
Lamellasoma, 25, *29*
    bacillaria, 29
lamellibranchs, 165
Lampsilis, 150
Larus argentatus, 174
larva: linguatulid, 317; red, 246, 269
larval migrans, 4
lateral: alae, 258, 262, 263; line, 13, 179, 187; lobes, 245
Lebouria, 157
Lecithaster, 135
    gibbosus, 136
    salmonis, 136
Lecithodendriidae, 159
leeches, 9, 15, 23, 28, 48, 166, *289*
Leighia, 179, *194*
Leiperia cincinalis, 318
lens, 13
Lentospora, 57
Lepeoptheirus, 11, *300*
Lepidauchen, 150, 151

Lepidotes, 83
Lepocreadidae, 154
Lepodermatidae, 140
Leptocleidus, 86, 88, 98
Leptomitus, 17
Leptorhynchoides, 272, *283*
    thecatum, 12, 271, *284*
Leptotheca, 54
Lernaea, 11, 15, 300, *307*
    anomala, 309
    carassii, 309
    catostomi, 309
    cruciata, 309
    cyprinacaea, 3, 309
    dolabrodes, 310
    elegans, 309, 310
    esocina, 309, 310
    gasterostei, 310
    insolens, 309, 310
    laterobrachialis, 310
    lophiara, 310
    pectoralis, 310
    pomotidis, 310
    ranae, 309, 310
    tenuis, 310
    tilapiae, 310
    tortua, 309, 310
    variabilis, 310
Lernaeidae, 307
Lernaeinae, 307
Lernaeocera, 11, 301, 307, 310
Lernaeopoda, 313, 314
Lerneopodidea, 310, 311, 313, 315
Leuceruthrus, 11, 12, *129*
    micropteri, 130
Leucocytozoon, 11, *48*
    salvelini, 49
Leuckartia, 226
lice, fish, 300
Ligula, 12, 225, *232*
    intestinalis, 232
limpet, 150
Lindane, 302
linguatulid larvae, 317
Linstowiella szidati, 177
Lintaxine, 99, 101
    cokeri, 101
lips, 245; dorsal and ventral, 258; lateral, 243, 246, 258, 262, 265
lips: four, 244, 246; papillae, 254; three, 246, 249; trilobed, 244, 262, 265; two, 259; two, four, six, 251
Lissorchiidae, 108, *123*, 165
Lissorchis, 124
    fairporti, 124
    gullaris, 124
liver, 4, 12
lizard, 318
location in fish, 11
loculi, 102
longevity, 171, 177, 182, 226
loons, 177, 191, 320
louse, salmon, 300

Lumbriculus, 269
Lymnaea, 171

Macrobdella decora, 289
macrocercous cercaria, 128
Macroderoides, 143, *152*, 181, *187*
   flavus, 152
   parva, 152
   spinifera, 152, 187
   typica, 154
Macroderoididae, 151
mammals, 187, 189–191, 194, 195, 197, 231, 250, 265, 269
man, 4, 189, 190, 231
marine, 4, 54
Marsipometra, 227
   confusa, 227
   hastata, 227
   parva, 227
Mastigophora, 24, *25*
mayfly, 145, 147, 261–263
Mazocraeoides, 99, *101*
   olentangiensis, 101
   megalocotyle, 101
   tennesseensis, 101
mechanical damage, 3
Megadistomum, 131
Megalogonia, 143, 145
Megathylacoides, 219
Meischer's tubes, 47
Melaniidae, 167
membranella, aboral, 39
merganser, 193, 231, 250, 320
merozoites, 48
mesenteries, 12
Mesoophorodiplostomum pricei, 176
Mesostephanus appendiculatoides, 5, 177
Mesostephanus appendiculatus, 177
Metabronema, 245, *263*
   canadense, 263
   harwoodi, 263
   prevosti, 263
   salvelini, 263
metacercariae, 11, *161*
metanauplius, 307, 311
Metorchis, 4, 181, *190*
   conjunctus, 190
mice, 171, 190, 193
Microcotyle, 99
   eriensis, 99
   macroura, 99
   spinicirrus, 99
Microcreadium, 154, *155*
   parvum, 155
Microphallidae, 108, *134*
Microphallus, 134
   medius, 134
   obstipes, 134
   opacus, 134
   ovatus, 134
Microsporidea, 11, 49, 65
Microtylidae, 99

migrans, larval, 6
migrating, 3
mink, 189, 190, 193, 195, 320
Mimodistomum, 133
mites, 12, 316
Mitraspora, 53, *64*
   elongata, 65
Mobilina, 34
Mollusca, 316
Monobothrium, 208, 213
   hunteri, 215
   ingens, 215
monogenetic trematodes, 8, 11, 14, *70*
Monopisthocotylea, 74
Monorchiidae, 108, *121*, 165
mouse, 171, 190, 193
mouth, 7, 11, 258; filament, 314
mucous envelopes, 49
mucus, 6, 7, 24, 73, 301
Mudalia, 178
Murraytrema (see *Pseudomurraytrema*)
muscle, 7, 13
Musculium, 128, 145
muskrats, 187, 190, 191
mycelium, 17
Myxidiidae, 56
Myxidium, 53, *56*
   americanum, 56
   aplodinoti, 56
   bellum, 56
   folium, 56
   gasterostei, 56
   illinoisense, 56
   kudoi, 56
   lieberkuhni, 56
   macrocapsulare, 56
   mellum, 56
   minteri, 56
   myoxocephali, 57
   oviforme, 57
   percae, 57
myxobacteria, 14
Myxobilatus, 53, *65*
   asymmetricus, 65
   caudalis, 65
   mictosporus, 65
   ohioensis, 65
   wisconsinensis, 65
Myxobolidae, 60
Myxobolus, 53, *60*
   angustus, 61
   aureatus, 61
   bellus, 61
   bilineatum, 61
   bubalis, 61
   capsulatus, 61
   catostomi, 61
   compressus, 61
   congesticius, 61
   conspicuus, 61
   couesii, 61
   dentium, 61

discrepans, 61
encephalica, 13
funduli, 61
gibbosus, 61
globosus, 61
grandis, 61
gravidus, 61
hyborhynchi, 61
inornatus, 61
insidiosus, 61
intestinalis, 61
iowensis, 61
kisutchi, 61
koi, 61
kostiri, 61
lintoni, 61
mesentericus, 61
moxostomi, 62
musculi, 62
mutabilis, 62
neurobius, 13, 62
nodosus, 62
notemigoni, 62
notropis, 62
obliquus, 62
oblongus, 62
okobojiensis, 62
orbiculatus, 62
osburni, 62
ovatus, 62
ovoidalis, 62
percae, 62
poecilichthidis, 62
rhinichthidis, 62
sparoidis, 62
squamae, 62
squamosus, 62
subcircularis, 62
symmetricus, 62
teres, 62
transovalis, 62
transversalis, 62
vastus, 62
Myxonema tenue, 17
Myxoproteus, 52, 54
Myxosoma, 53, 57
bibullatum, 59
cartilaginis, 13, 59
catostomi, 59
cerebralis, 2, 13, 59
commersonii, 59
cuneata, 59
diaphana, 59
ellipticoides, 59
endovasa, 59
funduli, 59
grandis, 59
hoffmani, 13, 59
hudsonis, 59
media, 59
microthecum, 59
mulleri, 59

multiplicatum, 60
neurophila, 60
notropis, 60
okobojiensis, 60
orbitalis, 60
ovalis, 60
parellipticoides, 60
pfrille, 60
procerum, 60
robustum, 60
rotundum, 60
scleroperca, 13, 60
squamalis, 60
subtecalis, 60
Myxosomatidae, 57
Myxosporidea, 2, 6, 8, 9, 11, 49
Myzobdella, 298

Nanophyetus, 181, 188
salmincola, 4, 12, 189
nares, 186
nasal capsule, 16
Natrix, 134, 285
nature, 2, 23
nauplius, 305, 307, 311
Neascus, 13, 164, 169, 174, 175
Cercaria flexicorpa, 176
Crassiphiala bulboglossa, 176
ellipticus, 176
grandis, 176
longicollis, 176
Mesoophorodiplostomum pricei, 176
nolfi, 176
Ornithodiplostomum ptychocheilus, 176
Posthodiplostomum minimum, 176
P. minimum centrarchi, 176
P. minimum minimum, 176
pyriformis, 176
rhinichthysi, 177
Uvulifer ambloplitis, 177
vancleavi, 176
wardi, 177
Necturus, 145
nematocysts, 318
nematode, 14; adult, 241; ceca, 243, 245; larval, 4, 8, 9, 245; red, 254; spines, 244, 246
Nematobothrium, 12, 14, 110
texomensis, 111, 159
Nembutal, 290
Neochasmus, 116, 117, 179, 202
ictaluri, 117
umbellus, 117
Neodactylogyrus, 79, 82
acus, 82
amblops, 82
apos, 82
attenuatus, 82
bifurcatus, 82

bulbus, 82
confusus, 82
cornutus, 82
distinctus, 82
duquesni, 82
fulcrum, 82
orchis, 82
perlus, 82
photogenis, 82
pyriformis, 82
rubellus, 83
scutatus, 83
simplex, 83
ursus, 83
vannus, 83
Neoechinorhynchidae, 273
Neoechinorhynchus, 272, *273*
    australe, 274
    crassum, 274
    cristatum, 274
    cylindratum, *274*, 287
    distractum, 274
    doryphorum, 274
    longirostris, 274
    paucihamatum, 274
    prolixoides, 275
    prolixum, 275
    rutili, 275, 281
    saginatum, 275
    strigosum, 275
    tenellum, 275
    tumidum, 275
Neogogatea kentuckiensis, 178
Nephelopsis obscura, 290, *293*
Nereis, 126
Neritina, 158
nervous system, 13
Neteridae, 320
newts, 171, 230
Nosema, 66, *67*
    pimephales, 67
Notonectidae, 320
nutrition, 2
Nyctotherus, 33, *36*

Octomacrum, 99, *102*
    lanceatum, 102
    microconfibula, 102
Octomitus, 31
Octospinifer, 272, *275*
    macilentus, 275
    torosus, 275
Octospiniferoides, 272, *277*
    chandleri, 277
Odonata, 265
oils for control of predators, 320
oligochaete, 124, 209, 217, 269
Onchocleidus, 79, 91, *98*
Ondatra, 187, 190, 191
oocyst, 47
Oodinium, 2, 11, 25, *27*
    limneticum, 16, 25, 27

    pillularis, 27
oogonia, 18
oospores, 17
Opecoelidae, 156
opercula, 16
Ophiotaenia, 218, *220*
    fragilis, 220
Opisthorchiida, 165, *189*
Opisthorchiidae, 165, *189*
opisthorchids, 4
Opisthorchis, 4, 181, *190*
    tonkae, 190
opossum, 195
oral papillae, 143, 144, 147, 178
Ornithodiplostomum, 13, 169, *176*
    ptychocheilus, 176
osmic acid fixative, 9
osprey, 187, 198, 320
ostracods, 271, 274, 277
otter, 320
ovary, fish, 12, 54, 160
owl, 193; screech, 170
Oxytrema, 110, 159, 189
Oxyuridea, 246
oxyurids, 246

pansporoblasts, 59
papillae: caudal, 259, 263, 265, 268,
    269; cephalic, 244; cervical, 263;
    head, 268, 269; lips, 254; oral,
    143, 147, 178; postanal, 250, 251,
    253, 257, 258, 261, 263, 267; pre-
    anal, 244, 250, 251, 253, 257, 258,
    261
Paramacroderoides, *151*, 179, *188*
    echinus, 152, 188
Parametorchis canadense, 190
Paramphostomidae, 107, *111*, 165
paramylon body, 28
parasite: accidental, 134, 159; as tag,
    234
Parastictodora, 13, 180, *197*
    hancocki, 198
pathology, 3
Paulisentis, 272, *277*
    fractus, 277
Paurorhynchus, 12, 113, *115*
    hiodontis, *115*, 159
peduncle, 83
pelecypoda, 113, 316
pelican, 170, 195, 231, 250
Pelichnibothrium, 239
    speciosum, 239
pellicle, 25
Pellucidhaptor, 79, 83
    pellucidhaptor, 83
Pentastomida, 317
    pepsin, 7
Peracreadium (see *Podocotyle*)
    perforatorium, 37
pericardial cavity, 174
Peritrichia, 39

permanent preparation, 9, 72
Petasiger, 179, *185*
    nitidus, 185
pH, 2, 9
Phagicola, 4, 179, *195*
    angrense, 195
    longa, 195
    molliensicola, 195
Pharyngobdellida, 290, *293*
Philometra, 13, 245, *268*
    carassii, 11, *268*
    cylindracea, 268
    nodulosa, 268
    obturans, 11
    sanguinea, 268
    translucida, 268
Philonema, 12, 13, 245, *267*
    agubernaculum, 267
    oncorhynchi, 267
    salvelini, 267
Phormidium mucicola, 17
Phycomycetes, 17, 18
Phyllobothrium, 239
    ketae, 239
    salmonis, 239
Phyllodistomum, 12, 13, 108, *127*
    americanum, 128
    brevicecum, 128
    carolini, 128
    caudatum, 128
    coregoni, 128
    etheostomae, 128
    fausti, 128
    folium, 128
    hunteri, 128
    lachancei, 128
    lacustri, 128
    lohrenzi, 128
    lysteri, 129
    nocomis, 129
    notropidus, 129
    pearsii, 129
    semotili, 129
    staffordi, 129
    superbum, 129
    undulans, 129
Physa, 176, 177
physiological saline, 7
pickling, 5
pigeon, 185, 191
pigment, black, 175–177, 181, 191, 197
Pisciamphistoma, 111
    reynoldsi, 111
Piscicola, 292, 297
    geometra, 292, *297*
    milneri, 292, *297*
    punctata, 292, *297*
    salmositica, 292, *297*
Piscicolaria, 292, 297
    reducta, 292, *297*
Piscicolidae, 291, 292, 295
Pisidium, 145, 148, 150

pithing, 6
Placobdella, 291, 294
    hollensis, 291, *294*, 295
    montifera, 291, 294
    ornata, 291, *295*
    parasitica, 291, *295*
    pediculata, 289, 291, *295*
    rugosa, 291, *295*
Plagiocirrus, 143, *149*
    primus, 149
    testeus, 149
Plagioporus, 157
    angusticollis, 158
    boleosomi, 158
    cooperi, 158
    lepomis, 158
    macrouterinus, 158
    serotinus, 158
    serratus, 158
    siliculus, 158
    sinitsini, 12, *158*
    truncatus, 158
    virens, 159
Plagiorchidae, 109, *140*, 141, 165
Plagiorchiida, 187
Plagiorchioidea, 165
Plagiorchis, 142
    corti, 142
    geminum, 141
planaria, 124
plant seeds, 318
Platysporea, 51, 52, *56*
plerocercoid, 219, 224, *225*
Plesiocreadium, 152
Plethodon, 134
Pleurocera, 131, 150
Pleurogenes, 109, *159*
Pliovitellaria, 207, *211*
    wisconsinensis, 211
Plistophora, 67
    cepedianae, 67
    ovariae, 68
    salmonae, 68
Podocotyle, 157
    shawi, 157
    simplex, 157
    virens, 157
Polamobius, 158
polar: capsule, 49; filament, 46, 65;
    plugs, eggs, 247, 249
Podilymbus, 185
"*Polymastigina*," 25, *30*
Polyopisthocotylea, 74, *98*
Polyphage, 320
Pomphorhynchidae, 284
Pomphorhynchus, 12, 272, *284*
    bulbocolli, 285
    tereticolli, 285
Porrocaecum, 246, *251*
Posthodiplostomum, 12, 13, *176*
    minimum, 168, *176*
    minimum centrarchi, 12, 175, *176*

minimum minimum, 175, *176*
  prosostomum, 176
preanal sucker, 244, 255
predators, 320; insect, 15; invertebrate, 320
preoral sting, 301
preservation, 3, 8
prevention, 3
Pristicola, 125
  sturionis, 125
Pristotrema, 125, *126*
proboscis, 9, 271
procercoid, 224
Procyon, 134
progenesis, 130, 194, 201
proglottides: campanulate, 229; craspedote, 229; serrate, 225
Prohemistomulum, 164, 169, *177*
  Linstowiella szidati, 177
  Mesostephanus appendiculatoides, 177
  Mesostephanus appendiculatus, 178
  Neogogatea kentuckiensis, 178
  Prohemistomum chandleri, 178
Protenteron, 141
  diaphanum, 141
Proteocephalidea, 218
Proteocephalidae, 218
Proteocephalus, 12, 218, *220*
  ambloplitis, 205, *220*
  arcticus, 221
  australis, 221
  cobraeformis, 221
  coregoni, 221
  elongatus, 221
  exiguus, 221
  filicollis, 221
  fluviatilis, 221
  laruei, 221
  leptosoma, 221
  luciopercae, 221
  macrocephalus, 221
  microcephalus, 221
  micropteri, 221
  nematosoma, 221
  osburni, 221
  parallacticus, 223
  pearsei, 223
  perplexus, 223
  pinguis, 223
  primaverus, 223
  ptychocheilus, 223
  pugitensis, 223
  pusillus, 223
  salmonidicola, 223
  salvelini, 224
  singularis, 224
  stizostethi, 224
  tumidocollus, 224
  wickliffi, 224
Proterometra, 12, *130*

catenaria, 131
  hodgesiana, 131
  macrostoma, 131
  sagittaria, 131
"Protomonadina," 25, *28*
Protozoa, 6, 8, 9, 11, 14, *23*
proventriculus, 187
Pseudoascocotyle, 195
Pseudolytocestus, 207, *216*
  differtus, 216
Pseudomurraytrema, 79, *87*
Pseudophyllidea, 224
pseudopodia, 24
pseudoscolex, 225
pseudosuckers, 169, 173
Psilostomum, 13
"*Pterocleidus*," 79, 91, *93*, 98
Ptychogonimidae, 109, *159*, 165
Ptychogonimus, 159
  fontanus, 159
public health, 4
Pulchrosomum, 186
PVA-AFA fix, 8
Pygidiopsoides, 179, *195*
  spindalis, 195
Pythium, 17

rabbit, 171, 193
raccoon, 189, 193, 195, 265
Rana, 134, 188, 265
Raphidascaris, 244, *251*
  alius, 251
  brachyurus, 251
  canadense, 251
  cayugensis, 251
  laurentianus, 251
rat, 171, 189, 190, 193, 195, 231
red intestine, 246
red larva, 245, 269
red nematode, 254
red worm, 244
relaxant, leech, 289
renal tubules, 12
reproduction impaired, 3
reptiles, 56, 191, 220, 231, 283
Rhabdochonidae, 258
Rhabdochona, 259
  cascadilla, 261
  cotti, 261
  decaturensis, 261
  laurentiana, 261
  milleri, 261
  ovifilamenta, 261
  pellucida, 261
rhabdocoele gut, 77
Rhadinorhynchidae, 285
Rhadinorhynchus, 272, *285*
  exilis, 286
  pristis, 286
Rhipidocotyle, 113, *114*, 178, *183*
  lepisostei, 114
  papillosum, 114, 183

septpapillata, 114, 183
Rhynchobdellida, 290, 293
rhynchus, 108, 113–115, 178, 183
Ribeiroia, 179, *186*
  thomasi, 187
rickettsia, 189
robin, 176
rodent, 240
Rossicotrema donicus, 191

salamanders, 134, 145, 171, 182, 189,
    262, 267, 309, 320
saline, physiological, 6, 7
Salmincola, 11, 15, 300, *313*, 315
  beani, 314
  bicauliculata, 314
  californiensis, 314
  carpenteri, 314
  edwardsi, 314
  extumescens, 314
  falculata, 315
  gibbera, 315
  gordoni, 315
  heintzi, 315
  inermis, 315
  omuli, 315
  oquassa, 315
  salvelini, 315
  siscowet, 315
  wisconsinensis, 315
Sanguinicola, 11, 14, 110, 159
  davisi, 110
  huronis, 110
  klamathensis, 110
  lophophora, 110
  occidentalis, 110
Sanguinicolidae, 107, *109*, 159, 164
Sanguinofilaria, 268
Saprolegnia, 11, 17, *18*
Saprolegniaceae, 18
Sarcocystis, 13, *47*
  salvelini, 47
Sarcodina, *24*
Scaphanocephalus, 180, *198*
  expansus, 198
Schistocephalus, 12, 225, *232*
  solidus, 232
  thomasi, 233
Schistosomatoidea, 164
Schizamoeba, 6, 12, 23, *24*
  salmonis, 25
sclera, 13
sclerites, 99
scolex: apical disc, 229; apical sucker,
    239; compressed laterally, 230,
    231; funnel-shaped, 235; hooks,
    234; pseudo-, 225, 233; triangu-
    lar, 232
scopula, 41
Scyphidia, 11, 34, *39*
  ameiuri, 39
  macropodia, 39
  micropteri, 41

tholiformis, 41
Scyphidiidae, 34
seal, 193
Sebekia oxycephalus, 31
secondary invader, 17
Sellacotyle, 181, *189*
Semichon's carmine, 9
Sessilina, 34
sexual sterility, 205
sigmoid trunk, 301
silver impregnation, 9
Sinistroporus, 157
Sinuolinea, 52, 55
sinuses, head, 12
skin, 11; blisters, 249
Skrjabinopsolus, 125, 126
  manteri, 126
skunk, 189
slime, blue, 23
smoking, 5
snail, 265
snake, 265, 267, 285, 318
Spartoides, 207, *208*
  wardi, 208
Spathebothriidae, 235
spawning, 18
sphaerid clams, 145, 148, 155, 316
Sphaerium, 145, 147
Sphaeromyxa, 53, 57
Sphaerospora, 52, 55
  carassii, 55
Sphaerosporea, 54
Sphaerosporidae, 54
spines, body, 108, 120, 121, 125–127,
    141, 151, 152, 178, 179, 187–188,
    261, 272
spines, circumoral, 115–117, 125, 178,
    179, 186, 188, 193–195
spines, head, 265
spines, nematode, 244, 246
Spinitectus, 244, *261*
  carolini, 262
  gracilis, 262
Spirorchidae, 164
Spirotrichia, 33, *36*
Spiroxys, 245, 246, *263*
  contorta, 265
Spiruridae, 262
Spiruridea, 243, 244, *251*
spleen, 12
sporangia, 17
sporangiospores, 17
spores, 3
sporoblast, 47
sporoplasm, 49
sporozoa, 24, *46*; uncertain classifica-
    tion, 68
sporozoites, 47
spurs, anchor, 79
squamodiscs, 79, 83
squirrel, 231
Stagnicola, 170
staining, 9

stalk, 41
starvation, 235
Stentoropsis, 33
Stephanoprora, 179, 186
  denticulata, 186
  polycestus, 186
sterility, sexual, 205
Stictodora, 197
Stigeoclonium, 16
stigma, 28
sting, preoral, 301
stocking, 2
stomach, 12, 24, 136–140, 149, 159,
  201, 211, 250, 253, 257, 258, 262
Strigeata, 166
Strigeatoidea, 165
Strigeidae, 164, *171*
Strigeoidea, 164, *169*
strobila, polyzoic, 205
Subulina, 182
sucker: apical, 239; preanal, 244, 255;
  pseudo-, 169, 173
sucking discs, 300, 301
Suctoria, 31, *35*
surface toxicants, 320
sutural line, 51
swans, 191
swim bladder, 12, 160, 245, 259
Syncoelidae, 108, *127*
Syncoelium, 11, *127*
  filiferum, 127
Szidatia joyeuxi, 167

tadpoles, 151, 152, 171, 188, 265, 309
tag, parasite, 234
tail, pointed, 246
Tanaorhamphus, 272, *279*
  ambiguus, 279
  longirostris, 279
Telosporea, 46
telotroch, 39
temperature: for control, 310; *Gyrodactylus*, 73; hatching, 234; lethal,
  234; optimum, 2
tentacles: appendages, 113, 114; protrusible, 225
tern, 193, 197, 231
testes, fish, 12
Tetracleidus, 86, 88, *98*
Tetracotyle, 164, 166, 167, 169, *171*,
  173, 174
  Apatemon gracilis pellucidus, 174
  Cotylurus communis, 174
  diminuta, 174
  intermedia, 174
  lepomensis, 174
  parvulum, 174
  tahoensis, 174
Tetrahymena, 41
Tetrahynchus, 287
Tetraonchinae, 79, *85*
Tetraonchus, 87
  alaskensis, 87

  monenteron, 87
  rauschi, 87
  variabilis, 87
Tetraphyllidea, 206, *239*
Thamnophis, 134
Thelohanellus, 52, *64*
  notatus, 64
  pyriformis, 64
Thelohania, 68
tissue-inhabiting trematode, 159
toad, 295
tobacco, 289
tomites, 37
toxicants, surface, 320
trachea, 4
transverse flagellum, 26
trash fish, 20
trauma, 18
treatment, 2
trematodes: bladder-inhabiting, 141;
  digenetic, 4, 8, 9, 14, 69, 106;
  monogenetic, 14
Triaenophoridae, 224, 234
Triaenophorus, 12, 13, 225, *234*
  crassus, 234
  lucii, 234
  nodulosus, 234
  robustus, 234
  stizostedionis, 234
  tricuspidatus, 234
Trianchoratus acleithrium, 85
Trichodina, 2, 9, 11, 23, 35, *43*, 46
  californica, 45
  discoidea, 45
  domerguei, 35, 45
  fultoni, 45
  guberleti, 46
  pediculus, 46
  platyformis, 46
  truttae, 46
  tumefaciens, 46
  vallata, 46
Trichodinella, 35, *44*
  myakkae, 44
Trichophrya, 11, 31, *35*
  ictaluri, 35
  intermedia, 35
  micropteri, 35
  piscium, 35
  sinensis, 35
Trichostomatida, 36
Trichuridea, 242, 243, *247*
Tricuspidaria, 234
Triganodistomum, 124
  attenuatum, 124
  crassicrurum, 124
  garricki, 124
  hypentelii, 124
  mutabile, 124
  polybatum, 124
  simeri, 124
  translucens, 124
Trilospora, 54

Tripartiella, 34, *43*
  bulbosa, 43
  bursiformis, 43
  symmetricus, 43
Triturus, 262, 267
Troglotrema salmincola, 189
Troglotrematidae, 165, 188
Trypanoplasma, 29
Trypanosoma, 6, 11, 25, *28*, 289
  myoxocephala, 28
  percae canadense, 28
  remaki, 28
Trypanosomatidae, 25
tubercle: dorsal, 254; dorso-cephalic, 244
Tubifex, 216
Tubificidae, 209, 216
turtles, 145, 171, 245, 258, 265, 274, 295, 320

Unicapsula, 56
  muscularis, 52, *56*
Unicapsulidae, 52
Unicauda, 53, *63*
  brachyura, 63
  clavicauda, 63
  crassicauda, 63
  fontinalis, 63
  plasmodia, 63
Unionidae clam, 154, 155
U.S. Food and Drug Administration, 5
Urceolariidae, 34, 43
ureters, 12, 77
urinary bladder, 6, 12, 44, 49, 72, 77, 108, 128
Urocleidoides, 79, *89*
  reticulatus, 91
Urocleidus, 79, 91
  acer, 95
  aculeatus, 91
  acuminatus, 95
  adspectus, 91
  affinis, 93
  angularis, 91
  attenuatus, 91
  banghami, 92
  biramosus, 92, 95
  chaenobryttus, 92
  chautauquaensis, 92
  chrysops, 92
  cyanellus, 92
  dispar, 93
  distinctus, 92
  doloresae, 92
  ferox, 92
  furcatus, 93
  grandis, 95
  interruptus, 92
  macropterus, 92
  malleus, 92
  mimus, 92
  miniatus, 95
  moorei, 92
  mucronatus, 92
  nigrofasciatus, 93
  parvicirrus, 93
  perdix, 93
  principalis, 93
  procax, 93
  seculus, 93
  similis, 93
  spiralis, 93
  torquatus, 93
  umbraensis, 93
  variabilis, 93
  wadei, 95
Urophagus, 27, *31*
uterus, ascending, 109
Uvulifer
  ambloplitis, 167, 176, *177*
  claviformis, 177
  magnibursiger, 177

vacuole, contractile, 25
Valvata, 110
Vauchomia, 12, 35, *44*
  nephritica, 44
  renicola, 44
Vietosoma, 143, *149*
  parvum, 149
virology, 2
virus, 289
viviparous, 267
Vorticella, 34, 41

Wardia, 12, *54*
  ovinocua, 54
Wardiidae, 51
water beetles, 320
water bugs, 320
water supply, 3
"whirling disease," 2, 3, 59
worm, red, 244
"wormy" halibut, 52, 56

zoospores, 17
Zoothamnium, 41
Zschokkella, 53, *57*
  salvelini, 57
zygospores, 17